Two-Dimensional Nanomaterials for Fire-Safe Polymers

This book provides an overview of the latest scientific developments and technological advances in two-dimensional (2D) nanomaterials for fire-safe polymers. It summarizes the preparation methods for diverse types of 2D nanomaterials and their polymer composites and reviews their flame-retardant properties, toxic gas and smoke emission during combustion, and inhibition strategies.

- Covers fundamental aspects like influence of size and dispersionof 2D nanomaterials to help readers develop efficient, multi-functional, and eco-friendly fire-safe polymer composites for a wide range of applications
- Discusses new-emerging 2D nanomaterials for fire-safe polymer applications, including MXenes, graphitic carbon nitride, boron nitride, and black phosphorus
- Introduces basic modes of flame retardant action of 2D nanomaterials, including smoke and toxic gas suppression, and the role of 2D nanomaterials in promoting char formation

This book is suitable for both scholars and engineers in the fields of polymer science and engineering. It is also aimed at graduate students in chemistry, materials, and safety science and engineering.

Emerging Materials and Technologies

Series Editor: Boris I. Kharissov

The *Emerging Materials and Technologies* series is devoted to highlighting publications centered on emerging advanced materials and novel technologies. Attention is paid to those newly discovered or applied materials with potential to solve pressing societal problems and improve quality of life, corresponding to environmental protection, medicine, communications, energy, transportation, advanced manufacturing, and related areas.

The series takes into account that, under present strong demands for energy, material, and cost savings, as well as heavy contamination problems and worldwide pandemic conditions, the area of emerging materials and related scalable technologies is a highly interdisciplinary field, with the need for researchers, professionals, and academics across the spectrum of engineering and technological disciplines. The main objective of this book series is to attract more attention to these materials and technologies and invite conversation among the international R&D community.

Smart Nanomaterials
Imalka Munaweera and M. L. Chamalki Madhusha

Nanocosmetics: Drug Delivery Approaches, Applications and Regulatory Aspects
Edited by Prashant Kesharwani and Sunil Kumar Dubey

Sustainability of Green and Eco-friendly Composites
Edited by Sumit Gupta, Vijay Chaudhary, and Pallav Gupta

Assessment of Polymeric Materials for Biomedical Applications
Edited by Vijay Chaudhary, Sumit Gupta, Pallav Gupta, and Partha Pratim Das

Nanomaterials for Sustainable Energy Applications
Edited by Piyush Kumar Sonkar and Vellaichamy Ganesan

Materials Science to Combat COVID-19
Edited by Neeraj Dwivedi and Avanish Kumar Srivastava

Two-Dimensional Nanomaterials for Fire-Safe Polymers
Yuan Hu and Xin Wang

For more information about this series, please visit: www.routledge.com/Emerging-Materials-and-Technologies/book-series/CRCEMT

Two-Dimensional Nanomaterials for Fire-Safe Polymers

Edited by
Yuan Hu
Xin Wang

CRC Press
Taylor & Francis Group
Boca Raton London New York

CRC Press is an imprint of the
Taylor & Francis Group, an **informa** business

First edition published 2024
by CRC Press
2385 NW Executive Center Drive, Suite 320, Boca Raton FL 33431

and by CRC Press
4 Park Square, Milton Park, Abingdon, Oxon, OX14 4RN

CRC Press is an imprint of Taylor & Francis Group, LLC

ISBN: 978-1-032-35268-8 (hbk)
ISBN: 978-1-032-35502-3 (pbk)
ISBN: 978-1-003-32715-8 (ebk)

DOI: 10.1201/9781003327158

Typeset in Times
by MPS Limited, Dehradun

Contents

Preface

With the increasing awareness of safety development, flame-retardant polymers have entered a new era of fire-safe polymers. Compared to conventional flame-retardant polymers, fire-safe polymers not only focus on the difficulty of ignition and the low heat release rate but also emphasize smoke and toxic gas suppression. Over the past few decades, two-dimensional nanomaterials have spawned a considerable interest in the development of fire-safe polymers.

Although there are a number of classic monographs on flame retardant polymer nanocomposites, including *Flame Retardant Polymer Nanocomposites* by Alexander B. Morgan and Charles A. Wilkie (2010), *Thermally Stable and Flame Retardant Polymer Nanocomposites* by Vikas Mittal (2011), and *Halogen-Free Flame-Retardant Polymers: Next-generation Fillers for Polymer Nanocomposite Applications* by Suprakas Sinha Ray and Malkappa Kuruma (2020), some newly emerging two-dimensional nanomaterials, including MXenes, graphitic carbon nitride, boron nitride and black phosphorus, for fire-safe polymers have not been covered. It is thereby imperative to introduce these latest scientific developments and technological advances to readers in this field.

This book aims to provide an overview of the latest scientific developments and technological advances in two-dimensional nanomaterials for fire-safe polymers. It summarizes the preparation methods for diverse types of two-dimensional nanomaterials and their polymer composites and specifically reviews their flame-retardant properties, toxic gas and smoke emissions during combustion, and inhibition strategies. In turn, the book discusses the application of clay, layered double hydroxides, MXenes, graphene, layered zirconium phosphate, graphitic carbon nitride, boron nitride, molybdenum disulfide, and black phosphorus in flame inhibition and addresses the flame-retardant properties of various two-dimensional nanomaterials containing polymers and their flame-retardant and smoke suppression mechanisms in detail. It also systematically covers the synergetic effects between two-dimensional nanomaterials and traditional flame-retardant compounds and explains the important role of two-dimensional nanomaterials in balancing the flame-retardant property and other behaviors (thermal stability, mechanical strength, glass transition temperature, etc.).

The fundamentals (Chapter 2) are also included to understand the flame-retardant properties of various two-dimensional nanomaterials and their polymer composites, and to develop efficient, multi-functional, eco-friendly fire-safe polymer composites for a wide range of applications. We hope that this book will be ideally suited for both scholars and engineers in the fields of polymer science and engineering, and for graduate students in chemistry, materials, and safety science and engineering.

We should express great acknowledgment to all those who helped make this book possible, especially the chapter contributors who have spent their valuable time to share their knowledge with readers. We would also like to thank the reviewers who have taken time out of their busy lives to review each chapter. Finally, we would like to express my gratitude to those at Taylor & Francis for their professional organizing and editing work.

Yuan Hu
Xin Wang

Editors

Yuan Hu, PhD, is the Director of the Institute of Fire-Safe Materials, University of Science and Technology of China (USTC). He earned a PhD at USTC in 1997. Prof. Hu has performed research on fire-safe materials for more than 30 years. His main research areas include polymer/inorganic compound nanocomposites, new flame retardants and their flame-retardant polymers, synthesis and properties of inorganic nanomaterials, combustion and decomposition mechanism of polymers. Prof. Hu has published more than 600 SCI papers (citations 30,000, H index = 98) in research areas covering the range of fire safety of polymer materials and has obtained over 70 invention patents. He was also awarded the Second Prize of the National Natural Science Award by the Chinese Government (2017).

Xin Wang, PhD, is an Associate Professor at the University of Science and Technology of China (USTC). He earned a PhD in safety science and engineering at USTC in 2013. His research interests focus on the synthesis and application of bio-based flame retardants and the preparation of layered nanomaterials and their use in fire-safe polymer composites. He has authored or co-authored more than 180 SCI-indexed papers in peer-reviewed international journals, 6 book chapters, and 2 monographs in this field. He was also awarded the Second Prize of the National Natural Science Award by the Chinese Government (2017).

Contributors

Wei Cai
State Key Laboratory of Fire Science
University of Science and Technology
  of China
Hefei, Anhui, PR China

Fukai Chu
State Key Laboratory of Fire Science
University of Science and Technology
  of China
Hefei, Anhui, PR China

Hong-Liang Ding
State Key Laboratory of Fire Science
University of Science and Technology
  of China
Hefei, Anhui, PR China

Bin Fei
School of Fashion and Textiles
Hong Kong Polytechnic University
Kowloon, Hong Kong, China

Xiaming Feng
Chongqing University
Chongqing, PR China

Wen-Wen Guo
Jiangnan University
Wuxi, Jiangsu, China

Yuan Hu
State Key Laboratory of Fire Science
University of Science and Technology
  of China
Hefei, Anhui, PR China

Ehsan Naderi Kalali
Department of Fire Safety Engineering
Faculty of Geosciences and
  Environmental Engineering
Southwest Jiaotong University
Chengdu, China
and
Institute of Macromolecular Chemistry
Czech Academy of Sciences
Prague, Czech Republic

Hiran Mayookh Lal
School of Energy Materials
Mahatma Gandhi University
Kottayam, Kerala, India

Ao Li
School of Mechanical and
  Manufacturing Engineering
University of New South Wales
Sydney, Australia

Yuchun Li
College of Materials Science and
  Engineering
Beijing University of Chemical
  Technology
Chaoyang, PR China

Zhaoxin Li
State Key Laboratory of Fire Science
University of Science and Technology
  of China
Hefei, Anhui, PR China
and
Hong Kong University of Science and
  Technology
Kowloon, Hong Kong, China

Miao Liu
Fuzhou University
Fuzhou, PR China

Giulio Malucelli
Department of Applied Science and
    Technology and Local INSTM Unit
Politecnico di Torino
Alessandria, Italy

Mariya Mathew
School of Energy Materials
Mahatma Gandhi University
Kottayam, Kerala, India

Yang Ming
School of Fashion and Textiles
Hong Kong Polytechnic University
Kowloon, Hong Kong, China

Bernhard Schartel
Bundesanstalt für Materialforschung
    und-prüfung (BAM)
Berlin, Germany

Marjan Entezar Shabestari
Department of Materials Science and
    Engineering and Chemical
    Engineering
Polytechnic School
Carlos III University of Madrid
Leganés, Madrid, Spain

Yong-Qian Shi
Fuzhou University
Fuzhou, PR China

Karen Ter-Zakaryan
TEPOFOL Ltd.
Moscow, Russia

Sabu Thomas
School of Energy Materials
and
International and Inter-University
    Center for Nanoscience and
    Nanotechnology
and
School of Nanoscience and
    Nanotechnology
Mahatma Gandhi University
Kottayam, Kerala, India

Arya Uthaman
School of Energy Materials
Mahatma Gandhi University
Kottayam, Kerala, India

Wei Wang
School of Mechanical and
    Manufacturing Engineering
University of New South Wales
Sydney, Australia

Xin Wang
State Key Laboratory of Fire Science
University of Science and Technology
    of China
Hefei, Anhui, PR China

Hongyu Yang
Chongqing University
Chongqing, PR China

Yu-Ting Yang
State Key Laboratory of Fire Science
University of Science and Technology
    of China
Hefei, Anhui, PR China

Bin Yu
State Key Laboratory of Fire Science
University of Science and Technology
   of China
Hefei, Anhui, PR China

Anthony Chun Yin Yuen
Department of Building Environment
   and Energy Engineering
The Hong Kong Polytechnic University
Hung Hom, Kowloon
Hong Kong SAR, PR China

1 Introduction to 2D Nanomaterials for Fire Safety of Polymers

Fukai Chu, Xin Wang, and Yuan Hu
State Key Laboratory of Fire Science, University of Science and Technology of China, Hefei, Anhui, PR China

CONTENTS

1.1 INTRODUCTION

Polymer materials, also known as macromolecular materials, are gradually becoming widely used in different fields due to their structural variability, diverse properties, good processability, simple molding process, and reasonable cost (Jeon and Baek 2010).

DOI: 10.1201/9781003327158-1

At present, human beings have entered the era of polymer materials. Polymers are widely used in various fields of production and life, and their end products cover many fields, such as construction, electrical appliances, transportation, household products, and aerospace. Depending on the thermal processing of the materials, polymers can be divided into thermoplastics and thermosets. Thermoplastic polymers have linear or branched structures. At elevated temperatures, it can be softened or melted by heat and then processed by molding; at lower temperatures, it becomes hard below the softening point and maintains the molded shape. For example, polyethylene (PE) and polypropylene (PP) are typical thermoplastic materials that have excellent electrical insulation, low-temperature resistance, easy processing, good molding chemical stability, and cost performance (Rong et al. 2001, Maddah 2016, Rezvani Ghomi et al. 2020). They have been widely used in the production of films, daily necessities, pipes, wire and cable insulation, and sheathing materials. In contrast to thermoplastic materials, thermosetting polymers are not easily softened by heating and dissolved by chemical reagents. They are generally cross-linked polymers formed by a large number of linear or branched macromolecules connected by chemical bonds (Rodriguez et al. 2014). Typical thermosetting epoxy (EP), unsaturated polyester resin (UPR) materials, etc., can be used as adhesives, coatings, and matrix resins for fiber-reinforced composite materials (Chu et al. 2019, Chu, Qiu, Zhang, et al. 2022).

However, most polymers contain a large amount of carbon and hydrogen elements, and flammable volatiles can be generated during the pyrolysis process when exposed to external heat sources (Rodriguez et al. 2014). After mixing with air, these flammable volatiles are ignited at temperatures above the ignition point. Thus, the inherently flammable nature of polymers and products produced from these materials presents serious fire hazards (Mouritz and Gibson 2007). These hazards are mainly manifested in thermal hazards and non-thermal hazards (Belpomme et al. 2018). A thermal hazard occurs when the polymer releases a large amount of heat when it burns. It spreads to the surrounding environment and expands the fire through heat conduction, convection, and radiation, thereby causing personal burns and thermal damage to buildings (Iimoto and Tanaka 2008, Nishino, Tanaka, and Hokugo 2012). Non-thermal hazards refer to the damage to life and the environment caused by the smoke and toxic gases released by the burning of polymers. Studies have shown that most casualties in fires are caused by non-thermal hazards (Nishino, Tanaka, and Hokugo 2012, Urashima and Chang 2000). For example, the released thick smoke will interfere with the line of sight and hinder rescue in the fire scene. The produced toxic and volatile products directly threaten the life safety of personnel. Irritant and corrosive combustion products can also harm the skin, eyes, and other organs of people.

The pyrolysis behavior of different polymers, as well as the heat release and smoke generation characteristics during combustion, are significantly different. For example, owing to the presence of highly reactive tertiary C-H, the pyrolysis sensitivity of PP in oxygen is 3.5 times greater than that of PE (Honus et al. 2018, Aboulkas and El Bouadili 2010); even when PP is degraded in an inert environment, the temperature at which PP begins to produce volatiles is 100°C lower than that of PE. However, the limiting oxygen index (LOI) of both PE and PP is approximately 17%, which makes them extremely flammable. Large amounts of heat released during combustion and high-temperature melt aggravate the fire spread (Wang, Liu, and Li 2017). Additionally, the toxicity of PP during combustion is mainly from the CO generated by combustion. Thermosetting EPs, such as diglycidyl ether of bisphenol-A (DGEBA), also show extremely flammable properties and complex decomposition mechanisms (Kandola, Magnoni, and Ebdon 2022). Common pyrolysis products of EP are CO, CO_2, methane, acetone, phenolic compounds, etc. (Giori and Yamauchi 1984, Grassie, Guy, and Tennent 1985). The LOI of EP is approximately 20%, and it continues to spontaneously ignite after removing the external ignition source. However, EP is a carbon-forming polymer, and the carbon layers generated by combustion can isolate the internal polymer from the flame zone, thereby slowing the heating rate of the underlying matrix and reducing its decomposition rate (Chu et al. 2020).

1.2 FLAME RETARDANTS FOR POLYMER MATERIALS

1.2.1 FIRE HAZARDS

Combustibles, sufficient heat, and oxygen are the three elements for the combustion of polymer materials. To prevent and control the fire hazard of polymer materials, it is necessary to carry out flame retardant modification of the polymer matrix. The flammability of polymers can be reduced by isolating oxygen, blocking the transfer of combustibles, and reducing fuel release. The thermophysical and chemical properties of combustible materials can be changed by adding suitable flame retardants or introducing flame retardant structural units (Lim et al. 2012).

1.2.2 HALOGEN-CONTAINING FLAME RETARDANTS

Halogen-containing flame retardants such as bromine-based and chlorine-based flame retardants are widely used in flame retardant polymer research (Bocchini and Camino 2010). During the combustion process, the produced hydrogen halide can dilute the air in the combustion zone and isolate the supplement of fresh air, suffocating the combustible. At the same time, hydrogen halide can also capture OH·, thereby reducing the flame and achieving flame retardancy. Although a small amount of halogen-containing flame retardants can impart a good flame-retardant effect to the polymer, more smoke, corrosive gases, and toxic products are generated during pyrolysis and combustion. Because of adverse effects on the environment and human health, halogenated flame retardants are gradually being constrained by relevant regulations, and alternative solutions are being sought (Georlette, Simons, and Costa 2000, Chen and Wang 2010).

1.2.3 PHOSPHORUS-BASED FLAME RETARDANTS

Among halogen-free flame retardants, phosphorus-based flame retardants have become the main research direction due to their excellent flame retardant performance, anti-dripping properties, and low price. Phosphorus-containing flame retardants can function in both condensed and gaseous phases (Chu, Zhou, et al. 2022). In the condensed phase, metaphosphoric acid and pyrophosphoric acid are generated from the degradation of phosphorus-containing flame retardants, which catalyze the dehydration, cyclization, and crosslinking of the polymer to form thermally stable carbon layers, delay the oxidative reaction and reduce the release of flammable volatiles (Zhang et al. 2021). In the gaseous phase, phosphorus-containing flame retardants such as dimethyl methylphosphonate (DMMP) and DOPO can pyrolyze to release free radicals such as PO· and HPO·(Krishnan, Kandola, and Ebdon 2021). After entering the gaseous phase, they capture the H· and HO·, cut off the combustion chain, and quench the flame. However, the thermal stability and compatibility between traditional phosphorus-containing flame retardants and polymer matrix need to be matched. During processing, flame retardants with low thermal stability will degrade, which has a negative influence on flame retardancy and cannot be used in engineering plastics that have high processing temperatures. At the same time, the poor compatibility between unsuitable flame retardants and the matrix makes it easy to migrate, resulting in decreased mechanical properties and flame retardant durability (Chu, Yu, et al. 2018). For example, although ammonium polyphosphate (APP), which was developed and applied by Monsanto Company in 1965, is widely used in the flame retardancy of polymer materials, its matrix compatibility and moisture absorption are still hot spots in current flame retardant research (Qin et al. 2015, Weil 1978). Additionally, reactive flame retardants are attracting extensive attention (Kandola et al. 2020, Bourbigot 2021, Seraji et al. 2022). This can introduce the flame retardant structures into the polymer molecular chain skeleton, thus improving the migration defects of additive flame retardants, and helping to enhance the flame retardant durability of materials.

1.2.4 SYNERGISTIC FLAME RETARDANTS

In the flame retardant system, phosphorus and other flame retardant elements, when used in a certain proportion, may produce good synergy, achieving better flame retardancy and improving mechanical properties. The N_2, NH_3, and H_2O produced by the nitrogen-containing structure during the pyrolysis and combustion process can dilute oxygen and combustibles in the gaseous phase and remove part of the heat through thermal convection (Jiang et al. 2015). The synergy of organophosphorus and nitrogen-containing flame retardant structures can become an effective intumescent system that produces a uniform carbonaceous foam layer when the polymer is burned (Xie et al. 2006, Liu et al. 2007, Nazir and Gaan 2020). In addition, silicon-containing flame retardants also attract attention due to their low hazards. When the material is burned, stable composite inorganic layers can be formed with produced silicon oxide, phosphoric acid, and pyrophosphoric acid, etc. (Sponton et al. 2008, Spontón et al. 2009, Qiu et al. 2018). However, the current phosphorus-containing synergistic flame retardant system still has the problem of low efficiency and needs to be added in large quantities to achieve a higher flame retardant level.

Thus, the need for further exploration of flame retardant technology can be attributed to the following points: (1) High flame retardant efficiency and cost performance, and the properties of polymer must be preserved to the greatest extent; (2) The thermal behaviors of flame retardant structures must match the processability of the polymer; and (3) The choice of flame retardants should meet the needs of environmental protection, recyclability, and sustainable development.

1.2.5 Two-Dimensional (2D) Nanosheet Flame Retardants

Nano flame retardant technology has recently become a hot spot in polymer flame retardant research (Innes and Innes 2011, Wang et al. 2020). Nanomaterials refer to materials with nanoscale (1–100 nm) and corresponding functional characteristics in structure. Due to various special effects, such as the small size effect, surface effect, quantum size effect, and macroscopic quantum tunneling effect, composite materials combining nanomaterials and polymers exhibit excellent physicochemical and multifunctional properties (Zhang et al. 2020, Liu et al. 2020). For the flame retardancy of composites, the small size effect and surface effect of nanomaterials help to greatly improve the flame retardant efficiency and dispersion. Through reasonable scale and structure design, the properties of polymer composites can be regulated.

During the combustion process of flame retardant polymers with 2D nanosheet flame retardants, the 2D nanosheet itself can act as a barrier layer, which can jointly block the heat and mass transfer processes with the catalyzed carbonaceous layer, thereby reducing the outward diffusion of heat and combustibles (Chu et al. 2021, Chu, Zhang, et al. 2018). The larger specific surface area of 2D nanosheets also facilitates the adsorption of pyrolysis and combustion products. At the same time, the 2D nanosheets containing transition metal elements can catalyze the formation of thermally stable carbon layers to form macroscopic physical barrier layers while using themselves as nanometer barrier layers. In addition, due to the large aspect ratio of 2D nanosheets, the penetration path of pyrolysis products of small molecules is prolonged, and the diffusion of gases is hindered so that the nanocomposites have a good barrier effect on smoke and flammable gases. In recent years, great progress has been made in the preparation of 2D nanosheet flame-retardant polymer composites. The 2D nanosheets used for flame retardancy mainly include clay, layered double hydroxides (LDHs), layered transition metal carbides (MXenes), graphene (oxide), graphitic carbon nitride (g-C_3N_4), layered boron nitride (BN), molybdenum disulfide (MoS_2), black phosphorus (BP), etc.

1.3 STRUCTURE AND PROPERTIES OF 2D NANOSHEETS

2D nanosheets have layered host structures, and the layers are stacked by specific structural units in a coplanar or co-edge manner. They can be divided into three types, cationic, anionic, and neutral inorganic layered materials, according to the charges of laminates. For example, the cationic type includes clay and polysulfide; the anionic type includes LDHs; and the neutral inorganic type includes single-element graphite, black phosphorus, and transition metal sulfide. The structural parameters of 2D nanomaterials determine their properties, specifically the composition, dimensions, and atomic arrangement. The differences in the composition, structure, and corresponding properties of 2D nanosheet materials also determine their effective application in flame retardant polymers. Some structural and performance parameters of different 2D nanosheets are shown in Table 1.1.

TABLE 1.1

Structural and Performance Parameters of Different 2D Nanosheets

2D Nanosheets	Thickness	Tensile Strength	Young's Modulus	Thermal Conductivity	Electrical Conductivity
MMT	0.96 nm	/	/	/	/
LDH	0.48 nm	/	/	/	/
Ti$_3$SiC$_2$	/	600 MPa	320 GPa	43 W/m.K	4.5×10^6 S/m
Graphene	0.335 nm	130±10 GPa	1 TPa	(4.84–5.30) $\times 10^3$ W/m.K	7200 S/m
g-C$_3$N$_4$	0.4–0.5 nm	40 GPa	320 GPa (Elastic)	/	/
h-BN	/	/	900 GPa	200 W/m·K	10^{18} Ω·cm (Resistivity)
MoS$_2$	0.6–0.7 nm	180 N/m (In-plane hardness)	270 GPa	52 W/m.K	1.37 S/m
BP	/	30% (tensile strain)	41.3 GPa (horizontal direction)	110 W/m.K (zigzag direction)	0.5–0.75 Ω·cm (Resistivity)

1.3.1 CLAY

Clay is an aluminosilicate produced by earth weathering or diagenesis. It can be divided into kaolin, bentonite, montmorillonite (MMT), illite, and so on, according to the different compositions. Among them, MMT is widely used in the preparation of polymer nanocomposites due to its low cost and easy availability, large cation exchange capacity, and easy intercalation. MMT is a plate-like particle with an average diameter of approximately 1 μm and a monolayer thickness of 0.96 nm. The Si^{4+} in the silicon-oxygen tetrahedron is often replaced by Al^{3+}, and the Al^{3+} in the aluminum-oxygen octahedron can be replaced by low-valent cations such as Mg^{2+} and Fe^{2+}, resulting in excess negative charge between the structural layers. To maintain electrical neutrality, cations such as K^+, Na^+, Ca^{2+}, and H^+ can be adsorbed between the crystal layers. These cations are present in a hydrated state and are exchangeable with each other. Therefore, the morphology, composition, and structure of MMT determine its excellent adsorption, cation exchange, dispersion and suspension, plasticity, and cohesion. At the same time, these features enable the intercalation of polymers between sheets to obtain nanocomposites with intercalated, exfoliated and intercalated-exfoliated hybrid structures. The mechanical strength, thermal properties, and flame retardant properties of the polymer have also been greatly improved (Kiliaris and Papaspyrides 2010, Gilman, Kashiwagi, and Lichtenhan 1997, Lu, Song, and Hu 2011).

1.3.2 LAYERED DOUBLE HYDROXIDES (LDHs)

LDHs are typical anionic 2D nanosheet materials with a common chemical composition of $Mg_6Al_2(OH)_{16}CO_3.4H_2O$. The smallest structural unit, the MgO_6

octahedron, is arranged in an orderly manner by sharing edges to form the main layer, where part of the Mg^{2+} located in the center of the octahedron is replaced by Al^{3+}, causing the layer to be positively charged. To balance the positive charge of the laminate, exchangeable anions are introduced between the layers to keep the overall structure electrically neutral. The changes in metal elements in lamellae and the exchangeable properties of anions between lamellae endow LDHs with controllability of the composition, structure, and properties, making LDHs promising functional materials in the field of flame retardants. When heated, the hydroxyl groups in the lamellar structure, interlayer anions, and water are removed, reducing the concentration of combustible gas and O_2 and decreasing the high temperature in the combustion zone through endothermic reactions. At the same time, due to the lamellar barrier effect, adsorption effect, and catalytic carbonization effect, LDHs can be used as halogen-free flame retardants and smoke suppressants, which are widely used in plastics, rubber, wood, coatings, and other fields (Liu et al. 2018, Matusinovic and Wilkie 2012).

1.3.3 LAYERED TRANSITION METAL CARBIDES (MXENES)

MXenes include 2D transition metal carbides, transition metal nitrides, and transition metal carbonitrides, with layered structures similar to graphene. The general chemical formula of MXenes is $M_{n+1}AX_n$ (n = 1, 2, 3), where M are transition metal elements, X are carbon or nitrogen elements, and A is from the 12–16th column of the element period (such as Al, Si, etc.). The crystal structure of bulk MAX is composed of alternating MX layers and A layers, which are the precursors for monolayer or few-layer MXenes. The weak bonding force between the precursor layers can be destroyed by chemical etching or mechanical lift-off. In 2011, Gogotsi et al. used hydrofluoric acid to selectively etch the Al atomic layer in Ti_3AlC_2, obtaining Ti_3C_2 nanosheets. Due to the large variety of MAX phases, a large number of MXenes with special properties can be prepared by etching the MAX phase, which would provide an important guarantee for the application of such 2D crystal materials. Additionally, the large specific surface area, good electrical conductivity, stability, magnetic properties, and mechanical properties of this 2D nanomaterial make it exhibit broad application prospects in multifunctional polymer composites (Chen et al. 2022).

1.3.4 GRAPHENE (OXIDE)

Graphene is composed of a layer of sp^2 hybridized carbon atoms, where the C-C length in the six-membered ring is 0.142 nm and the thickness of the nanosheet is approximately 0.34 nm. A single layer of graphite can be called graphene, and a stack of more than 10 layers of graphene is called graphite. Compared with graphite (10 m^2/g), graphene has a larger specific surface area (2630 m^2/g), as well as better mechanical, electrical, and thermal properties. Therefore, graphene can function as an excellent multifunctional reinforcement for polymer composites. However, the relatively strong van der Waals attraction between defect-free graphene sheets with perfect structures makes it easy for them to stack or aggregate with each other. Thus, it is difficult to disperse uniformly in the polymer matrix, which weakens the enhancement effect. After graphite is processed into graphite oxide, the interlayer spacing can be expanded to 0.65–0.75 nm according to different water contents. The van der Waals force between the layers is also weakened, and finally the graphene oxide (GO) with a thickness of about 0.78 nm can be exfoliated. The presence of functional groups such as carboxyl, hydroxyl, and epoxy groups on the edges and faces of GO facilitates the surface modification of GO, as well as the dispersion and strong interaction in the polymer matrix (Ahmad, Fan, and Hui 2018, Wang, Kalali, et al. 2017, Wang et al. 2022).

1.3.5 GRAPHITIC CARBON NITRIDE (g-C_3N_4)

C_3N_4 is a carbonaceous material with a certain absorption of visible light, excellent corrosion resistance, and good stability. There are several allotropes of C_3N_4, and different pressures on C_3N_4 in different directions can achieve the phase transition between the above allotropes. Thermodynamically, g-C_3N_4 is the most stable carbonitride under natural conditions, and it has attracted great attention due to its high chemical and thermal stability, excellent optical and photoelectrochemical properties. g-C_3N_4 contains -NH_2 and -NH groups remaining after thermal polycondensation, and the number of these groups varies with the degree of polycondensation. Notably, methods such as protonation, doping, loading of metals and inorganics, and chemical modification can be used to tune the performance of g-C_3N_4. Recently, researchers have prepared polymer composites with g-C_3N_4 to enhance the hydrophobicity, thermal stability, flame retardant properties, and mechanical properties (Myllymaa et al. 2009, Shi et al. 2017).

1.3.6 LAYERED BORON NITRIDE (BN)

BN has a variety of allotropes, which mainly include hexagonal BN (h-BN), rhombohedral BN (r-BN), cubic BN (c-BN), and wurtzite BN (w-BN) according to the different atomic arrangements. Thermodynamically, the h-BN structure is the most stable. The atomic structure of h-BN can be seen as the carbon atoms in graphene being replaced by boron atoms and nitrogen atoms, both of which present a six-membered ring honeycomb structure. The h-BN 2D nanosheets are called "white graphene" due to their lattice constant and interlayer spacing being similar to those of graphene. Due to their low density, high thermal conductivity, electrical insulation, excellent dielectric properties, and low friction coefficient, h-BN nanosheets greatly improve the corresponding properties of polymer composites, thereby meeting the needs of different fields. In terms of flame retardancy, the high thermal and chemical stability of h-BN, as well as the physical barrier effect of the 2D nanosheets, can effectively inhibit the mass and heat transfer during the combustion of polymers, thus improving the flame retardancy (Shen 2021, Yue et al. 2019).

1.3.7 MOLYBDENUM DISULFIDE (MoS$_2$)

MoS$_2$ is a 2D transition metal dichalcogenide (TMD). The layers of TMDs are stacked together by van der Waals forces, which are usually represented by the formula MX$_2$, where M represents transition metal elements such as Mo, W, Nb, and Re, and X represents chalcogen elements such as S, Se, and Te. TMDs are expected to be used in optoelectronic devices, photocatalysts, lithium-ion batteries, and supercapacitors due to their ultrathin monolayer atomic thickness, large specific surface area, abundant edge sites, good chemical stability, and tunable band gap. Among them, MoS$_2$ has good thermal stability and low thermal conductivity and can play a lamellar barrier effect in the process of pyrolysis and combustion of polymer materials. At the same time, as the main component of the smoke suppressant, molybdenum-containing compounds can reduce the amount of smoke generated during combustion through catalytic reactions, effectively reducing smoke toxicity and realizing efficient flame retardancy of the polymer material (Wang, Xing, et al. 2017).

1.3.8 BLACK PHOSPHORUS (BP)

BP is a typical single-element 2D layered material, and it is the most stable allotrope in phosphorus form. Different from layered graphite, each phosphorus atom in BP is connected with the other three surrounding phosphorus atoms through sp hybrid orbitals, with strong chemical covalent bonds, thereby forming a stable wrinkled layered structure. Among the layers of stacked BP nanosheets, the weak van der Waals interaction between layers makes it feasible for the exfoliation of BP to form monolayer or few-layer nanosheets. Therefore, the high mobility, appreciable direct band gap, and anisotropy make BP nanosheets significantly advantageous in nanoelectronics and optoelectronics. Recently, the excellent mechanical properties, reduced friction coefficient, and high thermal stability of BP nanosheets have also attracted extensive attention for their application in polymer composites. As an allotrope of red phosphorus, BP nanosheets are able to combine the properties of a 2D sheet structure and phosphorus-containing flame retardants. Compared with red phosphorus, BP has higher thermal stability; compared with ordinary 2D nanosheets, BP contains extremely high phosphorus contents, so it theoretically has higher flame retardant efficiency(Qiu 2021).

1.4 PREPARATION OF POLYMER-BASED 2D NANOSHEET COMPOSITES

1.4.1 SOL-GEL METHOD

By forming a homogeneous solution of the precursor in a certain organic solvent, the solute can be dissolved to form some nanoparticles and become a sol, and then the sol can be converted into a gel by solvent volatilization or heating (Pomogailo 2005, Wang et al. 2005, Guglielmi, Kickelbick, and Martucci 2014). According to the type of polymer and the interaction with nanomaterials, it can be divided into the following categories: (1) Dissolve the precursor in the polymer solution, and then sol and gel. (2) After the sol is generated, it is blended with the polymer and then gelled. (3) In the presence of the precursor, the monomer is first polymerized and then gelled. (4) The precursor and the monomer are dissolved in the solvent, and dissolution and polymerization are carried out simultaneously. The advantage of the sol-gel method is that it can be carried out under mild conditions, and the reaction conditions, the ratio of organic and inorganic components, and the dispersion can be easily controlled. For example, to prepare EP nanocomposites, Yu et al. first prepared hydroxylated h-BN (BNO) by heating h-BN in air at 1000°C, and then covalently bound it to the resin modified with (3-isocyanatopropyl) triethoxysilane by the sol-gel method. The results show that BNO can be dispersed in EP in the form of exfoliation and intercalation, with strong interfacial interactions (Yu et al. 2016). However, its disadvantage is that the volatilization of solvents and small molecules may cause shrinkage of the material and affect the mechanical properties of the material during the drying process of the gel (Chujo 1996).

1.4.2 BLENDING METHOD

The blending method is the simplest and most common method for preparing polymer-based nanocomposites and is suitable for particles of various shapes. Blending methods include solution blending, emulsion blending, melt blending, and other methods. In solution blending, the polymer resin is first dissolved in the solvent (Madaleno, Schjødt-Thomsen, and Pinto 2010, Kango et al. 2013). Then, the nanoparticles are added to the above system and stirred well. After film formation or casting into molds, the solvents were finally removed. Emulsion blending refers to dispersing the nanoparticles uniformly in the polymer emulsion first and then removing the solvent for processing. In melt blending, nanoparticles, and polymers are melt blended in an internal mixer, twin-screw extruder, and other equipment. During the process, the nanoparticles are easily agglomerated during the heating process, so surface treatment of the nanoparticles should be carried out before blending. The blending method can be used for the preparation of all polymer-based nanocomposites. However, according to the difference between thermoplastic and thermosetting polymers, the choice of blending methods is different. For example, solution blending is often used for thermosetting resin-based nanocomposites such as epoxy (Bari et al. 2017) and unsaturated polyester (Suh, Lim, and Park 2000). In the preparation of thermoplastic resin-based nanocomposites such as PP and polyurethane elastomers, solution blending and melt blending are both used (Deka and Maji 2010, Cruz and Viana 2015).

1.4.3 IN SITU POLYMERIZATION

In in-situ polymerization, nanoparticles are first uniformly dispersed in polymerized monomers. The polymer is then polymerized by an initiation reaction. The monomer can undergo free radical polymerization or polycondensation in the oil phase, so it is suitable for the preparation of most polymer-based nanocomposites. Due to the small molecules and low viscosity of polymer monomers, the nanoparticles are easily dispersed evenly after effective surface modification, which ensures the uniformity of the system and various physical properties. To uniformly disperse the stearate intercalated LDH in the poly(methyl methacrylate) (PMMA) matrix, Nogueira et al. dispersed the LDH in methyl methacrylate monomer and then conducted polymerization through an in-situ method. The results showed that the interaction between the two compounds was improved, and the mechanical and thermal properties of the nanocomposites were enhanced (Nogueira et al. 2011). At the same time, the reaction conditions of the in situ polymerization are mild, and only one polymerization process is performed during the polymerization process without thermal processing, which avoids the resulting degradation and ensures the nanometer characteristics of the particles, thereby maintaining the stability of their basic properties (Bhanvase and Sonawane 2014, Roghani-Mamaqani, Haddadi-Asl, and Salami-Kalajahi 2012).

1.4.4 INTERCALATION METHOD

Weak interactions exist between the layers in nanomaterials with a layered structure, which can be overcome by external action to exfoliate the lamellae. Single-layer or few-layer 2D nanomaterials are more helpful to improve their dispersion in the polymer matrix and realize their nano effects. Therefore, the intercalation and exfoliation of 2D nanosheets need to be considered during the preparation of polymer-based nanocomposites with 2D nanosheets. The intercalation method is specifically used for the preparation of polymer-based layered inorganic nanocomposites (Giannelis 1996). According to the different forms, intercalation can be divided into the following types: (1) By mixing polymerized monomers, initiators or their solutions with 2D nanosheets, the monomers are inserted between the sheets; alternatively, polymer monomers, initiators, emulsifiers and 2D nanosheets are mixed at high speed to form an emulsion, in which the intercalation of monomers is achieved, and then polymerization is carried out. For example, Ge et al. synthesized a new phosphorus-containing flame retardant copolyester/MMT nanocomposite by in-situ and interca-lation polycondensation of terephthalic acid, ethylene glycol, and 2-carboxyethyl (phenylphosphonic acid) with MMT (Ge et al. 2007). This can effectively improve the compatibility problem between inorganic and organic components and the difficulty in intercalation of organic macromolecular components. (2) After dissolving the polymer in the solvent, the macromolecular chains are intercalated into 2D nanosheets with the help of the solvent, and then the solvent is removed. For example, MgAl-LDH was first prepared through a homogeneous precipitation method and then intercalated and exfoliated with carboxymethyl cellulose in formamide solution at 80°C to prepare the corresponding layered nanocomposites (Kang et al. 2009). (3) The polymer is heated above the melting point or softening point, and the macromolecular chains are dif-fused and intercalated between the 2D nanosheets by mixing, shearing, etc. (4) In addition, the 2D nanosheet can be delaminated in a suitable dispersion first, and then polymers or polymerized monomers can be added to form nanocomposites directly or after polymerization. For example, Troutier-Thuilliez et al. exfoliated LDHs into nanosheets in solution, and then prepared nanocomposites by mixing LDH nanosheets with waterborne polyurethane (Troutier-Thuilliez et al. 2009).

1.5 FIRE SAFETY OF POLYMERS WITH 2D NANOMATERIALS

The 2D nanosheets are environmentally friendly flame-retardant materials. Their good catalytic charring capacity and nanosheet effects can effectively reduce the fire hazard of polymers, which have been used to compound with traditional flame retardants. Compared with traditional flame retardants, the 2D nanosheets can make the composites obtain good flame retardancy and smoke suppression effects at low dosages. Taking thermoplastic PP material as an example, compared with the same amounts of commercial APP flame retardants (Shao et al. 2014), phosphorus-silicon synergistic flame retardants (Si-APP) (Deng et al. 2014) and phosphorus-nitrogen intumescent flame retardants (PA-APP, APP/CFA/zeolite) (Shao et al. 2014, Li and Xu 2006) exhibited higher UL-94 and limiting oxygen index (LOI), lower peak heat release rate (PHRR) and peak smoke production rate (PSPR), and improved tensile and impact strengths (Table 1.2). The addition of a small amount of nanosheets such as MMT and GO cannot make PP reach higher UL-94 and LOI levels, but the reduction of PHRR and PSPR of modified PP is comparable to that of phosphorus-containing flame retardants with higher additions (Xiao et al. 2017, Huang et al. 2014, Chen et al. 2019). During the combustion of polymer-based nanocomposites, the 2D nanosheets can block the heat transfer to the underlying material and limit the entry of combustible volatiles into the flame, thereby slowing the combustion process of the composites. In addition, the PP nanocomposites exhibited different flame retardant efficiencies with different 2D nanosheets, which is related to the physical and chemical properties of the 2D nanosheets. This also indicates that different 2D nanomaterials should be selected according to the types of polymers and usage scenarios.

At the same time, the smaller size and higher specific surface area of the 2D nanosheets make them uniformly dispersed in the polymer matrix, which contributes to improving the mechanical properties of flame-retardant polymer materials. Compared with pure PP materials, the addition of a large amount of phosphorus-containing flame retardants reduces the tensile and impact strengths of PP to a certain extent. For example, compared with pure PP, the tensile and impact strengths of flame retardant PP with 30 wt% APP were approximately decreased by 47.5% and 69.6%, respectively (Wen et al. 2014). However, two-dimensional nanomaterials such as GO significantly enhance the mechanical properties of PP. For example, compared with pure PP, the tensile and impact strengths of the PP nanocomposite prepared by melt blending with 15.84 wt% of reduced GO (rGO) increased from 30 MPa and 2.6 kJ/M^2 to 43 MPa and 9.4 kJ/M^2, respectively (Huang et al. 2014, Chen et al. 2019). Although high-efficiency phosphorus-containing flame retardants such as PCH$_3$PG can reduce the content of 30 wt% for traditional flame retardant systems to 15 wt% to a certain extent, the mechanical properties of the corresponding flame retardant UPR materials have not been improved (Chu, Qiu, Zhou, et al. 2022). Compared with phosphorus-containing flame retardants, 2D nanosheets such as MoS$_2$ significantly improve the mechanical properties of UPR materials (Wang, Wen, et al. 2017, Wang1 et al. 2016). However, 2D sheets still exhibit some disadvantages, such as poor interfacial compatibility in polymer matrices, which limits their large-scale development in the

TABLE 1.2

Comparison of the Flame Retardant Efficiency of Different Types of Flame Retardants in Thermoplastic PP and Thermoset UPR

Flame Retardants	Polymer	Contents (wt%)	UL-94	LOI (%)	PHRR	PSPR	Tensile Strength	Impact Strength	References
APP	PP	25	NR	20.0	↓43.8%	↓8.4%	25.0 MPa	1.73 kJ/m^2	Shao et al. (2014)
PA-APP	PP	25	V0	32.5	↓80.7%	↓80.0%	25.6 MPa	1.70 kJ/m^2	Shao et al. (2014)
Si-APP	PP	25	NR	25.0	↓84.9%	↓64.8%	/	/	Deng et al. (2014)
APP/CFA/Zeolite	PP	16/8/1	V0	37.0	/	/	27.2 MPa	3.7 kJ/m^2	Li and Xu (2006)
IFR/MMT	PP	16/2	V0	28.8	↓74.6%	↓62.5%	/	/	Xiao et al. (2017)
GO	PP	2	NR	20.1	↓60.9%	/	/	/	Huang et al. (2014)
GO	PP	15.84	/	/	/	/	43.0 MPa	9.4 kJ/m^2	Chen et al. (2019)
DMMP/APP/ZB	UPR	10/12/2	V1	29.4	/	/	/	/	Jiang et al. (2019)
PCH$_3$PG	UPR	15	V0	29.0	↓66.3%	/	↓14.9% (20 wt%)	↓32.7% (20 wt%)	Chu et al. (2022)
MoS$_2$	UPR	5	/	/	↓12.7%	↓11.9% (CO)	30.0 GPa (storage modulus)	↑17°C (T$_g$)	Wang et al. (2017)
MoS$_2$@HPA	UPR	5	/	/	↓43.2%	↓38.1% (CO)	40.2 GPa (storage modulus)	↑23°C (T$_g$)	Wang et al. (2017)
MoS$_2$@GO	UPR	/	/	/	/	↓62.5% (CO)	/	/	Wang et al. (2016)

preparation of high-performance polymer composites. Efficient functional modification is the key to solving this problem. The functional modification of 2D nanosheets mainly includes organic modification and inorganic hybridization. Organic modification methods, such surface grafting, can introduce flame retardant surface modifiers on the surface of 2D nanosheets, simultaneously improving the flame retardant efficiency of 2D nanosheets and enhancing the interfacial compatibility. For example, the grafting of phosphorus-containing HPA on the surface of MoS_2 significantly enhanced the inhibition of 2D nanofillers on the heat release and toxic CO of UPR materials and further enhanced the mechanical properties of flame-retardant UPR materials. 2D nanomaterials and other inorganic substances are prepared by noncovalent bonds such as π-π to prepare hybrids, which can also enhance the flame retardancy efficiency of nanomaterials. This is because the nanohybrids can have synergistic flame retardant effects with different hybrid components (He et al. 2020, Visakh and Yoshihiko 2015, Wang, Kalali, et al. 2017).

Based on the above analysis, in order to function the performance enhancement effect of nanomaterials in polymers, the following conditions need to be met: (1) 2D nanomaterials with specific functions should be selected according to specific polymer materials and multifunctional modification requirements. For example, h-BN can be selected to simultaneously improve the flame retardancy and thermal conductivity of polymer materials; graphene or GO can be selected to simultaneously improve the flame retardant and mechanical properties of the polymer; and a variety of nanomaterials can also be used to produce anticorrosion, hydrophobic and oleophobic, and self-cleaning functions. (2) The realization of the nanoeffect of 2D nanosheets requires a smaller particle size and fewer nanosheets. The smaller the particle size of nanomaterials is, the more surface defects there are, the greater the activity, and the stronger the bonding ability with molecular chains, which is beneficial to improve the mechanical strength of polymer nanocomposites. However, there is a balance between particle size, performance, and cost. (3) The 2D nanomaterials should be uniformly dispersed in the polymer matrix. Nanomaterials exhibit a state of spontaneous aggregation in the matrix. By selecting a reasonable composite material preparation process and multifunctional surface modification methods, the tendency of easy aggregation can be reduced, and a uniform dispersion of 2D nanomaterials can be achieved. Nanomaterials exhibit a tendency to spontaneously aggregate in the matrix. By selecting a reasonable composite material preparation process and multifunctional surface modification methods, the aggregation tendency can be reduced, and a uniform dispersion of 2D nanomaterials can be achieved. (4) The interfacial interaction between 2D nanosheets and polymers should be improved. The stronger the interfacial interaction, the more energy required to break the noncovalent and covalent bonds between the interfaces, and the higher the mechanical strength of the polymer nanocomposite. The improvement of this interfacial interaction also comes from effective surface modification methods. According to the characteristics of polymer materials, surface modifiers of nanomaterials containing the same groups can be designed, or the surface roughness of nanomaterials can be enhanced by hybridization to enhance the interfacial bonding effect.

1.6 CONCLUSION

In recent years, research on polymer-based nanocomposites has developed rapidly and has attracted extensive attention in both academic research and industrial production. Due to the nanoeffect of nanoparticles, polymer-based nanocomposites not only have the common characteristics of ordinary composite materials but also have some special properties. Among them, research on the flame retardant properties of polymer composites with nanosheets such as MMT, LDHs, graphene, and GO has obtained relatively many reports. These composite materials exhibit unique properties due to the dispersed nanosheets in the polymer, as well as advantages unmatched by pure polymers and conventional filled polymers, such as improved mechanical strength, improved thermal stability, reduced gas permeability, reduced melt index, and improved flame retardant properties.

The improvement in the flame retardant properties of polymer composites with 2D nanosheets is first due to the lamellar barrier effect. This barrier effect effectively inhibits the release of heat and flue gas products during combustion. At the same time, 2D nanosheets containing metal elements, P, Si, and other elements are helpful for catalytic charring. This improves the charring capacity of polymers and the thickness of char layers. The thermally stable carbon layers formed by combustion not only reduce the heat transferred to the bottom layer but also isolate the oxygen required for combustion, thereby inhibiting combustion and achieving flame retardancy. In addition, for polymer composites with BP and other 2D nanomaterials modified with phosphorus-containing structures, free radical trapping reactions occur during the combustion process.

REFERENCES

Aboulkas, A., and A. El Bouadili. 2010. "Thermal degradation behaviors of polyethylene and polypropylene. Part I: Pyrolysis kinetics and mechanisms." *Energy Conversion and Management* 51 (7):1363–1369.

Ahmad, Hassan, Mizi Fan, and David Hui. 2018. "Graphene oxide incorporated functional materials: A review." *Composites Part B: Engineering* 145:270–280.

Bari, Pravin, Samrin Khan, James Njuguna, and Satyendra Mishra. 2017. "Elaboration of properties of graphene oxide reinforced epoxy nanocomposites." *International Journal of Plastics Technology* 21 (1):194–208.

Belpomme, Dominique, Lennart Hardell, Igor Belyaev, Ernesto Burgio, and David O. Carpenter. 2018. "Thermal and non-thermal health effects of low intensity non-ionizing radiation: An international perspective." *Environmental Pollution* 242:643–658.

Bhanvase, B.A., and S.H. Sonawane. 2014. "Ultrasound assisted in situ emulsion polymerization for polymer nanocomposite: A review." *Chemical Engineering and Processing: Process Intensification* 85:86–107.

Bocchini, Sergio, and Giovanni Camino. 2010. "Halogen-containing flame retardants." *Fire Retardancy of Polymeric Materials* 2: 75–106.

Bourbigot, Serge. 2021. "Intumescence-based flame retardant." In: "Non-Halogenated Flame Retardant Handbook." Alexander B. Morgan (Ed.). John Wiley & Sons. pp. 169–238.

Chen, Li, and Yu-Zhong Wang. 2010. "A review on flame retardant technology in China. Part I: development of flame retardants." *Polymers for Advanced Technologies* 21 (1):1–26.

Chen, Wenhua, Pengju Liu, Yuan Liu, and Zhuoxin Liu. 2022. "Recent advances in two-dimensional $Ti_3C_2T_x$ MXene for flame retardant polymer materials." *Chemical Engineering Journal* 446:137239.

Chen, Xiaosui, Yihan Ma, Yinjia Cheng, Aiqing Zhang, and Wei Liu. 2019. "Enhanced mechanical and flame-resistant properties of polypropylene nanocomposites with reduced graphene oxide-functionalized ammonium polyphosphate and pentaerythritol." *Journal of Applied Polymer Science* 136 (41):48036.

Chu, Fukai, Yanbei Hou, Longxiang Liu, Shuilai Qiu, Wei Cai, Zhoumei Xu, Lei Song, and Weizhao Hu. 2019. "Hierarchical structure: An effective strategy to enhance the mechanical performance and fire safety of unsaturated polyester resin." *ACS Applied Materials & Interfaces* 11(32):29436–29447.

Chu, Fukai, Chao Ma, Tao Zhang, Zhoumei Xu, Xiaowei Mu, Wei Cai, Xia Zhou, Shicong Ma, Yifan Zhou, and Weizhao Hu. 2020. "Renewable vanillin-based flame retardant toughening agent with ultra-low phosphorus loading for the fabrication of high-performance epoxy thermoset." *Composites Part B: Engineering* 190:107925.

Chu, Fukai, Shuilai Qiu, Shenghe Zhang, Zhoumei Xu, Yifan Zhou, Xiaoyu Luo, Xin Jiang, Lei Song, Weizhao Hu, and Yuan Hu. 2022. "Exploration on structural rules of highly efficient flame retardant unsaturated polyester resins." *Journal of Colloid and Interface Science* 608:142–157.

Chu, Fukai, Shuilai Qiu, Yifan Zhou, Xia Zhou, Wei Cai, Yulu Zhu, Lingxin He, Lei Song, and Weizhao Hu. 2022. "Novel glycerol-based polymerized flame retardants with combined phosphorus structures for preparation of high performance unsaturated polyester resin composites." *Composites Part B: Engineering* 233:109647.

Chu, Fukai, Zhoumei Xu, Yifan Zhou, Shenghe Zhang, Xiaowei Mu, Junling Wang, Weizhao Hu, and Lei Song. 2021. "Hierarchical core–shell $TiO_2@$ LDH@ Ni $(OH)_2$ architecture with regularly-oriented nanocatalyst shells: Towards improving the mechanical performance, flame retardancy and toxic smoke suppression of unsaturated polyester resin." *Chemical Engineering Journal* 405:126650.

Chu, Fukai, Xiaojuan Yu, Yanbei Hou, Xiaowei Mu, Lei Song, and Weizhao Hu. 2018. "A facile strategy to simultaneously improve the mechanical and fire safety properties of ramie fabric-reinforced unsaturated polyester resin composites." *Composites Part A: Applied Science and Manufacturing* 115:264–273.

Chu, Fukai, Dichang Zhang, Yanbei Hou, Shuilai Qiu, Junling Wang, Weizhao Hu, and Lei Song. 2018. "Construction of hierarchical natural fabric surface structure based on two-dimensional boron nitride nanosheets and its application for preparing biobased toughened unsaturated polyester resin composites." *ACS Applied Materials & Interfaces* 10 (46):40168–40179.

Chu, Fukai, Xia Zhou, Xiaowei Mu, Yulu Zhu, Wei Cai, Yifan Zhou, Zhoumei Xu, Bin Zou, Zhenzhen Mi, and Weizhao Hu. 2022. "An insight into pyrolysis and flame retardant mechanism of unsaturated polyester resin with different valance states of phosphorus structures." *Polymer Degradation and Stability* 202:110026.

Chujo, Yoshiki. 1996. "Organic—inorganic hybrid materials." *Current Opinion in Solid State and Materials Science* 1 (6):806–811.

Cruz, Silvia M., and Júlio C Viana. 2015. "Structure–properties relationships in thermo-plastic polyurethane elastomer nanocomposites: Interactions between polymer phases and nanofillers." *Macromolecular Materials and Engineering* 300 (11):1153–1162.

Deka, Biplab K., and T.K. Maji. 2010. "Effect of coupling agent and nanoclay on properties of HDPE, LDPE, PP, PVC blend and Phargamites karka nanocomposite." *Composites Science and Technology* 70 (12):1755–1761.

Deng, Cheng-Liang, Shuang-Lan Du, Jing Zhao, Zhen-Qi Shen, Cong Deng, and Yu-Zhong Wang. 2014. "An intumescent flame retardant polypropylene system with simultaneously improved flame retardancy and water resistance." *Polymer Degradation and Stability* 108:97–107.

Ge, Xin-Guo, De-Yi Wang, Chuan Wang, Ming-Hai Qu, Jun-Sheng Wang, Cheng-Shou Zhao, Xin-Ke Jing, and Yu-Zhong Wang. 2007. "A novel phosphorus-containing co-polyester/montmorillonite nanocomposites with improved flame retardancy." *European Polymer Journal* 43 (7):2882–2890.

Georlette, Pierre, Joseph Simons, and L. Costa. 2000. "Halogen-containing fire-retardant compounds." In: "Fire Retardancy of Polymeric Materials." A.F. Grand, C.A. Wilkie (Eds.). pp. Marcel Dekker Inc. pp. 245–284.

Giannelis, Emmanuel P. 1996. "Polymer layered silicate nanocomposites." *Advanced Materials* 8 (1):29–35.

Gilman, Jeffrey W., Takashi Kashiwagi, and Joseph D. Lichtenhan. 1997. "Nanocomposites: A revolutionary new flame retardant approach." *International Sampe Symposium and Exhibition*. Anaheim, California, USA.

Giori, C., and T. Yamauchi. 1984. "Effects of ultraviolet and electron radiations on graphite-reinforced polysulfone and epoxy resins." *Journal of Applied Polymer Science* 29 (1):237–249.

Grassie, Norman, Marilyn I. Guy, and Norman H. Tennent. 1985. "Degradation of epoxy polymers: 2—Mechanism of thermal degradation of bisphenol-A diglycidyl ether." *Polymer Degradation and Stability* 13 (1):11–20.

Guglielmi, Massimo, Guido Kickelbick, and Alessandro Martucci. 2014. *Sol-Gel Nanocomposites*. Springer.

He, Wentao, Pingan Song, Bin Yu, Zhengping Fang, and Hao Wang. 2020. "Flame retardant polymeric nanocomposites through the combination of nanomaterials and conventional flame retardants." *Progress in Materials Science* 114:100687.

Himoto, Keisuke, and Takeyoshi Tanaka. 2008. "Development and validation of a physics-based urban fire spread model." *Fire Safety Journal* 43 (7):477–494.

Honus, Stanislav, Shogo Kumagai, Vieroslav Molnár, Gabriel Fedorko, and Toshiaki Yoshioka. 2018. "Pyrolysis gases produced from individual and mixed PE, PP, PS, PVC, and PET—Part II: Fuel characteristics." *Fuel* 221:361–373.

Huang, Guobo, Shuqu Wang, Ping'an Song, Chenglin Wu, Suqing Chen, and Xu Wang. 2014. "Combination effect of carbon nanotubes with graphene on intumescent flame-retardant polypropylene nanocomposites." *Composites Part A: Applied Science and Manufacturing* 59:18–25.

Innes, Ann, and Jim Innes. 2011. "Flame retardants." In: *"Applied Plastics Engineering Handbook."* Myer Kutz. (Ed.). Elsevier. pp. 469–485.

Jeon, In-Yup, and Jong-Beom Baek. 2010. "Nanocomposites derived from polymers and inorganic nanoparticles." *Materials* 3 (6):3654–3674.

Jiang, Mengwei, Yunshu Zhang, Yuan Yu, Qingwu Zhang, Bujun Huang, Zhiquan Chen, Tingting Chen, and Juncheng Jiang. 2019. "Flame retardancy of unsaturated polyester composites with modified ammonium polyphosphate, montmorillonite, and zinc borate." *Journal of Applied Polymer Science* 136 (11):47180.

Jiang, Peng, Xiaoyu Gu, Sheng Zhang, Shende Wu, Qian Zhao, and Zhongwu Hu. 2015. "Synthesis, characterization, and utilization of a novel phosphorus/nitrogen-containing flame retardant." *Industrial & Engineering Chemistry Research* 54 (11):2974–2982.

Kandola, Baljinder K., Latha Krishnan, John R. Ebdon, and Peter Myler. 2020. "Structure-property relationships in structural glass fibre reinforced composites from unsaturated polyester and inherently fire retardant phenolic resin matrix blends." *Composites Part B: Engineering* 182:107607.

Kandola, Baljinder K., Federico Magnoni, and John R. Ebdon. 2022. "Flame retardants for epoxy resins: Application-related challenges and solutions." *Journal of Vinyl and Additive Technology* 28 (1):17–49.

Kang, Hongliang, Gailing Huang, Shulan Ma, Yongxiang Bai, Hui Ma, Yongliang Li, and Xiaojing Yang. 2009. "Coassembly of inorganic macromolecule of exfoliated LDH nanosheets with cellulose." *The Journal of Physical Chemistry C* 113 (21):9157–9163.

Kango, Sarita, Susheel Kalia, Annamaria Celli, James Njuguna, Youssef Habibi, and Rajesh Kumar. 2013. "Surface modification of inorganic nanoparticles for development of organic–inorganic nanocomposites—A review." *Progress in Polymer Science* 38 (8):1232–1261.

Kiliaris, Polymer, and CD Papaspyrides. 2010. "Polymer/layered silicate (clay) nanocomposites: An overview of flame retardancy." *Progress in Polymer Science* 35 (7):902–958.

Krishnan, Latha, Baljinder K. Kandola, and John R. Ebdon. 2021. "The effects of some phosphorus-containing fire retardants on the properties of glass fibre-reinforced composite laminates made from blends of unsaturated polyester and phenolic resins." *Journal of Composites Science* 5 (10):258.

Li, Bin, and Miaojun Xu. 2006. "Effect of a novel charring–foaming agent on flame retardancy and thermal degradation of intumescent flame retardant polypropylene." *Polymer Degradation and Stability* 91 (6):1380–1386.

Lim, W.K. Patrick, M. Mariatti, W.S. Chow, and K.T. Mar. 2012. "Effect of intumescent ammonium polyphosphate (APP) and melamine cyanurate (MC) on the properties of epoxy/glass fiber composites." *Composites Part B: Engineering* 43 (2):124–128.

Liu, Gang, Jianqing Zhao, Yonghua Zhang, Shumei Liu, and Hua Ye. 2007. "Synthesis and application in polypropylene of a novel nitrogen-containing intumescent flame-retardant." *Polymers and Polymer Composites* 15 (3):191–198.

Liu, Wenjun, Mengli Liu, Ximei Liu, Xiaoting Wang, Hui-Xiong Deng, Ming Lei, Zhongming Wei, and Zhiyi Wei. 2020. "Recent advances of 2D materials in nonlinear photonics and fiber lasers." *Advanced Optical Materials* 8 (8):1901631.

Liu, Yuan, Yanshan Gao, Qiang Wang, and Weiran Lin. 2018. "The synergistic effect of layered double hydroxides with other flame retardant additives for polymer nanocomposites: A critical review." *Dalton Transactions* 47 (42):14827–14840.

Lu, Hongdian, Lei Song, and Yuan Hu. 2011. "A review on flame retardant technology in China. Part II: flame retardant polymeric nanocomposites and coatings." *Polymers for Advanced Technologies* 22 (4):379–394.

Madaleno, Liliana, Jan Schjødt-Thomsen, and José Cruz Pinto. 2010. "Morphology, thermal and mechanical properties of PVC/MMT nanocomposites prepared by solution blending and solution blending+ melt compounding." *Composites Science and Technology* 70 (5):804–814.

Maddah, Hisham A. 2016. "Polypropylene as a promising plastic: A review." *American Journal of Polymer Science* 6 (1):1–11.

Matusinovic, Zvonimir, and Charles A. Wilkie. 2012. "Fire retardancy and morphology of layered double hydroxide nanocomposites: A review." *Journal of Materials Chemistry* 22 (36):18701–18704.

Mouritz, Adrian P., and Arthur Geoff Gibson. 2007. *Fire Properties of Polymer Composite Materials*. Vol. 143. Springer Science & Business Media.

Myllymaa, Katja, Sami Myllymaa, Hannu Korhonen, Mikko J. Lammi, Hanna Saarenpää, Mika Suvanto, Tapani A. Pakkanen, Virpi Tiitu, and Reijo Lappalainen. 2009. "Improved adherence and spreading of Saos-2 cells on polypropylene surfaces achieved by surface texturing and carbon nitride coating." *Journal of Materials Science: Materials in Medicine* 20 (11):2337–2347.

Nazir, Rashid, and Sabyasachi Gaan. 2020. "Recent developments in P (O/S)–N containing flame retardants." *Journal of Applied Polymer Science* 137 (1):47910.

Nishino, Tomoaki, Takeyoshi Tanaka, and Akihiko Hokugo. 2012. "An evaluation method for the urban post-earthquake fire risk considering multiple scenarios of fire spread and evacuation." *Fire Safety Journal* 54:167–180.

Nogueira, Telma, Rodrigo Botan, Fernando Wypych, and Liliane Lona. 2011. "Study of thermal and mechanical properties of PMMA/LDHs nanocomposites obtained by in situ bulk polymerization." *Composites Part A: Applied Science and Manufacturing* 42 (8):1025–1030.

Pomogailo, A.D. 2005. "Polymer sol-gel synthesis of hybrid nanocomposites." *Colloid Journal* 67 (6):658–677.

Qin, Zhaolu, Dinghua Li, Yanhua Lan, Qian Li, and Rongjie Yang. 2015. "Ammonium polyphosphate and silicon-containing cyclotriphosphazene: Synergistic effect in flame-retarded polypropylene." *Industrial & Engineering Chemistry Research* 54 (43):10707–10713.

Qiu, Shuilai. 2021. "Strengthening of Black Phosphorus/Nanofibrillar Cellulose Composite Film with Nacre-Inspired Structure and Superior Fire Resistance." In *Functionalized Two-Dimensional Black Phosphorus and Polymer Nanocomposites as Flame Retardant*, 85–110. Springer.

Qiu, Yong, Lijun Qian, Haisheng Feng, Shanglin Jin, and Jianwei Hao. 2018. "Toughening effect and flame-retardant behaviors of phosphaphenanthrene/phenylsiloxane bigroup macromolecules in epoxy thermoset." *Macromolecules* 51 (23):9992–10002.

Rezvani Ghomi, Erfan, Fatemeh Khosravi, Zahra Mossayebi, Ali Saedi Ardahaei, Fatemeh Morshedi Dehaghi, Masoud Khorasani, Rasoul Esmaeely Neisiany, Oisik Das, Atiye Marani, and Rhoda Afriyie Mensah. 2020. "The flame retardancy of polyethylene composites: From fundamental concepts to nanocomposites." *Molecules* 25 (21):5157.

Rodriguez, Ferdinand, Claude Cohen, Christopher K. Ober, and Lynden Archer. 2014. *Principles of Polymer Systems*. CRC Press.

Roghani-Mamaqani, Hossein, Vahid Haddadi-Asl, and Mehdi Salami-Kalajahi. 2012. "In situ controlled radical polymerization: A review on synthesis of well-defined nano-composites." *Polymer Reviews* 52 (2):142–188.

Rong, Min Zhi, Ming Qiu Zhang, Yong Xiang Zheng, Han Min Zeng, R. Walter, and K. Friedrich. 2001. "Structure–property relationships of irradiation grafted nano-inorganic particle filled polypropylene composites." *Polymer* 42 (1):167–183.

Seraji, Seyed Mohsen, Pingan Song, Russell J. Varley, Serge Bourbigot, Dean Voice, and Hao Wang. 2022. "Fire-retardant unsaturated polyester thermosets: The state-of-the-art, challenges and opportunities." *Chemical Engineering Journal* 430:132785.

Shao, Zhu-Bao, Cong Deng, Yi Tan, Ming-Jun Chen, Li Chen, and Yu-Zhong Wang. 2014. "An efficient mono-component polymeric intumescent flame retardant for poly-propylene: Preparation and application." *ACS Applied Materials & Interfaces* 6 (10):7363–7370.

Shen, Kelvin K. 2021. "Boron-based flame retardants in non-halogen based polymers." In: *"Non-Halogenated Flame Retardant Handbook."* Alexander B. Morgan. (Ed.). John Wiley & Sons. pp. 309–336.

Shi, Yongqian, Libi Fu, Xilei Chen, Jin Guo, Fuqiang Yang, Jingui Wang, Yuying Zheng, and Yuan Hu. 2017. "Hypophosphite/graphitic carbon nitride hybrids: preparation and flame-retardant application in thermoplastic polyurethane." *Nanomaterials* 7 (9):259.

Sponton, M., L.A. Mercado, J.C. Ronda, M. Galia, and V. Cadiz. 2008. "Preparation, thermal properties and flame retardancy of phosphorus-and silicon-containing epoxy resins." *Polymer Degradation and Stability* 93 (11):2025–2031.

Spontón, M., J.C. Ronda, M. Galià, and V. Cádiz. 2009. "Development of flame retardant phosphorus-and silicon-containing polybenzoxazines." *Polymer Degradation and Stability* 94 (2):145–150.

Suh, D.J., Y.T. Lim, and O.O.K. Park. 2000. "The property and formation mechanism of unsaturated polyester–layered silicate nanocomposite depending on the fabrication methods." *Polymer* 41 (24):8557–8563.

Troutier-Thuilliez, Anne-Lise, Christine Taviot-Guého, Joël Cellier, Horst Hintze-Bruening, and Fabrice Leroux. 2009. "Layered particle-based polymer composites for coatings: Part I. Evaluation of layered double hydroxides." *Progress in Organic Coatings* 64 (2–3):182–192.

Urashima, Kuniko, and Jen-Shih Chang. 2000. "Removal of volatile organic compounds from air streams and industrial flue gases by non-thermal plasma technology." *IEEE Transactions on Dielectrics and Electrical Insulation* 7 (5):602–614.

Visakh, P.M., and Arao Yoshihiko. 2015. *Flame Retardants: Polymer Blends, Composites and Nanocomposites*. Springer.

Wang, Dong, Weiyi Xing, Lei Song, and Yuan Hu. 2016. "Space-confined growth of defect-rich molybdenum disulfide nanosheets within graphene: Application in the removal of smoke particles and toxic volatiles." *ACS Applied Materials & Interfaces* 8 (50):34735–34743.

Wang, Baohe, Wenbo Zhang, Wei Zhang, Arun S. Mujumdar, and Lixin Huang. 2005. "Progress in drying technology for nanomaterials." *Drying Technology* 23 (1–2):7–32.

Wang, Dong, Panyue Wen, Jian Wang, Lei Song, and Yuan Hu. 2017. "The effect of defect-rich molybdenum disulfide nanosheets with phosphorus, nitrogen and silicon elements on mechanical, thermal, and fire behaviors of unsaturated polyester composites." *Chemical Engineering Journal* 313:238–249.

Wang, Xin, Wenwen Guo, Wei Cai, and Yuan Hu. 2022. "Graphene-based polymer composites for flame-retardant application." In *Innovations in Graphene-Based Polymer Composites*. pp. 61–89. Elsevier.

Wang, Xin, Wenwen Guo, Wei Cai, Junling Wang, Lei Song, and Yuan Hu. 2020. "Recent advances in construction of hybrid nano-structures for flame retardant polymers application." *Applied Materials Today* 20:100762.

Wang, Xin, Ehsan Naderi Kalali, Jin-Tao Wan, and De-Yi Wang. 2017. "Carbon-family materials for flame retardant polymeric materials." *Progress in Polymer Science* 69:22–46.

Wang, Xin, Weiyi Xing, Xiaming Feng, Lei Song, and Yuan Hu. 2017. "MoS$_2$/polymer nanocomposites: preparation, properties, and applications." *Polymer Reviews* 57 (3):440–466.

Wang, Zhijing, Yinfeng Liu, and Juan Li. 2017. "Regulating effects of nitrogenous bases on the char structure and flame retardancy of polypropylene/intumescent flame retardant composites." *ACS Sustainable Chemistry & Engineering* 5 (3):2375–2383.

Weil, Edward D. 1978. "Phosphorus-based flame retardants." In: "Flame-Retardant Polymeric Materials." Menachem Lewin, S.M. Atlas, Eli. M. Pearce. (Eds.). Springer. pp. 103–131.

Wen, Panyue, Xiaofeng Wang, Bibo Wang, Bihe Yuan, Keqing Zhou, Lei Song, Yuan Hu, and Richard K.K. Yuen. 2014. "One-pot synthesis of a novel s-triazine-based hyper-branched charring foaming agent and its enhancement on flame retardancy and water resistance of polypropylene." *Polymer Degradation and Stability* 110:165–174.

Xiao, Dan, Zhi Li, Xiaomin Zhao, Uwe Gohs, Udo Wagenknecht, Brigitte Voit, and De-Yi Wang. 2017. "Functional organoclay with high thermal stability and its synergistic effect on intumescent flame retardant polypropylene." *Applied Clay Science* 143:192–198.

Xie, Fei, Yu-Zhong Wang, Bing Yang, and Ya Liu. 2006. "A novel intumescent flame-retardant polyethylene system." *Macromolecular Materials and Engineering* 291 (3):247–253.

Yu, Bin, Weiyi Xing, Wenwen Guo, Shuilai Qiu, Xin Wang, Siuming Lo, and Yuan Hu. 2016. "Thermal exfoliation of hexagonal boron nitride for effective enhancements on thermal stability, flame retardancy and smoke suppression of epoxy resin nano-composites via sol–gel process." *Journal of Materials Chemistry A* 4 (19):7330–7340.

Yue, Xiaopeng, Chaofan Li, Yonghao Ni, Yongjian Xu, and Jian Wang. 2019. "Flame retardant nanocomposites based on 2D layered nanomaterials: A review." *Journal of Materials Science* 54(20):13070–13105.

Zhang, Peng, Song Han, Grzegorz Ludwik Golewski, and Xuhao Wang. 2020. *Nanoparticle-Reinforced Building Materials with Applications in Civil Engineering.* Sage Publications, Sage UK.

Zhang, Shenghe, Fukai Chu, Zhoumei Xu, Yifan Zhou, Weizhao Hu, and Yuan Hu. 2021. "Interfacial flame retardant unsaturated polyester composites with simultaneously improved fire safety and mechanical properties." *Chemical Engineering Journal* 426:131313.

2 Influence of the Size and Dispersion State of Two-Dimensional Nanomaterials on the Fire Safety of Polymers

Bernhard Schartel

Bundesanstalt für Materialforschung und-prüfung (BAM), Berlin, Germany

CONTENTS

2.1 INTRODUCTION

Nanotechnology has become a key technology of our time, as we approach the next level of miniaturization and attempt to exploit the surprising effects of nanostructures. Of all the groups these materials include, polymer nanocomposites have the greatest potential for industrial application at this time. Thermoplastic nanocomposites with modified layered silicates, for instance, have been prepared by melt blending for mechanical reinforcement for over three decades now [Kojima 1993, Usuki et al. 1993], and for over two decades to improve flame retardancy [Schall et al. 2000, Beyer 2001]. These applications entail industrial mass production and show a clear nanoscience character. In other words, the nano-scaled structure and dispersion of nanoparticles within the polymer matrix yield massive mechanical

DOI: 10.1201/9781003327158-2

reinforcement or fire retardancy, respectively [LeBaron et al. 1999, Alexandre and Dubois 2000], while the analogous microcomposites do not yield comparable property improvements. Furthermore, the dispersion of nanoplates often harbors multifunctional potential, combining flame retardancy [Gilman 1999, 1997, Zanetti et al. 2001a, Alexandre et al. 2001] and mechanical reinforcement, reducing gas permeation, and increasing viscosity. Graphene and graphene-related nanoplates are reported to increase the electrical conductivity and thermal conductivity of nano-composites and to work as antioxidants in nanocomposites [Frasca et al. 2015, 2016a]. Layered silicates are often discussed as environmentally friendly flame retardant adjuvants. However, it is usually the delamination of nanoplate particles from the microparticle consisting of thick stacks of nanoplates, the exfoliation of nanoplates in the polymer matrix, and their dispersion in the resulting nanocomposite that yield superior materials with the desired properties. The nanostructure and properties of the nanocomposites are dependent mainly on thermodynamic and kinetic factors during preparation [Mittal 2014]. The interfacial compatibility of polymer and nanoplates, or in other words, the polarity match between the nanoplate surface and the polymer chains, is important, as are the size, shape, time of mixing, and applied shear [Mittal 2014]. Thus, this chapter focuses on the dispersion of nanoplates, based mainly on studies of layered silicates and graphene/graphene-related nanoplates. Because the principles, phenomena, and approaches are general, the main conclusions are applicable to all kinds of nanoplates, including BN, WS_2, MoS_2, $MoSe_2$, $TaSe_2$, $NbSe_2$, $MoTe_2$, $NiTe_2$, Bi_2Te_3, and so on [Coleman et al. 2011]. Additionally, layered double hydroxides, an important family of nanoparticles applied for flame retardancy, are not discussed in detail; they show analogous results for modification, dispersion, characterization, mechanisms, combinations, and impact on fire behavior [Gao et al. 2014, Kalali et al. 2015, 2016, Babu et al. 2017].

2.2 DISPERSION AND CHARACTERIZATION

2.2.1 Adjusting Chemistry and Modification

The preparation of a nanocomposite entails mixing a polymer matrix and nano-particles. The formation of a nanocomposite occurs when a negative change in Gibbs free energy (ΔG) emerges due to mixing:

$$\Delta G = \Delta H - T\Delta S < 0 \tag{2.1}$$

Analogous to polymer blends, the thermodynamics of mixing nanocomposites are controlled by entropic ($T\Delta S$) and enthalpic factors (ΔH) [Manias et al. 2007]. Whereas entropy is a driving force for mixing low molecular weight molecules, polymers resist mixtures due to their long chains. The negative entropic contribution, which is thus the favorite for mixing, is reduced by division through the degree of polymerization. Polymer alloys are rare; phase-separated polymer blends usually govern our polymer world. The same is true for nanocomposites of a polymer and a nanoplate. Adapted to disk-like particles, the Onsager model for rigid rods dispersed in a polymer matrix was used to model phase diagrams that illustrate the function of the Flory-Huggins interaction parameter [Balazs et al. 1999, Lyatskaya and Balazs 1998]. The nanocomposite miscibility strongly decreases when the length of the polymer chains increases. The phase diagram also appears to be strongly dependent upon the aspect ratio of the particle. Consequently, it is always much easier to disperse the nanoparticle and polymer in a solvent or the nanoplates in the monomer, a method used for nanocomposite preparation in academic research. Nevertheless, this route is also promising for mass production in regard to thermosets and elastomers or for preparing master batches. Moreover, there are also entropic reasons for packing nanoplates in separated stacks. Dispersing nanoplates increases the amount of polymer chains or segments of the polymer that are restricted to a more linear conformation. These negative ΔS values result in a positive $(-T\Delta S)$ and thus unfavorable contribution to ΔG. The extensive amount of surface area imposes entropic penalties for adsorbed, physisorbed, or intercalated macromolecules [Manias et al. 2007].

Accepting that the entropic factor is never the chief factor in delaminating and dispersing nanoplates in a polymer matrix, favorable enthalpic contributions to the thermodynamics of mixing are crucial to overcome the apparently very high attractive forces between the plates in the stacks [Manias et al. 2007]. The same is true for intercalation. Clever or sophisticated preparation of nanocomposites is often the key to achieving dispersed nanoplates in the polymer. Nevertheless, these impacts of mixing kinetics can rarely compensate for unfavorable thermodynamics. In the case of layered silicates, intercalated small counterions yield small layer-to-layer distances, accompanied by strong ion-ion attractive forces. Graphene stacks together to form graphite due to $\pi-\pi$ interactions. For many nanoplate/polymer systems, mixing is unlikely to be effective without crucial adjustments of the chemistry of the nanoparticle and polymer, modification of the nanoparticle, or modification of the polymer matrix.

Layered silicates such as montmorillonite are hydrophilic and only miscible with hydrophilic polymers, such as poly(ethylene oxide) (PEO) or poly(vinyl alcohol) (PVA); the most common modification to enable the preparation of nanocomposites using all kinds of polymers is to replace interlayer cations (Na+ or K+) with quaternized ammonium or phosphonium cations with alkyl chains [Pavlidoua and Papaspyrides 2008, Sinha Ray 2003]. The normally hydrophilic silicate surface is converted to an organophilic surface, enabling the intercalation of many engineering polymers. The ion-exchange reactions with cationic surfactants lower the surface energy of the inorganic host and improve the wetting characteristics of the polymer matrix [Giannelis 1996, Zanetti et al. 2000]. Additionally, intercalation with primary, secondary, tertiary, and quaternary alkyl ammonium or alkylphosphonium cations results in a larger interlayer spacing [Kim et al. 2001, Ogawa and Kuroda 1997]. Because of the negative charges of the silicate layer, the cationic head group of the surfactants prefers to reside at the layer surface; the nonpolar organic tail radiates away from the surface. The cation-exchange capacity (CEC) measures the excess negative charge of layered silicates and their capability to exchange ions [Manias et al. 2001]. The CEC depends on the silicate structure; thus, for example, montmorillonites from different origins show different CEC values [Manias et al. 2001, Kornmann and

Lindberg 2001]. A higher CEC is often key to improved nanocomposite formation [Chavarria et al. 2007].

The higher cation exchange capacity of the layered silicate, which increases the number of intercalated surfactants, together with the length or volume of organic tail(s), controls the increase in interlayer spacing [Lagaly 1986, Vaia et al. 1994]. For example, interlayer spacing increases with the increase in the length of the alkylamine chain [LeBaron et al. 1999]. Some works have proposed different hypotheses regarding how surfactants may be ordered in the interlayer, positing that the available surface area per molecule might result in a different ordered structure [Hacket et al. 1998]. Several studies have demonstrated that applying different organic cations yields different nano-scaled morphologies, and the better the nanocomposite formation is, the higher the efficiency in terms of fire retardancy efficiencies [Bartholmai and Schartel 2004, Duquesne et al. 2003, Gilman et al. 2000, Wu et al. 2012a, Zanetti et al. 2001c].

Often, research proceeds from and goes ahead with such a limited empirical approach and easy models because only one certain or limited extent of chemical modifications can be applied in each project. Nevertheless, when the modification of the nanoparticle is varied or must be optimized, a set of different approaches for matrix/nanoparticle combinations are compared; in regard to phase-selective filling, the focus should shift to more detailed descriptions of the thermodynamics. To investigate the surface/interfacial tension or solubility parameters, the adhesive energies and adhesive pressure can be calculated, and enthalpic and entropic contributions can be quantified [Manias et al. 2007]. A tractable approach to calculate the entropic and enthalpic contributions has been proposed by Vaia et al. [Manias et al. 2007, Vaia and Giannelis 1997]. The miscibility of polystyrene with alkyl-ammonium-modified montmorillonite and fluorohectorite was predicted. The application of the surfactant reduces the unfavorable entropic contributions by increasing conformational freedom and gallery spacing. Favorable enthalpic inter-actions involve excess enthalpy, analogous to the χ parameter in the Flory-Huggins theory for polymer mixing [Manias et al. 2007, Vaia and Giannelis 1997]·

$$\Delta H \sim \varepsilon_{ps} + \varepsilon_{pa} - (\varepsilon_{aa} + \varepsilon_{as}) \qquad (2.2)$$

where ε_{ij} is the interaction between components i and j, and i and j refer to silicate (s) modified by a surfactant (a) and a polymer (p). Thus, the enthalpic interaction can be approximated by solubility parameters, cohesive energy densities, or inter-facial tension. The commonly used organic modification by alkyl-cationic surfac-tants is sufficient to promote nanocomposite formation with montmorillonite and most polymers. Nevertheless, considering the solubility parameters of different polymers, pristine or organically modified clays, and different surfactants, it has been shown that the higher the solubility parameter of the polymer is, the better the mixing that can be achieved with layered silicate [Jang et al. 2005]. For the systems investigated in this study, the authors conclude that the modification of the sur-factants may be less important than the modification of the polymer. It should be noted that this main message affirms the scientific understanding of why it is so much easier to obtain a polyamide/layered silicate nanocomposite than a polypropylene (PP)/layered silicate nanocomposite, why adding a small portion of maleic acid modified olefin polymer is such a strong game changer when dispersing layered silicate in nonpolar olefins, and why the in situ polymerization of monomer/ nanoplates shows better dispersion than does the melt-blending of nanoplate and polymer chains. Realizing that small portions of polyolefin-grafted maleic anhydride crucially change polymer properties, such as solubility parameters, has established the practice of grafting maleic anhydride as an intrinsic or additive compatibilizer to efficiently disperse the organoclay in PP or other nonpolar polyolefins. Furthermore, different interesting attempts at organically modifying the interlayer ions in layered silicates have been proposed, including ammonium, phosphonium, and imidazolium, as well as different organic groups on these ions [Jang et al. 2005, Wu et al. 2012a]. In the case of montmorillonite, inorganic cations can be easily exchanged with organic ions [Jasmund and Lagaly 1993, LeBaron et al. 1999, Mittal 2009, Theng 1974]. In theory, for every polymer, there are tailored organic modifiers, namely, specific cationic surfactants suitable to compatibilize the given polymer during nanocomposite preparation. Surfactants can potentially be engineered to optimize their compatibility with a given polymer [Xie et al. 2001, Vaia and Giannelis 1997]. Bentonite was modified with either an ammonium or a phosphonium salt and applied in epoxy resins [Hartwig et al. 2003]. In this case, phosphonium-based modification was proposed not only to achieve better dispersion and thus a greater reduction in fire risks, such as peak heat release rate (PHRR) but also to reduce the worsening of the time to ignition because the thermal stability of phosphonium salts is higher than that of ammonium salts. Sometimes, even functional groups have been reported to react with the polymer or initiate polymerization of monomers [Chin et al. 2001]. In general, organic cations trigger organophilic and hydrophobic behavior, lowering the surface energy of silicates and increasing the interlayer spacing (Figure 2.1). Wetting, swelling, intercalation, delamination, and exfoliation of the aluminosilicate plates are enhanced in the polymer matrix [Pinnavaia 1983, Krishnamoorti and Giannelis 1997, Theng 1974]. Montmorillonite modified with different interlayer alkyl ammonium ions has been successfully commercialized as a filler for polymer nanocomposites. Usually, excess organic cations are used in these products as an additional compatibilizer [LeBaron et al. 1999], as indicated in Figure 2.1.

Adjusting the chemistry of the matrix and filler, adding compatibilizer and modifying the filler surface are extremely versatile approaches to improve not only layered silicates but also all kinds of nanocomposites. In the case of graphene and graphene-related nanoplates, the number of functional groups (such as carboxyl, epoxide, hydroxyl, etc.) on the surface are varied to achieve compatibility with different polymer matrices [Mittal 2014]. Graphite oxide is produced first by treating graphite with strong mineral acids and oxidizing agents [Potts et al. 2011, Brodie 1859, McAllister et al. 2007, Hummers and Offeman 1958]. Graphite oxide is proposed to contain hydroxyl and epoxy groups on the basal plane, whereas carboxylic groups are present around the sheet periphery. Not only surface chemistry and thus solubility but also all of the mechanical properties and the thermal and electrical conductivity of graphite oxide strongly differ from those of graphene. However, a wide variety

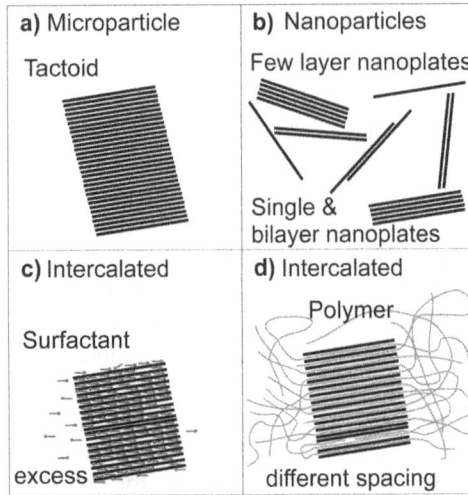

FIGURE 2.1 Illustration of the different kinds of particles discussed; (a) tactoid belonging to the microparticles, (b) single, bi, and few-layer nanoplates, (c) with surfactant-intercalated particles showing increased interlayer distances and some excess surfactants, one layer not intercalated, and (d) with polymer chain-intercalated particles with increased interlayer distance, one layer not intercalated, another layer intercalated with two chains.

of chemical and tailored thermal reduction methods have been applied to graphite oxide to form exfoliated graphene [Mittal 2014]. Sophisticated rapid thermal reduction of graphene oxide generates CO_2 due to the decomposition of functional groups on the surface and, consequently, pressure within the stack, causing the exfoliation of graphene sheets. The reactive functional surface groups of graphite oxide and the remaining few reactive functional groups after reduction are starting points for tailored modification, such as grafting tailored compatibilizer or flame retardant groups [Yang et al. 2009, Fang et al. 1982, Wang et al. 2011, Xu et al. 2011, Deng et al. 2011, Zhuang et al. 2010, Salavagione et al. 2011, Lin et al. 2011, Yuan et al. 2017, Liao et al. 2012, Shi et al. 2018]. Whereas graphene lacks usable functionality, graphite oxide or reduced graphite oxide platelets offer crucial possibilities for modification; for instance, even covalent bonds between graphite oxide platelets and polymers (=grafting) have been demonstrated.

How thermodynamic considerations are used to control advanced nanostructured composites is shown in works on the phase-selective filling of polymer blends. Contact angle measurements of three liquids were applied to determine the total surface tension of the two polymers of the blend and the plate-like nanoparticle [Viretto et al. 2016]. Total surface tensions consist of dispersive and polar contributions and were calculated according to the Owens-Wendt approach [Owens and Wendt 1969]:

$$\gamma_L = 1 + \cos\theta = 2(\gamma_S^d \gamma_L^d)^{-0.5} + 2(\gamma_S^p \gamma_L^p)^{-0.5} \qquad (2.3)$$

where θ = contact angle, γ = surface tension, index L and S for liquid and surface, index p and d for dispersive and polar. The interfacial tension γ_{ij} between the two polymers was calculated using the harmonic mean equation (equation 2.4) of Wu [Steimann et al. 2002, Wu 1987]; for the interfacial tension between nanofiller and the polymers, the geometric mean equation (equation 2.5) of Wu was used [Steimann et al. 2002, Wu 1987]:

$$\gamma_{ij} = \gamma_i + \gamma_j - (4\gamma_i^d \gamma_j^d)/(\gamma_i^d + \gamma_j^d) - (4\gamma_i^p \gamma_j^p)/(\gamma_i^p + \gamma_j^p) \tag{2.4}$$

$$\gamma_{ij} = \gamma_i + \gamma_j - (\gamma_i^d \gamma_j^d)^{-0.5} - (\gamma_i^p \gamma_j^p)^{-0.5} \tag{2.5}$$

The wetting parameter ω_{12} of a solid particle (index I, the indices j and k stand for the two polymers)

$$\omega_{12} = \cos \theta = (\gamma_{ij} - \gamma_{ik})/\gamma_{jk} \tag{2.6}$$

marks the most favorable position of the nanoparticle in a phase-selective filled polymer blend [Fenouillot et al. 2009]. When $\omega_{12} > 1$ or < -1, the nanoplates are dispersed in one or the other polymer phase, whereas $-1 < \omega_{12} < 1$ results in the localization of the nanoplates in the interface (Figure 2.2). A series of studies have tried to work out the best dispersion in blends by comparing the fire behavior

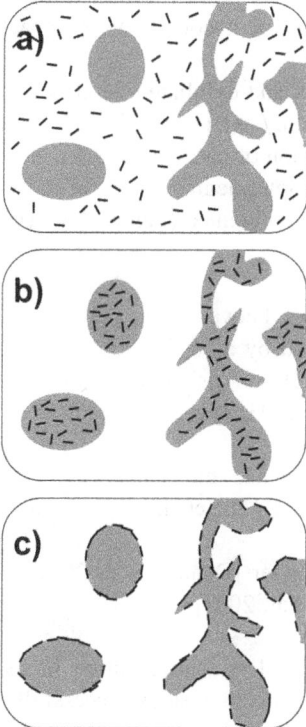

FIGURE 2.2 Phase-selective filled nanocomposites; (a) nanocomposite as a matrix hosting a polymer phase, (b) nanocomposite phase within a polymer matrix or phase-selective cocontinuous structure, and (c) nanoplates localized in the interface between the two polymer phases.

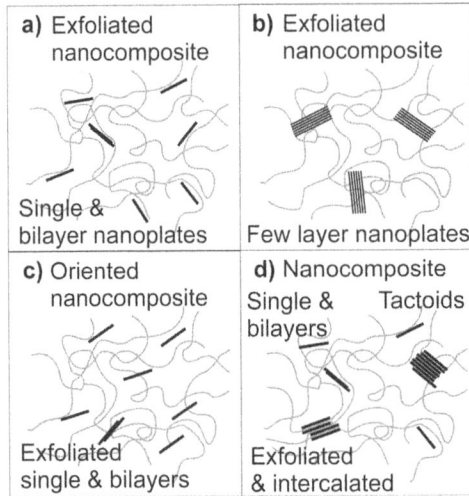

FIGURE 2.3 Typical dispersion states of nanoplates in nanocomposites; (a) statistically oriented, exfoliated single and bilayers, (b) statistically oriented, exfoliated few layers, (c) oriented, exfoliated single and bilayers, and (d) mixture of exfoliated single and bilayers, intercalated stack, and tactoids.

of the filling of nanoplates in the different phases of various ammonium polyphosphate-flame retarded polymer blends [Lu et al. 2015, Lu et al. 2018, Lu et al. 2014]. Overall, filling that supports the formation of a continuous compact residue is proposed to be the best choice.

Although tactoid (= microcomposites), intercalated, and exfoliated describe the preferred thermodynamic states for nanoplates in a polymer matrix (Figure 2.3), the mixing of the various compounds and the distribution of particles in the matrix are phenomena that are controlled kinetically. Often, intermediate states of nanocomposites or partially nanostructured morphologies are obtained (Figure 2.3). The nanostructure is controlled by:

- the mixing enthalpy, the chemical interactions between the polymer, nanofiller, and surfactants, which can be boosted by adjusting the nanoparticle/polymer pair, modifying the nanoparticle, or using the optimal surfactant [Vaia and Giannelis 1997, Yoon et al. 2003, Zanetti et al. 2001a, 2001b],
- the physical structure of the components (particle size, molecular weight, etc.) [Schartel et al. 2006a, Fornes et al. 2001, 2002, Vaia et al. 1995, Berta et al. 2006],
- processing, as will be discussed in the next chapter [Dennis et al. 2001, Vaia et al. 1995, Bendaoudi et al. 2005, Kim et al. 2002].

It should not escape our notice that the factor physical structure, buried in the middle of this list, is sometimes the key to preparing nanocomposites successfully. For instance, the influence of different drying methods of phosphonium bentonite on the distribution of the layered silicate in the epoxy matrix, and thus the flame retardancy, has been reported [Schartel et al. 2006a, 2011]. Phosphonium-modified bentonite was dried in an oven, and the formed product was ball-milled and sifted, dried via spray-drying, or dried via freeze-drying. Neither the interlayer distance nor the chemical composition was influenced by these drying methods, whereas the specific surface area (BET, specific surface area determined according the theory of Brunauer, Emmett, and Teller) of the obtained particles was $45 \ m^2 \ g^{-1}$, $100 \ m^2 \ g^{-1}$ and $175 \ m^2 \ g^{-1}$, respectively. The deviations in powder morphology before preparing the nanocomposite crucially control the nanoparticle dispersion and flame retardancy efficiency [Schartel et al. 2006a, 2011]. Very similar dependencies of the reduction in PHRR on the BET (specific surface area determined according to the theory of Brunauer, Emmett, and Teller) of the applied nanoplate-based nanoparticles have been reported in comparison of graphene with few-layer and multilayer graphene [Dittrich et al. 2013a, 2013b], and may constitute a quite general structure–property relationship.

2.2.2 DISPERSION TECHNIQUES

Beyond thermodynamics, addressed by modification of the nanoplates and polymers in the previous chapter, mechanisms such as the intercalation of layered structures by polymer chains, the delamination of nanoplate-based nanoparticles or single layers, and the dispersion of nanoparticles in the matrix are controlled mainly by kinetics. Designing the correct interfacial chemistry between nanoplates and the polymer matrix is crucial but often not sufficient to obtain a well-dispersed polymeric nanocomposite [Vaia and Wagner 2004]. Nanocomposite formation depends strongly on the preparation conditions, such as time, temperature, and shear stress. In practice, nanocomposite preparation rarely produces a thermodynamically stable state or perfectly homogeneous materials (Figure 2.3 and Figure 2.4). Intermediate states of nanocomposites or partially nanostructured morphologies occur. Nanocomposites are often inhomogeneous materials: Regions of different concentrations of exfoliated layers, single tactoids, or separated regions of intercalated stacks are observed at the same time. The nanostructure becomes very sensitive to the nanocomposite preparation.

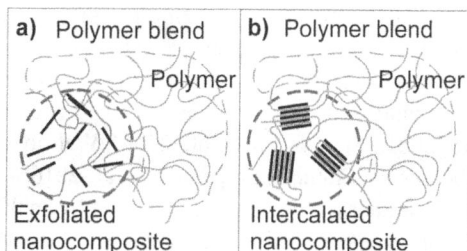

FIGURE 2.4 Polymer blends of the polymer phase and (a) exfoliated nanocomposite and (b) intercalated nanocomposite phases.

Different routes of nanocomposite preparation, such as solvent mixing, latex methods, in situ polymerization and melt blending, have been reported and reviewed [Sinha Ray and Okamoto 2003, Pavlidoua and Papaspyrides 2008, Paul and Robeson 2008]. In situ polymerization exploits the superior exfoliation of nanoplates within the low molecular weight monomer or the easy intercalation of monomers within the nanoplate-stack particle. The polymerization fixes the nanostructure. In situ polymerization is often proposed for emulsion polymerization. The curing (polymerization and crosslinking) of all kinds of elastomer and thermoset nanocomposites may be seen as a special kind of in situ polymerization [Guo et al. 2011]. In template synthesis, the desired nanoparticle is finally formed within a nanodispersed nanoparticle precursor/polymer composite. While the melt blending approach provides a more environmentally friendly and industrially viable route, filler dispersion is generally less optimal. Nevertheless, the greatest interest has been directed toward melt processing because this method is generally considered more economical, as it takes place in compounding and fabrication facilities that are commonly used in commercial practice. Usually, double screw extrusion is applied to disperse nanofillers in thermoplastics [Wu et al. 2002, Zou et al. 2008, Fawaz and Mittal 2014]. The shear force, temperature, and residence time must be optimized because of the obvious dependency of delamination and exfoliation on the preparation parameters. Nanoplate/polymer nanocomposites prepared by extrusion and injection molding exhibited better exfoliation and thus improved flame retardancy than those prepared by extrusion and hot melt pressing [Bartholmai and Schartel 2004]. The dispersion and thus the flame retardancy are reported to depend on extrusion parameters such as the temperature. Groundbreaking investigations on the influence of processing parameters on nanocomposite formation, such as temperature, shear stress and residence time, can be found in the literature [Dennis et al. 2001]. The nature of the extruder and its screw configuration are important [Paul and Robeson 2008]. Even details such as where the components are introduced into the extruder can make a difference [Chavarria et al. 2007]. Although higher molecular weights of the polymer chain are unfavorable when considering the thermodynamics of the mixing, nanocomposites based on higher molecular weight polyamides yielded higher degrees of clay exfoliation [Fornes et al. 2001]. The higher melt viscosity results in higher shear stresses, favoring the mixing kinetics.

Furthermore, because the nanostructured morphologies are not equilibrium morphologies, the morphology becomes very sensitive to all kinds of improvements to mixing during preparation. Not only optimizing the parameters has an effect but also adding measures or processing steps, such as auxiliary premixing, the deployment of masterbatches, applying ultrasonication, and ball milling [Teng et al. 2021]. Ultrasonically assisted solution mixing is widely used in academic nanocomposite research [Teng et al. 2021, Stankovich et al. 2006, Yousefi et al. 2013a, 2014, Shen et al. 2016] whereas it is rather rare [Frasca et al. 2016b], in regard to nanocomposites for flame retardancy, which are dominated by melt-processed thermoplastics. The solvent increases the miscibility of the components, and ultrasonic energy is used to overcome the energy of interaction between the nanoparticles or nanoplates in the aggregates. Using monomers as solvents, this process is used for in situ polymerization at high temperatures [Yousefi et al. 2013b]. Compared to ultrasonically assisted solution mixing, simple shear mixing is used only to separate aggregates of nanofillers [Teng et al. 2021].

Roll milling or three-roll milling (calendering) [Teng et al. 2021] are methods of dispersing nanofillers through the shearing force between rolls in a high-viscosity matrix and have been proposed for anisotropic nanoplates, such as graphene nanosheets [Ahmadi-Moghadam and Taheri 2014, Chandrasekaran et al. 2014, Li et al. 2016a] and nanoclays [Dalir et al. 2012, Park and Lee 2013, Kothmann et al. 2015] For elastomer/graphene nanocomposites, direct two-roll milling of the elastomer and particles was compared with the preparation of the final nanocomposite via two-roll milling of the elastomer and an elastomer/graphene nanocomposite masterbatch. An ultrasonically assisted solution mixing procedure was applied to obtain the nanocomposite masterbatch. Multifunctional nanocomposites were obtained. The combination of an ultrasonically assisted solution mixed masterbatch and subsequent two-roll milling resulted in better dispersion and clearly outperformed the directly mixed nanocomposite [Frasca et al. 2016b]. A two-step preparation was proposed for all kinds of elastomers as a possible drop-in technology [Frasca et al. 2015]. The existing companies specializing in solvent-assisted rubber mixing produce nanocomposite masterbatches, whereas the final production can be performed at any typical elastomer production site.

All kinds of grinding followed by the premixing of powders can be used as a crucial preliminary step for the melt mixing of nanocomposites. Even the simple mechanical premixing of polymer granulates and nanoparticle powder before feeding the extruder can make a crucial difference. Different PP nanocomposites with graphene, few-layer graphene, and multilayer graphene nanocomposites were obtained via ordinary extrusion after premixing [Dittrich et al. 2013]. The carbon nanoplate particles were dispersed in acetone by means of ultrasonication and used to coat PP powder. The extrusion of the nanoparticle-coated PP powder resulted in perfect dispersion of the nanoparticle so that changes in the fire behavior and any other properties became a function of the size and shape of the particles used.

Ball milling grinds and mixes powders while applying large shear and compression forces. Graphene nanoparticles may be peeled off; furthermore, reaction sites are activated for chemomechanical modification or grafting [Wu et al. 2011a, Jiang and Drzal 2012, Tang et al. 2013, Gu et al. 2015, Han et al. 2021].

Mixing by latex technology may be understood as a special kind of solution mixing. Latex is a colloidal dispersion of discrete particles, usually in an aqueous medium. The latex approach may offer some benefits when applied to polymer matrices that are produced by emulsion polymerization or that can be prepared as emulsions. Exfoliation-adsorption: Layered silicate is exfoliated into single layers using a solvent in which the polymer (or a prepolymer in the case of insoluble polymers such as polyimide) is soluble. Layered silicates can be easily dispersed in an adequate solvent. Nanocomposites can be obtained through emulsion polymerization, which disperses the layered silicate in the aqueous phase. Latex-based mixing methods are also proposed for elastomer/graphene nanocomposites [Strommer 2022a, 2022b].

The solution-assisted mixing of nanoplates in polymers followed by either vaporization of the solvent or mixing within the monomer followed by polymerization is considered a method to obtain well-dispersed exfoliated nanocomposites. Exfoliated well-dispersed nanoplatelets may be favored when composites are processed by solution mixing or in situ polymerization compared with composites produced by melt mixing [Kim et al. 2010]. However, these methods are not as straightforward as desired. Vaporization of the solvent can result in agglomeration of the nanoplates, polymerization, and crosslinking with demixing; crystallization is also known to accompany segregation. However, even trivial liquid nanocomposites that are homogenous under stirring may start to demix or sediment when poured into a mold. For instance, although they started with a homogenous mix of nanoparticles in the uncured epoxy system under stirring, epoxy resin nanocomposites were reported to settle into a blend of a nanoplate-rich nanocomposite phase and an epoxy phase, as shown in Figure 2.4 [Hartwig et al. 2003, Schartel et al. 2006a]. On the other hand, melt temperature-enhanced mixing and pyrolysis-induced delamination and dispersion of nanoplates may occur during pyrolysis [Schartel et al. 2011]. As the nanostructure of the pyrolyzing melt is reported to control the flame retardancy effects, the nanocomposite morphology in the solid-state may be understood as a prerequisite or morphological precursor [Schartel 2007].

2.2.3 CHARACTERIZATION OF THE NANOSTRUCTURED MORPHOLOGIES

The quantitative assessment of the degree of nanocomposite formation is quite a challenge, along with the consideration of their homogeneity. The oversimplification of frequently used models, which merely distinguish among microcomposites, intercalated nanocomposites, and exfoliated nanocomposites, fails to offer a satisfactory description [Vaia 2000] Nanocomposite preparation rarely produces homogenous, thermodynamically stable morphologies. Partially exfoliated or strongly inhomogeneous materials, such as blends of polymer and polymer/nanocomposites, are obtained (Figure 2.4). Tactoids within exfoliated nanoplates were observed. Nevertheless, the better the nanocomposite formation, such as greater dispersion of nanofiller, the higher the efficiency in terms of property improvement, such as flame retardancy. Thus, apart from methods targeting nanostructured morphology directly, such as transmission electron microscopy (TEM) and X-ray investigations, properties strongly influenced by the dispersion of nanoplates, such as melt viscosity, can be used to characterize nanocomposite formation.

X-ray diffraction and TEM are the most common methods used to gain insight into the morphology of nanocomposites [Vaia 2000, Alexandre and Dubois 2000, Sinha Ray and Okamoto 2003]. Wide angle X-ray diffraction (WAXD) is used to determine the Bragg peak of the layer spacing at a small angle resolution. Tactoids, aggregates of tactoids, and particles consisting of aggregates of tactoids show the same X-ray pattern [Paul and Robeson 2008]. Shifting the X-ray d-spacing to larger interlayer distances is interpreted as intercalation, and the vanishing d-spacing peaks are interpreted as exfoliation. The latter interpretation is somewhat questionable because the loss of electron density periodicity may be caused by delamination without any convincing dispersion of the nanoplates. Strictly speaking, this kind of X-ray diffraction investigation evaluating the Bragg peak fails to assess the dispersion of delaminated nanoplates [Morgan and Gilman 2003]. The Bragg peak is not a direct measure for exfoliation and is reported to be insufficient when comparing rather good nanocomposites with each other [Morgan and Gilman 2003]. There is crucial but usually unused potential for more comprehensive X-ray investigations, such as small angle X-ray scattering (SAXS), 2D SAXS, and online WAXD, for the hierarchical dispersion, orientation, and kinetics of formation [Sinha Ray and Okamoto 2003, Bafna et al. 2003, Tidjani et al. 2003]. A far more direct way to visualize nano-composite morphology is TEM. However, because each TEM image covers only a limited small area, the best-looking image is not always representative. Only a series of TEM images with different magnifications taken at different sites provide good insight into the morphology [Paul 2008]. A series of TEM images were taken to prove the absence of tactoids. Only high-resolution TEM provides the accuracy to characterize single-, bi, and few-layer nanoplate nanoparticles. TEM images covering larger areas deliver information on homogenous dispersion and orientation. X-ray scattering and TEM techniques are usually applied for the qualitative evaluation of the degree of plate dispersion, whereas quantitative assessment demands specific and extended efforts. Quantitative analysis based on TEM image analysis is extremely rare and falls into the following two categories [Gou et al. 2014]: distance measurement-based methods [Eckel et al. 2004, Luo and Koo 2008, Tyson et al. 2011, Bray et al. 2012] and particle analysis-based methods [Dennis et al. 2001, Fornes et al. 2001, Nam et al. 2001, Carastan et al. 2010, Vermogen et al. 2005, Lee et al. 2005, Shah et al. 2007, Kim et al. 2007, Spencer et al. 2010]. Atomic force microscopy (AFM) as well as all kinds of spectroscopic methods, such as FTIR [Loo and Gleason 2003, Wu et al. 2001], Raman spectroscopy [do Nascimento et al. 2002], and nuclear magnetic resonance (NMR), have been proposed to address the nano-composite structure [Sinha Ray and Okamoto 2003]. However, only a few studies use these methods, perhaps because their prevalence cannot compete with X-ray and TEM, and they are regarded as too sophisticated. For instance, it has been 20 years since, solid-state NMR was proposed for tackling the problem of quantitative char-acterization of both delamination and exfoliation (dispersion) in flame-retarded na-nocomposites [VanderHart 2001a, 2001b], but this technique has rarely been adopted by the community [Sinha Ray and Okamoto 2003].

Many studies believe that using scanning electron microscopy (SEM) images of cold fracture surfaces can prove nanocomposite formation. Investigation of the cold fracture surface of any composite is worth performing; indeed, it is one of the standard methods used to characterize composites, and images always work well to convince humans of all sorts of arguments. Nevertheless, strictly speaking, increases in roughness and in the number of protuberances deliver no direct or directly related measure of the nanostructure. Often, only a qualitative comparison of the particle dispersion and adhesion between the matrix and particles is indicated. Despite their value as a supplementary method [Xing et al. 2014, Kang et al. 2014, Frasca et al. 2016c], SEM images of cold fracture surfaces are not recommended as an exclusive method to characterize the nanostructure.

Melt rheology is already a standard method to characterize thermoplastics; at the same time, the nanocomposite structure shows a pronounced impact on the melt vis-cosity for lower shear rates. Several works have proposed melt rheology for the quantitative characterization of nanocomposite structures [Wagener and Reisinger 2003,

Solomon et al. 2001, Zhang et al. 2008]. For instance, the dispersion of differently modified layered silicate nanoparticles correlates with the difference in shear viscosity. The shear thinning exponent n for polymer–clay nanocomposites has been proposed as a quantitative measure to compare the extent of dispersion based on a routine polymer characterization experiment [Wagener and Reisinger 2003, Krishnamoorti et al. 2001a]. A strong correspondence between the reduction in PHRR and the melt viscosity for low shear rates has been reported because both shear viscosity and protective layer formation are controlled by the quality of nanocomposite formation [Bartholmai and Schartel 2004, Vermant et al. 2007].

2.3 NANOPARTICLE AND DISPERSION STATE

The improvement in a given property depends strongly on the effective nanoparticles and their properties. It is important to consider that even for composites that are believed to be the best nanocomposites, such as polyamide 6/modified layered silicate, exfoliation into single plates is usually not complete [Fornes and Paul 2003]. Often, bilayer nanoplates or stacks of few platelets are common, as can be visualized by high-resolution TEM, which shows well-defined bilayers and stacks but sometimes also plates skewed relative to one another [Chavarria and Paul 2004]. Bilayer and few-layer nanoplates may bring more complex changes than just thicker nanoplates [Fornes and Paul 2003]. Usually, multiple properties of polymer nanocomposites are expected to be enhanced when small amounts of single-layer graphene are added due to the outstanding properties of single-layer graphene, including a specific surface area of 2600 m^2 g^{-1} [Peigney 2001]. an extraordinary Young's modulus of 1 TPa [Gong 2012, Jiang 2009, Li 2006, Yu 2000], and remarkable thermal and electrical conductivities [Wu et al. 2007]. Nevertheless, scalability in production, handling in nanocomposite processing, and the cost of perfect single-layer graphene present unresolved challenges [Zhong et al. 2015, Zhu et al. 2018]. In practice, graphene oxide, thermally reduced graphene oxide, few layer graphene, multilayer graphene, and even submicron graphite particles are used in flame-retarded polymer/graphene nanocomposites [Araby et al. 2017, 2018, 2019, 2021, Attia et al. 2021, Dittrich et al. 2013, 2015, Frasca et al. 2016, Wang et al. 2019, Yang et al. 2018, Zhong et al. 2015]. Although several of these nanoparticles are fascinating multifunctional fillers in nanocomposites, their efficacy is as much as an order of magnitude lower than perfect single-layer graphene [Gómez-Navarro et al. 2008, Gong et al. 2012, Suk et al. 2010, Young et al. 2012]. Nevertheless, few or multilayer graphene with a BET of 250–350 m^2 g^{-1} [Frasca et al. 2015], meaning nanoplate particles consisting of stacks with an average of <15 graphene sheets, are reported to function as very convincing multifunctional nanofillers enhancing flame retardancy when they are well dispersed in elastomers [Battig et al. 2021, Strommer et al. 2022c]. The BET specific surface area is a highly effective parameter to describe the degree of exfoliation and thus the number of layers in graphene stacks [Steurer et al. 2009, Boehm et al. 1962a]. Commercialized multilayer graphene well dispersed on nanoplates offers flame-retarded nanocomposites at reasonable costs. Furthermore, a few composites with well-dispersed submicron or few-micron particles are reported to show all the effects desired from a nanocomposite; put simply, it turns out that the best polymer/10 wt. % microparticle microcomposites may easily outperform the worst polymer/5 wt. % nanoplate nanocomposites. For instance, well-dispersed fine microplate particles such as talc are reported to strongly influence the pyrolysis, melt flow and dripping behavior, and hence the flammability, and increase the efficiency of fire behavior [Schartel et al. 2012a], in a very similar way to nanoplate nanocomposites. Melt flow and dripping were prevented, char yield and char properties improved, the PHRR was reduced, and the limiting oxygen index (LOI) increased [Schartel et al. 2012a]. At the same time, nanocomposites display crucially different behavior than the corresponding typical microcomposites, and these characteristics continuously change the particle size. Thus, valuably synergistic flame retardancy action is reported for both microcomposites and nanocomposites [Eckel et al. 1994, Bödiger et al. 1997]. Nevertheless, when the flame retardancy effect of stacked carbon nanoplates is compared with nanocomposites consisting of different numbers of sheets, the strong influence of the nanoparticle becomes clear [Dittrich et al. 2013a, 2013b]. The reduction in PHRR is nonlinear with respect to the specific surface area (BET) of the graphene particles, strongly favoring few and multilayer graphene nanoparticles. The effective nanoplate and its properties demand our attention. Many of the surprising effects in nanocomposites follow the same rules as other kinds of composites and thus can be described by the same mathematical formulas. For instance, mechanical reinforcement is controlled by the mechanical properties of the matrix and the filler, the aspect ratio of the filler, the adhesion/attraction between filler and matrix, the orientation of the fillers, and the volume fraction of the fillers regardless of molecular reinforcement, nanocomposites. or short-fiber composites used. This is described by the Halpins Tsai equation and similar approaches that estimate the properties of the composites somewhere between the Voigt and Reuss average [Ward and Hadley 1993, Schartel and Wendorff 1999, Halpin and Kordos 1976, Papageorgiou et al. 2017, Yang et al. 2013, Eduljee and McCullough 1993]. Similarly, valid overall particle dimensions are reported for permeation using the tortuous pathway model and its corresponding equations [Yano 1997], such as the model of Nielsen [1967] based on the factors permeability of the matrix and filler, aspect ratio of the filler, orientation, and volume fraction. A key to understanding and thus utilizing nanocomposites is to describe the nanoparticle properties and their impact. The drastic increase in the surface-to-volume ratio and thus the extremely large effective specific surface area of the nanoplates is crucial for the extraordinary nanocomposite performance. The aspect ratio also plays a major role, although strictly speaking, it is a dimensionless factor. Moreover, nanoparticles often show superior properties because they are a perfect kind of particle. Usually, particles are agglomerates of several smaller particles, with their properties governed by the boundaries of the grain. For plate-like chemical structures, this means there are three property regimes. The mechanical properties of an agglomerate of graphite stacks are first determined by the adhesion between the different stacks; second, those of a particle consisting of a single stack of graphene sheets are determined by the way graphene sheets are skewed in relation to each other; and third, the mechanical properties of a single graphene sheet are determined by the deformation of the hexagonal structure.

Similar regimes exist for the thermal and electrical conductivity, where the behavior of a particle consisting of several stacks is dominated by different behavior than is a stack of plates or a single plate [Allen et al. 2009, Wu et al. 2007]. The formation of single-layer graphene nanoplates via chemically or thermally reduced graphene oxide was reported in the 1960s [Boehm 1962a, 1962b], but the graphene hype of the last 20 years did not start until the extraordinary properties of graphene sheets had been investigated, addressed, and utilized.

Summing up what we have discussed in this chapter thus far, it is apparent that the common classification: (A) exfoliated nanocomposites, meaning that the single nanoplates are statistically distributed within the polymer matrix, (B) intercalated nanocomposites characterized by polymer chains between the layers of stacks, and (C) microcomposites based on tactoids is merely a rough and idealized picture. Depending on the starting particle, on the affinity between the polymer and the modified nanoplate, and on the preparation procedure, the composites formed differ in terms of the effective nanoparticle or often exhibit a variety of different nanoparticles, as well as different dispersion of these nanoparticles, including inhomogeneity at different scales such as intercalation or partly intercalation, regions with different nanoplate concentrations, partly exfoliated or partly dispersed, or phase-selective filled systems [Vaia 2000]. It is nonsensical and nonproductive to bemoan the loss of easy, thermodynamically stable states of dispersion. Taking all kinds of percolation-controlled properties into account, prestructuring in well-dispersed systems is preferable to perfectly homogenous dispersed systems. For instance, the lowest threshold for the electric conductivity of anisotropic carbon nanofiller nanocomposites is achieved through some degree of prestructuring [Alig et al. 2012]. The desired properties include phenomena that depend on the formation of a network. The formation of a network during pyrolysis was proposed by Kashiwagi to be one of the main mechanisms in flame-retarded nanocomposites [Kashiwagi et al. 2005, 2008]. Furthermore, many typical polymer preparation methods, such as injection molding, pressing, and film production, result in efficient alignment of the nanoplates. During injection molding, for instance, the polymer chains and particles are highly aligned due to the shear rates typically applied. Nanoplates, with their pronounced anisotropic shape and high stiffness, are easy to orient as discotic or sanidic liquid crystals [Demus 1994, Marrucci 1991]. Complex anisotropic morphologies known from smectic systems were proposed to explain the rheological behavior of polymer/nanoplate nanocomposites [Krishnamoorti et al. 2001b, Schmidt et al. 2002a, 2002b]. Because the kinetics of disorientation are very different for polymer chains and anisotropic nanoplate particles [Ren et al. 2003, Malwitz et al. 2004] common temperature–time profiles during cooling in the preparation of thermoplastics yield highly oriented nanocomposites. In fact, it is extremely difficult to obtain polymer/nanoplate nanocomposites without any preferential orientation of the nanoplates. The aspect ratio and orientation of the nanoplates play a major role in any kind of mechanical reinforcement [Papageorgiou et al. 2017, Li et al. 2015, 2016b] as well as in any kind of barrier property, such as reducing the permeation. The processing technique can be used to induce orientation beneficial for reinforcement [Fornes and Paul 2003] but may raise the concentration of nanoplates demanded to meet the percolation threshold [Kim 2009]. Randomly oriented nanoplatelets may be preferable when nanocomposites are processed by solution mixing or in situ polymerization compared with a more oriented structure for composites produced by melt mixing [Kim et al. 2010]. Nevertheless, an obvious orientation and the resulting anisotropic properties have also been reported for elastomer/graphene nanocomposites prepared via latex premixing followed by conventional two-roll milling [Strommer et al. 2022a, 2022b].

Several studies on different polymer/nanoplate nanocomposites show the strong influence of the degree of nanodispersion on fire behavior [Bartholmai and Schartel 2004, Duquesne et al. 2003, Dittrich et al. 2013a, Wu et al. 2012a]. Nevertheless, this picture is an oversimplification and does not address one of the most important points. Indeed, the details of this structure–property relationship are reported to become shaky when nanostructured materials are compared, and the impact of the degree of dispersion on flame retardancy differs from system to system. For instance, no convincing significant advantage was found when inter-calated or exfoliated systems were compared [Gilman et al. 1998, Gilman 1999, Morgan et al. 2001, Wilkie 2005]. One reason for this is that fire behavior is defined by the response of a test specimen or a component in a fire scenario, which depends on the interaction of a slew of various material properties. Changing several properties in a nanocomposite leads to complex changes in the fire behavior. The other main reason is that the nanocomposite morphology and properties at room temperature do not appear to control fire behavior directly. Flame retardant mechanisms are proposed, such as the formation of an inorganic-carbonaceous protective layer or changed melt flow and dripping during pyrolysis [Schartel et al. 2006b]. Although the dispersion of nanoplates in nanocomposites at room tem-perature is the essential starting point for fire retardancy, this is only a prerequisite, as the nanocomposite does the main job during pyrolysis, when the composition of the materials and their morphology change strongly due to ablative reassembly [Gilman 1999, Zanetti et al. 2002]. The enrichment of nanoparticles at the surface was shown by XPS [Du et al. 2002, Wang et al. 2002]. Only a few studies have addressed the different stages of protective residue formation, mainly through visual observations [Kashiwagi et al. 2004a]. Other results suggest that increased nano-dispersion occurs at higher temperatures and during pyrolysis and is exploited. It becomes clear that protective layer formation at elevated temperatures and during pyrolysis is key to understanding the relationship between the nanostructure and fire retardancy [Schartel et al. 2011]. Exfoliation induced by polymer decomposition and heat takes place, as well as changes in orientation in the molten nanocomposite due to increased mobility and the tendency to mix at very high temperatures [Schartel et al. 2011]. The nanoplates migrate through the melt [Sturm et al. 2012, Lewin 2003, 2006, Schartel et al. 2011, Tang et al. 2006, Tang and Lewin 2007, Zammarano et al. 2006], or are transported via bubbles [Schartel et al. 2011]. Burning specimens were quenched in liquid nitrogen for subsequent advanced SEM/EDX imaging to enable a quasi-online analysis of the formation of a residual protection layer in layered silicate epoxy resin nanocomposites. Different zones in the condensed phase were identified as well as distinct phenomena occurring within these zones, such as the melting and ablation of organic material, agglomeration, and the depletion, exfoliation and reorientation of the layered silicate nanoplates [Sturm et al. 2012]. All these changes in the nanostructure are restricted by the limited time period available before ablation and solidification finalize the formation of the protective surface layer [Schartel et al. 2011]. The agglomeration of nanoparticles within the residue is done mainly by the ablation of the polymer, not the migration of the nanoplates within the polymer melt. Thus, it is commonly concluded that better nano-dispersion at room temperature favors the formation of a homogenous closed or continuous protective layer with no cracks or holes on the macro- or micro-level.

2.4 INTERPLAY WITH FLAME RETARDANT MECHANISMS AND FIRE PROPERTIES

2.4.1 IGNITION

Ignition occurs when the critical conditions are reached for a fuel/air gas mixture in contact with an ignition source. Reaching critical conditions in fire tests can often be summarized by when the material-specific ignition temperature is reached at the surface of the specimen [Babrauskas 2003, Lyon 2004]. Ignition is a complex phenomenon controlled by the interplay of many different properties and processes, including thermal and thermo-oxidative decomposition, heat absorption and thermal conductivity, and thermal inertia, which can be influenced by adding and dispersing nanoplates into nanocomposites.

The influence of nanoparticles on thermal decomposition and thermo-oxidative decomposition differs strongly from one nanocomposite system to the next, varying between enhanced decomposition, no significant influence, and strong improvement [Zanetti et al. 2001c, Bourbigot 2004a, Burnside 1995, Tidjani 2003]. The necessary molecular movements during decomposition and product release are believed to be hindered in nanocomposites. Even large effects were reported for a few intercalated systems; poly(methyl methacrylate) or polydimethylsiloxane intercalated in montmorillonite, for instance, showed a decomposition temperature increase of up to 50 K and 140 K, respectively [Blumstein 1965, Burnside 1995]. On the other hand, most nanocomposites exhibit earlier times to ignition than their corresponding polymers [Zhu 2000] presumably due to decomposition through Hofmann elimination of modifiers used to modify bentonite, such as alkyl ammonium salt [Zanetti et al. 2001c]. Acidic sites formed on the surface of the clay particles may also accelerate the decomposition of the polymer [Tang et al. 2002, Fina et al. 2012]. The decomposition of organic modifiers or the increase in impurities such as water may trigger decomposition. However, the change in thermal decomposition does not seem to be a general mechanism and is usually of minor importance. Furthermore, nanoplate nanocomposites in the solid-state show a reduction in permeability, often by approximately 50%, even for small gas molecules such as nitrogen and oxygen [Lan et al. 1994, Messerschmith and Giannelis 1995, Frasca et al. 2016c]. This pronounced reduction is explained by a tortuous pathway model for gas molecules through the nanocomposite. Nanocomposites may show some increase in residence time under pyrolyzing conditions; for instance, a changed ratio between monomers and oligomers for polystyrene has been reported [Su and Wilkie 2004]. However, the bubbling pyrolyzing melt dominating the burning of a polymer is hardly affected by this reduced diffusion of gas molecules through the solid state. Nevertheless, the beginning of decomposition, when the first volatiles must diffuse through a rather intact polymer matrix, may be influenced, as may thermo-oxidation, which depends on the permeation of oxygen into the nanocomposite. A relevant change in thermal decomposition temperature is often observed only at the beginning of pyrolysis. The two model conceptions explaining the reduction in permeability, a prolonged tortuous pathway or a nanoplate network increasing the viscosity, are rather exchangeable. For instance, carbon nanoparticles shifted the onset temperature of PP decomposition to higher temperatures, depending on the nanofiller dispersion achieved [Dittrich et al. 2013a, 2013b]. Whereas composites with tactoid graphene stacks of approximately 40 or 60 graphene sheets started to decompose at an onset temperature similar to that of PP, well-dispersed few-layer graphene achieved an improvement of up to 30°C. The release of the first volatile products was concluded to be hindered by the effect of the particle network. The beginning of PP decomposition increased with increasing exfoliation of graphene layers [Dittrich et al. 2013a, 2013b]. Most polymers in most fire tests burn based on anaerobic pyrolysis, with thermo-oxidation playing a negligible role [Lyon 2004, Schartel et al. 2017]. Nevertheless, for several polymers, ignition marks the point in time when systems switch from thermo-oxidation on their surface to anaerobe pyrolysis of the bulk. Thus, the reduction in oxygen diffusion harbors the potential to delay ignition in systems where the thermo-oxidative decomposition of the polymer occurs at significantly lower temperatures and is delayed in the nanocomposite. For polypropylene nanoplate nanocomposites, for instance, a shift in decomposition temperature up to 100°C is reported in TGA under air [Tidjani et al. 2003, Zanetti et al. 2001b]. However, most nanocomposites show no significant change in time to ignition or even a reduced time to ignition, since the latter was often ascribed to early decomposition of the modifiers.

Beyond the ignition temperature, which is influenced by the decomposition temperature, the time it takes to reach this critical temperature may be the key to understanding the influence of nanoplate dispersion on the ignition of nanocomposites. The time needed for the top layer of the polymer to reach the ignition temperature under external irradiance is controlled by the in-depth heat absorption and material physical properties (thermal conductivity, melting temperature, melt viscosity, etc.) [Fina et al. 2013]. Clear changes in absorption have been shown to be key for controlling ignition [Schartel et al. 2012b]. Carbon nanoplates and even nanoclays are reported to reduce the in-depth heat absorption via additional infrared absorbance. As heat absorption and thus heating are somewhat concentrated in the top layer, the time to ignition decreases as well [Fina et al. 2013, Dittrich et al. 2013a, 2013b]. The effect is very pronounced for carbon nanoplates and increases with increasing dispersion of graphene nanoplates [Dittrich et al. 2013a, 2013b] thus, well-dispersed few-layer graphene nanocomposites are characterized by complete heat absorption in their top layer, namely, the pyrolysis zone for ignition. Thermal inertia, the product of specific heat capacity, material density and thermal conductivity, is another important material characteristic influencing ignition [Thomson and Drysdale 1987]. Carbon nanomaterials significantly increase the thermal conductivity of nanocomposites compared to their polymer matrix

[Kashiwagi et al. 2004b, Dittrich et al. 2013a, 2013b]. The increase in heat conduction away from the surface layer reduces the heating rate and delays ignition. Thus, carbon nanoplates' pronounced effects of increased heat absorption and increased conductivity counteract each other in regard to ignition.

Convective heat transfer outperforms thermal conduction in polymer melts. During the ignition or burning of a polymer specimen, significant convective flows are observed depending on the viscosity of the molten polymer [Fina and Camino 2011]. Nanocomposites are characterized by a dramatic increase in melt viscosity and thus drastically reduced convective heat transport in the pyrolyzing melt. Less heat is transported away from the top layer; a general mechanism is proposed to reduce the time to ignition via reduced convective heat conduction [Fina and Camino 2011].

2.4.2 FLAMMABILITY

Protective residual layer formation and increasing the viscosity of the pyrolyzing melt are accepted as the two main general fire retardancy mechanisms in polymer nanocomposites [Schartel et al. 2006b, Schartel 2007]. Their distinct impacts on fire properties result in varying flame retardancy effects and efficiencies in different fire tests. The formation of a nanoparticle network is the reason for the completely changed rheological behavior, as a flow limit is introduced at low shear rates [Wawrzyn et al. 2014, Kempel et al. 2015]. Increased melt viscosity reduces melt flow and prevents dripping, which are the main factors that fix the molten polymer in reaction to fire tests using a small ignition source [Schartel et al. 2006b, Dittrich et al. 2015]. Nanoplates in polymer nanocomposites are reported to work as efficient anti-dripping agents to achieve a V-0 rating in vertical UL 94 testing, yet at the same time, a worsening from V-2 to HB rating was reported for several nanoplate nanocomposites [Bartholmai and Schartel 2004, Dittrich et al. 2013a, 2013b]. This effect is based on the increase in melt viscosity and thus depends strongly on the degree of dispersion.

Most of the applied nanoparticles work mainly as an inert filler during the burning of the nanocomposite. Although some of the nanoplates deliver some catalytic sites (layered silicate) or radical scavenging properties (graphene), the increase in the amount of carbonaceous char and thus the decrease in fire load is usually limited [Bartholmai and Schartel 2004, Schartel et al. 2006b]. Indeed, distinct modification of the nanoplates to enhance flame retardant properties or their combination with conventional flame retardants is needed. The main flame retardancy mode of action of nanocomposites is not the reduction of the amount of fuel by charring but the reduction in the fuel release rate achieved by improved properties of the residue. An efficient protective layer is formed. Inorganic-carbonaceous fire residues are reported to show increased thermal, thermo-oxidative and mechanical stability. Moreover, the formation of homogeneous residues with a continuous, closed surface and high structural integrity is proposed to be crucial for providing the most effective protective layer [Gilman 2007, Wu et al. 2012a]. The better the dispersion of the nanoplates within the nanocomposite, the higher the tendency to form such a protective layer [Duquesne 2003, Bartholmai 2004].

The ability of a nanocomposite to form an efficient protective layer causes flame retardancy but not the nanocomposite morphology at room temperature. Release channels, cracks, and openings must be avoided on all of the length scales [Gilman 2007, Wu et al. 2012a]. Nevertheless, some works show correlations between the dispersion of the nanoplates in the nanocomposite and a reduction in PHRR [Bartholmai and Schartel 2004, Dittrich 2013a, 2013b]. A large interphase between nanoparticles and the polymer may be a crucial prerequisite that enhances the formation of an active fire protection layer. Additionally, an essential influence of oxygen was reported for the formation of a char-silicate layer [Pastore et al. 2004]. The most plausible mechanism in polymer nanocomposites based on noncharring or nearly noncharring polymers is the introduction of a heat-shielding effect caused by the inorganic-carbonaceous residue layer. In charring polymers, nanoplates mainly improve the properties of fire residues, while the relative reduction in PHRR may often be smaller, but only the low absolute values count. The protective layer reduces heat and mass transport. Detailed analysis and quantitative assessment of the protective layer effect are rare [Schartel and Weiß 2010, Schartel et al. 2011, Wu et al. 2012b]. The significantly enhanced reradiation of heat by the hot surface has been identified as the most important factor. This heat shield explains the size of the reduction in PHRR as well as the reduction in PHRR for different external heat fluxes [Wu et al. 2012b, Schartel and Weiß 2010]. Similar descriptions were obtained by simulating the burning of nanocomposites in a cone calorimeter [Zhang et al. 2009a, Zhang and Delichatsios 2010].

For noncharring or hardly charring polymers, nanocomposites yield a fundamental change in fire behavior from noncharring to charring/residue-forming polymers. The characteristic curve pattern of the HRR is changed in the cone calorimeter [Schartel and Kebelmann 2020]. Hereby, the initial increases in HRR show no significant difference, but a residual protective layer is formed during combustion. The PHRR at the end typical for noncharring polymers vanishes, the HRR curves become plateau-like, and the burning time is prolonged [Schartel 2007]. The better the dispersion of the nanoplates, the more pronounced the transformation of fire behavior. The role of nanocomposites as a prerequisite for the formation of a protective layer, as well as the formation of the protective layer after ignition during burning, is demonstrated in a study on aged nanocomposites [Colonna et al. 2014]. Aging the polymer layered silicate nanocomposites at high temperatures under air gave the nanoplate enough time for migration toward the surface. During aging, the protective layer is formed due to migration before ignition, shifting the time to ignition and changing the HRR curve pattern. The burning behavior of the aged material crucially differs from the burning behavior of the nanocomposite, for which the protective layer is formed after ignition.

In charring polymers, nanoplates can interact with char formation, enhancing the amount or stabilizing the final fire residue [Bourbigot et al. 2004b]. Furthermore, well-dispersed nanoplates change the polymeric melt viscosity as well as the viscosity of the liquid decomposition products. Nanocomposites showed significantly changed char deformation during burning, and thus, a crucial impact on intumescent systems is expected. Synergistic effects are usually reported when nanoplates are used as adjuvants in intumescent materials [Bourbigot et al. 2004b,

Bourbigot and Duquesne 2007]. Nevertheless, increased viscosity may also inhibit intumescence [Schartel et al. 2010]. Two effects must be considered when combining nanocomposites with the intumescent flame retardant mode of action. Since extremely small concentrations of nanoplates crucially change the viscosity of the pyrolyzing melt, nanoplates may be used to adjust the viscosity, and solidification may be used to optimize the intumescence. Since nanoplates strengthen the fire residue, the thermal and mechanical stability of the multicellular structure can be significantly improved [Dittrich et al. 2014, Bourbigot et al. 2004b, Bourbigot and Duquesne 2007].

Successfully commercialized flame retardant nanocomposites combine nanoparticles with conventional flame retardants [Bödiger et al. 1997, Schall et al. 2000]. Furthermore, synergistic effects of nanoplates with other nanoparticles, such as nanotubes or nanospheres, have been reported [Beyer 2005, Beyer 2006, Schartel and Schmaucks 2017]. It should be noted that well-dispersed nanoplates may increase the dispersion of the other components because increased viscosity results in higher shear rates during compounding. Combination with a glass-forming adjuvant may improve the closed surface of the protective layer [Wu et al. 2012, Yu et al. 2011, Wu et al. 2011b]. The combination of nanoplate nanocomposites with metal hydrates in different polymers is proposed for a large variety of systems to crucially improve the protective properties of the metal oxide/carbonaceous fire residue [Beyer 2001, Zhang et al. 2009b, Dittrich 2014, Schartel et al. 2006a, Zirnstein 2018, Hull 2003, Strommer 2022c, Kong 2009]. Summing up, the interplay and interactions of nanoplates with other fillers in multicomponent systems are influenced by the dispersion.

2.5 CONCLUSION

The fire behavior and thus the active flame retardancy mechanisms in nanoplate nanocomposites show a strong structure–property relationship. Improving nanodispersion often directly improves flame retardancy. Therefore, the modification of the nanoplates as well as the preparation of nanocomposites becomes very important to control this dispersion. Physical properties influencing fire behavior, such as thermal conductivity and heat absorption, are crucially changed by well-dispersed nanoplates. The increase in pyrolyzing melt viscosity influences fire behavior in different ways, such as changing the melt flow, dripping, and/or deformation of the fire residue during charring. The dispersion of nanoplates functions as a prerequisite for the formation of an efficient protective layer or improvement of the char properties.

REFERENCES

B. Ahmadi-Moghadam, F. Taheri. 2014. "Effect of processing parameters on the structure and multifunctional performance of epoxy/GNP-nanocomposites." *J. Mater. Sci.* 49:6180–6190. doi: 10.1007/s10853-014-8332-y.
M. Alexandre, P. Dubois. 2000. "Polymer-layered silicate nanocomposites: preparation, properties and uses of a new class of materials." *Mater. Sci. Engin.* 28:1–63. doi: 10.1016/S0927-796X(00)00012-7.

M. Alexandre, G. Beyer, C. Henrist, R. Cloots, A. Rulmont, R. Jerome, P. Dubois. 2001. "Preparation and properties of layered silicate nanocomposites based on ethylene vinyl acetate copolymers." *Macromol. Rapid Comm.* 22:643–646. doi: 10.1002/1521-3927 (20010501)22:8<643::AID-MARC643>3.0.CO;2-%23.

I. Alig, P. Pötschke, D. Lellinger, T. Skipa, S. Pegel, G.R. Kasaliwal, T. Villmow. 2012. "Establishment, morphology and properties of carbon nanotube networks in polymer melts." *Polymer* 53:4–28. doi: 10.1016/j.polymer.2011.10.063.

M.J. Allen, V.C. Tung, R.B. Kaner. 2009. "Honeycomb carbon: A review of graphene." *Chem. Rev.* 110:132–145. doi: 10.1021/cr900070d.

S. Araby, J. Li, G. Shi, Z. Ma, J. Ma. 2017. "Graphene for Flame- Retarding Elastomeric Composite Foams Having Strong Interface." *Compos. Part A* 101:254–264. doi: 10.1016/j.compositesa.2017.06.022.

S. Araby, C.-H. Wang, H. Wu, Q. Meng, H.-C. Kuan, N.K. Kim, A. Mouritz, J. Ma. 2018. "Development of flame-retarding elastomeric composites with high mechanical performance." *Compos. Part A* 109:257–266. doi: 10.1016/j.compositesa.2018.03.012.

S. Araby, X. Su, Q. Meng, H.-C. Kuan, C.-H.Wang, A. Mouritz, A. Maged, J. Ma. 2019. "Graphene platelets versus phosphorus compounds for elastomeric composites: flame retardancy, mechanical performance and mechanisms." *Nanotechnology* 30:385703. doi: 10.1088/1361-6528/ab2a3d.

S. Araby, B. Philips, Q. Meng, J. Ma, T. Laoui, C. H. Wang. 2021. "Recent advances in carbon-based nanomaterials for flame retardant polymers and composites." *Compos. Part B* 212:108675. doi: 10.1016/j.compositesb.2021.108675.

N. F. Attia, S. E. A. Elashery, A. M. Zakria, A. S. Eltaweil, H. Oh. 2021. "Recent advances in graphene sheets as new generation of flame retardant materials." *Mater. Sci. Eng. B* 274:115460. doi: 10.1016/j.mseb.2021.115460.

V. Babrauskas. 2003. "*Ignition Handbook.*" Fire Science Publishers and SFPE, Issaquah.

H. V. Babu, C. Coluccini, D.-Y. Wang. 2017. "Functional layered double hydroxides and their use in fire retardant polymer materials." In: "Novel Fire Retardant Polymers and Composite Materials." D.-Y. Wang, (Ed.). Woodhead publishing. Chap 8: 201–228.

A. Bafna, G. Beaucage, F. Mirabella, S. Mehta. 2003. "3D hierarchical orientation in polymer–clay nanocomposite films." *Polymer* 44:1103–1115. doi: 10.1016/S0032-3 861(02)00833-9.

A.C. Balazs, C. Singh, E. Zhulina, Y. Lyatskaya. 1999. "Modeling the phase behavior of polymer/clay nanocomposites." *Acc. Chem. Res.* 8:651–657. doi: 10.1021/ar970336m.

M. Bartholmai, B. Schartel. 2004. "Layered silicate polymer nanocomposites: new approach or illusion for fire retardancy? Investigations of the potentials and the tasks using a model system." *Polym. Adv. Technol.* 15:355–364. doi: 10.1002/pat.483.

A. Battig, N. Abdou-Rahaman Fadul, D. Frasca, D. Schulze, B. Schartel. 2021. "Multifunctional graphene nanofiller in flame retarded polybutadiene/chloroprene/ carbon black composites." *e-Polymers* 21:244–262. doi: 10.1515/epoly-2021-0026.

A. Bendaoudi, S. Duquesne, C. Jama, M. Le Bras, R. Delobel, P. Recourt, J-M. Gloaguen, J-M. Lefebvre, A. Addad. 2005. "Effect of the processing conditions on the fire retardant and thermomechanical properties of PP-clay nanocomposites." In: "Fire Retardancy of Polymers: New Applications of Mineral Fillers." M. Le Bras, C. A. Wilkie, S. Bourbigot, S. Duquesne, C. Jama (Eds.). Royal Society of Chemistry, Cambridge. chap. 8:114–125.

M. Berta, C. Lindsay, G. Pans, G. Camino. 2006. "Effect of chemical structure on combustion and thermal behaviour of polyurethane elastomer layered silicate nanocomposites." *Polym. Degrad. Stab.* 91:1179–1191. doi: 10.1016/j.polymdegradstab.2005.05.027.

G. Beyer. 2001. "Flame retardant properties of EVA-nanocomposites and improvements by combination of nanofillers with aluminium trihydrate." *Fire Mater.* 25:193–197. doi: 10.1002/fam.776.

G. Beyer. 2005. "Filler blend of carbon nanotubes and organoclays with improved char as a new flame retardant system for polymers and cable applications." *Fire Mater.* 29:61–69. doi: 10.1002/fam.866.

G. Beyer. 2006. "Flame retardancy of nanocomposites based on organoclays and carbon nanotubes with aluminium trihydrate." *Polym. Adv. Technol.* 17:218–225. doi: 10.1002/pat.696.

A. Blumstein. 1965. "Polymerization of adsorbed monolayers. II. Thermal degradation of the inserted polymer." *J. Polym. Sci. Part A Polym. Chem.* 3:2665–2673. doi: 10.1002/pol.1965.100030721.

M. Bödiger, T. Eckel, D. Wittmann, H. Alberts, German Patent DE 195 30 200 A 1, 1997.

H. P. Boehm, A. Clauss, G. O. Fischer, U. Hofmann. 1962a. "Das Adsorptionsverhalten sehr dünner Kohlenstoff-Folien." *Z. Anorg. Allg. Chem.* 316:119–127. doi: 10.1002/zaac.19623160303.

H. P. Boehm, A. Clauss, G. O. Fischer, U. Hofmann. 1962b. "Dünnste Kohlenstoff-Folien." *Z. Naturforsch.* 17b:150–153. doi: 10.1515/znb-1962-0302.

S. Bourbigot, J. W. Gilman, C. A. Wilkie. 2004a. "Kinetic analysis of the thermal degradation of polystyrene–montmorillonite nanocomposite." *Polym. Degrad. Stabil.* 84:483–492. doi: 10.1016/j.polymdegradstab.2004.01.006.

S. Bourbigot, M. Le Bras, S. Duquesne, M. Rochery. 2004b. "Recent advances for intumescent polymers." *Macromol. Mater. Eng.* 289:499–511. doi: 10.1002/mame.200400007.

S. Bourbigot, S. Duquesne. 2007. "Intumescence and nanocomposites: A novel route for flame-retarding polymeric materials." In: "Flame Retardant Polymer Nanocomposites." A.B. Morgan, C.A. Wilkie (Eds.).Wiley, Hoboken. chap 6:131–162.

D. J. Bray, S. G. Gilmour, F. J. Guild, A. C. Taylor. 2012. "Quantifying nanoparticle dispersion by using the area disorder of Delaunay triangulation." *J. Royal Statistical Soc. Ser. C Appl. Statistics* 61:253–275. doi: 10.1111/j.1467-9876.2011.01009.x.

B. C. Brodie. 1859. "XIII. On the atomic weight of graphite." *Philos. Trans. R. Soc. London* 149:249. doi: 10.1098/rstl.1859.0013.

S. D. Burnside, E. P. Giannelis. 1995. "Synthesis and properties of new poly(dimethylsiloxane) nanocomposites." *Chem. Mater.* **7**:1597–1600. doi: 10.1021/cm00057a001.

D. J. Carastan, A. Vermogen, K. Masenelli-Varlot, M. R. Demarquette. 2010. "Quantification of clay dispersion in nanocomposites of styrenic polymers." *Polym. Eng. Sci.* 50:257–267. doi: 10.1002/pen.21527.

S. Chandrasekaran, N. Sato, F. Tölle, R. Mülhaupt, B. Fiedler, K. Schulte. 2004. "Fracture toughness and failure mechanism of graphene based epoxy composites." *Compos. Sci. Technol.* 97:90–99. doi: 10.1016/j.compscitech.2014.03.014.

C. Lu, X.-p. Gao, D. Yang, Q.-q. Cao, X.-h. Huang, J.-c. Liu, Y.-q. Zhang. 2014. "Flame retardancy of polystyrene/nylon-6 blends with dispersion of clay at the interface." *Polym. Degrad. Stab.* 107:10–20. doi: 10.1016/j.polymdegradstab.2014.04.028.

C. Lu, L. Liu, N. Chen, X. Wang, D. Yang, X.-h. Huang, D.-h. Yao. 2015. "Influence of clay dispersion on flame retardancy of ABS/PA6/APP Blends." *Polym. Degrad. Stab.* 114:16–29. doi: 10.1016/j.polymdegradstab.2015.01.024.

C. Lu, X.-p. Gao, D.-h. Yao, C.-l. Cao, Y.-j. Luo. 2018. "Improving flame retardancy of linear low-density polyethylene/nylon 6 blends via controlling localization of clay and intumescent flame retardant." *Polym. Degrad. Stab.* 153:75–87. doi: 10.1016/j.polymdegradstab.2018.04.022.

F. Chavarria, D.R. Paul. 2004. "Comparison of nanocomposites based on nylon 6 and nylon 66." *Polymer* 45:8501–8515. doi: 10.1016/j.polymer.2004.09.074.

F. Chavarria, K. Nairn, P. White, A.J. Hill, D.L. Hunter, D.R. Paul. 2007. "Morphology and properties of nanocomposites from organoclays with reduced cation exchange capacity." *J. Appl. Polym. Sci.* 105:2910–2924. doi: 10.1002/app.26362.

I.-J. Chin, T. Thurn-Albrecht, H.C. Kim, T.P. Russell, J. Wang. 2001. "On exfoliation of montmorillonite in epoxy." *Polymer* 42:5947–5952. doi: 10.1016/S0032-3861(00)00898-3.

J. N. Coleman, M. Lotya, A. O'Neill, S. D. Bergin, P. J. King, U. Khan, K. Young, A. Gaucher, S. De, R. J. Smith. 2011. "Two-dimensional nanosheets produced by liquid exfoliation of layered materials." *Science* 42:568–571. doi: 10.1126/science.1194975.

S. Colonna, F. Cuttica, A. Frache. 2014. "Aging of EVA/organically modified clay: effect on dispersion, distribution and combustion behavior." *Polym. Degrad. Stab.* 107:184–187. doi: 10.1016/j.polymdegradstab.2014.05.019.

H. Dalir, R. D. Farahani, V. Nhim, B. Samson, M. Lévesque, D. Therriault. 2012. "Preparation of highly exfoliated polyester-clay nanocomposites: process-property correlations." *Langmuir* 28:791–803. doi: 10.1021/la203331h.

D. Demus. 1994. "Phase types, structures, and chemistry of liquid crystals." In: "Liquid Crystals." H. Stegemeyer, (guest Ed.). In: "Topics in Physical Chemistry." H. Baumgartel, E.V. Franck, W. Grunbein (Eds.). Steinkopff Verlag, Darmstadt, Germany. pp. 1–50.

Y. Deng, Y. Li, J. Dai, M. Lang, X. Huang. 2011. "An efficient way to functionalize graphene sheets with presynthesized polymer via ATNRC chemistry." *J. Polym. Sci., Part A: Polym. Chem.* 49:1582–1590. doi: 10.1002/pola.24579.

H. R. Dennis, D. L. Hunter, D. Chang, S. Kim, J. L. White, J. W. Cho, D. R. Paul. 2001. "Effect of melt processing conditions on the extent of exfoliation in organoclay-based nanocomposites." *Polymer* 42:9513–9522. doi: 10.1016/S0032-3861(01)00473-6.

B. Dittrich, K.-A. Wartig, D. Hofmann, R. Mülhaupt, B. Schartel. 2013a. "Flame retardancy through carbon nanomaterials: carbon black, multiwall carbon nanotubes, expanded graphite, multi-layer graphene and graphene in polypropylene." *Polym. Degrad. Stab.* 98:1495–1505. doi: 10.1016/j.polymdegradstab.2013.04.009.

B. Dittrich, K.-A. Wartig, D. Hofmann, R. Mülhaupt, B. Schartel. 2013b. "Carbon black, multiwall carbon nanotubes, expanded graphite and functionalized graphene flame retarded polypropylene nanocomposites." *Polym. Adv. Technol.* 24:916–926. doi: 10.1002/pat.3165.

B. Dittrich, K.-A. Wartig, R. Mülhaupt, B. Schartel. 2014. "Flame-retardancy properties of intumescent ammonium poly(phosphate) and mineral filler magnesium hydroxide in combination with graphene." *Polymers* 6:2875–2895. doi: 10.3390/polym6112875.

B. Dittrich, K.-A. Wartig, D. Hofmann, R. Mülhaupt, B. Schartel. 2015. "The Influence of layered, spherical, and tubular carbonnanomaterials' concentration on the flame retardancy of polypropylene." *Polym. Compos.* 36:1230–1241. doi: 10.1002/pc.23027.

G. M. do Nascimento, V. R. L. Constantino, M. L. A. Temperini. 2002. "Spectroscopic characterization of a new type of conducting polymer–clay nanocomposite." *Macromolecules* 35:7535–7537. doi: 10.1021/ma0255711.

J. X. Du, J. Zhu, C. A. Wilkie, J. Q. Wang. 2002. "An XPS study of the thermal degradation and flame retardant mechanism of polystyrene-clay nanocomposites." *Polym. Degrad. Stab.* 77:377–381. doi: 10.1016/S0141-3910(02)00055-1.

S. Duquesne, C. Jama, M. Le Bras, R. Delobel, P. Recourt, J. M. Gloaguen. 2003. "Elaboration of EVA–nanoclay systems—characterization, thermal behaviour and fire performance." *Compos. Sci. Technol.* 63:1141–1148. doi: 10.1016/S0266-3538(03)00035-6.

T. Eckel, P. Ooms, D. Wittmann, H.-J. Buysch, German patent. DE 42 31 774 A 1, 1994.

D. F. Eckel, M. P. Balogh, P. D. Fasulo, W. R. Rodgers. 2004. "Assessing organo-clay dispersion in polymer nanocomposites." *J. Appl. Polym. Sci.* 93:1110–1117. doi: 10.1002/app.20566.

R. F. Eduljee, R. L. McCullough. 1993. "Elastic Properties of composites." In: "Structure and Properties of Composites." Vol. 13, T. W. Chou, (Ed.). from the series Materials Science and Technology, R. W. Cahn, P. Haasen, E. J. Kramer, (Eds.). VCH. chap 9:381–474

M. Fang, K. Wang, H. Lu, Y. Yang, S. Nutt. 2010. "Single-layer graphene nanosheets with controlled grafting of polymer chains." *J. Mater. Chem.* 20:1982–1992. doi: 10.1039/B919078C.

J. Fawaz, V. Mittal. V. 2014. "Synthesis of Polymer Nanocomposites: Review of Various Techniques." Wiley-VCH.

F. Fenouillot, P. Cassagnau, J. C. Majest. 2009. "Uneven distribution of nanoparticles in immiscible fluids: morphology development in polymer blends." *Polymer* 50:1333–1350. doi: 10.1016/j.polymer.2008.12.029.

A. Fina, G. Camino. 2011. "Ignition mechanisms in polymers and polymer nanocomposites." *Polym. Adv. Technol.* 22:1147–1155. doi: 10.1002/pat.1971.

A. Fina, F. Cuttica, G. Camino. 2012. "Ignition of polypropylene/montmorillonite nanocomposites." *Polym. Degrad. Stab.* 97:2619–2626. doi: 10.1016/j.polymdegradstab.2012.07.017.

A. Fina, J. Feng, F. Cuttica. 2013. "In-depth radiative heat transmittance through polypropylene/nanoclay composites." *Polym. Degrad. Stab.* 98:1030–1035. doi: 10.1016/j.polymdegradstab.2013.02.003.

T. D. Fornes, P. J. Yoon, H. Keskkula, D. R. Paul. 2001. "Nylon 6 nanocomposites: the effect of matrix molecular weight." *Polymer* 42:9929–9940. doi: 10.1016/S0032-3861(01)00552-3.

T. D. Fornes, P. J. Yoon, D. L. Hunter, H. Keskkula, D. R. Paul. 2002. "Effect of organoclay structure on nylon 6 nanocomposite morphology and properties." *Polymer* 43:5915–5933. doi: 10.1016/S0032-3861(02)00400-7.

T. D. Fornes, D. R. Paul. 2003. "Modeling properties of nylon 6/clay nanocomposites using composite theories." *Polymer* 44:4993–5013. doi: 10.1016/S0032-3861(03)00471-3.

D. Frasca, D. Schulze, V. Wachtendorf, C. Huth, B. Schartel. 2015. "Multifunctional multilayer graphene/elastomer nanocomposites." *Eur. Polym. J.* 71:99–113. doi: j.eurpolymj.2015.07.050.

D. Frasca, D. Schulze, V. Wachtendorf, B. Krafft, T. Rybak, B. Schartel. 2016a. "Multilayer graphene/carbon black/chlorine isobutyl isoprene rubber nanocomposites." *Polymers* 8:95. doi: 10.3390/polym8030095

D. Frasca, D. Schulze, V. Wachtendorf, M. Morys, B. Schartel. 2016b. "Multilayer graphene/chlorine-isobutene-isoprene rubber nanocomposites: the effect of dispersion." *Polym. Adv. Technol.* 27:872–881. doi: 10.1002/pat.3740.

D. Frasca, D. Schulze, M. Böhning, B. Krafft, B. Schartel. 2016c. "Multilayer graphene chlorine isobutyl isoprene rubber nanocomposites: influence of the multilayer graphene concentration on physical and flame-retardant properties." *Rubber Chem. Technol.* 89:316–334. doi: 10.5254/rct.15.84838.

Y. S. Gao, J. W. Wu, Q. Wang, C. A. Wilkie, D. O'Hare. 2014. "Flame retardant polymer/layered double hydroxide nanocomposites." *J. Mater. Chem. A* 2:10996–11016. doi: 10.1039/c4ta01030b.

E. P. Giannelis. 1996. "Polymer layered silicate nanocomposites." *Adv. Mater.* 8:29–35. doi: 10.1002/adma.19960080104.

J. W. Gilman, T. Kashiwagi, J.D. Lichtenhan. 1997. "Nanocomposites: a revolutionary new flame retardant approach." *SAMPE J.* 33:40–46.

J. W. Gilman, T. Kashiwagi, E. P. Giannelis, E. Manias, S. Lomakin, J. D. Lichtenham, P. Jones. 1998. "Nanocomposites: radiative gasification and vinyl polymer flammability." In: "Fire Retardancy of Polymers: The Use of Intumescence." M. Le Bras,

G. Camino, S. Bourbigot, R. Delobel (Eds.). Royal Society of Chemistry, Cambridge. pp. 203–221.

J. W. Gilman. 1999. "Flammability and thermal stability studies of polymer layered-silicate (clay) nanocomposites." *Appl. Clay Sci.* 15:31–49. doi: 10.1016/S0169-1317(99)0001 9-8.

J. W. Gilman, C. L. Jackson, A. B. Morgan, R. Harris, E. Manias, E. P. Giannelis, M. Wuthenow, D. Hilton, S. H. Phillips. 2000. "Flammability properties of poly-mer–layered-silicate nanocomposites. polypropylene and polystyrene nanocompo-sites." *Chem. Mater.* 12:1866–1873. doi: 10.1021/cm0001760.

J.W. Gilman. 2007. "Flame retardant mechanism of polymer-clay nanocomposites." In: "Flame Retardant Polymer Nanocomposites." A. B. Morgan, C. A. Wilkie, (Eds.). Wiley, Hoboken. chap 3: pp. 67–87.

C. Gómez-Navarro, M. Burghard, K. Kern. 2008. "Elastic properties of chemically derived single graphene sheets." *Nano Lett.* 8:2045–2049. doi: 10.1021/nl801384y.

L. Gong, R. J. Young, I. A. Kinloch, I. Riaz, R. Jalil, K. S. Novoselov. 2012. "Optimizing the reinforcement of polymer-based nanocomposites by graphene." *ACS Nano* 6:2086–2095. doi: 10.1021/nn203917d.

Q. Gou, M. D. Wetzel, B. A. Ogunnaike. 2014. "Quantification of layered silicate dispersion in polymer nanocomposites." *Polymer* 55:4216–4225. doi: 10.1016/j.polymer.2014.05.047.

J. Gu, N. Li, L. Tian, Z. Lv, Q. Zhang. 2015. "High thermal conductivity graphite nano-platelet/UHMWPE nanocomposites." *RSC Adv.* 5:36334–36339. doi: 10.1039/C5 RA03284A.

Y. Q. Guo, C. L. Bao, L. Song, B. H. Yuan, Y. Hu. 2011. "In situ polymerization of gra-phene, graphite oxide, and functionalized graphite oxide into epoxy resin and com-parison study of on-the-flame behavior." *Indust. Engin. Chem. Res.* 50:7772–7783. doi: 10.1021/ie200152x.

E. Hackett, E. Manias, E. P. Giannelis. 1998. "Molecular dynamics simulations of organi-cally modified layered silicates." *J. Chem. Phys.* 108:7410–7415. doi: 10.1063/1.4761 61.

J. C. Halpin, J. L. Kordos. 1976. "The Halpin-Tsai equations: a review." *Polym. Eng. Sci.* 16:344–352. doi: 10.1002/pen.760160512.

G. J. Han, X.Y. Zhao, Y.Z. Feng, J.M. Ma, K.Q. Zhou, Y.Q. Shi, C.T. Liu, X.L. Xie. 2021. "Highly flame-retardant epoxy-based thermal conductive composites with functiona-lized boron nitride nanosheets exfoliated by one-step ball milling." *Chem. Engin. J.* 401:127099. doi: 10.1016/j.cej.2020.127099.

A. Hartwig, D. Pütz, B. Schartel, M. Bartholmai, M. Wendschuh-Josties. 2003. "Combustion behaviour of epoxide based nanocomposites with ammonium and phosphonium ben-tonites." *Macromol. Chem. Phys.* 204:2247–2257. doi: 10.1002/macp.200300047.

T. R. Hull, D. Price, Y. Liu, C. L. Wills, J. Brady. 2003. "An investigation into the decomposition and burning behaviour of ethylene-vinyl acetate copolymer nano-composite materials." *Polym. Degrad. Stab.* 82:365–371. doi: 10.1016/S0141-3910 (03)00214-3.

W. S. Hummers, R. E. Offeman. 1958. "Preparation of graphitic oxide." *J. Am. Chem. Soc.* 80:1339. doi: 10.1021/ja01539a017.

B. N. Jang, D. Wang, C. A. Wilkie. 2005. "Relationship between the solubility parameter of polymers and the clay dispersion in polymer/clay nanocomposites and the role of the surfactant macromolecules." 38:6533–6543. doi: 10.1021/ma0508909.

K. Jasmund, G. Lagaly. 1993. "Tonminerale und Tone Struktur Eigenschaften, Anwendungen und Einsatz in Industrie und Umwelt." Steinkopff, Darmstadt, Germany.

J.-W. Jiang, J.-S. Wang, B. Li. 2009. "Young's modulus of graphene: a molecular dynamics study." *Phys. Rev. B* 80:113–405. doi: 10.1103/PhysRevB.80.113405.

X. Jiang, L. T. Drzal. 2012. "Reduction in percolation threshold of injection molded high-density polyethylene/exfoliated graphene nanoplatelets composites by solid state ball milling and solid state shear pulverization." *J. Appl. Polym. Sci.* 124:525–535. doi: 10.1002/app.34891.

E. N. Kalali, X. Wang, D. Y. Wang. 2015. "Functionalized layered double hydroxide-based epoxy nanocomposites with improved flame retardancy and mechanical properties." *J. Mater. Chem. A* 3:6819–6826. doi: 10.1039/c5ta00010f.

E. N. Kalali, X. Wang, D. Y. Wang. 2016. "Multifunctional intercalation in layered double hydroxide: toward multifunctional nanohybrids for epoxy resin." *J. Mater. Chem. A* 4:2147–2157. doi: 10.1039/c5ta09482h.

H. Kang, K. Zuo, Z. Wang, L. Zhang, L. Liu, B. Guo. 2014. "Using a green method to develop graphene oxide/elastomers nanocomposites with combination of high barrier and mechanical performance." *Compos. Sci. Technol.* 92:1–8. doi: 10.1016/j.compscitech.2013.12.004.

T. Kashiwagi, R. H. Harris, X. Zhang, R. M. Briber, B. H. Cipriano, S. R. Raghavan, W. H. Awad, J. R. Shields. 2004a. "Flame retardant mechanism of polyamide 6–clay nanocomposites." *Polymer* 45:881–891. doi: 10.1016/j.polymer.2003.11.036.

T. Kashiwagi, E. Grulke, J. Hilding, K. Groth, R. Harris, K. Butler, J. Shields, S. Kharchenko, J. Douglas. 2004b. "Thermal and flammability properties of polypropylene/carbon nanotube nanocomposites." *Polymer* 45:4227–4239. doi: 10.1016/j.polymer.2004.03.088.

T. Kashiwagi, F. M. Du, J. F. Douglas, K. I. Winey, R. H. Harris, J. R. Shields. 2005. "Nanoparticle networks reduce the flammability of polymer nanocomposites." *Nature Mater.* 4:928–933. doi: 10.1038/nmat1502.

T. Kashiwagi, M. F. Mu, K. Winey, B. Cipriano, S. R. Raghavan, S. Pack, M. Rafailovich, Y. Yang, E. Grulke, J. Shields. 2008. "Relation between the viscoelastic and flammability properties of polymer nanocomposites." *Polymer* 49:4358–4368. doi: 10.1016/j.polymer.2008.07.054.

F. Kempel, B. Schartel, J. M. Marti, K. M. Butler, R. Rossi, S. R. Idelsohn, E. Oñate, A. Hofmann. 2015. "Modelling the Vertical UL 94 test: competition and collaboration between melt dripping, gasification and combustion." *Fire Mater.* 39:570–584. doi: 10.1002/fam.2257.

C.-M. Kim, D.-H. Lee, B. Hoffmann, J. Kressler, G. Stoppelmann. 2001. "Influence of nanofillers on the deformation process in layered silicate/polyamide 12 nanocomposites." *Polymer* 42:1095–1100. doi: 10.1016/S0032-3861(00)00468-7.

S. W. Kim, W. H. Jo, M. S. Lee, M. B. Ko, J. Y. Jho. 2002. "Effects of shear on melt exfoliation of clay in preparation of nylon 6/organoclay nanocomposite." *Polym. J.* 34:103–111. doi: 10.1295/polymj.34.103.

D. H. Kim, P. D. Fasulo, W. R. Rodgers, D. R. Paul. 2007. "Structure and properties of polypropylene-based nanocomposites: effect of PP-g-MA to organoclay ratio." *Polymer* 48:5308–5323. doi: 10.1016/j.polymer.2007.07.011.

H. Kim, C.W. Macosko CW. 2009. "Processing-property relationships of polycarbonate/graphene composites." *Polymer* 50:3797–3809. doi: 10.1016/j.polymer.2009.05.038.

H. Kim, Y. Miura, C. W. Macosko. 2010. "Graphene/polyurethane nanocomposites for improved gas barrier and electrical conductivity." *Chem Mater* 22:3441–3450. doi: 10.1021/cm100477v.

Y. Kojima, A. Usuki, M. Kawasumi, Y. Fukushima, A. Okada, T. Karauchi, O. Kamigaito. 1992. "Mechanical properties of nylon 6-clay hybrid." *J. Mater. Res.* 8:1185–1189. doi: 10.1557/JMR.1993.1185.

Q. H. Kong, Y. Hu, L. Song, C. W. Yi. 2009. "Synergistic flammability and thermal stability of polypropylene/aluminum trihydroxide/Fe-montmorillonite nanocomposites." *Polym. Adv. Technol.* 20:404–409. doi: 10.1002/pat.1285.

X. Kornmann, H. Lindberg, L. A. Berglund. 2001. "Synthesis of epoxy–clay nanocomposites: influence of the nature of the clay on structure." *Polymer* 42:1303–1310. doi: 10.1016/S0032-3861(00)00346-3.

M. H. Kothmann, M. Ziadeh, G. Bakis, A. R. de Anda, J. Breu, V, Altstädt. 2015. "Analyzing the influence of particle size and stiffness state of the nanofiller on the mechanical properties of epoxy/clay nanocomposites using a novel shear-stiff nano-mica." *J. Mater. Sci.* 50:4845–4859. doi: 10.1007/s10853-015-9028-7.

R. Krishnamoorti, E. P. Giannelis. 1997. "Rheology of end tethered polymer layered silicate nanocomposites." *Macromolecules* 30:4097–4102. doi: 10.1021/ma960550a.

R. Krishnamoorti, J. X. Ren, A. S. Silva. 2001a. "Shear response of layered silicate." *J. Chem. Phys.* 114:4968–4973. doi: 10.1063/1.1345908.

R. Krishnamoorti, K. Yurekli. 2001b. "Rheology of polymer layered silicate nanocomposites." *Curr. Opin. Colloid Interfac. Sci.* 6:464–470. doi: 10.1016/S1359-0294(01)00121-2.

G. Lagaly. 1986. "Interaction of alkylamines with different types of layered compounds." *Solid State Ionics* 22:43–45. doi: 10.1016/0167-2738(86)90057-3.

T. Lan, P. D. Kaviratna, T. J. Pinnavaia. 1994. "On the nature of polyimide-clay hybrid composites." *Chem. Mater.* 6:573–577. doi: 10.1021/cm00041a002.

P. C. LeBaron, Z. Wang, T. J. Pinnavaia. 1999. "Polymer-layered silicate nanocomposites: an overview." *Appl. Clay Sci.* 15:11–29. doi: 10.1016/S0169-1317(99)00017-4.

H. Lee, P. D. Fasulo, W. R. Rodgers, D. R. Paul. 2005. "TPO based nanocomposites. Part 1. Morphology and mechanical properties." *Polymer* 46:11673–11689. doi: 10.1016/j.polymer.2005.09.068.

M. Lewin. 2003. "Some comments on the modes of action of nanocomposites in the flame retardancy of polymers." *Fire Mater.* 27:1–7. doi: 10.1002/fam.813.

M. Lewin. 2006. "Reflections on migration of clay and structural changes in nanocomposites." *Polym. Adv. Technol.* 17:758–763. doi: 10.1002/pat.762.

J.-L. Li, K. N. Kudin, M. J. McAllister, R. K. Prud'homme, I. A. Aksay, R. Car. 2006. "Oxygen-driven unzipping of graphitic materials." *Phys. Rev. Lett.* 96:176101. doi: 10.1103/PhysRevLett.96.176101.

Z. Li, R. J. Young, I. A. Kinloch, N. R. Wilson, A. J. Marsden, A. P. A. Raju. 2015. "Quantitative determination of the spatial orientation of graphene by polarized Raman spectroscopy." *Carbon* 88:215–224. doi: 10.1016/j.carbon.2015.02.072.

Y. Li, H. Zhang, E. Bilotti, T. Peijs. 2016a. "Optimization of three-roll mill parameters for in-situ exfoliation of graphene." *MRS Adv.* 1:1389–1394. doi: 10.1557/adv.2016.191.

Z. Li, R. J. Young, N. R. Wilson, I. A. Kinloch, C. Vallés, Z. Li. 2016b. "Effect of the orientation of graphene-based nanoplatelets upon the Young's modulus of nanocomposites." *Compos. Sci. Technol.* 123:125–133. doi: 10.1016/j.compscitech.2015.12.005.

S. H. Liao, P. L. Liu, M. C. Hsiao, C. C. Teng, C. A. Wang, M. D. Ger, C. L. Chiang. 2012. "One-step reduction and functionalization of graphene oxide with phosphorus-based compound to produce flame-retardant epoxy nanocomposite". *Ind. Engin. Chem. Res.* 51:4573–4581. doi: 10.1021/ie2026647.

Y. Lin, J. Jin, M. Song. 2011. "Preparation and characterisation of covalent polymer functionalized graphene oxide." *J. Mater. Chem.* 21:3455–3461. doi: 10.1039/c0jm01859g.

L. S. Loo, K. K. Gleason. 2003. "Fourier transforms infrared investigation of the deformation behavior of montmorillonite in nylon 6/nanoclay nanocomposite." *Macromolecules* 36:2587–2590. doi: 10.1021/ma0259057.

Z. P. Luo, J. H. Koo. 2008. "Quantification of the layer dispersion degree in polymer layered silicate nanocomposites by transmission electron microscopy." *Polymer* 49:1841–1852. doi: 10.1016/j.polymer.2008.02.028.

Y. Lyatskaya, A. C. Balazs. 1998. "Modeling the phase behavior of polymer-clay composites." *Macromolecules* 31:6676–6680. doi: 10.1021/ma980687w.

R. E. Lyon. 2004. "Plastics and rubber." In: "Handbook of Building Materials for Fire Protection." C. A. Harper, (Ed.).McGraw-Hill, New York. chap 3:pp. 3.1–3.51.

M. M. Malwitz, P. D. Butler, L. Porcar, D. P. Angelette, G. Schmidt. 2004. "Orientation and relaxation of polymer-clay solutions studied by rheology and small-angle neutron scattering." *J. Polym. Sci. B Polym. Phys.* 42:3102–3112. 10.1002/polb.20175.

E. Manias, A. Touny, L. Wu, K. Strawhecker, B. Lu, T. C. Chung. 2001. "Polypropylene/montmorillonite nanocomposites. Review of the synthetic routes and materials properties." *Chem. Mater.* 13:3516–3523. doi: 10.1021/cm0110627.

E. Manias, G. Polizos, H. Nakajima†, M. J. Heidecker. 2007. "Fundamentals of polymer nanocomposite technology." In: "Flame Retardant Polymer Nanocomposites." A. B. Morgan, C. A. Wilkie (Eds). John Wiley & Sons, Inc. chap 2:31–66

G. Marrucci. 1991. "Rheology of nematic polymers." In: "Liquid Crystallinity in Polymers: Principles and Fundamental Properties." A. Ciferri (Ed.).VCH Publishers, New York. chap 11:395–422.

M. J. McAllister, J. L. Li, D. H. Adamson, H. C. Schniepp, A. A. Abdala, J. Liu, M. Herrera-Alonso, D. L. Milius, R. Car, R. K. Prud'homme, I. A. Aksay. 2007. "Single sheet functionalized graphene by oxidation and thermal expansion of graphite." *Chem. Mater.* 19:4396–4404. doi: 10.1021/cm0630800.

P. B. Messersmith, E. P. Giannelis. 1995. "Synthesis and barrier properties of poly(epsilon-caprolactone)-layered silicate nanocomposites." *J. Polym. Sci. A Polym. Chem.* 33:1047–1057. doi: 10.1002/pola.1995.080330707.

V. Mittal. 2009. "Polymer layered silicate nanocomposites: a review." *Materials* 2:992–1057. doi:10.3390/ma2030992.

V. Mittal. 2014. "Functional polymer nanocomposites with graphene: a review." *Macromol. Mater. Eng.* 299:906–931. doi: 10.1002/mame.201300394.

A. B. Morgan, T. Kashiwagi, R. H. Harris, J. R. Campbell, K. Shibayama, K. Iwasa, J. W. Gilman. 2001. "Flammability properties of polymer-clay nanocomposites: polyamide-6 and polypropylene clay nanocomposites." In: "Fire and Polymers, Materials and Solutions for Hazard Prevention." G. L. Nelson, C. A. Wilkie (Eds.). ACS Symposium Series 797, American Chemical Society, Washington. chap. 2:9–23.

A. B. Morgan, J. W. Gilman. 2003. "Characterization of polymer-layered silicate (clay) nanocomposites by transmission electron microscopy and X-ray diffraction: a comparative study." *J. Appl. Polym. Sci.* 87:1329–1338. doi: 10.1002/app.11884.

P. H. Nam, P. Maiti, M. Okamoto, T. Kotaka, N. Hasegawa, A. Usuki. 2001. "A hierarchical structure and properties of intercalated polypropylene/clay nanocomposites." *Polymer* 42:9633–9640. doi: 10.1016/S0032-3861(01)00512-2.

L. E. Nielsen. 1967. "Models for the permeability of filled polymer systems." *J. Macromol. Sci. Chem.* 1:929–942. doi: 10.1080/10601326708053745.

M. Ogawa, K. Kuroda. 1997. "Preparation of inorganic-organic nanocomposites through intercalation of organoammonium ions into layered silicates." *Bull. Chem. Soc. Jpn.* 70:2593–2618. doi: 10.1246/bcsj.70.2593.

D. K. Owens, R. C. Wendt. 1969. "Estimation of the surface free energy of polymers." *J. Appl. Polym. Sci.* 13:1741–1747. doi: 10.1002/app.1969.070130815.

D. G. Papageorgiou, I. A. Kinloch, R. J. Young. 2017. "Mechanical properties of graphene and graphene-based nanocomposites." *Prog. Mater. Sci.* 90:75–127. doi: 10.1016/j.pmatsci.2017.07.004.

J.-J. Park, J.-Y. Lee. 2013. "Effect of nano-sized layered silicate on AC electrical treeing behavior of epoxy/layered silicate nanocomposite in needle-plate electrodes." *Mater. Chem. Phys.* 141:776–780. doi: 10.1016/j.matchemphys.2013.06.003.

H. O. Pastore, A. Frache, E. Boccaleri, L. Marchese, G. Camino. 2004. "Heat induced structure modifications in polymer-layered silicate nanocomposites." *Macromol. Mater. Eng.* 289:783–786. doi: 10.1002/mame.200400109.

D. R. Paul, L. M. Robeson. 2008. "Polymer nanotechnology: nanocomposites." *Polymer* 49:3187–3204. doi: 10.1016/j.polymer.2008.04.017.

S. Pavlidoua, C.D. Papaspyrides. 2008. "A review on polymer–layered silicate nanocomposites." *Prog. Polym. Sci.* 33:1119–1198. doi: 10.1016/j.progpolymsci.2008.07.008.

A. Peigney, C. Laurent, E. Flahaut, R. R. Bacsa, A. Rousset. 2001. "Specific surface area of carbon nanotubes and bundles of carbon nanotubes". *Carbon* 39:507–514. doi: 10.1016/S0008-6223(00)00155-X.

T. J. Pinnavaia. 1983. "Intercalated clay catalysts." *Science* 220:365–371. doi: 10.1126/science.220.4595.365.

J. R. Potts, D. R. Dreyer, C. W. Bielawski, R. S. Ruoff. 2011. "Graphene-based polymer nanocomposites." *Polymer* 52:5–25. doi: 10.1016/j.polymer.2010.11.042.

J. Ren, B. F. Casanueva, C. A. Mitchell, R. Krishnamoorti. 2003. "Disorientation kinetics of aligned polymer layered silicate nanocomposites." *Macromolecules* 36:4188–4194. doi: 10.1021/ma025703a.

H. J. Salavagione, G. Martinez. 2011. "Importance of covalent linkages in the preparation of effective reduced graphene oxide–poly(vinyl chloride) nanocomposites." *Macromolecules* 44:2685–2692. doi: 10.1021/ma102932c.

N. Schall, T. Engelhardt, H. Simmler-Hübenthal, G. Beyer. German Patent. DE 199 21 472 A 1. 2000.

B. Schartel, J. H. Wendorff. 1999. "Molecular composites for molecular reinforcement: A promising concept between success and failure." *Polym. Engin. Sci.* 39:128–151. doi: 10.1002/pen.11403.

B. Schartel, U. Knoll, A. Hartwig, D. Pütz. 2006a. "Phosphonium modified layered silicate epoxy resins nanocomposites and their combinations with ATH and organo-phosphorus fire retardants." *Polym. Adv. Technol.* 17:281–293. doi: 0.1002/pat.686.

B. Schartel, M. Bartholmai, U. Knoll. 2006b. "Some comments on the main fire retardancy mechanisms in polymer nanocomposites." *Polym. Adv. Technol.* 17:772–777. doi: 10.1002/pat.792.

B. Schartel. 2007. "Considerations regarding specific impacts of the principal fire retardancy mechanisms in nanocomposites." In: "Flame Retardant Polymer Nanocomposites." A. B. Morgan, C. A. Wilkie (Eds.), John Wiley & Sons, Hoboken. chap 5:107–129.

B. Schartel, A. Weiß. 2010. "Temperature inside burning polymer specimens: Pyrolysis zone and shielding." *Fire Mater.* 34:217–235. doi: 10.1002/fam.1007.

B. Schartel, A. Weiß, F. Mohr, M. Kleemeier, A. Hartwig, U. Braun. 2010. "Flame retarded epoxy resins by adding layered silicate in combination with the conventional protection-layer-building flame retardants melamine borate and ammonium polyphosphate." *J. Appl. Polym. Sci.* 118:1134–1143. doi: 10.1002/app.32512.

B. Schartel, A. Weiß, H. Sturm, M. Kleemeier, A. Hartwig, C. Vogtand, R. X. Fischer. 2011. "Layered silicate epoxy nanocomposites: formation of the inorganic-carbonaceous fire protection layer." *Polym. Adv. Technol.* 22:1581–1592. doi: 10.1002/pat.1644.

B. Schartel, K. H. Richter, M. Böhning. 2012a. "Synergistic use of talc in halogen-free flame retarded polycarbonate/acrylonitrile- butadiene-styrene blends." In: "Fire and Polymers VI: New Advances in Flame Retardant Chemistry and Science." A. B. Morgan, C. A. Wilkie, G. L. Nelson (Eds.). ACS Symposium Series, Vol. 1118, American Chemical Society, Washington, DC. chap 2:15–36. doi: 10.1021/bk-2012-1118.ch002.

B. Schartel, U. Beck, H. Bahr, A. Hertwig, U. Knoll, M. Weise. 2012b. "Short communication: sub-micrometer coatings as an IR mirror: a new route to flame retardancy." *Fire Mater.* 36:671–677. doi: 10.1002/fam.1122.

B. Schartel, C. A. Wilkie, G. Camino. 2017. "Recommendations on the scientific approach to polymer flame retardancy: Part 2 – concepts." *J. Fire Sci.* 35:3–20. doi: 10.1177/0734904116675370.

B. Schartel, G. Schmaucks. 2017. "Flame Retardancy Synergism In Polymers Through Different Inert Fillers' Geometry." *Polym. Engin. Sci.* 57:1099–1109. doi: 10.1002/pen.24485.

B. Schartel, K. Kebelmann. 2020. "Fire testing for the development of flame retardant polymeric materials." In: "Flame Retardant Polymeric Materials A Handbook." Y. Hu, X. Wang (Eds.). CRC Press, Taylor & Francis Group, Boca Raton, Fl, US. chap 3:35–55.

G. Schmidt, A. I. Nakatani, C. C. Han. 2002a. "Rheology and flow-birefringence from viscoelastic polymer-clay solutions." *Rheol. Act.* 41:45–54. doi: 10.1007/s003970200004.

G. Schmidt, A. I. Nakatani, P. D. Butler, C. C. Han. 2002b. "Small-angle neutron scattering from viscoelastic polymer-clay solutions." *Macromolecules* 35:4725–4732. doi: 10.1021/ma0115141.

R. K. Shah, D. H. Kim, D. R. Paul. 2007. "Morphology and properties of nanocomposites formed from ethylene/methacrylic acid copolymers and organoclays." *Polymer* 48:1047–1057. doi: 10.1016/j.polymer.2007.01.002.

X. Shen, Z. Wang, Y. Wu, X. Lui, Y.-B. He, J.-K. Kim. 2016. "Multilayer graphene enables higher efficiency in improving thermal conductivities of graphene/epoxy composites." *Nano Lett.* 16:3585–3593. doi: 10.1021/acs.nanolett.6b00722.

Y. Q. Shi, B. Yu, Ý. Y. Zheng, J. Yang, Z. P. Duan, Y. Hu, Y. 2018. "Design of reduced graphene oxide decorated with DOPO-phosphanomidate for enhanced fire safety of epoxy resin." *J. Coll. Interfac. Sci.* 521:160–171. doi: 10.1016/j.jcis.2018.02.054.

S. Sinha Ray, M. Okamoto. 2003. "Polymer/layered silicate nanocomposites: a review from preparation to processing." *Prog. Polym. Sci.* 28:1539–1641. doi: 10.1016/j.progpolymsci.2003.08.002.

M. J. Solomon, A. S. Almusallam, K. F. Seefeldt, A. Somwangthanaroj, P. Varadan. 2001. "Rheology of polypropylene/clay hybrid materials." *Macromolecules* 34:1864–1872. doi: 10.1021/ma001122e.

M. W. Spencer, L. Cui, Y. Yoo, D. R. Paul. 2010. "Morphology and properties of nanocomposites based on HDPE/HDPE-g-MA blends." *Polymer* 51:1056–1070. doi: 10.1016/j.polymer.2009.12.047.

S. Stankovich, D. A. Dikin, G. H. Dommett, K. M. Kohlhaas, E. J. Zimney, E. A. Stach, R. D. Piner, S. T. Nguyen, R. S. Ruoff. 2006. "Graphene-based composite materials." *Nature* 442:282–286. doi: 10.1038/nature04969.

S. Steinmann, W. Gronski, C. Friedrich. 2002. "Influence of selective filling on rheological properties and phase inversion of two-phase polymer blends." *Polymer* 43:4467–4477. doi: 10.1016/S0032-3861(02)00271-9.

P. Steurer, R. Wissert, R. Thomann, R. Mülhaupt. 2009. "Functionalized graphenes and thermoplastic nanocomposites based upon expanded graphite oxide." *Macromol. Rapid Commun.* 30, 316–327. doi: 10.1002/marc.200800754.

B. Strommer, D. Schulze, B. Schartel, M. Böhning. 2022a. "Networking skills: the effect of graphene on the crosslinking of natural rubber nanocomposites with sulfur and peroxide systems." *Polymers* 14:4363. doi: 10.3390/polym14204363.

B. Strommer, A. Battig, D. Schulze, L. Agudo Jácome, B Schartel, M. Böhning. 2022b. "Shape, orientation, interaction or dispersion: valorization of the influence factors in natural rubber nanocomposites." *Rubber Chem. Technol.* Early view. doi: 10.5254/rct.23.77961.

B. Strommer, A. Battig, D. Frasca, D. Schulze, C. Huth, M. Böhning, B. Schartel. 2022c. "Multifunctional property improvements by combining graphene and conventional fillers in chlorosulfonated polyethylene rubber composites." *ACS Appl. Polym. Mater.* 4:1021–1034. doi: 10.1021/acsapm.1c01469.

H. Sturm, B. Schartel, A. Weiß, U. Braun. 2012. "SEM/EDX: Advanced investigation of structured fire residues and residue formation." *Polym. Test.* 31:606–619. doi: 10.1016/j.polymertesting.2012.03.005.

S.P. Su, C. A. Wilkie. 2004. "The thermal degradation of nanocomposites that contain an oligomeric ammonium cation on the clay." *Polym. Degrad. Stabil.* 83:347–362. doi: 10.1016/S0141-3910(03)00279-9.

J. W. Suk, R. D. Piner, J. An, R. S. Ruoff. 2010. "Mechanical properties of monolayer graphene oxide." *ACS Nano* 4:6557–6564. doi: 10.1021/nn101781v.

Y. Tang, Y. Hu, S. F. Wang, Z. Gui, Z. Chen, W. C. Fan. 2002. "Preparation and flammability of ethylene-vinyl acetate copolymer/montmorillonite nanocomposites." *Polym. Degrad. Stab.* 78:555–559. doi: 10.1016/S0141-3910(02)00231-8.

Y. Tang, M. Lewin, E. M. Pearce. 2006. "Effects of annealing on the migration behavior of PA6/clay nanocomposites." *Macromol. Rapid Commun.* 27:1545–1549. doi: 10.1002/marc.200600356.

Y. Tang, M. Lewin. 2007. "Maleated polypropylene OMMT nanocomposite: annealing, structural changes, exfoliated and migration." *Polym. Degrad. Stab.* 92:53–60. doi: 10.1016/j.polymdegradstab.2006.09.013.

L.-C. Tang, Y.-J. Wan, D. Yan, Y.-B. Pei, L. Zhao, Y.-B. Li, L.-B. Wu, J.-X. Jiang, G.-Q. Lai. 2013. "The effect of graphene dispersion on the mechanical properties of graphene/epoxy composites." *Carbon* 60:16–27. doi: 10.1016/j.carbon.2013.03.050.

T. Li, G. Ding, S.-T. Han, Y. Zhou. 2021. "Introduction of polymer nanocomposites." In: "Polymer Nanocomposite Materials: Applications in Integrated Electronic Devices." Y. Zhou, G. Ding (Eds.). WILEY-VCH. Chap 1:pp. 1–20.

B. K. G. Theng. 1974. "The Chemistry of Clay-Organic Reactions." Wiley, New York, NY, USA.

H. E. Thomson, D. D. Drysdale. 1987. "Flammability of plastics I: ignition temperatures." *Fire Mater.* 11:163–172. doi: 10.1002/fam.810110402.

A. Tidjani, O. Wald, M.-M. Pohl, M. P. Hentschel, B. Schartel. 2003. "Polypropylene-graft-maleic anhydride-nanocomposites: I - characterization and thermal stability of nanocomposites produced under nitrogen and under air." *Polym. Degrad. Stab.* 82:133–140. doi: 10.1016/S0141-3910(03)00174-5.

B. M. Tyson, R. K. Abu Al-Rub, A. Yazdanbakhsh, Z. Grasley. 2011. "A quantitative method for analyzing the dispersion and agglomeration of nano-particles in composite materials." *Compos. Part B: Eng.* 42:1395–1403. doi: 10.1016/j.compositesb.2011.05.020.

A. Usuki, Y. Kojima, M. Kawasumi, A. Okada, Y. Fukushima, T. Karauchi, O. Kamigaito. 1993. "Synthesis of nylon 6-clay hybrid." *J. Mater. Res.* 8:1179–1184. doi: 10.1557/JMR.1993.1179.

D. L. VanderHart, A. Asano, J. W. Gilman. 2001a. "Solid-state NMR investigation of paramagnetic nylon-6 clay nanocomposites. 2. Measurement of clay dispersion, crystal stratification, and stability of organic modifiers." *Chem. Mater.* 13:3796–3809. doi: 10.1021/cm011078x.

D. L. VanderHart, A. Asano, J. W. Gilman. 2001b. "NMR measurements related to clay-dispersion quality and organic-modifier stability in nylon-6/clay nanocomposites." *Macromolecules* 34:3819–3822. doi: 10.1021/ma002089z.

R.A. Vaia, R.K. Teukolsky, E.P. Giannelis. 1994. Interlayer structure and molecular environment of alkylammonium layered silicates." *Chem. Mater.* 6:1017–1022. doi: 10.1021/cm00043a025.

R. A. Vaia, K. D. Jandt, E. J. Kramer, E. P. Giannelis. 1995. "Kinetics of polymer melt intercalation." *Macromolecules* 28:8080–8085. doi: 10.1021/ma00128a016.

R. A. Vaia, E.P. Giannelis. 1997. "Polymer melt intercalation in organically modified layered silicates: model predictions and experiment." *Macromolecules* 30:8000–8009. doi: 10.1021/ma9603488.

R. A. Vaia. 2000. "Structural characterization of polymer- layered silicate nanocomposites." In: "Polymer-Clay Nanocomposites." T. J. Pinnavaia, G. W. Beall (Eds.). John Wiley & Sons, Chichester. chap. 12:229–266.

R. A. Vaia, H. D. Wagner. 2004. "Framework for nanocomposites." *Mater. Today*, 7:32–37. doi: 10.1016/S1369-7021(04)00506-1.

J. Vermant, S. Ceccia, M. K. Dolgovskij, P. L. Maffettone, C. W. Macosko. 2007. "Quantifying dispersion of layered nanocomposites via melt rheology." *J. Rheol.* 51:429–450. doi: 10.1122/1.2516399.

A. Vermogen, K. Masenelli-Varlot, S. Roland, J. Duchet-Rumeau, S. Boucard, P. Prele. 2005. "Evaluation of the structure and dispersion in polymer-layered silicate nano-composites." *Macromolecules* 38:9661–9669. doi: 10.1021/ma051249+.

A. Viretto, A. Taguet, R. Sonnier. 2016. "Selective dispersion of nanoplatelets of MDH in a HDPE/PBT binary blend: Effect on flame retardancy." *Polym. Degrad. Stab.* 126:107–116. doi: 10.1016/j.polymdegradstab.2016.01.021.

R. Wagener, T. J. G. Reisinger. 2003. "A rheological method to compare the degree of exfoliation of nanocomposites." *Polymer* 44:7513–7518. doi: 10.1016/j.polymer.2003.01.001.

J. Q. Wang, J. X. Du, J. Zhu, C. A. Wilkie. 2002. "An XPS study of the thermal degradation and flame retardant mechanism of polystyrene-clay nanocomposites." *Polym. Degrad. Stab.* 77:249–252. doi: 10.1016/S0141-3910(02)00055-1.

B. Wang, D. Yang, J. Z. Zhang, C. Xi, J. Hu. 2011. "Stimuli-responsive polymer covalent functionalization of graphene oxide by Ce(IV)-induced redox polymerization." *J. Phys. Chem. C* 115:24636. doi: 10.1021/jp209077z.

C. Wang, J. Wang, Z. Men, Y. Wang, Z. Han. 2019. "Thermal degradation and combustion behaviors of polyethylene/alumina trihydrate/graphene nanoplatelets." *Polymers* 11:772. doi: 10.3390/polym11050772.

I. M. Ward, D. W. Hadley. 1993. "An Introduction to the Mechanical Properties of Solid Polymers." Wiley&Sons, Chichester, England.

E. Wawrzyn, B. Schartel, A. Karrasch, C. Jäger. 2014. "Flame-retarded bisphenol A poly-carbonate/silicon rubber/bisphenol A bis(diphenyl phosphate): adding inorganic ad-ditives." *Polym. Degrad. Stab.* 106:74–87. doi: 10.1016/j.polymdegradstab.2013.08.006.

C. A. Wilkie. 2005. "An introduction to the use of fillers and nanocomposites in fire retardancy." In: "Fire Retardancy of Polymers: New Applications of Mineral Fillers." M. Le Bras, C. A. Wilkie, S. Bourbigot, S. Duquesne, C. Jama (Eds.). Royal Society of Chemistry, Cambridge. chap 1:3–15.

S. Wu. 1987. "Formation of dispersed phase in incompatible polymer blends: interfacial and rheological effects." *Polym. Eng. Sci.* 27:335–343. doi: 10.1002/pen.760270506.

H. D. Wu, C. R. Tseng, F. C. Chang. 2001. "Chain conformation and crystallization behavior of the syndiotactic polystyrene nanocomposites studied using Fourier transform infrared analysis." *Macromolecules* 34:2992–2999. doi: 10.1021/ma991897r.

C. L. Wu, M. Q. Zhang, M. Z. Rong, K. Friedrich. 2002. "Tensile performance improvement of low nanoparticles filled-polypropylene composites." *Compos. Sci. Technol.* 62:1327–1340. doi: 10.1016/S0266-3538(02)00079-9.

J. Wu, W. Pisula, K. Müllen. 2007. "Graphenes as potential material for electronics." *Chem. Rev.* 107:718–747. doi: 10.1021/cr068010r.

H. Wu, W. Zhao, H. Hu, G. Chen. 2011a. "One-step in situ ball milling synthesis of polymer-functionalized graphene nanocomposites." *J. Mater. Chem.* 21:8626–8632. doi: 10.1039/c1jm10819k.

G. M. Wu, B. Schartel D. Yu, M. Kleemeier, A. Hartwig. 2011b. "Synergistic fire retardancy in layered-silicate nanocomposite combined with low-melting phenysiloxane glass." *J. Fire Sci.* 30:69–87. doi: 10.1177/0734904111422417.

G. Mei Wu, B. Schartel, M. Kleemeier, A. Hartwig. 2012a. "Flammability of layered silicate epoxy nanocomposites combined with low-melting inorganic Ceepree glass." *Polym. Engin. Sci.* 52:507–517. doi: 10.1002/pen.22111.

G. M. Wu, B. Schartel, H. Bahr, M. Kleemeier, D. Yu, A. Hartwig. 2012b. "Experimental and quantitative assessment of flame retardancy by the shielding effect in layered silicate epoxy nanocomposites." *Combust. Flame* 159:3616–3623. doi: 10.1016/j.combustflame.2012.07.003

W. Xie, Z. Gao, K. Liu, W.-P. Pan, R. Vaia, D. Hunter, A. Singh. 2001. "Thermal characterization of organically modified montmorillonite." *Thermochim. Act.* 367/368: 339–350. doi: 10.1016/S0040-6031(00)00690-0.

W. Xing, J. Wu, G. Huang, H. Li, M. Tang, X. Fu. 2014. "Enhanced mechanical properties of graphene/natural rubber nanocomposites at low content." *Polym. Int.* 63:1674–1681. doi: 10.1002/pi.4689.

X. Xu, Q. Luo, W. Lv, Y. Dong, Y. Lin, Q. Yang, A. Shen, D. Pang, J. Hu, J. Qin, Z. Li. 2011. "Functionalization of graphene sheets by polyacetylene: convenient synthesis and enhanced emission." *Macromol. Chem. Phys.* 212:768–773. doi: 10.1002/macp.201000608.

Y. Yang, J. Wang, J. Zhang, J. Liu, X. Yang, H. Zhao. 2009. "Exfoliated graphite oxide decorated by PDMAEMA chains and polymer particles." *Langmuir* 25:11808–11814. doi: 10.1021/la901441p.

Y. Yang, W. Rigdon, X. Huang, X. Li. 2013. "Enhancing graphene reinforcing potential in composites by hydrogen passivation induced dispersion." *Sci. Rep.* 3:2086. doi: 10.1038/srep02086.

Z. Yang, Z. Xu, L. Zhang, B. Guo. 2018. "Dispersion of graphene in chlorosulfonated polyethylene by slurry compounding." *Compos. Sci. Technol.* 162:156–162. doi: 10.1016/j.compscitech.2018.04.030.

K. Yano, A. Usuki, A. Okada. 1997. "Synthesis and properties of polyimide-clay hybrid films." *J. Polym. Sci. A: Polym. Chem.* 35:2289–2294. doi: 10.1002/(SICI)1099-0518(199708)35:11<2289::AID-POLA20>3.0.CO;2-9.

P. J. Yoon, D. L. Hunter, D. R. Paul. 2003. Polycarbonate nanocomposites. Part 1. Effect of organoclay structure on morphology and properties." *Polymer* 44:5323–5339. doi: 10.1016/S0032-3861(03)00528-7.

R. J. Young, I. A. Kinloch, L. Gong, K. S. Novoselov. 2012. "The mechanics of graphene nanocomposites: a review." *Compos. Sci. Technol.* 72:1459–1476. doi: 10.1016/j.compscitech.2012.05.005.

N. Yousefi, M. M. Gudarzi, Q. Zheng, X. Lin, X. Shen, J. Jia, F. Sharif, J.-K. Kim. 2013a. "Highly aligned, ultralarge-size reduced graphene oxide/polyurethane nanocomposites: mechanical properties and moisture permeability." *Compos. Part A: Appl. Sci. Manuf.* 49:42–50. doi: 10.1016/j.compositesa.2013.02.005.

N. Yousefi, X. Lin, Q. Zheng, X. Shen, J. R. Pothnis, J. Jia, E. Zussman, J.-K. Kim. 2013b. "Simultaneous in situ reduction, self-alignment and covalent bonding in graphene oxide/epoxy composites." *Carbon* 59:406–417. doi: 10.1016/j.carbon.2013.03.034.

N. Yousefi, X. Sun, X. Lin, X. Shen, J. Jia, B. Zhang, B. Tang, M. Chan, J.-K. Kim. 2014. "Highly aligned graphene/polymer nanocomposites with excellent dielectric properties for high-performance electromagnetic interference shielding." *Adv. Mater.* 26:5480–5487. doi: 10.1002/adma.201305293.

M. F. Yu, O. Lourie, M. J. Dyer, K. Moloni, T. F. Kelly, R. S. Ruoff. 2000. "Strength and breaking mechanism of multiwalled carbon nanotubes under tensile load." *Science* 287:637–640. doi: 10.1126/science.287.5453.6.

D. Yu, M. Kleemeier, G. M. Wu, B. Schartel, W. Q. Liu, A. Hartwig. 2011. "Phosphorus and silicon containing low- melting organic–inorganic glasses improve flame retardancy of epoxy/clay composites." *Macromol. Mater. Eng.* 296:952–964. doi: 10.1002/mame.201100014.

B. H. Yuan, Y. Hu, X. F. Chen, Y. Q. Shi, Y. Niu, Y. Zhang, S. He, H. M. Dai. 2017. "Dual modification of graphene by polymeric flame retardant and Ni (OH)(2) nanosheets for improving flame retardancy of polypropylene." *Compos Part A- Appl. Sci. Manufac.* 100:106–117. doi: 10.1016/j.compositesa.2017.04.012.

M. Zammarano, J. W. Gilmann, M. Nyden, E. M. Pearce, M. Lewin. 2006. "The role of oxidation in the migration mechanism of layered silicate in poly(propylene) nanocomposites." *Macromol. Rapid Commun.* 27:693–696. doi: 10.1002/marc.200600068.

M. Zanetti, S. Lomakin, G. Camino. 2000. "Polymer layered silicate nanocomposites." *Macromol. Mater. Eng.* 279:1–9. doi: 10.1002/1439-2054(20000601)279:1<1::AID-MAME1>3.0.CO;2-Q.

M. Zanetti, G. Camino, R. Mülhaupt. 2001a. "Combustion behaviour of EVA/fluorohectorite nanocomposites." *Polym. Degrad. Stabil.* 74:413–417. doi: 10.1016/S0141-3910(01) 00178-1.

M. Zanetti, G. Camino, R. Thomann, R. Mülhaupt. 2001b. "Synthesis and thermal behaviour of layered silicate–EVA nanocomposites." *Polymer* 42:4501–4507. doi: 10.1016/S0032-3861(00)00775-8.

M. Zanetti, G. Camino, P. Reichert, R. Mülhaupt. 2001c. "Thermal behaviour of poly(propylene) layered silicate nanocomposites." *Macromol. Rapid Commun.* 22:176–180. doi: 10.1002/1521-3927(200102)22:3<176::AID-MARC176>3.0.CO;2-C.

M. Zanetti, T. Kashiwagi, L. Falqui, G. Camino. 2002. "Cone calorimeter combustion and gasification studies of polymer layered silicate nanocomposites." *Chem. Mater.* 14:881–887. doi: 10.1021/cm011236k.

Q. Zhang, F. Fang, X. Zhao, Y. Li, M. Zhu, D. Chen. 2008. "Use of dynamic rheological behavior to estimate the dispersion of carbon nanotubes in carbon nanotube/polymer composites." *J. Phys. Chem. B* 112:12606–12611. doi: 10.1021/jp802708j.

J. Zhang, M. Delichatsios, S. Bourbigot. 2009a. "Experimental and numerical study of the effects of nanoparticles on pyrolysis of a polyamide 6 (PA6) nanocomposite in the cone calorimeter." *Combust. Flame* 156:2056–2062. doi: 10.1016/j.combustflame.2009.08.002.

J. Zhang, J. Hereid, M. Hagen, D. Bakirtzis, M. A. Delichatsios, A. Fina, A. Castrovinci, G. Camino, F. Samyn, S. Bourbigot. 2009b. "Effects of nanoclay and fire retardants on fire retardancy of a polymer blend of EVA and LDPE." *Fire Safety J*. 44:504–513. doi: 10.1016/j.firesaf.2008.10.005.

J. Zhang, M. Delichatsios. 2010. "Further validation of a numerical model for prediction of pyrolysis of polymer nanocomposites in the cone calorimeter." *Fire Technol.* 46:307–319. doi: 10.1007/s10694-008-0073-5.

Y. L.Zhong, Z. Tian, G. P. Simon, D. Li. 2015. "Scalable production of graphene via wet chemistry: progress and challenges." *Mater. Today* 18:73–78. doi: 10.1016/j.mattod.2014.08.019.

J. Zhu, C. A. Wilkie. 2000. "Thermal and fire studies on polystyrene–clay nanocomposites." *Polym. Int.* 49:1158–1163. doi: 10.1002/1097-0126(200010)49:10<1158::AID-PI505>3.0.CO;2-G.

Y. Zhu, H. Ji, H.-M. Cheng, R. S. Ruoff. 2018. "Mass production and industrial applications of graphene materials." *Natl. Sci. Rev.* 5:90–101. doi: 10.1093/nsr/nwx055.

X. D. Zhuang, Y. Chen, G. Liu, P. P. Li, C. X. Zhu, E. T. Kang, K. G. Noeh, B. Zhang, J. H. Zhu, Y. X. Li. 2010. "Conjugated-polymer-functionalized graphene oxide: synthesis and nonvolatile rewritable memory effect." *Adv. Mater.* 22:1731–1735. doi: 10.1002/adma.200903469.

B. Zirnstein, W. Tabaka, D. Frasca, D, Schulze, B. Schartel. 2018. "Graphene/hydrogenated acrylonitrile-butadiene rubber nanocomposites: dispersion, curing, mechanical reinforcement, multifunctional filler." *Polym. Test.* 66:268–279. doi: 10.1016/j.polymertesting.2018.01.035.

H. Zou, S. Wu, J. Shen. 2008. "Polymer/silica nanocomposites: preparation, characterization, properties, and applications." *Chem. Rev.* 108:3893–3957. doi: 10.1021/cr068035q.

3 Synthesis and Characterization of Flame Retardant Polymer/Clay Nanocomposites

Mariya Mathew
School of Energy Materials, Mahatma Gandhi University, Kottayam, Kerala, India

Karen Ter-Zakaryan
TEPOFOL Ltd., Moscow, Russia

Sabu Thomas
School of Energy Materials, Mahatma Gandhi University, Kottayam, Kerala, India

International and Inter-University Center for Nanoscience and Nanotechnology, Mahatma Gandhi University, Kottayam, Kerala, India

School of Nanoscience and Nanotechnology, Mahatma Gandhi University, Kottayam, Kerala, India

Hiran Mayookh Lal and Arya Uthaman
School of Energy Materials, Mahatma Gandhi University, Kottayam, Kerala, India

CONTENTS

DOI: 10.1201/9781003327158-3

59

3.1 INTRODUCTION

Many fire injuries and accident hazards always accompany the steady demand for plastics and polymers. The consumption of plastics has increased dramatically from 1.5 million metric tons to 299 million metric tons (Troitzsch 2016). The massive demand for combustible materials as a part of the changes in the living standards of the public and global developments has resulted in several fire tragedies around the world. Moreover, the increasing consumer needs for electric devices, apparel, upholstery furniture, floor and wall coverings, modern transportation, and building materials and their rising fire outbreaks advocate the importance of flame retardants (FRs) in the present scenario. FRs are chemicals or compounds capable of pre-venting fire propagation by forming a protective mask on the material's surface or degrading the chemical reactions in the flame itself (Speight 2017). Several chemical reagents are added to the polymers to minimize the flammability of the polymer backbone (Camino and Costa 1988). The inclusion of FR materials reduces the flammability of the polymers and the toxic emissions and thereby considers them a milestone in the development of new materials (Hodgson and Gras 2002). The commonly used flame retardant fillers include halogenated flame retardants, phosphorous and nitrogen-based compounds, and inorganic and mineral com-pounds. The fire spread and explosions were the essential roots of the FR in the initial times. In contrast, innovations have been made in flame retardant systems, focusing on environmental hazards, smoke toxicity, and thermal stability (Horacek and Grabner 1996). Studies have reported that flammability control might also be the reason for adverse health and environmental impacts.

Furthermore, fire deaths also result from inhaling irritant gases, including carbon monoxide (Hodgson and Gras 2002). Halogen-based FRs are commonly employed in most industrial or electrical applications. The carcinogenic furans and dioxins generated upon the combustion of halogenated FR hindered halogens from their flame retardant applications (Chen and Wang 2010). Halogenated flame-resistant materials not only cause the emission of irritant gases but also behave as global contaminants due to their massive spread in the air, water, soil, dust, crustaceans, fish, amphibians, reptiles, birds, mammals and human tissues. Although halogenated FR has received much attention in the household furnishings and construction sectors, its fire effluent toxicity limits its application as an FR additive. Inorganics and phosphorous-, silicon-, and nitrogen-based compounds have acquired importance in flame resistance due to the absence of halogenated compounds (Troitzsch 2016). Nitrogen-based compounds are attracting attention because of their low toxicity and lack of poisonous gases and dioxins once a fire occurs. Nitrogen-based FR compounds also decrease the mechanical behaviour of polymers. However, it was found that the deterioration was less than that of metal hydroxides (Horacek and Grabner 1996). Phosphorous-based FRs were also industrialized from the background of elaborating environmentally safe flame retardant reagents (Salmeia et al. 2018).

Developing a cost-effective, environmentally friendly, non-halogenated FR filler with excellent flame retardancy and mechanical properties remains a research gap. The search for a potential replacement for toxic FR agents opened the door to nanoparticles, as they possess excellent mechanical, thermal, physicochemical and fire resistance properties. Furthermore, nanotechnologies and nanostructures have recently received massive interest in several technologies. Nanocomposites are a two-phase system consisting of polymers and dispersed particles in the nanometer range (Choudalakis and Gotsis 2009). Nanocomposites are categorized based on the number of nanosized particles dispersed in the polymer matrix. For some nanocomposites, the three dimensions will be in the nanometer range, whereas for other nanocomposites, only two or one dimension will be in the nanometer range. According to Manzi-Nshuti et al., nanocomposite-loaded polymers show enhanced flexural modulus, improved heat distortion temperature (HDT), low permeability and better fire properties. Layered silicate-containing nanocomposites have gained more attention because they substantially lessen the flammability of polymers (Tai et al. 2012).

Nanoclay is a conventionally used nanoparticle in flame retardancy (Misra et al. 2011). Based on the structure, clays are classified into allophane, kaolinite, halloysite, smectite, illite, chlorite, vermiculite, attapulgite–palygorskite–sepiolite and mixed layered minerals. Clay nanocomposites are those in which nanosized layers of clay are randomly and uniformly dispersed in the polymer matrix. Smectic clays are most commonly employed for polymer clay composites due to their excellent swelling properties resulting from their ability to accommodate water and organic molecules between the silicate layers, higher aspect ratio, high cation exchange capacities and large surface area (Chen et al. 2008). Most polymer clay nanocomposites are fabricated only after modifying either clay or polymer. Depending on the layered silicates, there are three categories for polymer clay nanocomposites: conventional composites, intercalated nanocomposites and exfoliated nanocomposites. Basal plane spacing (d001) is the distance between one layer and the next layer. The d001 of clay remains unaffected if the polymer material does not enter the galleries; then, such composites are defined as conventional.

In some cases, an increase in d001 is observed even when the clay mineral is stacked. Such a type of composite is known as intercalated. The composite is named exfoliated whenever the clay layers seem to be pushed apart, creating a disordered array (Chen et al. 2008). Clay-reinforced nanocomposites have attracted public attention due to their minimum gas permeability and flammability. Upon incorporation into the polymer matrix, their mechanical, thermal and barrier properties could be enhanced when compared with neat polymer. Enhanced flame retardancy is due to the char formation of clay particles and the clay's barrier properties, which restrict the diffusion of volatile components formed via thermal degradation (Chen et al. 2008). It was noted that the water absorption capacity of nylon 6 decreases upon the addition of nanoparticles. Sinha et al. found that the oxygen permeability of polylactic acid was decreased upon the inclusion of organically treated clay (Sinha Ray et al. 2003). Studies have reported that several clay nanocomposites maintain toughness, whereas others do not when dispersed in the polymer (Chen et al. 2008). Clay has received significant focus as a reinforcement tool in polymer-based composites due to its ability to resist structural modifications due to exposure to solvents. Additionally, clay nanocomposites support biodegradability (Misra et al. 2011). The enhanced thermal properties of polymer–clay nanocomposites equipped them for high temperature and fire resistance applications. Polymer–clay composites are also well known for better flame retardancy. It was observed that adding a small amount of clay to a polymer matrix can appreciably minimize the heat release rate. It also causes a change in char structure and a reduction in mass loss during the combustion stage (Qin et al. 2005). From this standpoint, polymer–clay nanocomposites have become important for flame retardancy.

The mechanism to enhance the flammability and thermal stability of the polymer matrix using nanoclay as a filler still needs to be explored. Various mechanisms have been suggested for the flame retardancy of polymer clay nanocomposites. Radical trapping is one of the highlighted mechanisms among them, which was proposed in the early 1990s by Wilkie et al. According to them, transition metals, mainly Fe, can accept electrons from radicals at the time of polymer decomposition. Fullerenes, graphene and carbon nanotubes are also used for realizing the trapping effect. Polymer degradation may be delayed because of the scavenging of radicals by the trappers. Radical scavenging leads to the quenching of polymer degradation centres, which in turn leads to a delay in polymer decomposition (Carvalho et al. 2013).

Zhang et al. developed polyethylene/clay nanocomposites. The authors found an enhancement effect on flammability and thermal stability. In addition, maleic anhydride improved the intercalation between the polymer and clay particles. The authors observed a reduction in the peak heat release rate (Zhang and Wilkie 2003). Yang et al. reported that the nanosized clay in a wood fibre-plastic blend enhanced the composites' mechanical and flame retardant properties. They presented the effect of clay and clay dispersion on the mechanical and fire-resistant behaviour. They inferred that a high degree of dispersion is crucial for enhancing the mechanical and FR properties (Yang et al. 2010). A ternary blend wherein sepiolite nanoclay (Sep) along with multiwalled nanotubes (MWNT) were loaded in

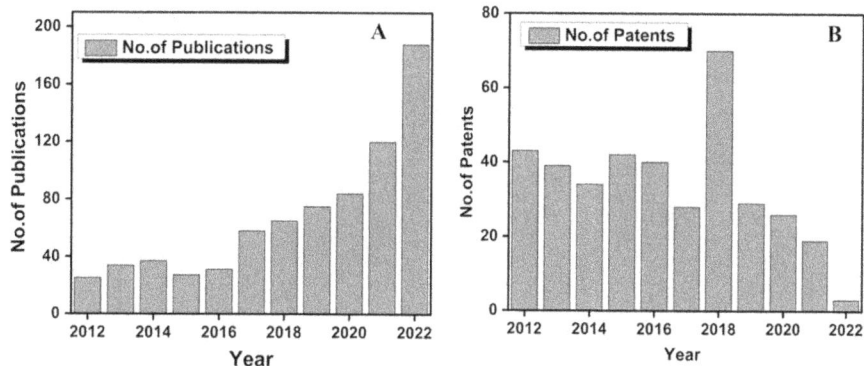

FIGURE 3.1 Graphical representations of (A) publications and (B) patents in the past 10 years on fire-safe polymer clay nanocomposites.

bio-based polylactic acid was conducted by Hapuarachchi et al. The authors found a reduction in the flammability of the resultant nanocomposites. The reduction was characterized by a 45% decrease in the peak heat release rate (PHRR) with cone calorimetry (Hapuarachchi and Peijs 2010). Thirumal et al. also fabricated a polyurethane foam in which organically modified clay (ONC) was used as the flame retardant filler. It was noted that adding three parts per hundred of ONC could improve flame retardancy performance (Thirumal et al. 2009). The study concluded that the prepared nanocomposites could be used as a halogen-free flame retardant filler for a wide range of polymeric compounds (Tai et al. 2012).

Figure 3.1 shows the number of publications and patents reported in fire-safe polymer/clay nanocomposites. This suggests that clay-incorporated polymer composites are an emerging area in fire retardancy applications, and significant reviews and research articles on this topic still need to be reported. Therefore, this review assesses various preparation techniques, in-depth structural and morphological analyses, and flame retardancy.

3.2 PREPARATION TECHNIQUES FOR THE FABRICATION OF POLYMER/CLAY NANOCOMPOSITES

The flow chart representing the different techniques for the preparation of polymer clay nanocomposites is represented in Figure 3.2.

The main techniques utilized for the fabrication of polymer clay nanocomposites are as follows.

3.2.1 MELT MIXING

A two-step method commonly used for the elaboration of polymer clay nanocomposites. In this technique, the nanoparticle is dispersed by the mechanical action of shearing the polymer backbone in the molten state. Many studies have employed this melt mixing technique with the help of an internal mixer. Shah et al. worked on nylon 6 nanocomposites, where they approached melt mixing to prepare the same

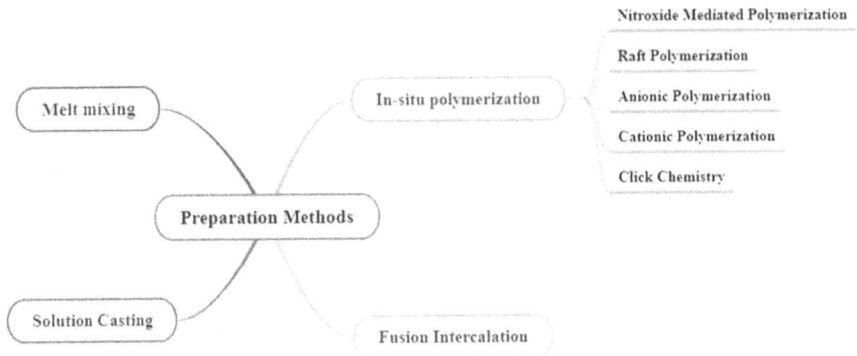

FIGURE 3.2 Flow chart diagram of preparation methods of fire-resistant polymer–clay nanocomposites.

(Shah and Paul 2004). Among the two steps, the first consists of fabricating a masterbatch of organoclay in high molecular weight nylon 6. This step is followed by diluting the masterbatch with low molecular weight nylon 6 to the preferred montmorillonite (MMT) to reduce melt viscosity. The Brabender Torque Rheometer measures the melt viscosity during mixing.

A study on a flame-resistant polyethylene clay nanocomposite was conducted in which melt mixing was used to prepare the composite (Zhang and Wilkie 2003). In this procedure, they mixed the polymer and maleic anhydride in a beaker and added the mixed sample into the brander chamber. Clay was then added in small portions into the chamber. The molten composite will be detached from the chamber and cooled to room temperature.

3.2.2 IN SITU POLYMERIZATION

The mixing of the nanomaterials or the dispersion of monomers was followed by polymerization in the presence of dispersed nanomaterials (Figure 3.3) (Mrah and Meghabar 2021). It is a highly preferable technique for the expansion of polymer/clay nanocomposites.

3.2.2.1 Nitroxide-Mediated Polymerization (NMP)

In this technique, stable free radicles, including nitroxides, might be used as reversible terminating agents for monitoring the polymerization process (Tasdelen, Kreutzer, and Yagci 2010). The fabrication of polystyrene/clay nanocomposites with ammonium cations as the nitroxide mediator was the first composite using this method. This composite has received much attention owing to the polymer/clay composite with controlled molar mass using this vinyl group. The nanocomposite's molecular mass also varied according to the concentration of the nitroxide initiator. Another study in the same area was performed by Konn et al., who employed an SG-type alkoxyamine as the initiator. The initiator possesses a terminal quaternary ammonium functional group for electrostatic anchoring to the clay layers. To study the kinetics

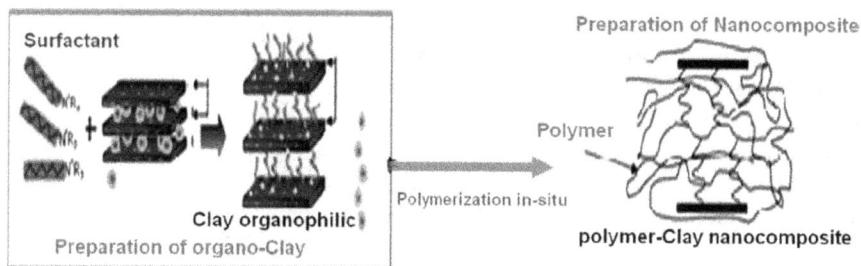

FIGURE 3.3 In situ polymerization method for the preparation of polymer clay nanocomposites.

Source: Reprinted from Mrah, L. and Rachid M., *Polym. Bull.*, 78, 3509–3526, 2021. With permission.

and reaction rates of pristine and clay-bonded polymerization, they adopted trials with a free initiator in the solution (Shen et al. 2010).

3.2.2.2 RAFT Polymerization

Reversible addition-fragmentation transfer (RAFT) belongs to the group of living polymerization, where a RAFT agent mediates the common radical polymerization. It is an independent metal polymerization owing to its comprehensive approach for a large number of polymers. The polymerization tolerates reaction conditions such as solvent and temperature. This RAFT polymerization method uses thiocarbonyls for chain transfer for transport between active and dormant chains. As RAFT polymerization is an adaptable approach, it is considered a worthy polymerization technique for a wide range of polymer clay nanocomposites (Ding et al. 2007). In RAFT polymerization, polymer growth occurs from the clay itself.

Moreover, the anchoring of RAFT agents inside the clay will also support RAFT polymerization. Salem et al. developed polystyrene-, poly(methyl methacrylate)-, and poly(n-butyl acrylate)-layered silicate nanocomposites in the presence of organo-modified clays by utilizing the RAFT method (Salem and Shipp 2005). The polymerization was conducted in an oil bath, and the ratio of the monomer to the RAFT agent taken was 300:1 assuming one mole of RAFT agent will initiate one polymer chain.

3.2.2.3 Anionic Polymerization

Anionic polymerization is a living polymerization widely used to establish polymers that verify the bond between structure and properties (Hadjichristidis et al. 2000). In anionic polymerization, the propagating chain is reactive or living. For polymer–clay nanocomposites, the bonded H_2O molecule in the clay particle terminates the living anion in the polymerization. A study investigated the preparation of nylon 6/clay nanocomposites by exploiting anionic polymerization (Yoonessi et al. 2005). The polymerization was carried out in a three-neck flask with nitrogen gas purging to avoid contact with moisture. The study presented two feeding times of clay to understand the importance of the intercalation of clay layers. The anionic polymerization made use of a ring-opening mechanism of caprolactam. Moisture prohibits the anionic polymerization of caprolactam by eliminating the active primary amine. Using n-BuLi as an initiator, Zhang et al. developed styrene–butadiene rubber/montmorillonite (SBR/MMT) nanocomposites with in situ living anionic polymerization (Zhang et al. 2005). According to the kinetic study, it is verified that the presence of organophilic montmorillonite (OMMT) did not affect living copolymerization and the microstructure of the nanocomposite.

3.2.2.4   Cationic Polymerization

A type of in situ living polymerization in which the monomer becomes reactive only after a cation transfers its charge to the monomer. Tasdelen et al. adopted cationic polymerization for formulating poly tetrahydrofuran (PTHF)/clay nanocomposites (Tasdelen et al. 2008). Using cationic ring-opening polymerization (CROP) of tetrahydrofuran, the study establishes the growth of PTHF chains from the clay surface with the help of trifluoromethanesulfonic anhydride provided 2,6- di-tert-butyl pyridine as the proton trap and dichloromethane as the solvent.

Exploiting an initiator trapped inside the silicate coverings of clay particles, Bayram et al. prepared poly(cyclohexene oxide) (PCHO)/clay nanocomposites using cationic polymerization (Bayram, Oral, and Şirin 2013). Through the cation exchange process, diphenyliodonium molecules were intercalated between Cloisite Ca and diphenyliodonium. The homogenous nanosized distribution of the clay layers in the polymer backbone resulted from the polymerization of cyclohexene oxide among the interlayer galleries of the clay particles. This results in the formation of the PCHO/clay nanocomposites.

A study showed the usage of clay as an activator for photoinitiated cationic-activated monomer polymerization of cyclohexene oxide. In this study, the clay layers propagate individually, and terminal oxonium ions react with the intercalant. Accordingly, clay layers are homogeneously distributed in the polymer matrix, which leads to the formation of polymer/clay nanocomposites (Boyer et al. 2008).

3.2.2.5   Click Chemistry

Click chemistry is an idea of several chemical reactions that share standard features, including increased reaction yields, mild reaction conditions and easily removable byproducts, stereospecific reaction coordinates and tolerance to different functional groups. It has been reported that emerging click chemistry can enhance the properties of polymer clay nanocomposites (Arslan and Tasdelen 2017). A study examined the elaboration of poly(epsilon-caprolactone)/clay nanocomposites wherein an azide/alkyne cycloaddition (CuAAC) click reaction was employed. Moreover, ring-opening polymerization occurred with an initiator such as propargyl alcohol (Tasdelen 2011). Purut et al. expanded thermoset/clay nanocomposites in a high yield using thiol-epoxy click reactions provided with ambient reaction conditions and investigated the effect of adding clays by examining thermal and mechanical properties. They found that there is an improvement in the thermal behaviour upon the addition of clay particles (Purut Koc et al. 2018).

3.2.3 SOLUTION CASTING

Solution casting is also an approachable method for preparing polymer clay nanocomposites. Al-Mulla et al. employed this technique for formulating polylactic acid (PLA)/clay nanocomposites (Jaffar Al-Mulla 2011). A particular amount of PLA was dissolved in chloroform, followed by continuous stirring for 1 hour. The needed clay was added into the dissolved PLA in tiny portions and refluxed for an hour. It was then ultrasonically stirred until the clay was dispersed. The mixture was then poured into a petri dish and kept in an oven until it was dried.

3.2.4 FUSION INTERCALATION

Vaia et al. developed another method for preparing polymer clay composites in 1993. In this procedure, thermoplastic and clay are blended in a molten condition (Vaia, Ishii, and Giannelis 1993). The polymer backbone is then pushed to the interlamellar space, providing a nanocomposite. The driving force for these steps is the heat produced upon the interaction between the polymer matrix and clay nanoparticles.

3.3 CHARACTERIZATION TECHNIQUES OF FIRE-SAFE POLYMER CLAY NANOCOMPOSITES

3.3.1 FOURIER TRANSFORM INFRARED SPECTROSCOPY (FTIR)

FTIR spectra are used to identify the structure of the prepared materials (Papadopoulos et al. 2020). FTIR spectra of the polymer/clay composites will be recorded in FTIR spectroscopy using a KBr pellet. Das et al. (Das and Karak 2009) revealed the interaction of clay with epoxy resins from the FTIR spectra of epoxy clay nanocomposites. The spectra showed bands at 3627 cm^{-1} indicative of OH stretching vibrations. Moreover, a shift in the spectra is from 3437 to 3422 cm^{-1} when we go from flame-resistant epoxy resin to flame-resistant epoxy resin clay nanocomposites, suggestive of the intermolecular hydrogen bond between epoxy and clay particles (Das and Karak 2009).

Research on the mechanical, thermal and flammability properties of polyethylene/clay nanocomposites exhibited FTIR spectra that confirmed the char structure during combustion (Zhao et al. 2005). The spectra demonstrated many peaks corresponding to stretching, bending and rocking of the CH2 group, stretching of Si-O and absorption of Al-O. It is confirmed from the peaks of the FTIR spectra that the char residue consists of a huge portion of clay particles. Interestingly, it was found that the residue of char has almost no hydrocarbon absorption.

3.3.2 X-RAY DIFFRACTION (XRD)

X-ray diffractogram is mainly used to analyse the dispersion nature of clay particles and the crystallinity of the prepared nanocomposites (Mehrabzadeh and Kamal 2002,

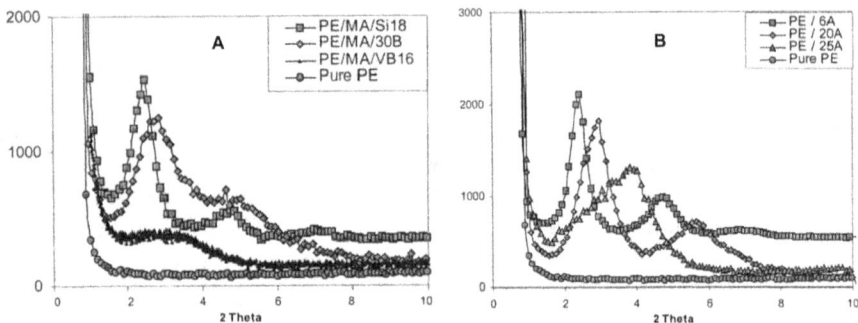

FIGURE 3.4 X-ray diffractogram of PE/clay and PE/MA/clay nanocomposites.

Source: Reprinted from Zhang, J. and Wilkie, C. A., *Polym. Degrad. Stabil.*, 80, 163–169, 2003. With permission.

Uthaman, Lal, and Thomas 2021b). Polymer clay nanocomposites are formed by including polymers between the clay layers. Upon the inclusion of the polymer, the gallery space increases and forces the clay layers to separate (Zhang and Wilkie 2003). Here, the X-ray diffractogram gained more importance. For clay nanocomposites, XRD gives the idea of nanostructures if the peaks are found in the low-angle zone. Such diffraction peaks indicate the d-spacing of the ordered-intercalated and ordered-delaminated nanocomposites. If the nanocomposites are not so ordered, no peaks will be obtained owing to the large d-spacing (Gilman et al. 2000). However, the absence of an XRD peak is not enough to confirm the nanostructure, while a combination of intercalated and exfoliated structures can also be obtained (Zhu et al. 2002). TEM images and XRD will provide more information regarding the nanostructures in such cases.

Tai et al. showed an XRD spectrum of flame-resistant and exfoliated polymer clay nanocomposites (Tai et al. 2012). The polymer/clay nanocomposites can be distinguished as intercalated and/or exfoliated depending on the peak changing with the gallery height of the clay particles. X-ray diffraction of the novel nano-composite comprising polymeric flame retardant (PDEDP) and clay revealed that the clay layers are almost isolated and evenly dispersed in the PDEPD matrix (Tai et al. 2012).

Zhang et al. fabricated flame retardant polyethylene/clay nanocomposites wherein XRD proved the exfoliated structure of the composite. There is only a slight variation in d-spacing for several clay-added composites, whereas d-spacing was found to be large in others. The effect of clay in the presence or absence of maleic anhydride (MA) can be easily understood from the d-spacing exposed in XRD spectra (Figure 3.4) (Zhang and Wilkie 2003).

3.3.3 Transmission Electron Microscopy (TEM)

Transition electron micrographs are used to obtain the morphology of the developed nanocomposite (Uthaman, Lal, and Thomas 2021a, Uthaman et al. 2021). The

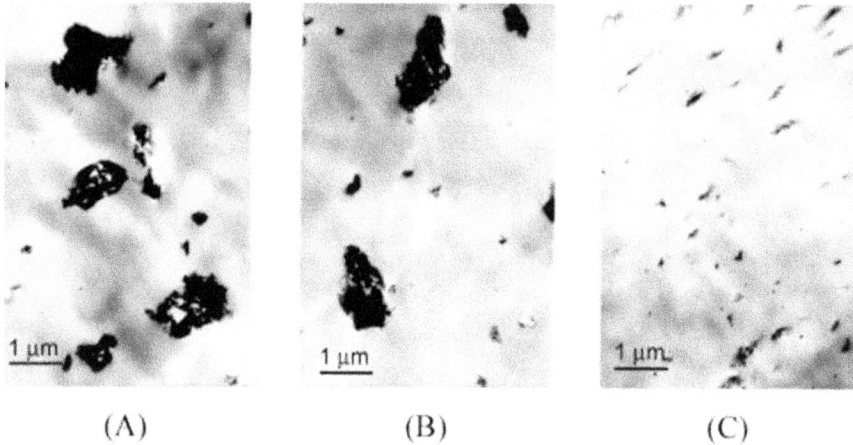

(A)   (B)   (C)

FIGURE 3.5   TEM images of (A) PP/Na-MMT, (B) PP/H-MMT, and (C) PP/OMMT nanocomposites.

morphological study of various polypropylene (PP) clay nanocomposites, such as polypropylene sodium montmorillonite (PP/Na-MMT), polypropylene organic clay (PP/OMMT), and polypropylene protonic clay PP/H-MMT, was studied by Qin et al. (Qin et al. 2005). The TEM images presented a vast and unevenly immersed nanoclay in the PP/Na-MMT composite. Interestingly, the dispersion state of PP/H-MMT is almost parallel to that of PP/Na-MMT, while the clay particles in the PP/OMMT composites were found to be comparatively small (Figure 3.5).

Zhu et al. elaborated poly(methyl methacrylate)-clay nanocomposites wherein the TEM images confirmed that the nanocomposites were intercalated, exfoliated or mixed intercalated-exfoliated (Zhu et al. 2002). According to the study, the high polymerization degree of poly(methyl methacrylate) may be due to the enhanced rate of propagation and the prominent gel effect. The gel effect may increase the polymerisation and molecular mass because of the remarkable decrement in the termination rate.

A study on vegetable oil-based flame retardant epoxy/clay nanocomposites by Das et al. (Das and Karak 2009) displayed a TEM image that proved the behaviour of clay dispersion in the polymer surface. The dark spots in the TEM micro-photographs are indicative of silicate nanolayers. The TEM image established the exfoliated nature of the clay nanocomposite. Moreover, the nanoclay appeared to be evenly dispersed in the polymer matrix (Figure 3.6).

3.3.4   THERMOGRAVIMETRIC ANALYSIS (TGA)

TGA analysis is mostly employed for a better understanding of the thermal deg-radation of prepared composites. Clay layers possess excellent barrier properties, which contribute to the enhanced thermal stability of the polymer–clay blends. The alkyl ammonium cations in the clay decompose to a product that stimulates polymer matrix degradation (Zhao et al. 2005). Additionally, the clay itself causes the

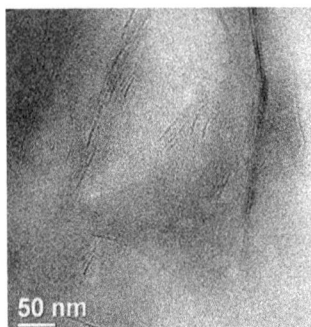

FIGURE 3.6 TEM microphotograph of epoxy/clay nanocomposites. (Reprinted from Das, G. & Karak, N., Polym. Degrad. Stabil. 94, 1948–1954, 2009. With permission.)

degradation of the polymer (Zanetti et al. 2001). Unfortunately, these processes reduce the thermal stability of polymer clay nanocomposites (Zhao et al. 2005).

TGA analysis of different polypropylene (PP)/clay nanocomposites, such as polypropylene (PP)/sodium montmorillonite (PP/Na-MMT), polypropylene/octa-decyltrimethyl ammonium chloride (symbolized as C18) (PP/C18), polypropylene/organic clay (PP/OMMT), and polypropylene/protonic clay (PP/H-MMT) blends, was examined by Qin et al. (2005). The results showed that the thermal decomposition temperature of pure PP seemed lower than that of PP/Na-MMT and PP/C18. However, the TGA curve of PP/H-MMT was almost parallel to that of PP/OMMT and was higher than that of PP/Na-MMT. The study inferred that the thermal stability increases with the number of acid sites in the clay particles. On the other hand, the initial decomposition of these composites occurred prior to pristine PP.

Zhu et al. conducted an investigation on the thermal stability and flame retardancy of poly(methyl methacrylate) (PMMA)-clay and clay, including *N,N*-dimethyl-*n*-hexadecyl-(*p*-vinylbezyl) ammonium chloride (VB16), *N,N*-dimethyl-*n*-hexadeylallyl ammonium chloride (Allyl16) and *N,N*-dimethyl-*n*-hexadecylbenzyl ammonium chloride (Bz16)-modified nanocomposites, where TGA was employed for thermal studies (Zhu et al. 2002). Blumstein reported the reasons for the improved thermal stability of PMMA intercalated within clay. According to the research of Blumstein, the changes in the chemical structure and the controlled thermal mobility of polymer chains in the silicate interlayers might be the reasons for the same. The TGA curves implied that the thermal stability of pure PMMA is lower than that of these nanocomposites (Figure 3.7). Furthermore, the temperature at which 10% and 50% degradation occurred was noted. The results suggested that the temperature at which 10% degradation occurs is a measure of the onset temperature of the degradation, and this is higher for both allyl16 and VB16, yet it is lower for Bz16. The temperature at which 50% degradation of all the samples was observed was 40°C, which is also considered a measure of thermal stability.

The TGA data of polyethylene (PE)-clay nanocomposites by Zhang et al. consist of temperatures at which 10% and 50% degradation occurs, the midpoint of the degradation, and the fraction of material that is nonvolatile at 600°C (Zhang and Wilkie 2003). The outcomes suggested that the changes in degradation temperature and amount of nonvolatile material are indicative of clay particles that have been

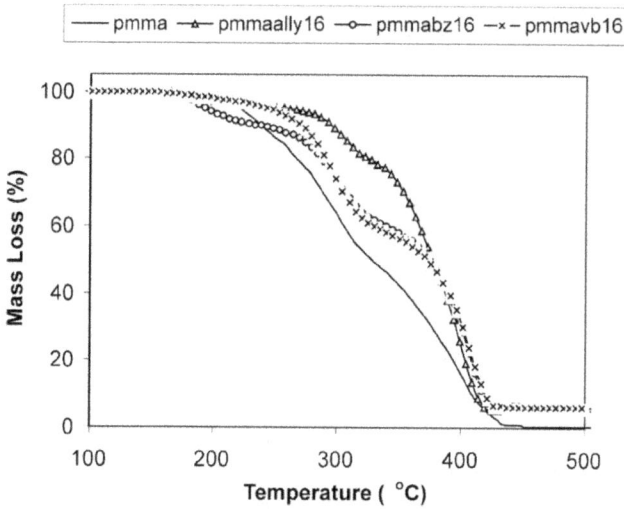

FIGURE 3.7 TGA curves of poly(methyl methacrylate) (PMMA) and various clay-loaded PMMA nanocomposites.

Source: Reprinted from Zhu, J. et al., *Polym. Degrad. Stabil.* 77, 253–258, 2002. With permission.

added. It was also reported that nanocomposite fabrication has no significance in the thermal stability of PE.

3.3.5 FLAMMABILITY PROPERTIES

3.3.5.1 Cone Calorimetry

Cone calorimetry is considered the touchstone for analysing the flammability properties. Polymer clay nanocomposites are characterized under high temperatures to test their flammability properties (Gilman 1999). The cone calorimeter defines several combustion factors, including time to ignition (TTI), heat release rate (HRR), PHRR, total heat release (THR), total smoke release (TSR), carbon mon-oxide yield and mean specific extinction area (mean SEA) (Zhang et al. 2014). The HRR is considered the single most crucial variable in determining the fire safety of a particular compound (Babrauskas and Peacock 1992).

The cone calorimeter works on the oxygen consumption principle, which is relatively simple and offers high resolution (Babrauskas 1984). Because of the oxygen consumption principle, this technique is considered an accurate and efficient method that has yet to be developed. This technique can test specimens horizontally or vertically and uniformly exposed to radiation up to 100 kW/m². The heater design chosen prevents frequent mistakes brought on by uncontrollable heating fluxes found in other systems. A user-friendly approach may be used to calibrate both irradiance and heat release rate. Measurements of mass loss and adequate heat of combustion have been included in the operation.

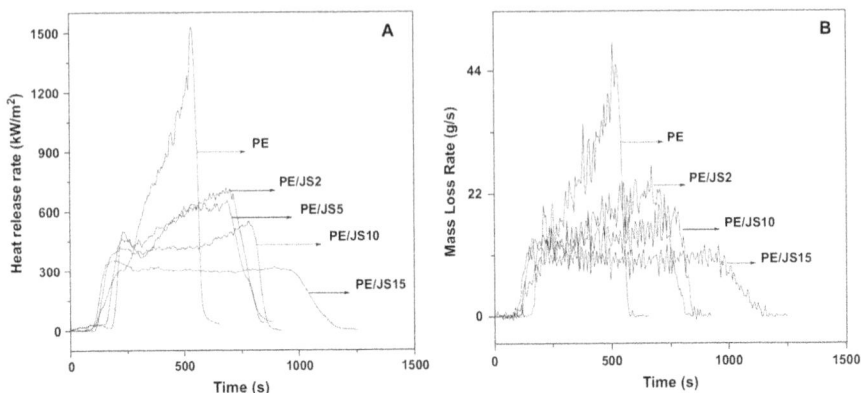

FIGURE 3.8 The influence of time on (A) heat release rate and (B) mass loss rate of neat PE and clay-loaded PE nanocomposites.

Source: Reprinted from Zhao, C. et al., *Polym. Degrad. Stabil.* 87, 183–189, 2005. With permission.

Zhao et al. explored the flammability of polyethylene (PE)/clay nanocomposites. The study used a cone calorimeter to characterize the flammability of the fabricated nanocomposites (Zhao et al. 2005). Figure 3.8A displays the heat release rate (HRR) curves of PE and PE/clay nanocomposites. It was clear that the peak HRR values of the PE/clay nanocomposites were lower than those of pure PE. The peak value decreases with clay loading, and the flame retardancy of the PE/clay nanocomposites is enhanced. As a result of the catalysis of organically modified clay, the HRR of PE/clay nanocomposites improved more quickly as a function of time than that of PE. The PE/clay nanocomposites turn into the plateau stage due to specific improvement of flame resistance soon after ignition. The plot of the mass loss rate (Figure 3.8B) is also the same as that of the HRR, and the lower HRR of the nanocomposites is due to the lowering of the MLR and thereby the improved flame resistivity of the PE/clay nanocomposites. This is because of the variations in the condensed phase and not in the gaseous phase.

The flame retardant behaviour of polypropylene (PP)/clay composite was studied by Qin et al. with a cone calorimeter (Qin et al. 2005). An improvement in the fire resistance was noted owing to the condensed-phase flame retardant mechanism. The delay in thermal-oxidative degradation and decrease in the HRR of the nano-composites were due to the barrier effect of exfoliated layered silicates for volatiles (Figure 3.9). It was found that the ignition time for PP nanocomposites was lower even though the initial HRR was superior for PP/clay composites. The release of the thermal degradation products of the organic treatment of the clay would be the reason for the higher HRR in the initial stage as well as the lower ignition time in the polymer nanocomposites.

The enhanced fire-retardant properties of polymer clay nanocomposites are attributed to their large interfacial area. The decrease in the PHRR of a polymer clay nanocomposite is mainly because of the polymer we used as the base material.

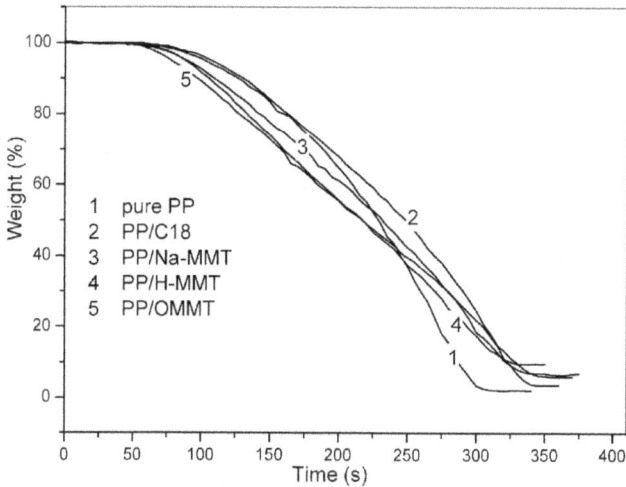

FIGURE 3.9 Mass loss curves obtained via cone calorimeter for pure PP, PP/C18, PP/Na-MMT, PP/H-MMT and PP/OMMT.

Source: Reprinted from Qin, H. et al., *Polymer* 46, 8386–8395, 2005. With permission.

Reports have shown that there is a significant reduction in the PHRR rate upon the addition of clay into polyamide 6 (PA6), polystyrene (PS) and ethylene-covinyl acetate co-polymers (EVA) nanocomposites. The interchain interactions, including intermolecular aminolysis/acidolysis, radical recombination and hydrogen abstraction, might be the reasons for the substantial decrease in the PHHR (Jang, Costache, and Wilkie 2005).

Clay composites of poly(styrene-*co*-acrylonitrile) (SAN) and acrylonitrile-butadiene-styrene terpolymer (ABS) displayed a mild decrease in PHRR upon the inclusion of clay. However, poly(methyl methacrylate) (PMMA)/clay nanocomposites do not show any reduction in the PHRR upon the incorporation of clay nanoparticles. The study inferred that the intermolecular interactions upon the clay addition of the polymer, which degrades at several temperature conditions, are a crucial factor in lowering the PHRR. In addition, fire resistance is not achieved if the degradation of the polymer is single-stage, as no thermally stable products will be formed.

3.3.5.2 Limited Oxygen Index (LOI)

LOI is also considered a tool for determining the flammability properties of polymer nanocomposites. It is expressed as the minimum percentage of oxygen required to maintain flaming combustion in a flowing mixture of nitrogen and oxygen (Devi and Maji 2012, Funar-Timofei, Iliescu, and Suzuki 2014). It is regarded as a standard of the feasibility of the elimination of material.

The limited oxygen index was used to understand the flammability of styrene-acrylonitrile copolymer SAN- and SAN/clay/SiO$_2$-treated samples. The network structure formed because inorganic silica reduced the flammability properties.

Interestingly, the addition of nanoclay enhanced the fire retardancy of the nano-composite (Devi and Maji 2012). However, the fire-retardant behaviour falls at a higher concentration of clay due to the tortuous path created due to the addition of clay, leading to a decrement in the LOI value.

Yeh et al. researched the burning behaviour of poly(*o*-methoxy aniline) (PMA) and polymeric clay nanocomposites (PCN) by focusing on the LOI value (Yeh and Chin 2003). Montmorillonite (MMT) clay is the precursor for fire resistance. The LOI value improved substantially upon the addition of MMT. The dispersed clay in the polymer matrix accelerates the fire resistance, which is proven by the increased LOI value at higher MMT loadings.

LOI was considered a standard to characterize the flammability of poly-propylene/clay nanocomposites (Wagenknecht, Kretzschmar, and Reinhardt 2003). The study examined the effect of natural and synthetic clay and talc on the flame retardancy reflected in the parameter LOI. The results demonstrated a higher LOI for synthetic clay than for natural clay. Even though the polymer matrix was the same for all the composites, the study claimed that the composition of gaseous products and the temperature at which gaseous substances and fillers move out of the combustion cycle may cause a difference in the ignition time.

3.3.5.3 Underwriters Laboratories Test Standard UL-94 (UL-94) Test

This technique is also utilized for determining the fire-retardant properties of polymer nanocomposites. This test is conventionally used to evaluate plastic material ignitability and flame spread (Wang et al. 2017). However, this technique alone is not suitable for the proper analysis of the flammability characteristics of a composite. It is known that the fire test is almost the same as the American Society for Testing and Materials (ASTM) or National Fire Protection Association (NFPA) standards. This method will carry out a flame test, and the flame property will be evaluated from V0 to V2. If the flame stops within an interval of 10 seconds after two applications of a flame to a test bar lasting 10 seconds each and no burning drips are allowed, then it is said that flame retardancy is quite good and rated V0 (Hodgson and Gras 2002).

The flame retardant behaviour of ethylene vinyl acetate (EVA) copolymer/or-ganoclay/alumina trihydrate (ATH) was employed using this UL-94 test (Cárdenas et al. 2008). The effect of organoclay on the burning properties of the polymer was investigated using the UL-94 test by Araújo et al. They reported that clay forms a thermal insulation barrier that acts as a barrier for both energy and transport (Araújo et al. 2007). To analyse the flammability, a study on vegetable oil-based flame retardant epoxy/clay nanocomposites exploited this technique (Das and Karak 2009). The study concluded that flame retardant epoxy resin displayed a flamma-bility rate of V1, whereas the nanocomposites exhibited a rate of V0 using the UL-94 test.

3.3.5.4 Pyrolysis Combustion Flow Calorimetry (PCFC) test

This method is considered one of the tools for determining the flammability characteristics of polymer clay nanocomposites. This procedure works on the principle that the experiment's net oxygen consumed is related to the heat of combustion of the polymer (Das and Karak 2009). The time-integrated heat release rate is the measured net heat of PC total combustion of PC based on oxygen consumption.

This approach was employed to analyse the flammability properties of poly-propylene/clay/carbon nanotube composite systems. The study also compared the flame test results with a cone calorimeter and PCFC. Changes in the HRR were noticed when measured in both the cone calorimeter and PCFC (Hapuarachchi, Peijs, and Bilotti 2013). The study concluded that due to its tiny size and arbitrary form, testing in the PCFC might restrict the behaviour of the material, whereas specimen dimensions are standardized when testing in the cone calorimeter. Compared to the PCFC, the cone calorimeter offers a much greater surface area where flame retarding devices can work to their full potential. The testing environment in the microcalorimeter, where these tests are carried out in an enclosed system with regulated gas fluxes, is another potential source of worry. Anaerobic settings have been proven to reduce a material's ability to resist flame, particularly in regard to char formation, which is a crucial component of such resistance. On the other hand, the cone calorimeter is an open device that more closely matches real-life situations.

3.4 CONCLUSIONS

This chapter overviews the importance of clay as a fire-retardant tool in polymer composites. Additionally, the chapter focused on various fabrication techniques, characterization techniques and flammability properties. As clay nanoparticles are biodegradable, they are a suitable alternative to halogenated FRs. Moreover, it possesses several advantages, including low cost, the feasibility of processing and a minimal quantity needed for the elaboration of the composites.

Regarding flame retardant behaviour, studies have proven that incorporating clay reduces the peak heat release rate. Clay-enriched composites seemed to be thermally stable, as polymer nanocomposites have several applications, such as packaging, aircraft, space, automotive, sporting goods, marine, biomedical, energy, infrastructures and electronic devices. The applications can be fulfilled only if the clay used ensures the fire safety of the fabricated composites. The use of clay is a strategic approach for the expansion of a wide variety of polymer nanocomposites with improved mechanical, thermal and fire-resistant characteristics.

REFERENCES

Araújo, E. M., R. Barbosa, A. W. B. Rodrigues, T. J. A. Melo, and E. N. Ito. 2007. "Processing and characterization of polyethylene/Brazilian clay nanocomposites." *Materials Science and Engineering: A* no. 445–446:141–147. doi: 10.1016/j.msea.2 006.09.012.

Arslan, Mehmet, and Mehmet Atilla Tasdelen. 2017. "Polymer nanocomposites via click chemistry reactions." *Polymers* no. 9 (10):499.

Babrauskas, Vytenis. 1984. "Development of the cone calorimeter—A bench-scale heat release rate apparatus based on oxygen consumption." *Fire and Materials* no. 8 (2):81–95. doi: 10.1002/fam.810080206.

Babrauskas, Vytenis, and Richard D. Peacock. 1992. "Heat release rate: The single most important variable in fire hazard." *Fire Safety Journal* no. 18 (3):255–272. doi: 10.101 6/0379-7112(92)90019-9.

Bayram, Işıl, Ayhan Oral, and Kamil Şirin. 2013. "Synthesis of poly(cyclohexene oxide)-montmorillonite nanocomposite via in situ photoinitiated cationic polymerization with bifunctional clay." *Journal of Chemistry* no. 2013:617498. doi: 10.1155/2013/617498.

Boyer, Cyrille, Jingquan Liu, Lingjiun Wong, Michael Tippett, Volga Bulmus, and Thomas P. Davis. 2008. "Stability and utility of pyridyl disulfide functionality in RAFT and conventional radical polymerizations." *Journal of Polymer Science Part A: Polymer Chemistry* no. 46 (21):7207–7224. doi: 10.1002/pola.23028.

Cárdenas, M. A., D. García-López, I. Gobernado-Mitre, J. C. Merino, J. M. Pastor, J. de, D. Martínez, J. Barbeta, and D. Calveras. 2008. "Mechanical and fire retardant properties of EVA/clay/ATH nanocomposites – Effect of particle size and surface treatment of ATH filler." *Polymer Degradation and Stability* no. 93 (11):2032–2037. doi: 10.101 6/j.polymdegradstab.2008.02.015.

Camino, G., and L. Costa. 1988. "Performance and mechanisms of fire retardants in polymers—A review." *Polymer Degradation and Stability* no. 20 (3):271–294. doi: 10.1016/0141-3910(88)90073-0.

Carvalho, Hudson W. P., Celso V. Santilli, Valérie Briois, and Sandra H. Pulcinelli. 2013. "Polymer–clay nanocomposites thermal stability: Experimental evidence of the radical trapping effect." *RSC Advances* no. 3 (45):22830–22833. doi: 10.1039/C3RA44388D.

Chen, Biqiong, Julian R. G. Evans, H. Christopher Greenwell, Pascal Boulet, Peter V. Coveney, Allen A. Bowden, and Andrew Whiting. 2008. "A critical appraisal of polymer–clay nanocomposites." *Chemical Society Reviews* no. 37 (3):568–594. doi: 10.1039/B702653F.

Chen, Li, and Yu-Zhong Wang. 2010. "A review on flame retardant technology in China. Part I: Development of flame retardants." *Polymers for Advanced Technologies* no. 21 (1):1–26. doi: 10.1002/pat.1550.

Choudalakis, G., and A. D. Gotsis. 2009. "Permeability of polymer/clay nanocomposites: A review." *European Polymer Journal* no. 45 (4):967–984. doi: 10.1016/j.eurpolymj.2 009.01.027.

Das, Gautam, and Niranjan Karak. 2009. "Vegetable oil-based flame retardant epoxy/clay nanocomposites." *Polymer Degradation and Stability* no. 94 (11):1948–1954. doi: 10.1016/j.polymdegradstab.2009.07.028.

Devi, Rashmi Rekha, and Tarun K. Maji. 2012. "Effect of nano-SiO_2 on properties of wood/polymer/clay nanocomposites." *Wood Science and Technology* no. 46 (6):1151–1168. doi: 10.1007/s00226-012-0471-1.

Ding, Peng, Ming Zhang, Jing Gai, and Baojun Qu. 2007. "Homogeneous dispersion and enhanced thermal properties of polystyrene-layered double hydroxide nanocomposites prepared by in situ reversible addition–fragmentation chain transfer (RAFT) polymerization." *Journal of Materials Chemistry* no. 17 (11):1117–1122. doi: 10.1039/B615334H.

Funar-Timofei, Simona, Smaranda Iliescu, and Takahiro Suzuki. 2014. "Correlations of limiting oxygen index with structural polyphosphoester features by QSPR approaches." *Structural Chemistry* no. 25 (6):1847–1863. doi: 10.1007/s11224-014-0474-7.

Gilman, Jeffrey W. 1999. "Flammability and thermal stability studies of polymer layered-silicate (clay) nanocomposites. This work was carried out by the National Institute of Standards and Technology (NIST), an agency of the U. S. government, and by statute is not subject to copyright in the United States.1." *Applied Clay Science* no. 15 (1):31–49. doi: 10.1016/S0169-1317(99)00019-8.

Gilman, Jeffrey W., Catheryn L. Jackson, Alexander B. Morgan, Richard Harris, Evangelos Manias, Emmanuel P. Giannelis, Melanie Wuthenow, Dawn Hilton, and Shawn H. Phillips.

2000. "Flammability properties of polymer–layered-silicate nanocomposites. Polypropylene and polystyrene nanocomposites." *Chemistry of Materials* no. 12 (7):1866–1873. doi: 10.1 021/cm0001760.

Hadjichristidis, Nikos, Hermis Iatrou, Stergios Pispas, and Marinos Pitsikalis. 2000. "Anionic polymerization: High vacuum techniques." *Journal of Polymer Science Part A: Polymer Chemistry* no. 38 (18):3211–3234. doi: 10.1002/1099-0518(20000915) 38:18<3211::AID-POLA10>3.0.CO;2-L.

Hapuarachchi, T. Dhanushka, and Ton Peijs. 2010. "Multiwalled carbon nanotubes and sepiolite nanoclays as flame retardants for polylactide and its natural fibre reinforced composites." *Composites Part A: Applied Science and Manufacturing* no. 41 (8):954–963. doi: 10.1016/j.compositesa.2010.03.004.

Hapuarachchi, Tharindu Dhanushka, Ton Peijs, and Emiliano Bilotti. 2013. "Thermal degradation and flammability behavior of polypropylene/clay/carbon nanotube composite systems." *Polymers for Advanced Technologies* no. 24 (3):331–338. doi: 10.1002/ pat.3087.

Hodgson, David M., and Emmanuel Gras. 2002. "Recent developments in the chemistry of lithiated epoxides." *Synthesis* no. 2002 (12):1625–1642. doi: 10.1055/s-2002-33635.

Horacek, H., and R. Grabner. 1996. "Advantages of flame retardants based on nitrogen compounds." *Polymer Degradation and Stability* no. 54 (2–3):205–215. doi: 10.1016/ S0141-3910(96)00045-6.

Jaffar Al-Mulla, Emad A. 2011. "Preparation of new polymer nanocomposites based on poly (lactic acid)/fatty nitrogen compounds modified clay by a solution casting process." *Fibers and Polymers* no. 12 (4):444–450. doi: 10.1007/s12221-011-0444-2.

Jang, Bok Nam, Marius Costache, and Charles A. Wilkie. 2005. "The relationship between thermal degradation behavior of polymer and the fire retardancy of polymer/clay nanocomposites." *Polymer* no. 46 (24):10678–10687. doi: 10.1016/j.polymer.2005 .08.085.

Mehrabzadeh, M., and M. R. Kamal. 2002. "Polymer-clay nanocomposites based on blends of polyamide-6 and polyethylene." *Canadian Journal of Chemical Engineering* no. 80 (6):1083–1092. doi: 10.1002/cjce.5450800610.

Misra, M., S. Vivekanandhan, A. K. Mohanty, and J. Denault. 2011. "4.09 - Nanotechnologies for Agricultural Bioproducts." In *Comprehensive Biotechnology (Second Edition)*, edited by Murray Moo-Young, 111–119. Burlington: Academic Press.

Mrah, Lahouari, and Rachid Meghabar. 2021. "In situ polymerization of styrene–clay nanocomposites and their properties." *Polymer Bulletin* no. 78 (7):3509–3526. doi: 10.1007/s00289-020-03274-5.

Papadopoulos, Lazaros, Zoi Terzopoulou, Antonios Vlachopoulos, Panagiotis A. Klonos, Apostolos Kyritsis, Dimitrios Tzetzis, George Z. Papageorgiou, and Dimitrios Bikiaris. 2020. "Synthesis and characterization of novel polymer/clay nanocomposites based on poly (butylene 2,5-furan dicarboxylate)." *Applied Clay Science* no. 190:105588. doi: 10.1016/j.clay.2020.105588.

Purut Koc, Ozlem, Seda Bekin Acar, Tamer Uyar, and Mehmet Atilla Tasdelen. 2018. "In situ preparation of thermoset/clay nanocomposites via thiol-epoxy click chemistry." *Polymer Bulletin* no. 75 (11):4901–4911. doi: 10.1007/s00289-018-2306-1.

Qin, Huaili, Shimin Zhang, Chungui Zhao, Guangjun Hu, and Mingshu Yang. 2005. "Flame retardant mechanism of polymer/clay nanocomposites based on polypropylene." *Polymer* no. 46 (19):8386–8395. doi: 10.1016/j.polymer.2005.07.019.

Salem, Nuha, and Devon A. Shipp. 2005. "Polymer-layered silicate nanocomposites prepared through in situ reversible addition–fragmentation chain transfer (RAFT) polymerization." *Polymer* no. 46 (19):8573–8581. doi: 10.1016/j.polymer.2005.04.082.

Salmeia, K. A., A. Gooneie, P. Simonetti, R. Nazir, J. P. Kaiser, A. Rippl, C. Hirsch, S. Lehner, P. Rupper, R. Hufenus, and S. Gaan. 2018. "Comprehensive study on flame retardant polyesters from phosphorus additives." *Polymer Degradation and Stability* no. 155:22–34. doi: 10.1016/j.polymdegradstab.2018.07.006.

Shah, Rhutesh K., and D. R. Paul. 2004. "Nylon 6 nanocomposites prepared by a melt mixing masterbatch process." *Polymer* no. 45 (9):2991–3000. doi: 10.1016/j.polymer.2004.02.058.

Shen, Yongqian, Yunpu Wang, Jiucun Chen, Hongjun Li, Zhiwen Li, and Chenglin Li. 2010. "Nitroxide-mediated polymerization of styrene initiated from the surface of montmorillonite clay platelets." *Journal of Applied Polymer Science* no. 118 (2):1198–1203. doi: 10.1002/app.32498.

Sinha Ray, Suprakas, Kazunobu Yamada, Masami Okamoto, Youhei Fujimoto, Akinobu Ogami, and Kazue Ueda. 2003. "New polylactide/layered silicate nanocomposites. 5. Designing of materials with desired properties." *Polymer* no. 44 (21):6633–6646. doi: 10.1016/j.polymer.2003.08.021.

Speight, James G. 2017. "Chapter 4 – Sources and types of organic pollutants." In *Environmental Organic Chemistry for Engineers*, edited by James G. Speight, 153–201. Butterworth-Heinemann.

Tai, Qilong, Richard K. K. Yuen, Lei Song, and Yuan Hu. 2012. "A novel polymeric flame retardant and exfoliated clay nanocomposites: Preparation and properties." *Chemical Engineering Journal* no. 183:542–549. doi: 10.1016/j.cej.2011.12.095.

Tasdelen, Mehmet Atilla. 2011. "Poly(epsilon-caprolactone)/clay nanocomposites via "click" chemistry." *European Polymer Journal* no. 47 (5):937–941. doi: 10.1016/j.eurpolymj.2011.01.004.

Tasdelen, Mehmet Atilla, Johannes Kreutzer, and Yusuf Yagci. 2010. "In situ synthesis of polymer/clay nanocomposites by living and controlled/living polymerization." *Macromolecular Chemistry and Physics* no. 211 (3):279–285. doi: 10.1002/macp.200900590.

Tasdelen, Mehmet Atilla, Wim Van Camp, Eric Goethals, Philippe Dubois, Filip Du Prez, and Yusuf Yagci. 2008. "Polytetrahydrofuran/clay nanocomposites by in situ polymerization and "click" chemistry processes." *Macromolecules* no. 41 (16):6035–6040. doi: 10.1021/ma801149x.

Thirumal, M., Dipak Khastgir, Nikhil K. Singha, B. S. Manjunath, and Y. P. Naik. 2009. "Effect of a nanoclay on the mechanical, thermal and flame retardant properties of rigid polyurethane foam." *Journal of Macromolecular Science, Part A* no. 46 (7):704–712. doi: 10.1080/10601320902939101.

Troitzsch, Juergen H. 2016. "Fires, statistics, ignition sources, and passive fire protection measures." *Journal of Fire Sciences* no. 34 (3):171–198. doi: 10.1177/0734904116636642.

Uthaman, Arya, Hiran Mayookh Lal, Chenggao Li, Guijun Xian, and Sabu Thomas. 2021. "Mechanical and water uptake properties of epoxy nanocomposites with surfactant-modified functionalized multiwalled carbon nanotubes." *Nanomaterials* no. 11 (5):1234.

Uthaman, Arya, Hiran Mayookh Lal, and Sabu Thomas. 2021a. "Fundamentals of silver nanoparticles and their toxicological aspects." In *Polymer Nanocomposites Based on Silver Nanoparticles: Synthesis, Characterization and Applications*, edited by Hiran Mayookh Lal, Sabu Thomas, Tianduo Li and Hanna J. Maria, 1–24. Cham: Springer International Publishing.

Uthaman, Arya, Hiran Mayookh Lal, and Sabu Thomas. 2021b. "Silver nanoparticle on various synthetic polymer matrices: Preparative techniques, characterizations, and applications." In *Polymer Nanocomposites Based on Silver Nanoparticles: Synthesis, Characterization and Applications*, edited by Hiran Mayookh Lal, Sabu Thomas, Tianduo Li and Hanna J. Maria, 109–138. Cham: Springer International Publishing.

Vaia, Richard A., Hope Ishii, and Emmanuel P. Giannelis. 1993. "Synthesis and properties of two-dimensional nanostructures by direct intercalation of polymer melts in layered silicates." *Chemistry of Materials* no. 5 (12):1694–1696. doi: 10.1021/cm00036a004.

Wagenknecht, Udo, Bernd Kretzschmar, and Gerd Reinhardt. 2003. "Investigations of fire retardant properties of polypropylene-clay-nanocomposites." *Macromolecular Symposia* no. 194 (1):207–212. doi: 10.1002/masy.200390084.

Wang, Xin, Ehsan Naderi Kalali, Jin-Tao Wan, and De-Yi Wang. 2017. "Carbon-family materials for flame retardant polymeric materials." *Progress in Polymer Science* no. 69:22–46. doi: 10.1016/j.progpolymsci.2017.02.001.

Yang, Zhe, Hongdan Peng, Weizhi Wang, and Tianxi Liu. 2010. "Crystallization behavior of poly(ε-caprolactone)/layered double hydroxide nanocomposites." *Journal of Applied Polymer Science* no. 116 (5):2658–2667. doi: 10.1002/app.31787.

Yeh, Jui-Ming, and Chih-Ping Chin. 2003. "Structure and properties of poly(o-methoxyaniline)–clay nanocomposite materials." *Journal of Applied Polymer Science* no. 88 (4):1072–1080. doi: 10.1002/app.11829.

Yoonessi, Mitra, Hossien Toghiani, Tyrone L. Daulton, Jar-Shyong Lin, and Charles U. Pittman. 2005. "Clay delamination in clay/poly(dicyclopentadiene) nanocomposites quantified by small angle neutron scattering and high-resolution transmission electron microscopy." *Macromolecules* no. 38 (3):818–831. doi: 10.1021/ma048663e.

Zanetti, Marco, Giovanni Camino, Peter Reichert, and Rolf Mülhaupt. 2001. "Thermal behaviour of poly(propylene) layered silicate nanocomposites." macromolecular rapid communications no. 22 (3):176–180. doi: 10.1002/1521-3927(200102)22:3<1 76::AID-MARC176>3.0.CO;2-C.

Zhang, Jinguo, and Charles A. Wilkie. 2003. "Preparation and flammability properties of polyethylene–clay nanocomposites." *Polymer Degradation and Stability* no. 80 (1):163–169. doi: 10.1016/S0141-3910(02)00398-1.

Zhang, Wenchao, Xiangdong He, Tinglu Song, Qingjie Jiao, and Rongjie Yang. 2014. "The influence of the phosphorus-based flame retardant on the flame retardancy of the epoxy resins." *Polymer Degradation and Stability* no. 109:209–217. doi: 10.1016/j.polymdegradstab.2014.07.023.

Zhang, Zhenjun, Lina Zhang, Yang Li, and Hongde Xu. 2005. "New fabricate of styrene–butadiene rubber/montmorillonite nanocomposites by anionic polymerization." *Polymer* no. 46 (1):129–136. doi: 10.1016/j.polymer.2004.11.008.

Zhao, Chungui, Huaili Qin, Fangling Gong, Meng Feng, Shimin Zhang, and Mingshu Yang. 2005. "Mechanical, thermal and flammability properties of polyethylene/clay nanocomposites." *Polymer Degradation and Stability* no. 87 (1):183–189. doi: 10.1016/j.polymdegradstab.2004.08.005.

Zhu, Jin, Paul Start, Kenneth A. Mauritz, and Charles A. Wilkie. 2002. "Thermal stability and flame retardancy of poly(methyl methacrylate)-clay nanocomposites." *Polymer Degradation and Stability* no. 77 (2):253–258. doi: 10.1016/S0141-3910(02)00056-3.

4 Application of Layered Double Hydroxides in Fire-Safe Polymers

Ehsan Naderi Kalali

Department of Fire Safety Engineering, Faculty of
Geosciences and Environmental Engineering, Southwest
Jiaotong University, Chengdu, China

Institute of Macromolecular Chemistry, Czech Academy of
Sciences, Prague, Czech Republic

Marjan Entezar Shabestari

Department of Materials Science and Engineering and
Chemical Engineering, Polytechnic School, Carlos III
University of Madrid, Leganés, Madrid, Spain

CONTENTS

DOI: 10.1201/9781003327158-4

4.1 INTRODUCTION

In recent years, fire safety and environmental compatibility have gained great attention and are known as one of the main concerns for selecting suitable compounds to develop flame retardants for polymers to safeguard people's lives. Recently, vast investigations in green chemistry have been performed, and a variety of novel bio-safe and eco-friendly materials and methods have already been designed to eliminate or diminish the side effects of chemical compounds. Due to their individual behaviors as eco-friendly and bio-safe nanomaterials, they have been selected as potential substances for advanced applications (Chen et al. 2013, Doğan et al. 2021). The applications and final properties of nano-architectured materials depend on different parameters, such as their composition, shape, phase, size distribution, and structure (Dong et al. 2019, Zhang and Horrocks 2003). Since the last decade, in both academia and industry, layered double hydroxide (LDH), which is an anionic clay, has become an attractive compound due to its hydrotalcite-like structure (Becker et al. 2012, Wang, Li, Zhao, et al. 2012, Wu et al. 2018). These materials are cost-effective and demonstrate different features, such as tunable interior structure, thermal resistivity, and pH variations. They also have a high surface area and great anion-exchange capability (Mallakpour and Hatami 2017, Wang, Zhang, and Zhou 2017). In addition, LDHs are able to form multifunctional nano-sized compounds due to their novel crystalline clustered compounds (Kalali, Wang, and Wang 2016, Kang and Wang 2013). However, there are some restrictions that limit their applications, such as high hydrophilicity and large charge density. Therefore, such materials do not extensively tend to interact with other materials with a hydrophobic nature; hence, their intramolecular interactions increase their agglomeration (Huang et al. 2010, Ma et al. 2006). Moreover, pristine LDH does not demonstrate remarkable applications in advanced technology (Zubair et al. 2017, Yin and Tang 2016). Therefore, to promote their organophilic nature, to expand their interlayer gallery spacing and to develop their applications in industrial and technical fields, LDHs need to be well modified with convenient organic and/or inorganic modifying agents (Anto Jeffery, Nethravathi, and Rajamathi 2014). Accordingly, in recent years, modification of LDHs has gained significant attention from scientists. Based on the Scopus database, more than 500 documents have been issued during the last decade (Figure 4.1a) (Mallakpour, Hatami, and Hussain 2020), which are grouped into different types, such as articles (89.4%), reviews (5.4%), conference papers (3.2%), book chapters (1.2%), and conference reviews (0.8%) (Figure 4.1b) (Mallakpour, Hatami, and Hussain 2020). In this chapter, we will explain the synthesis methods, morphological aspects, characterization, and flame-retardant properties. Furthermore, we will briefly clarify the major applications and the advantages and disadvantages of the modified LDHs.

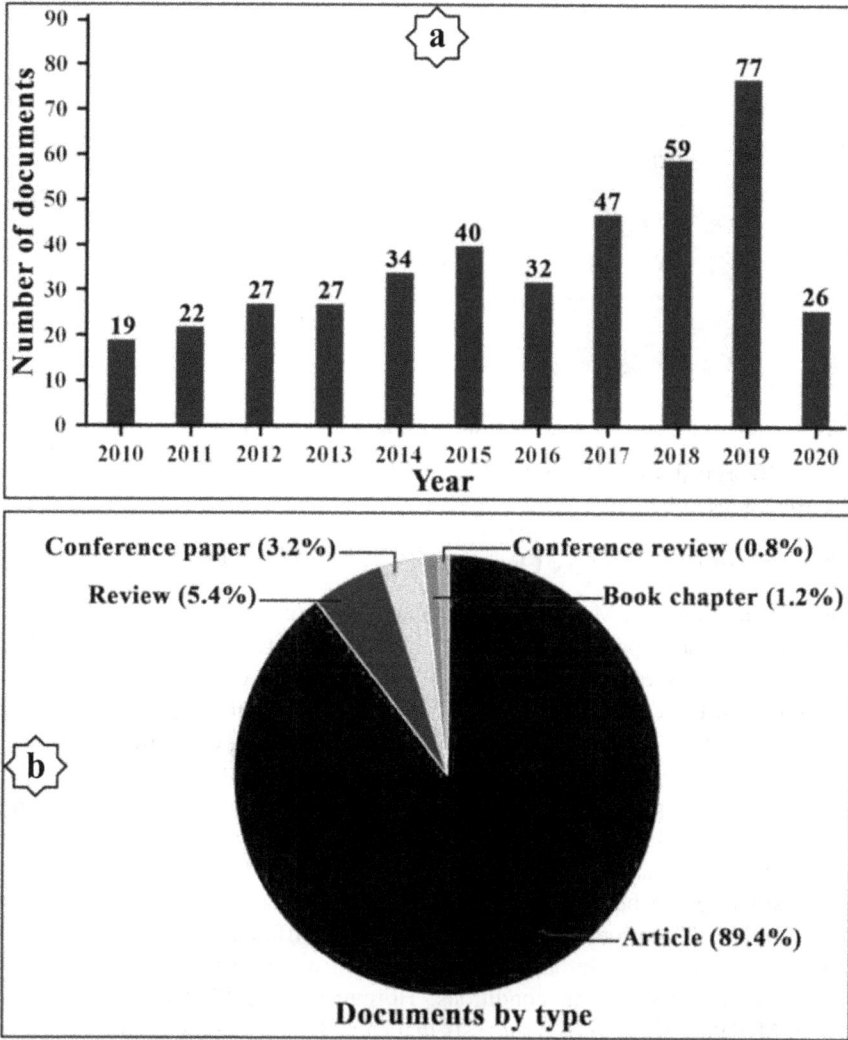

FIGURE 4.1   Diagrams observed about (a) modification of LDH in the range of 2010–2020 and (b) documents by type for modified LDH based on the Scopus database.

Source: These data were obtained on April 14, 2020. Reprinted from Mallakpour, S., Hatami, M., and Hussain, C. M., *Adv. Colloid Interface Sci.*, 283, 102216, 2020.

4.2   STRUCTURAL CHARACTERISTICS OF LDHS

The general formula of LDHs is the following: $[M_{1-x}{}^{2+} M_x{}^{3+}(OH)_2]^{x+} \cdot [A_{x/n}]^{n-} \cdot mH_2O$, where M^{2+} and M^{3+} are the divalent and trivalent metal ions, respectively, A^{n-} are inorganic or organic anions, "m" is the number of interlayer water, and $x=M^{3+}/(M^{2+}+M^{3+})$, which is the layer charge density or molar ratio (Wijitwongwan, Intasa-ard, and Ogawa 2019). Figure 4.2 is a schematic illustration of the general structure of LDH.

Brucite-like sheet

A^{n-} Interlayer anion

Water molecule

M^{2+} or M^{3+}

Basal spacing

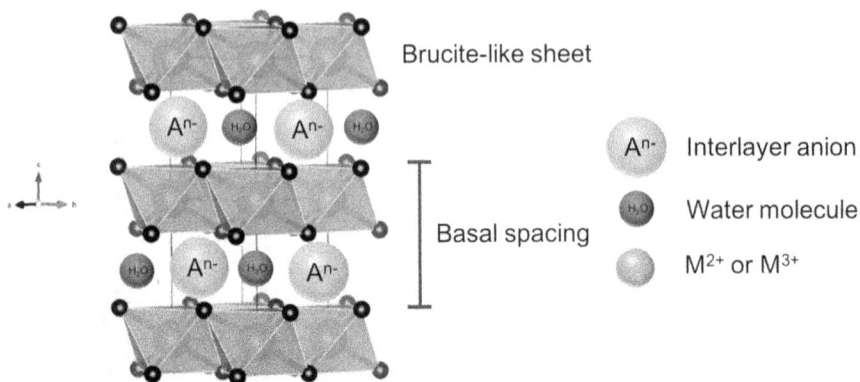

FIGURE 4.2 Schematic illustration of the general structure of LDH.

Source: Reprinted from Wijitwongwan, R., Intasa-ard, S., and Ogawa, M., *Chem. Engineer.*, 3, 68, 2019.

4.2.1 CATION AND ANION ORDERING

Vucelic et al. (Vucelic, Moggridge, and Jones 1995, Vucelic, Jones, and Moggridge 1997) showed the importance of order within LDH sheets for catalytic applications where large surface areas merged with a high degree of metal dispersion are demanded. In the case of (LiAl$_2$–OH) (Besserguenev et al. 1997, Thiel, Chiang, and Poeppelmeier 1993) and pyroaurite (Mg$_{2.3}$Fe-NO$_3$) (Vucelic, Jones, and Moggridge 1997), structural cation ordering is shown by the existence of additional diffraction peaks that were related to the super lattices. For the former compound, the ordering of the cations is illustrated by the migration of Li$^+$ ions into the hollow octahedral sites. Except for the mentioned cases, a general lack of cation ordering can be observed due to the large difference in the size between M(II) and M(III) cations (Bellotto et al. 1996). Different LDHs with trivalent cations (Al^{3+}) indicate similar properties. Cation ordering is strongly dependent on the particular structural conditions. Hofmeister and Platen (1992) showed that an M(II)/M(III) ratio of 2 suggests that each divalent cation is surrounded by 3 M(II) and 3 M(III) and that the trivalent cations are surrounded by 6 M(II). A change in the mentioned ratio also resulted in a change in the conceptual local order. The interlayer anions may also exhibit an ordered distribution, epitomized by the highly ordered two-dimensional actual superlattice in (Zn$_2$Al–SO$_4^{2-}$) (Bookin, Cherkashin, and Drits 1993).

4.2.2 STACKING

There are various articles in which the intercalation chemistry of layered double hydroxides (LDHs) describes fully exchanged anions. However, there are only a few reports on mixed ion-exchanged forms. Previous studies (Mendiboure and Schoellhorn 1987) showed the competitive intercalation of NO^{3-} and ClO^{4-} into [ZnCr] LDH. They demonstrated the coexistence of the two phases, which implies that the generation of phases with mixed anions in the interlayer distance is inadequate energetically. Schon, Adler, and Dresselhaus (1988) showed that some theoretical models suggested that regular stacking, referred to as "staging", which is commonly observed in graphite, does not lie in the LDH due to the solidity of the layers. However, Fogg et al. (Fogg, Dunn, and O'Hare 1998) used in situ time-resolved X-ray diffraction (XRD) to demonstrate that $[LiAl_2(OH)_6Cl \cdot 2H_2O]$ can generate second-stage intermediates. Every second layer is filled by dicarboxylate anions (l-malate, tartrate, fumarate, maleate, adipate, phthalate, succinate and terephthalate). Khan and O'Hare (2002) used a similar technique to study the intercalation chemistry of layered materials. Kaneyoshi and Jones (1998) observed the interstratification of the layers during the interlayer exchange of terephthalate with Cl^- and NO^{3-} anions. This phe-nomenon is ascribed to the two orientations of terephthalate anions, vertical and horizontal, which can be changed by varying the extent of drying and the layer charge density. Moreover, Iyi, Kurashima, and Fujita (2002) investigated the direct synthesis of mixed azobenzene/hydroxide intercalates and suggested a separation of hydrophilic inorganic anions and hydrophobic organic anions, similar to staged fluorohectorites reported (Ijdo and Pinnavaia 1998). Layered nanoparticles in the form of colloids are generated into a solution that subsequently transforms into staging structures during the drying process. These cases demonstrate that staging in LDH can occur either by direct synthesis or during the exchange process and depends on different orientations of the same molecule or different interlayer contents.

4.2.3 GUEST–HOST INTERACTIONS

Interactions between the host (the brucite-like layer) and guests (anions) can be controlled by hydrogen bonding and electrostatic interactions between the hydroxyl surface groups and the anions. Such interactions strongly depend on the order of the metal cations in the layer and the chemistry of the anions. Fogg and O'Hare (1999) showed that for ordered structure LDHs such as [Li–Al], the anions are well aligned perpendicularly to the Li^+–Li^+ axis, exhibiting the effect of electrostatic interactions in this structure.

In general, oxo-anions tend to interact with the hydroxyl surface groups of the layers, which leads to dense packing of the anions. It is suggested that the sym-metric compatibility between the octahedral layers and the oxoanion polyhedra of the host structure is the main reason. The various rhombohedral or hexagonal stacking of the LDH layers describe the prismatic, octahedral, and tetrahedral interlayer crystallographic sites that can ideally embed monomeric oxoanions. In the case of face-to-face or corner-to-corner arrangements, the Td or Oh units are formed as optimal hydrogen bonds. Interlayer domains of $[Zn–Cr–SO_4]$ are an example of the ordered arrangement structure of the alternate inverse interlayer SO_4 tetrahedral. The tetrahedral arrangement retains their C_3-axis perpendicular to the layers, in which one oxygen points toward a metal cation of the octahedral layer and the other three oxygens point toward three hydroxyl (OH) groups of the opposite layer. Strong interactions between sulfate anions and the layers are indicated by the short hydrogen bond lengths of 0.293 and 0.271 nm, which lead to a 0.892 nm basal spacing (Khaldi et al. 1997). The XO_4^- containing series of LDHs ($[Zn–Al–CrO_4]$

[Zn–Al–SO$_4$], [Cu–Cr–SO$_4$], [Zn–Cr–SO$_4$], [Cu–Cr–Cr$_2$O$_7$], [Zn–Cr–Cr$_2$O$_7$], and [Cu–Cr–CrO$_4$]) shows basal distances close to the d value of [Zn–Cr–SO$_4$]. Bigey et al. (1997) showed that in all mentioned phases, the oxoanions possess similar orientations to the SO$_4^{2-}$ ions within the reference material. Similar d values were achieved even with the more pliable Cr$_2$O$_7$, as the anions are able to lie flat and stay parallel to the layers.

In the case of strong interaction of the interlayer anions with the layers, modification of molecular dynamics can be expected. Intercalation of SO$_4^{2-}$, CO$_3^{2-}$, and ClO$_4^-$ into [Mg–Al] LDH reduces the symmetry of the anions caused by the hydrogen bonding associated with the OH groups and intercalated water molecules, as shown by Raman spectroscopy and Fourier transform infrared (FTIR) spectroscopy (Kloprogge et al. 2002). However, the NO$_3^-$ anions tended to be randomly distributed within the interlayer distance due to their vibration bands, which remained unaffected during intercalation.

In the case of intercalated silicate anions [Zn–Cr–SiO$_4$], cell parameter reformation gives a hexagonal 3R lattice similar to that for [Zn–Cr–Cl] (Schutz and Biloen 1987, Depège et al. 1996). The other reflections of the [Zn–Al–SiO$_4$] phase are ascribed to the super structural order of the silicate species in the interlayer space of LDH.

Intercalation of large-scale anions such as V$_{10}$O$_{28}^{6-}$ resulted in very short basal spacing due to the plane orientation of the decaoctahedra building block of V$_{10}$O$_{28}^{6-}$. Due to the limitation in size of the oxometallate anions and their small charge density, no interlayer microporosity can be generated. In fact, the basal distancing of LDH-containing oxopolyanions cannot surpass 1.4 nm, while the intercalation of Al$_{13}$ hydroxocations into the clay minerals resulted in significantly greater interlayer expansion (Pinnavaia 1987).

Investigations have shown that organic anions interact with their anionic groups, such as –CO$_2^-$, –OPO$_3^{2-}$, –PO$_3^{2-}$, –SO$_3^-$, and –OSO$_3^-$, which are bonded strongly with hydrogen bonds to the surface hydroxyl groups (Meyn, Beneke, and Lagaly 1990). It is demonstrated that α, ω-Dianionic molecules possess the capability of bridging two adjacent layers, providing a basal spacing appropriate to the length of the hydrocarbon chain. It is also shown that with the incorporation of monovalent long-chain anions, both interdigitated arrangements or bilayers can be formed. Such an arrangement leads to the formation of a basal spacing larger than 1.5 nm (Sanz and Serratosa 1984, Meyn, Beneke, and Lagaly 1990, Bonnet et al. 1996, Costantino et al. 1999, Prevot et al. 1998).

In particular, the equivalent area of the layers and the relation between the size and charge of the organic anions are the factors that determine the interlayer packing. Optimal packing can be achieved when the cross-sectional area of the anions is the same as the equivalent area. Such a phenomenon can be observed in the case of dodecylsulfate-containing [Zn$_2$Al] LDH. The anions embrace a packing structure similar to that in their sodium salt, with hydrogen bonding and electrostatic interactions between the surface OH groups and anionic head groups and between the lateral hydrocarbon chains caused by van der Waals bonding (Boehm, Steinle, and Vieweger 1977, Kopka, Beneke, and Lagaly 1988).

4.3 FUNDAMENTAL PROPERTIES OF LDHS

4.3.1 CHEMICAL STABILITY

One of the main LDH features is their chemical stability considering their applications. For example, in the case of LDH usage as a sink for radioactive metal ions existing in nuclear waste repositories, such a parameter needs to be measured (Allada et al. 2002). The solubility of LDHs in water must be evaluated in geochemical processes (Ford et al. 1999, Scheckel et al. 2000). pH titration is a suitable method to estimate the solubility of metal hydroxides and LDHs (Boclair and Braterman 1998, 1999, Boclair et al. 1999). The stability of LDHs shows an increment in the order $Al^{3+}<Fe^{3+}$ for trivalent cations and $Mg^{2+}<Mn^{2+}<Co^{2+} \approx Ni^{2+}<Zn^{2+}$ for divalent cations (Boclair and Braterman 1999). This trend complies with the pK_{sp} values recorded for the corresponding metal hydroxides (K_{sp} is the solubility constant). Due to the smaller pK_{sp} value of $Mg(OH)_2$ compared to $Zn(OH)_2$, Mg-based LDHs exhibit higher solubility. Therefore, the aqueous solution of Mg-containing LDH shows more basic properties than the corresponding aqueous solution of a Zn-containing LDH (Shaw et al. 1990).

Allada et al. (2002) demonstrated that thermochemical data can also be used to calculate the aqueous solubility of an LDH. The free energy recorded during the formation of the carbonate form of an LDH and the heat of formation are identical to the values recorded for a physical blend of the binary compounds, i.e., carbonates and the metal hydroxides. Their ordinary physical blend model demonstrates that the interlayer anions are the key factors determining the aqueous suspension ability of LDHs. Similar to carbonate, borate, and silicate can also reduce the suspension ability, while sulfate and nitrate cause an increase.

4.3.2 THERMAL STABILITY

As the thermal decomposition products of LDH are defined as catalysts, investigation of the thermal stability of such materials was necessary. Despite the variety in the composition of LDHs, they usually exhibit similar thermal decomposition behavior. By increasing the temperature, LDH liberates its interlayer water up to ca. 250°C, leads to dehydroxylation of the hydroxide layers and hence decomposition of the interlayer anions at elevated temperatures. In the case of $Mg–Al–CO_3$ LDH, the differential thermal analysis (DTA) curves exhibit two endothermic peaks. The first peak appears in the range of 150 to 250°C and corresponds to the elimination of interlayer water, and the second peak in the range of 250 to 450°C shows the decomposition of the carbonate anions and hydroxide layers (Miyata 1980). Similar weight-loss behavior can be observed in their corresponding thermal gravimetric analysis (TGA) curve. Moreover, the XRD patterns indicated that with the elimination of interlayer water, the layer structure remained unchanged. Valente et al. (2000) evaluated the decomposition temperatures of carbonate-containing LDH composed of different metal cations under similar experimental situations. They discovered that the thermal stability can be elevated in the order Co–Al<Zn–Al≈Cu–Al<Mg–Fe≈Ni–Al<Mg–Al≈Mg–Cr. The greatest decomposition

temperature was recorded at 400°C for Mg–Cr LDH and Mg–Al LDH, and the lowest decomposition temperature was observed at 220°C for Co–Al LDH. The changes in the XRD pattern when Mg–Al–CO$_3$ LDH with Mg/Al=2:1 is heated at various temperatures. The calcination temperatures and Miller indices are indicated.

The thermal decomposition progression of Mg-containing LDH, as specified by Mg–Al–CO$_3$ LDH, is well defined. The MgO phase was observed in the range of 400 to 850°C and exhibits a smaller lattice constant in comparison with pure MgO because of the incorporation of Al^{3+} ions. Above 900°C, Al ions are liberated, and spinel (MgAl$_2$O$_4$) is generated. Pure MgAl$_2$O$_4$ and MgO are obtained at approximately 1000°C (Miyata 1980). MgFe$_2$O$_4$ spinel is generated together with the MgO phase at 350°C in the case of Mg–Fe–CO$_3$ LDH (Maria Fernández et al. 1998, Hibino and Tsunashima 2000).

Zinc-based layered double hydroxides demonstrate lower thermal stability than Mg-based LDH (Bertoldi et al. 1988, Kiyoshi et al. 1993). On the other hand, nickel-containing LDHs show an intermediate thermal stability between Mg- and Zn-based layered double hydroxides (Clause et al. 1991, 1992), and cobalt-containing LDH starts the decomposition process at reduced temperatures compared with other LDHs (Kannan and Swamy 1992, Kannan et al. 1995, Xu and C. Zeng 1998, Pérez-Ramírez et al. 2001, 2001).

4.4   APPLICATION OF LAYERED DOUBLE HYDROXIDES IN FIRE-SAFE POLYMERS

4.4.1   Application of LDHs in Fire-Safe Thermoplastic Polymers

4.4.1.1   Polyethylene (PE)

One of the most widely used polymers for commodity purposes is known as PE. The flame retardant efficiency of LDHs in PE has been widely investigated. Achieving a high dispersion stage of polar additives such as silica or layered silicates through the PE matrix is a challenging task, which contains a nonpolar hydrocarbon molecular backbone structure. Therefore, as mentioned before, obtaining a high level of dispersion of LDHs in PE is the initial challenge in dealing with the flammability of PE-LDH composites. Costantino et al. (Costantino et al. 2005) synthesized Zn/AleCO$_3$ LDH by conducting urea hydrolysis, in which the carbonate anions exchanged with nitrate to enlarge the interlayer distance, and then, the prepared LDH intercalated with sodium stearate to enhance mixing with PE. The stearate intercalated LDHs had a basalt spacing of 30.9 A° according to the XRD characterization. PE nanocomposites containing modified LDHs were prepared by a Brabender internal mixer via melt mixing. At low-angle XRD, no diffraction peaks were observed, demonstrating a partial exfoliation of LDHs through the PE matrix. It has been recorded that incorporation of 5 wt% (calculated from inorganic content) of modified LDHs into the PE resulted in a 55% reduction in the peak heat release rate (PHRR).

In another investigation by Costa et al. (2005), sodium dodecylbenzene sulfonate (DBS) was used to modify Mg/AleCO$_3$ LDH via solution intercalation. Then, they prepared a master batch of a 1:1 ratio of maleic anhydride grafted PE (PE-g-MAH) and organically modified LDHs (LDH-DBS) by melt compounding. Subsequently,

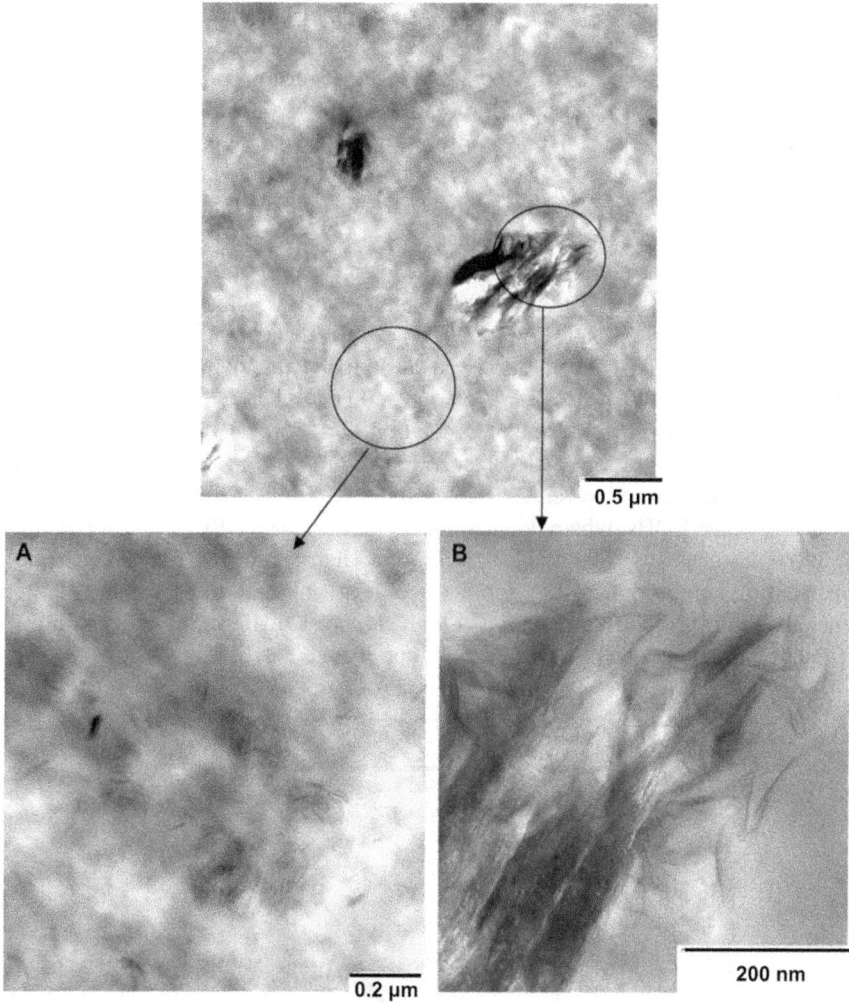

FIGURE 4.3   TEM micrographs showing that exfoliated LDH particles exist both (a) in the bulk matrix of LDPE and (b) also in the vicinity of the originating bigger particles.

Source: Reprinted from Costa, F. R., Wagenknecht, U., and Heinrich, G., *Polym. Degrad. Stabil.,* **92, 1813–1823, 2007.**

the prepared master batch was mixed with the desired amounts of low-density PE (LDPE) via a co-rotating twin-screw extruder. It has been reported that the employment of such intensive mixing equipment resulted in significant enhancement in dispersion of the nanoparticles through the polymer matrix (Costa, Wagenknecht, and Heinrich 2007). It is indicated that the vigorous shear force while mixing facilitates and promotes the intercalation of the polymer chains into the interlayers, which results in further rupture and delamination of layers, hence resulting in more exfoliation and size reduction of the particles, as illustrated by the TEM images (Figure 4.3).

Costa, Wagenknecht, and Heinrich (2007) reported that the thermal stability of LDPE/LDH-polymer composites was significantly influenced by the melt mixing method, as explained before. An increase in the onset decomposition temperature of LDPE by 20°C was observed by incorporation of only 2.43 wt% of LDHs. It is also demonstrated that loading of 6.89 to 16.20 wt% of LDHs resulted in significant weight loss only above 400°C and a great suppression effect on the low-temperature decomposition of LDPE. The presence of LDHs also resulted in significant residual char at 600°C for the LDPE composites.

Under 30 kW/m^2 radiant heat flux in a cone calorimeter, LDPE/LDH-DBS composites showed significant flame retardant properties. In the case of neat LDPE, the PHRR was approximately 800 kW/m^2, while the composites with approximately 16.2 wt% filler concentration showed a PHRR as low as 300 kW/m^2. The existence of LDHs also resulted in a lower total heat release (THR) and a delay in the ignition time (t_{ig}). Moreover, incorporation of 16.2 wt% LDHs also leads to a significant reduction in the THR value to approximately 44% compared with the efficiency of LDHs and layered silicates on the THR values and highlights the advantage of the LDHs, where no significant change is generally observed regarding the layered silicates (Costa, Wagenknecht, and Heinrich 2007). LDHs can release water during combustion, which can act as a heat sink and decrease the total heat generated during combustion. On the other hand, layered silicates act predominantly in a physical mode, which helps generate a protective char layer.

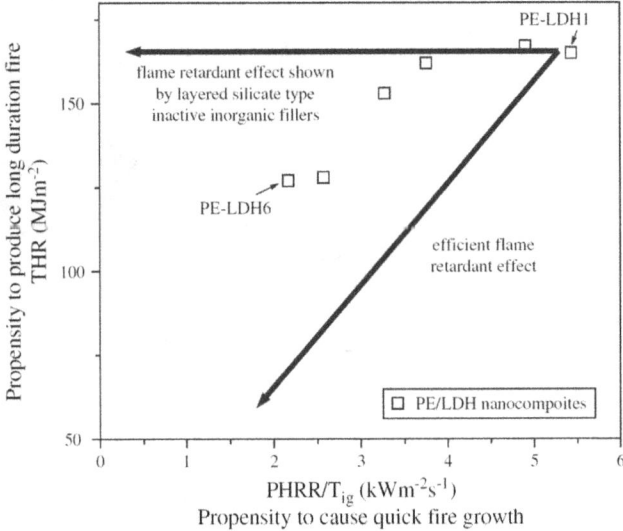

FIGURE 4.4 A plot of THR against PHRR/t_{ig} for different LDH concentrations indicating fire risk associated with LDPE/LDH nanocomposites (the LDH concentration increases from the right to the left as shown, PE-LDH1-2.43 wt%, PE-LDH2-4.72 wt%, PE-LDH3-6.89 wt%, PE-LDH4-8.95 wt%, PE-LDH5-12.75 wt%, PE-LDH6-16.2 wt%).

Source: Reprinted from Costa, F. R., Wagenknecht, U., and Heinrich, G., *Polym. Degrad. Stabil.*, 92, 1813–1823, 2007.

Regarding the flammability of materials, several aspects can be considered, such as the tendency to ignite, propensity for sustained combustion, dripping of burning material, and fire growth. Therefore, translation of different results recorded during cone calorimeter tests into a meaningful value is advantageous to define the tendency to cause fast fire growth. The ratio of PHRR to ignition time (PHRR/t_{ig}) is known as an index for comparing the flammability of materials in texts. According to the cone calorimetry results, LDPE/LDH-DBS nanocomposites demonstrated a reduced trend of the PHRR/t_{ig} ratio by increasing the concentration of LDH-DBS. Figure 4.4 shows the PHRR/t_{ig} ratio against THR, in which less flammability of the composites can be concluded, as the position of the LDPE/LDH-DBS nanocomposite shifted toward the origin. It is worth noting that a variety of testing conditions, such as heat flux, source of heat flux and ignition, orientation and topography of the sample surface and its dimensions, and ambient conditions, indicate the flammability of a material.

4.4.1.2 Polypropylene (PP)

One of the most important polymers is PP, which is commercially used in many different applications due to its great performance and low cost. However, the main drawback of polypropylene is its thermo-oxidative degradation and flammability. Hence, finding non-halogenated flame retardants has become of significant interest for PP. For this purpose, a variety of inorganic particulates, including carbon nanotubes (CNTs), montmorillonite (Qin et al. 2005), MH (Mishra et al. 2004), and silica (Kashiwagi et al. 2000), in combination with other compounds, such as melamine phosphate (Lv et al. 2005) and APP (Chiu and Wang 1998), have been studied as flame retardant additive packages.

In recent years, layered double hydroxides (LDHs) have also been investigated as potential candidates for use as flame retardants for PP. LDHs show great chemical versatility as "anionic clays" and possess hydrated structures. The thermal degradation behavior of PP-incorporated montmorillonite and LDHs was compared by Ding and Qu (2006). According to the TGA records, the PP/LDH composites demonstrated better thermal stability than the PP/montmorillonite composites based on the last stages of decomposition, in which LDHs resulted in better char formation. For example, 80% of the weight of PP/LDH nanocomposites was lost at approximately 416°C, and in the case of PP/montmorillonite composites, this phenomenon occurred at 399 °C. Similar results have also been reported by other researchers (Shi et al. 2010, Zhang, Ding, and Qu 2009), which exhibited a higher decomposition temperature and a significant enhancement in the thermal stability of PP in the presence of LDHs. Co/Al-NO$_3$ LDHs synthesized by Wang et al. (2011) and dodecylbenzene sulfonate were used to modify it. A master batch containing modified maleic anhydride grafted PP (PP-g-MA) and LDH was first melt compounded via a compounder and later mixed with non-maleated PP. The prepared PP/LDH composites demonstrated an improved thermal stability in TGA characterization. The Freidman method (Starink 2003) or Flynn-Wall-Ozawa method was used to calculate the activation energies in different decomposition stages of polypropylene. The results showed an increment of 10–30% with incorporation of 1.5–6 wt% of LDHs.

In general, different compatibilizers, such as PP-*g*-MA, have been used to improve the dispersion of LDHs into hydrophobic PP. However, the effect of such chemical compatibilizers on the flame retardant properties of the final PP composites should be considered.

In this regard, Manzi-Nshuti et al. (2009) employed a series of oleate-modified Zn/Mg/Al LDHs composed of different Zn/Mg contents to study the influence of the presence of PP-*g*-MA and LDH composition on the dispersion state and flame retardant properties of the PP/LDH system. In this investigation, incorporation of 1 to 4 wt% (inorganic mass fraction) LDH did not show any significant reduction in the PHRR of nonmaleated PP (0–38%). Poor dispersion of LDHs through nonmaleated PP is known as the main reason for such weak efficiency. Therefore, different amounts of compatibilizer (PP-*g*-MA) were added to the mixture to facilitate the dispersion of these oleate-modified LDHs within the PP matrix.

Cone calorimeter tests demonstrated a comprehensive correlation between the amount of compatibilizer (PP-*g*-MA) and the flame retardant properties of the mentioned PP/LDH systems containing maleated PP. In this study, it was demonstrated that incorporation of 4 wt% (inorganic mass fraction) LDH into the maleated PP matrix with a PP/PP-*g*-MA ratio of 1:1 resulted in a 68% reduction in the PHRR values compared with neat PP. It was also observed that by loading 4 wt% oleate-modified LDH and reducing the amount of PP-*g*-MA from a PP/PP-*g*-MA ratio of 1:1 to 4:1 and 8:1, the PHRR values were reduced to 57% and 51%, respectively. According to the TEM and XRD analysis, the authors showed that PP-*g*-MA greatly enhances the dispersion state of LDH nanoparticles through the PP matrix, hence yielding improved flame retardant properties.

Recent studies have also shown that the chemical composition of LDHs can also significantly affect the flame retardant behavior of PP composites. In one study, it was demonstrated that incorporation of 4 wt% Zn/Al LDH into PP/PP-g-MA (1:1) led to a 68% decrease in the PHRR value, while only a 52% reduction was recorded for Mg/Al LDH under similar conditions. Such a difference might be attributed to their char morphologies that are generated during combustion. Zn/Al LDH formed a compact char during combustion in comparison with Mg/Al LDH. Until now, all LDH-related investigations regarding polypropylene systems have been related to organically modified LDHs. The main reason for this was that PP is a nonpolar macromolecule that has intrinsically negligible compatibility with intercalated LDHs, which are hydrophobic. The studies also propose that using LDHs alone is not sufficient to obtain desirable flame retardant properties and generally requires other compounds to improve its efficiency.

4.4.1.3 Poly(Methyl Methacrylate) (PMMA)

PMMA is a polymer that has been widely used in many applications because it possesses excellent mechanical, optical, electric, and thermal properties. However, its main drawback is its high combustibility, and flame retardation is required to overcome this issue (Wang et al. 2010). For PMMA, different types of LDHs, including MgAl–DBS, MgAl–alkyl carboxylates, MgAl–oleate, ZnAl–oleate, MgAl–$CH_3(CH_2)$ $NCOO^-$, CaAl–undecenoate, ZnAl–undecenoate, CuAl–undecenoate, MgAl–N-(2-(5,5-dimethyl-1,3,2-dioxaphosphinyl-2-ylamino)-*N*- hexyl) formamide-2-propenyl acid (MgAl-DPHPA), MgAl–undecenoate, CoAl–undecenoate, CaFe–undecenoate, NiAl–undecenoate, MgAl–CO_3, MgAl–PO_4, and MgAl–NO_3, have been studied. Most of them are intercalated with a variety of organic anions containing long hydrocarbon chains. Nyambo, Songtipya, et al. (2008) synthesized MgAl-LDH via a co-precipitation method, modified it with different alkyl carboxylate modifiers with various chain lengths, and investigated their flame retardant effects on poly (methylmethacrylate) (PMMA). Mg/Al-NO_3 LDHs were intercalated with decanoate ($C_{10}H_{19}O_2$ or C10), laurate (C12), myristate (C14), palmitate (C16), and stearate (C18) through an anion-exchange process.

A similar study was conducted by Nyambo and Wilkie (2009), in which the effect of organo-modified LDHs on the combustion properties of LDH–PMMA composites was investigated. In this study, the recorded cone calorimetry results showed that the PHRR was significantly decreased to 51% in the polymer composites containing organically modified MgAl LDH compared with unmodified MgAl–CO_3, MgAl–NO_3, and calcined LDH-PMM composites, which showed only a 30% reduction.

Enhanced thermal and combustion properties might be attributed to the improvement in the physical interactions and good nano-dispersion. Nyambo, Songtipya, et al. (2008) used long-chain linear alkyl carboxylates ($CH_3(CH_2)_nCOO^-$) with alkyl numbers of n= 1, 4, 8, 10, 12, 14, 16, and 20 to intercalate MgAl LDHs as flame retardant additives for PMMA. They discovered that the PHRR decrement improved by increasing the loadings of the LDH. A similar result was exhibited in that the PHRR of the composite was reduced by 46% at 10 wt% LDH loading of MgAl–DBS LDH. The flame retardant properties of undecenoate intercalated ZnAl, MgAl, CoAl, NiAl, CuAl, and CaAl LDHs were qualitatively investigated to compare the role of the divalent metal elements on the fire properties of PMMA. The reduction in PHRR at 6 wt% LDH loading can be summarized as follows: CaAl (36%) ¼ NiAl (36%) > MgAl (24%) > ZnAl (16%) > CoAl (7%) > CuAl (0%). Therefore, it can be concluded that both the trivalent and divalent metals used for the synthesis of LDH influence the combustion of PMMA, and higher flame retardancy efficiency can be achieved via organic modification.

The lack of an IR absorbance peak of NO_3 groups at 1384 cm^{-1} confirmed the completion of the anion-exchange process. An increase in the "d" spacing by increasing the alkyl chain lengths of the carboxylate anion in the XRD studies on the PMMA/modified LDH composites made by melt compounding revealed that more organophilic LDHs lead to greater expansion of the interlayer galleries to a greater distance by enhancing polymer intercalation.

LDHs modification resulted in an enhancement to the thermal stability and residual char formation of PMMA at any filler loading. However, it was observed that the residual char formation was greater than the expected amount based on the loaded LDH. From the above, it can be concluded that residual char was not only the residue of mixed oxide but also may lead to additional mechanisms of generation of carbonaceous char formation in the PMMA matrix. Cone calorimeter records demonstrate a decrease between 49% and 58% in the PHRR of PMMA for all types of alkyl carboxylate-modified LDHs at 10 wt% incorporation.

In some investigations, it has been demonstrated that the PHRR reduction with LDHs is higher than that with mineral fillers such as MH and ATH. However, some studies showed that in the case of PMMA, incorporation of 10 wt% nanodispersed LDH gives a similar amount of reduction in the PHRR as the 10 wt% loading of ATH or MH. This phenomenon indicates that further investigation is required to explain such similarities.

Manzi-Nshuti et al. (2008) studied the effect of divalent metal cations in LDHs on the flame retardant properties of PMMA by using various LDHs with aluminum as the trivalent cation and cobalt, zinc, or nickel as divalent cations. The mentioned LDHs were synthesized via co-precipitation and mixed through melt blending with PMMA. Cobalt-containing LDH demonstrated the best flame retardant efficiency among the LDHs in the PMMA matrix. The PHRR values showed 41%, 26%, and 16% reductions for Co/Al-LDH, Zn/Al-LDH, and Ni/Al-LDH, respectively. In addition, a better dispersion state was reported for cobalt-containing LDH/PMMA nanocomposites compared with other nanocomposites. However, the superior flame retardant efficiency of Co/Al-LDH over other LDHs has not been clearly explained. In another study, the role of trivalent metals in LDHs was compared. In this investigation, incorporation of 10 wt% Ca/Al-LDH leads to a 54% reduction in the PHRR of PMMA, while in the case of Ca/Fe-LDH, only a 34% reduction has been reported (Manzi-Nshuti et al. 2009), which is attributed to the different dispersion states of these LDHs. Nyambo et al. synthesized intercalated Mg/Al-LDH by using various benzyl anions of differing functionalities (i.e., phosphonate, sulfonate, and carboxylate) via the co-precipitation method, and their flame retardant properties in PMMA were characterized (Nyambo et al. 2009b). The highest decrement in the PHRR of PMMA was achieved by incorporation of 10 wt% benzoic acid-modified Mg/Al-LDH (46%), while benzene sulfonic acid-modified Mg/Al-LDH demonstrated only a 26% reduction in the PHRR. However, even a comprehensive characterization of all the available results does not provide a clear correlation between different factors, such as LDH dispersion state through the polymer matrix, its chemistry, PHRR, and char structure.

4.4.1.4 Ethylene-Vinyl Acetate Copolymers (EVA)

EVA has been extensively used in a broad range of applications, including hot melt adhesives, wire and cable insulation, carpeting, and packaging. In general, EVA copolymer resins can also meet the requirements of electrical applications that require great processing ability, excellent heat resistance, and good resistance to environmental stress cracking. Similar to many other polymeric materials, the flammability of EVAs is known as their major drawback. Therefore, flame retardancy is an important requirement for EVA in most of the abovementioned applications, specifically in cable insulation, which is usually obtained through the addition of conventional flame retardants. LDHs with a variety of chemical compositions have been used to improve the thermal stability and flame retardancy of EVA and have been characterized extensively. In one of the earliest reports, Camino et al. (2001) demonstrated the flame retardant efficiency of LDHs on EVA. In this study, it is reported that Mg/Al LDHs (which are naturally occurring hydrotalcites) have a significant effect on the ignitability and combustion behavior

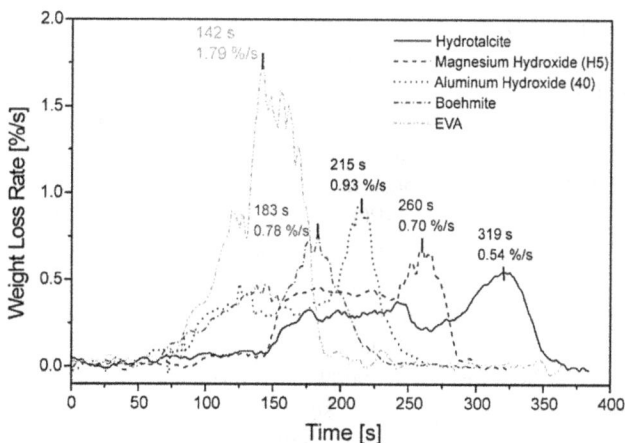

FIGURE 4.5 Mass loss calorimeter curves of EVA and EVA composites obtained with an incident heat flux of 30.3 kW/m^2.

Source: Reprinted from Camino, G., et al. *Polym. Degrad. Stabil.*, 74, 457–464, 2001.

of EVA. Figure 4.5 exhibits a comparative study on the mass loss rate among hydrated fillers, including aluminum trihydrate (ATH), magnesium hydroxide (MH), and hydrotalcites. The hydrotalcites showed the lowest mass loss rate during the mass combustion calorimetry tests, which were performed at an incident heat flux of 30.3 kW/m^2. Moreover, the time to ignition (t_{ig}) of the EVA/hydrotalcite composites was recorded at 150 s, which is almost two times that of neat EVA (78 s). However, it should be noted that, regardless of the filler type, at least 50 wt% filler loading was required to achieve significant flame retardant properties on EVA.

Conventional fillers such as MH and ATH can endothermally release water during decomposition. Apart from significant heat absorption, the released water vapor also dilutes the flammable gases during pyrolysis, thereby inhibiting flame spread in burning materials. LDHs have been reported to have significant quantities of water in the interlayer and can act in a similar way to retard flame spread. In addition, they can also form a mixed oxide residue upon burning, leading to the formation of an insulative char layer that serves as further protection for the underlying polymer. The enhanced flame retardant effect of LDHs can be attributed to this combined physical and chemical action. Ye and Qu (2008) investigated the flame retardant effects of Mg/Al-PO$_4$ LDH in EVA and compared it with Mg/Al-CO$_3$. Phosphate-intercalated Mg/Al LDH was prepared by anion exchange of Mg/Al-NO$_3$ LDH and incorporated into EVA by melt compounding. Mg/Al-PO$_4$ significantly influenced the thermal degradation and residual char formation in EVA. The thermal degradation of EVA is a two-step process. The first weight loss step occurs at 250–420°C and is due to the evolution of acetic acid, while the second weight loss step occurs between 420 and 600°C and corresponds to the degradation of ethylene chains in EVA. Incorporation of 60 wt% Mg/Al-PO$_4$ did not enhance the thermal stability of EVA but resulted in 40–50% residual char. Phosphorous-containing compounds are known to catalyze the conversion of organic matter to charred layers by the formation of P-O-P and P-O-C complexes during combustion. This enhanced char formation of Mg/Al-PO4 LDH can be expected to confer better flame retardance to EVA when compared to Mg/AlCO$_3$ LDH. As expected, phosphate intercalated LDH resulted in a more compact char layer compared to Mg/AlCO$_3$, thereby efficiently protecting the underlying polymer material from burning rapidly. The PHRR of EVA during combustion in a cone calorimeter was reduced by 76% by incorporating 60 wt% Mg/Al-PO$_4$, while incorporating a similar amount of Mg/AleCO$_3$ LDH caused a reduction in the PHRR of only approximately 66%. The heat released during burning of a material is a driving force for further burning and can make the combustion self-sustaining. Hence, PHRR has been found to be the most important parameter to evaluate fire safety (Gilman 1999) and represents the point in a fire where the heat generation is sufficient to propagate the flame or ignite adjacent material. Therefore, a significant reduction in the PHRR by the addition of Mg/Al-PO$_4$ LDH indicates inhibition of the combustion process and factors associated with rapid flame spread. Phosphate intercalated LDH also revealed a better flame retardant performance than Mg/AleCO$_3$ LDH in terms of reduction in mass loss rate and delayed ignition. LDHs with intercalated borate anions were also found to be better flame retardants in EVA when compared to conventional flame retardant fillers such as ATH, MH, or zinc hydroxide (ZH). Nyambo and Wilkie (2009) found that at 40 wt% loading, borate-intercalated Mg/Al and Zn/Al LDHs showed a significantly higher reduction of the PHRR (74% and 77%) in EVA/LDH composites than those consisting of ZH (47%), MH (65%), and their combinations. In another study, Jiao (2006) compared the flame retardant effects of nano MH and nanohydrotalcite (both treated with stearic acid with a median particle size of 50 nm) in EVA. While EVA composites with 150 phr of MH nanoparticles had a limiting oxygen index (LOI) of 40, hydrotalcite nanoparticles at a similar loading had an LOI of approximately 43. Therefore, we can infer that ZnAl, MgAl, and ZnMgAl LDHs are all promising flame retardant additives for EVA.

LOI is the minimum concentration of oxygen (in volume percent) that will just support combustion in a flowing mixture of oxygen and nitrogen and lead to downward burning of a vertically mounted test specimen (ASTM D2863). Since air comprises approximately 20.95% oxygen by volume, materials with LOI values below 20.95 are considered "flammable". It is suggested that materials with LOI values between 20.95 and 28 be classified as "slow burning" and those that have values >28 as "self-extinguishing" (Horrocks, Tune, and Cegielka 1988).

4.4.1.5 Polyurethane (PU)

PU is known as a versatile polymeric material that can be modified to meet the large demands of modern technologies, such as foams, coatings, composites, thermoplastic elastomers, adhesives, and fibers. PU has a broad range of chemical and physical properties. It also has very high elongation at break together with a high Young's modulus. Abrasion resistance, excellent hardness, low water absorption, and low thermal conductivity are known as other superior properties of PU. However, the major drawback of polyurethane is its flammability, as it decomposes rapidly and generates a large amount of heat during combustion. Therefore, in

TABLE 4.1

Different Types of LDHs Used to Impart Flame Retardancy to PUs

Type of LDH	References
NO_3-MgAl–LDH	(Xu, Wang, et al. 2016)
Heptamolybdate (Mo_7O_{24})-intercalated MgAl–LDH	(Xu, Wang, et al. 2016)
MgAl–LDH with nitrate in its interlayer	(Xu, Xu, et al. 2016)
MgAl–LDH-containing tripolyphosphate (P-LDH)	(Xu, Xu, et al. 2016)
Aminopropyltriethoxysilane (APTS) grafted onto the surface of P_3O_{10}–MgAl–LDH	(Xu, Xu, et al. 2016)
MgAl–LDH-loaded graphene hybrid	(Xu, Zhang, et al. 2016)
NO_3MgAl–LDH	(Li et al. 2020)
Sodium dodecyl sulfate (SDS)–modified NO_3–MgAl–LDH	(Li et al. 2020)
SC4A-modified MgAl–LDH	(Mohammadi et al. 2019)
SuBC4A-modified MgAl–LDH	(Mohammadi et al. 2019)
Lanthanum-doped MgAl–LDH (La–MgAl–LDH)	(Qian et al. 2020)
Sodium dodecyl sulfate–modified CoAl–LDH	(Gao et al. 2014)
Ammonium polyphosphate (APP)@CoAl–LDH	(Huang et al. 2019)
Stearate-intercalated MgAl–LDH	(Kotal et al. 2013)

recent years, flame retardants have been used in polyurethane (PU) to overcome its flammability (Visakh et al. 2019, Soni and Bhatt 2022, Yadav et al. 2022). To date, different types of LDH have been employed in the preparation of flame retardant polymers. Table 4.1 shows different types of LDHs used to impart flame retardancy to PUs. One of the most common LDHs used for FR purposes is magnesium–aluminum LDH (MgAl–LDH). In fact, its accessibility and ease of synthesis are the main reasons for the abundant use of LDH in different studies. The effect of LDH on the decrease in PHRR is significant even at a low loading. However, to achieve a higher reduction in the PHRR values, it is necessary to modify LDH with some chemicals as well as its physical combination with conventional flame retardants, particularly phosphorus ones. A large number of investigations have been conducted for this purpose. The mentioned works will be summarized in two main sections: chemically modified LDHs and combinations of LDH and conventional FR systems to enhance the flame behavior of PU (Vahabi et al. 2021).

- **Chemically modified LDH used in PUs**

Chemical modification of LDH is usually employed to enhance its dispersion state in the PU matrix or enhance the flame retardant properties. For the latter, the chemicals used for the intercalation of LDHs are commonly nitrogen-, phosphorus-, or silicon-based molecules. Xu, Wang, et al. (2016) showed that the incorporation of 10 wt% MgAl–LDH and Mo–MgAl–LDH into pure PU significantly reduced the PHRR values to approximately 52 and 69%, respectively, in the cone calorimeter test. Similar to other polymers, the barrier effect and dilution of volatiles in the gas phases clarified the enhancement in flame retardant properties of PU in the presence of the mentioned LDHs. Xu, Wang, et al. (2016) synthesized modified LDHs following three different synthesis techniques to improve the flame retardancy of PU elastomers: (1) the ion exchange technique to prepare tripolyphosphate ($P_3O_{10}^{5-}$) modified MgAl–LDH (P-LDH), (2) the hydrolysis silylation method to graft aminopropyltriethoxysilane (APTS) onto the surface of P-LDH and to prepare APTS grafted onto the surface of P-LDH (S-LDH), and (3) the hydrothermal treatment method to obtain nitrogen-containing MgAl–LDH with nitrate in its interlayers. 1, 3, and 5 wt% of these synthesized LDHs were used to prepare in situ polymerization of PU elastomer nanocomposites. According to the cone calorimeter test results, 58 and 70% reductions in the PHRR values were achieved when 5 wt% of P-LDH and S-LDH were loaded into PU elastomers. The analysis of residual char of the previously prepared PU elastomer nanocomposites after the cone calorimeter test showed the formation of a graphitic crystalline char that acted as a barrier during combustion instead of a disordered graphitic structure. Moreover, the presence of some stable structures, such as –P(=O)–O–Si– and –P(=O)–O–C–, had a beneficial effect on the thermal oxidative resistance of the residual char layer and enhanced the flame retardant properties of PUE.

Huang et al. (Huang et al. 2020) studied the effect of incorporation of a phosphoryl polyethyleneimine amide-modified LDH (PPEIA)–LDH on the flame retardant properties of TPU. It has been demonstrated that the incorporation of 6 wt% PPEIA–LDH resulted in an increase in the LOI of 29 and a dramatic reduction in the PHRR to 228 kW/m². The main reason for the enhancement in the flame retardant properties was the synergistic effect between LDH and PPEIA in the condensed phase mechanism, which improved the barrier effect.

- **Combination of LDH and conventional FRs in PUs**

Gao et al. (2014) prepared a flame retardant package composed of diethyl ethyl phosphonate (DEEP), EG, and montmorillonite (MMT) or LDH and studied the flame retardant effect of these packages on the flame retardant properties of rigid polyisocyanurate-PU (PIR) foams. The authors changed the weight ratio between DEEP and EG (fixed at 45 php) and changed the percentage of MMT or LDH from 1 to 5 php. According to the LOI test results, the samples that showed the highest LOI percentage were selected for cone calorimeter testing. The HRR curves for pure PIR and PIR containing EG and DEEP at 45 php (ratio 2:1) and LDH at 2 php or MMT at 3 php clearly demonstrate the higher efficiency of a combination of DEEP and EG on the flame retardant properties of PIR by lowering the PHRR. The comparison of the HRR curves of PIR-FR45 and PIR-FR45-LDH indicated that LDH had almost no effect on flame retardancy, as the PHRR values were similar and the THR values were shifted to higher amounts for the sample containing LDH.

Gómez-Fernández et al. studied the effect of the combination of lignin and Exolit OP560 with LDH (EX) on the flame retardant properties of synthetic flexible PU foams (FPUFs) (Gómez-Fernández et al. 2018). The LOI results showed that incorporation of LDH in combination with lignin had no effect on the flammability of the FPUF. However, the mentioned combination demonstrated some efficiency during the cone calorimetry tests, in which incorporation of LDH at 3 php in combination with lignin and EX both at 5 php resulted in the reduction of PHRR to 645 kW/m^2, compared with the sample containing lignin and EX, which was 851 kW/m^2.

Ammonium polyphosphate (APP) is a well-known conventional flame retardant (Lim et al. 2016). To date, several authors have reported its synergistic effect in combination with LDH on the flame retardant properties of a variety of polymers (Liu et al. 2021, Nyambo, Kandare, et al. 2008). However, it is also shown that the incorporation of APP into polymers in general results in deterioration of the mechanical properties due to the poor interfacial adhesion between the polymer matrix and APP. To overcome this problem, Huang et al. (2019) prepared a modified APP in which LDH nanosheets were deposited electrostatically on the surface of micronized APP (APP@LDH). In this manner, the author achieved the benefit of the synergistic effect while maintaining the mechanical properties. In the first step, CoAl–LDH was dispersed in water and then loaded into a suspension containing APP under stirring. Afterward, the precipitated particles were centrifuged and dried (Figure 4.6). APP@ LDHs were prepared by combining different ratios of APP and LDH: 2.6:0.4 (APP2.6@LDH0.4), 4.3:0.7 (APP4.3@LDH0.7), and 6:1 (APP6@LDH1). Finally, the modified APP (APP@LDH) and the physical mixture of APP and LDH without treatment, with a mixing ratio of 6:1 [APP6/LDH1], were incorporated into TPU at 190°C via melt blending, and the obtained samples were prepared via an injection-molding machine.

The vertical burning UL-94 test showed that all the samples achieved a V-2 rating, and in the case of the TPU sample containing APP6@LDH1, V-0 was obtained. The LOI value of pure TPU was 22, and incorporation of APP@LDH additives led to an increase in the LOI values to 28% for TPU/APP2.6@LDH0.4, 28.7 for TPU/APP4.3@ LDH0.7, and 29 for TPU/APP6.0@LDH1.0. The sample containing physically mixed APP/LDH showed an LOI value of 28%. Based on the results recorded by cone calorimeter tests, the best performance was observed for

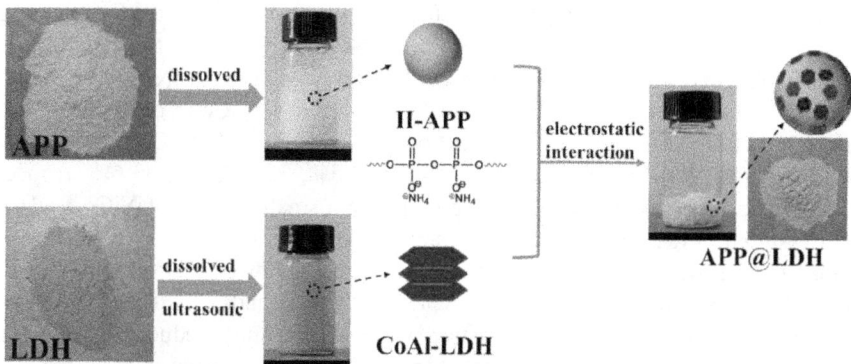

FIGURE 4.6 Preparation method of APP coated at the surface with CoAl–LDH (APP@LDH).

Source: Reprinted from Huang, S. C., et al. *Polym. Degrad. Stabil.*, 165, 126–136, 2019.

the sample containing APP6.0@LDH1.0, in which the PHRR was reduced to 219 kW/m^2, compared with 282 kW/m^2 for the TPU/APP6.0/LDH1.0 sample. These results indicated that APP modified with LDH showed better performance than did its simple mixing to enhance the flame retardant properties due to the formation of a more stable residual char layer during combustion by the accumulation of LDH nanosheets on the surface of the polymeric sample.

4.4.1.6 Others

Zhang et al. (2004) investigated the effects of MgAl, ZnAl, and CaAl LDHs on polystyrene (PS). The flame retardant performance of PS after incorporation of MgAl–CO$_3$ was evaluated. The results showed that incorporation of 50 wt% LDH led to an increase in the LOI values from 19 to 27% and a significant decrease in smoke production compared to neat PS. Costache et al. (2007) prepared a nanocomposite by combining undecenoate intercalated ZnAl–LDH and PS and demonstrated that a 5 wt% loading of LDH showed a 35% reduction in the PHRR values. In another project, benzoate intercalated CaAl–LDH was synthesized, and 10 wt% of the prepared LDH was loaded into the PS matrix via two different processing methods: in situ bulk polymerization and melt-blending. The presence of the modified LDH enhanced the flame retardant properties, with PHRR reductions of 42% and 24%, respectively (Matusinovic, Lu, and Wilkie 2012).

ABS resin is a famous member of the thermoplastic polymer family that has a wide range of industrial applications; however, similar to other thermoplastics, it also severely suffers from flammability and, if ignited, can generate a great amount of toxic gases and smoke. In some studies, MgAl-LDH and ZnMgAl-LDHs have been incorporated into the ABS matrix as flame retardants, leading to a significant decrease in the PHRR results. In addition, a higher LOI value was obtained by using ZnMgAl LDHs and a longer combustion time, demonstrating that ZnMgAl LDHs possess a better flame retardant effect in the ABS matrix (Xu et al. 2012).

In another project, a hierarchical 3D NiCo-LDH@PZS hollow dodecahedral structure was prepared to improve the fire safety of EP composites by lowering their flammability, smoke, and emission of toxic gases during the combustion process. The PHRR and THR of EP/NiCo-LDH@PZS4.0 are decreased by 30.9% and 11.2%, respectively, compared to those of pure EP. Cone calorimeter test results demonstrated that the total yield of CO$_2$ and CO and the O$_2$ consumption of EP/NiCoLDH@PZS4.0 were reduced by 64.5%, 32.4%, and 33.6%, respectively. It was also observed that incorporation of a small amount of NiCo-LDH@PZS resulted in achieving high efficiency of the flame retardant material in the smoke suppression of the EP composite. The results showed that the amounts of aromatic compounds, nitric oxides, carbonyl compounds, hydrocarbons, and oxycarbide released during combustion were significantly reduced during the combustion of EP/NiCo-LDH@PZS. Moreover, the maximum release concentrations of HCN of EP/NiCo-LDH4.0 decreased by 87.8%. From the obtained results, it can be concluded that the fire retardant properties of the EP/NiCo-LDH@PZS composites greatly improved due to the coefficient effect of NiCo-LDH and PZS (Zhou et al. 2019).

Apart from the polymers discussed above, many other polymers, such as polyvinyl chloride (PVC) (Bao et al. 2008), polylactic acid (PLA) (Shan et al. 2012), ethylene propylene diene monomer (EPDM) (Huang et al. 2012), and poly(butyl acrylate-vinyl acetate) (P(BA-VAc)) (Zhao et al. 2011), have also been investigated by using different LDHs, and all exhibited enhanced flame retardant properties.

4.4.2   APPLICATION OF LDHs IN FIRE-SAFE THERMOSETTING POLYMERS

4.4.2.1   Epoxy Resins

In a systematic investigation, Zammarano et al. (2005) compared ATH, montmorillonite (MMT), and Mg/Al-LDH modified with different organic anions, such as 4-toluenesulfonic acid monohydrate (TS), aminobenzene sulfonic acid (AS), or 4-hydroxybenzenesulfonic acid. In this study, he showed for the first time that incorporation of 5 wt% LDHs resulted in a self-extinguishing behavior to epoxy in the UL 94 horizontal burning test, while montmorillonite or ATH only demonstrated a slight reduction to the flame spread rate in epoxy. The above fillers were mixed with epoxy under different conditions and cured with the proper curing agents. Among the different fillers studied, only LDHs modified with organic anions and organically modified montmorillonite (Cloisite 30B) provided an intercalated nanocomposite structure, while ATH, unmodified LDHs and APP formed microcomposites. The flame retardant properties of such materials are revealed to be a complex interaction of the dispersion state of fillers, intrinsic properties of the additives, their mode of action, etc. The excellent flame retardant properties of organo-modified LDHs in polymer matrices are due to the effect of mixed metal oxide residue (confirmed by X-ray diffraction analysis) and the release of water during combustion. The improved flame retardant efficiency of organo-LDH/epoxy was also recorded through cone calorimeter tests, where a 40–51% decrease in the PHRR was observed by loading 5 wt% organo-LDH into the epoxy, while at a similar loading, organically modified MMTs exhibited only a 27% reduction in the PHRR values. The difference in the performances of the above-mentioned samples has been proven with a more intumescent and compact residual char obtained from LDH/epoxy compared with fragmented char residues obtained in the case of the montmorillonite/epoxy system (Figure 4.7). Therefore, epoxy- and organic sulfonate (TS and AS)-modified LDHs evolve into an intumescent system while burning, where the epoxy matrix acts as a char source, CO_2 and water (evolved during the thermal decomposition of hydroxyl layers) act as blowing agents, and sulfonate serves as the charring agent.

4.4.2.2   Unsaturated Polyester Resin (UPR)

UPR is the most commonly used thermoset resin for the manufacture of a wide assortment of products, naval constructions, offshore applications, pipes, tanks, sanitary ware, chemical containers and high-performance components for the building, marine, electric and automotive industries. High thermal and heat stability, low shrinkage, and excellent mechanical properties are typical for their polymers. Those applications, in addition to the mechanical and electrical properties, also

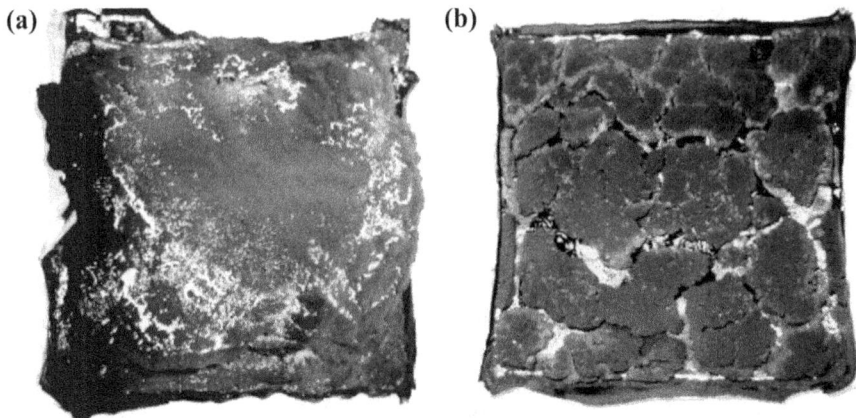

FIGURE 4.7 Images of carbonaceous residue obtained after cone calorimeter tests: (a) Compact and intumescent char of epoxy-LDH nanocomposite; (b) fragmented char of epoxy- montmorillonite nanocomposite at 5 wt% loading.

Source: Reprinted from Zammarano, M., et al., *Polymer*, 46, 9314–9328, 2005.

require good flame retardants (Haq 2007). Therefore, suppressing the flammability of polyester and epoxy resins is currently one of the most serious and difficult challenges for state-of-the-art technologies for the production of these materials (Pereira 2009).

Xie, Wang, and Zhang (2011) synthesized *p*-toluene sulfonate- pillared layered double hydroxide (PTS-LDH) by a co-precipitation method, and unsaturated polyester resin/PTS-LDH composites were prepared by solution blending and solidification. The addition of PTS-LDH improved the flame retardancy of the unsaturated polyester resin. The limit oxygen index of the unsaturated polyester resin/PTS-LDH composite increased to 23.30% from 21% of the pure resin, the char residue rate increased by 3.7%, and the speed of horizontal burning decreased by 20.24%. The vertical burning test (UL-94) results indicated that the materials achieved V-2 grade.

4.4.3 LDHs for Fire-Safe Coating Applications

Almost all the reviewed studies during the recent years mentioned in this chapter deal with LDH as an additive loaded into bulk polymer matrices. However, it should be noted that LDH was also employed in the form of coating flame retardant systems for different polymers. The flame retardancy of PU foam was investigated by Yang et al. (2015) using a layer-by-layer (LbL) assembly technique via a multicomponent system containing poly(acrylic acid), branched polyethyleneimine, sodium montmorillonite (Na-MMT), and layered double hydroxide (LDH). Several samples were prepared employing the mentioned materials and by bi-, tri-, and quad-layer LbL coating progression. The authors discovered that a five-bilayer poly (acrylic acid)/branched polyethyleneimine/LDH coating reduced both the PHRR and the THR by 40% and 70%, respectively. They also demonstrated that the combination of LDH and Na-MMT decreased the flammability of PU and resulted in a decrease in the mass of the coating materials. Liu, Wang, and Hu (2015) also used the LbL deposition method to improve the flame retardancy of flexible PU foams (FPUFs). They impregnated FPUF sequentially in three solutions containing chitosan, MgAl–LDH, or NiAl–LDH, and alginate. These three layers were replicated 3, 6, or 12 times. The cone calorimetry results demonstrated that the PHRR was reduced by increasing the number of layers. Therefore, without considering the type of LDH, the best performance was obtained for the samples containing 12 layered coatings. It was also investigated that NiAl–LDH showed a greater effect on the reduction of PHRR than MgAl–LDH, and the physical barrier effect was proposed regarding their mechanism of action.

APP, pentaerythritol (PER) and melamine (EN) are the most common flame retardants used in conventional intumescent flame retardant coating systems. The APP/PER/EN system generates thick protective char layers; however, generally due to their mechanically weak char structure in fire, they easily detach from the substrate, hence losing their efficiency. Although conventional APP/PER/EN coatings showed great intumescent properties, their chemical and physical properties and their char layer structure must be improved (Reshetnikov, Yablokova, and Khalturinskij 1997). A recent study has shown that nano-additives can enhance the chemical and physical properties of coatings. However, fewer studies have been performed on the effect of nano-size additives or flame retardants on intumescent flame-retardant coating systems. Magnesium aluminum-layered double hydroxide (MgAl-LDH) nanosheets are able to absorb heat and release H_2O and CO_2 during combustion, which decreases the temperature of the substrate and improves the char structure. Thermal decomposition products of nano-LDHs form a porous structure possessing a large specific surface area, which provides smoke suppression effects by absorbing volatile gases and smoke during combustion.

Wang, Han, and Ke (2005) demonstrated that incorporation of only 1.5% of LDHs resulted in the formation of efficient interpenetrating polymer networks and could significantly enhance the flame retardant properties of the coating and improve the mechanical properties. The resistance of the substrate to fire is the key factor in the transference speed of heat through the char layer. Therefore, the char formation process, foam structure, insulation efficiency and effect of the intumescence of the char layer can greatly influence the general flame-retardant properties of the coating. The internal and external surfaces of the intumescent char layer of the flame-retardant coating generate well-distributed small close cells, with an internal diameter of 10–30 μm and a wall thickness of 2–3 μm. During the combustion of the coating in the second stage, high-pressure gas charged the char layer formed in the first stage. The process above blows the char layer and results in the formation of an inflated, thick, and holey structure. Although the intumescent char layer of the flame-retardant coating can also generate some close holes without nano-particles, many open holes and cracks in the char layer of the coating can appear due to the low viscosity of the molten coating and more volatile gas in some parts of the coating. In such a manner, air in small holes can decrease the efficiency of heat transference, but the air convection phenomenon can increase the efficiency of heat transfer if the volume of the small hole is too large. Heat degradation of resin and flame retardant can be accelerated by increased heat transference, which damages the expansion process. It is demonstrated that the size of honeycomb-like small holes should not go beyond 50 μm to achieve a good char layer.

Wang et al. (2018) prepared multifunctional coatings with superhydrophobicity and flame retardancy on medium-density wood fiber composites (commercially known as MDFs) by deposition of polydimethylsiloxane (PDMS) and 1H, 1H, 2H, 2H-perfluoro decyltrichlorosilane (FDTS)-intercalated Mg/Al layered double hydroxide (LDH) particles via a stepwise procedure. The PDMS@FDTS-Mg/Al LDH coating demonstrated superhydrophobicity possessing a water contact angle of ~155° and superior self-cleaning properties. The limiting oxygen index (LOI) value of the superhydrophobic MDFs showed a 60.4% increase compared to that of the pristine MDFs. In addition, the PHRR and THR of MDFs coated with PDMS@ FDTS-Mg/Al LDH were reduced by 24.7% and 11.2%, respectively, in comparison with MDFs alone, exhibiting the positive effect of the flame retardant coating on the MDFs.

4.5 FLAME RETARDANT AND SMOKE SUPPRESSION MECHANISMS OF LDHS

In fact, the layered structure and unique chemical composition of LDHs are the origin of their excellent flame retardancy and smoke suppression properties. In general, the barrier properties of two-dimensional layered nanomaterials are widely proven to be their main flame retardant mechanism for polymeric nanocomposites. Moreover, the incorporation of layered nanomaterials results in the formation of carbonaceous residual char due to their carbon donor properties, hence increasing the residual char and therefore generating an insulating char layer between the unburnt and burnt structures during the combustion process (Dong et al. 2016). LDHs can enhance the flame retardancy of polymeric materials by generating refractory oxide residuals on the external surface of the material and releasing carbon dioxide and water vapor during polymer combustion (Liu et al. 2016). In addition, the incorporation of LDHs can not only catalyze the formation of residues in the thermal decomposition process of carbonaceous materials but also enhance the thermal stability of the polymer composites (Wang et al. 2017).

As mentioned previously, solvent blending (Edenharter et al. 2016), melt blending (Nevare et al. 2014, Wang et al. 2014) and in situ polymerization (Hou et al. 2018, Benaddi, Benachour, and Grohens 2016) are common incorporation methods of LDHs into the polymer matrix. However, the high surface energy of LDHs results in a great tendency to aggregate (Liu et al. 2017). On the other hand, the poor hydrophobicity of LDHs leads to severe aggregation, which greatly influences the mechanical and flame retardant properties of nanocomposites (Guo et al. 2017).

Similar to the cationic clays, intercalation of organic anions into the interlayer gallery distances of LDH layers leads to modification of LDHs. Such organic-modified LDHs can be employed as nanofillers for the preparation of polymer–LDH nanocomposites (Manzi-Nshuti et al. 2009, Leroux and Besse 2001). The barrier mechanism and radical trapping are known as the major flame retardancy mechanisms for silicate clay-containing nanocomposites. However, no specified mechanism has been proposed for the flame retardant mechanism of LDH nanocomposites (Zhu et al. 2001, Nyambo et al. 2009a), and based on a recent study, it is suggested that the flame retardant mechanism of LDHs is not similar to that of silicate clay-based nanocomposites (Nyambo, Songtipya, et al. 2008). LDHs can release their interlayer water during combustion, while decomposition of the intercalated anions (e.g., NO_3^- and CO_3^{2-}, etc.) and the metal hydroxide network will generate water and other gases (such as CO_2). The existence of the mentioned gases for these processes will dilute the volatiles and the "fuels" available for combustion. In such a way, they can decrease the heat release and ultimately lead to disruption of propagation reactions due to lack of the fuel and, hence, stop the combustion and promote the generation of a continuous and compacted carbonaceous char. The aforementioned residual char can significantly protect the underneath bulk polymer from exposure to oxygen and heat while hindering supplying the fuel to the combustion zone, thus minimizing the heat release and suppressing smoke production (Zammarano et al. 2005, Chen et al. 2002). In general, the flame retardant mechanism of layered double hydroxides can be described by the combination of the following three functions: (1) absorption of heat, (2) dilution of volatile gas, and (3) formation of char. As a result, the mass loss rate (MLR) will be reduced by a combination of the abovementioned functions. MgAl LDH was used as a flame retardant for ethylene vinyl acetate (EVA) (Camino et al. 2001). According to the results, he observed good flame-resistant efficiency. Thermogravimetric analysis results demonstrated three clear endothermic incidents at 207°C, 291°C, and 416°C during the thermal decomposition of the LDH, which can be attributed to the decomposition of interlayer CO_3^{2-}, the loss of interlayer H_2O, and the decomposition of hydroxyl groups of the pristine LDH layers. Moreover, MgO and $MgAl_2O_4$ were generated at 500°C and 850°C, respectively, which enhanced char formation (Becker et al. 2011). Zhao et al. (2008) prepared and characterized residual poly (vinyl alcohol)/ammonium polyphosphate/LDH (PVA/APP/LDH) nanocomposites and demonstrated that LDH can improve char formation. The abovementioned mechanism is proposed in Figure 4.8a. One hypothesis is that the LDH layers might react with poly(phosphoric acid), releasing H_2O molecules and forming bridges between APP chains. The generation of the mentioned bridges will lead to a stabilization of APP, hence a reduction in the volatility of phosphorus, and thus, retention of more APP, which is required for the phosphorylation and char formation process. Such crosslinks can increase the polymerization degree of poly (phosphoric acid), which results in an increase in the viscosity of the melt during pyrolysis and combustion and hence improves the formation of a dense and compact char layer. Another hypothesis (Figure 4.8b) is that during the combustion process, PVA chains undergo dehydrogenation, consequently forming double bonds on the backbone, and LDHs can act as catalysts for the cyclization reaction to produce char precursors.

In fact, LDH can enhance both flame retardant and smoke suppression while increasing the rheological and mechanical properties, photo-stability and IR absorption properties of the final nanocomposites (Chai et al. 2008, 2009, Guo et al. 2010, Zhang et al. 2007, Zhu et al. 2011). Moreover, the chemical structure of LDHs can be precisely adjusted, which provides great purity levels for advanced

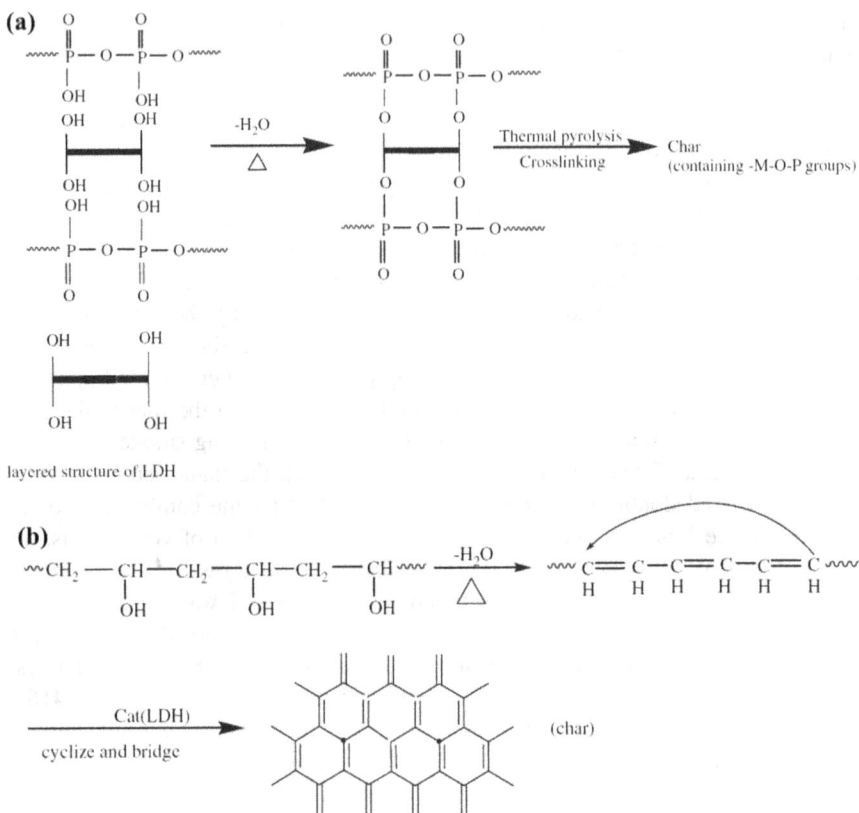

FIGURE 4.8 Schematic of a possible mechanism for char formation using an LDH.

Source: Reprinted from Zhao, C. X., et al., *Polym. Degrad. Stabil.*, 93, 1323–1331, 2008.

applications. In addition, LDHs are increasingly considered as a new potential green flame retardant for polymer applications.

There are many anions that exist in the interlayer gallery distance of the LDH layers due to their positive charge on its surface, hence providing the possibility for modification. Co-precipitation, reconstruction and ion exchange are different methods to intercalate and modify LDHs (Wang, Zhang, and Zhou 2017). Anionic surfactants are the most widely used organic modifiers, such as dodecylsulfate (DDS) (Kumar et al. 2018, Qiu et al. 2018) and sodium dodecyl benzene sulfonate (SDBS) (Guan et al. 2014, Jiang et al. 2018). The interlayer spacing of LDHs can be enlarged via the solution intercalation method, which leads to changing the surface properties of LDHs from hydrophilic to hydrophobic, which can enhance the compatibility between polymer matrices and LDHs. As an example, DDS-intercalated NiAl–LDH nanofillers were synthesized by following the co-precipitation method coupled with the microwave hydrothermal treatment by Wang, Li, Zhang, et al. (2012), and the prepared nanomaterials were compounded with EVA. DDS-intercalated Ni/Al–LDH demonstrated a high dispersion state in the EVA matrix, and the polymer nanocomposite

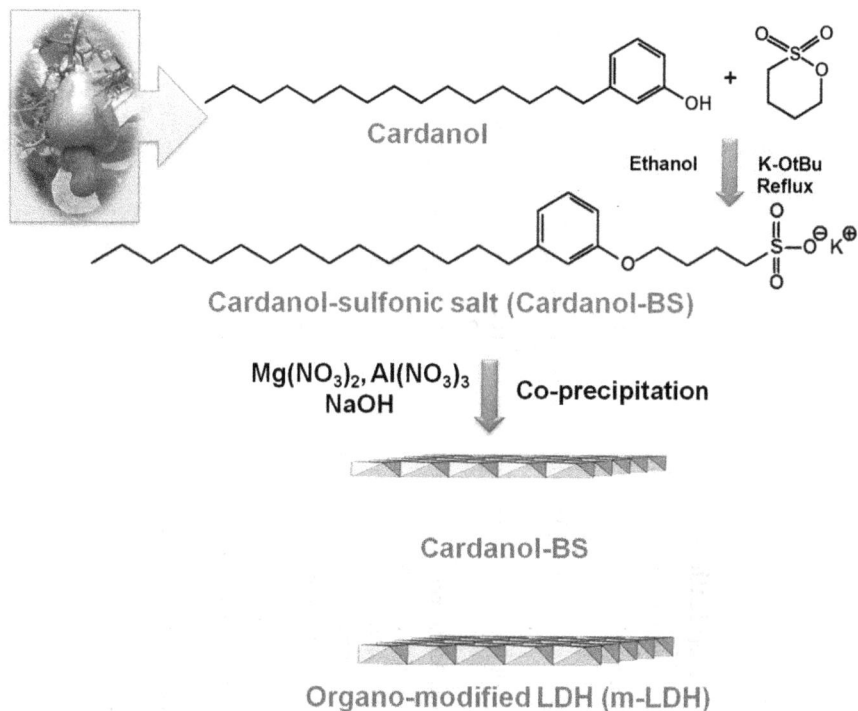

FIGURE 4.9 Diagrammatic illustration of the synthetic route of cardanol-BS modified LDH.

Source: Reprinted from Wang, X., Kalali, E. N., and Wang, D. Y., *ACS Sustain. Chem. Eng.*, 3, 3281–3290, 2015.

showed improved tensile strength and reduced PHRR compared to the specimens prepared by Na_2-CO_3-intercalated Ni/Al-LDH. In another project, cardanol as a kind of renewable resource was employed to synthesize a bio-based modifier (cardanol-BS) through the ring-opening of 1,4-butane sulfonate (BS) (as indicated in Figure 4.9). The EP composite exhibited a UL-94 V0 rating and reached a limiting oxygen index (LOI) of 29.2% with only the incorporation of 6 wt% of modified LDH. The PHRR, total smoke production (TSP) and total heat release (THR) values were reduced by 62%, 45%, and 19%, respectively, compared to those of pure EP (Wang, Kalali, and Wang 2015).

Furthermore, other methods were developed without surfactants. It has been reported that the microencapsulation technique can be conducted to improve the compatibility between the polymer matrix and LDHs. For example, LDHs were recently microencapsulated with melamine formaldehyde (MF) resin. This microencapsulation resulted in a sharp reduction in the particle size of the modified LDHs, and the related EP-based nanocomposites showed significant improvement in flame retardant properties (Zhu et al. 2018). The synergistic effect is known as the effect of combining two or more components, which together are larger than the

TABLE 4.2

Synergistic Effect of LDHs and Other Flame Retardants and Their Applications

Matrix	LDHs and Synergistic Flame Retardant Loading	FR Properties	References
PVA	Total 15 wt% (APP = 14.7 wt%, NiAl-LDH = 0.3 wt%)	33.9% LOI value, UL-94 V0 rating	(Ding et al. 2015)
PC	Total 10 wt% (BSR = 5 wt%, NiAl-LDH = 5 wt%)	44.0% reductions in PHRR, 29.7% LOI value, UL-94 V-0 rating	(Jiang et al. 2018)
PE	16.2% Mg/Al-LDH (including SDS)	35% LOI value, UL-94 V0 rating, 44% reduction in THR	(Costa, Wagenknecht, and Heinrich 2007)
SBS	Total 15 wt% (LDH = 10 wt%, ATH = 10 wt%)	33.9% LOI value, 34.8% reduction in PHRR	(Li, Pang, et al. 2018)
P(BA-VAc)	Total 15 wt% (LDH = 0.5 wt%, APP = 14.7 wt%)	30.7% LOI value, UL-94 V0 rating	(Zhao et al. 2011)
PMMA	Total 5 wt% (including DPHPA)	45.0% and 25.4% reductions in PHRR and THR, respectively. 28.2% LOI value, UL-94 V-1 rating	(Huang et al. 2013)
PS	Total 10 wt% (5 wt%MgAl-LDH, 5 wt%APP)	42.2% and 42.8% reduction in PHRR and average MLR, respectively.	(Nyambo, Kandare, et al. 2008)
PLA	Total 20 wt% (2 wt%PWA- LDH, 18 w:% IFR)	48.3% LOI value, UL-94 V-0 rating	(Zhang et al. 2017)
PLA	Total 15 wt% (1 wt% LDH, 14 wt% DTM)	62.5% and 43% reduction in PHRR and THR, respectively, 19% LOI value, UL-94 V-0 rating	(Jin et al. 2017)
LLDPE	20% Ni/Al-LDH (including SDS)	26.5% LOI value, UL-94 V-0 rating, 50.7% reduction in THR.	(Wang, Tan, and Song 2019)
PP	Total 30 phr (APP: PER: ZnAl- LDH = 2:1:2)	31.0% LOI value, UL-94 V-1 rating	(Zhang et al. 2008)

PP	Total 20 wt%(5 wt% NiAl-OLDH, 15 wt% IFR)	27.5% LOI value, UL-94 V-0 rating, 49% and 19% reductions in PHRR and THR, respectively	(Zheng et al. 2017)
EVA	Total 30 wt%(10 wt% EG, 20 wt% LDH)	43.0% and 27% reductions in PHRR and THR, respectively. 29.7% LOI value, UL-94 V-0 rating	(Pang, Tian, and Shi 2017)
EVA	Total 20 wt% (19 wt% IFR, 1 wt% MgAl-LDH)	26.9% LOI value, UL-94 V-2 rating, 41.6% reduction in PHRR.	(Zhang et al. 2018)
EP	Total 20 wt% (15 wt% MPP, 5 wt% MgAl-LDH)	24.4% LOI value, UL-94 V-1 rating	(Kaul et al. 2017)
EP	Total 20 wt% (5 wt% GNS, 2.5 wt% LDH)	23.6% LOI value rating, 64% and 33% reductions in PHRR and THR, respectively	(Liu et al. 2016)

PVA, poly(vinyl alcohol); P(BA-VAc), poly(butyl acrylate–vinyl acetate); PS, polystyrene; PMMA, polymethyl methacrylate; APP, ammonium polyphosphate; PER, pentaerythritol; DPHPA, N-(2-(5,5-dimethyl-1,3,2-dioxaphosphinyl-2-ylamino)-N-hexylformamide-2-propenyl acid; MH, magnesium hydroxide; RP, red phosphorus; MP, melamine polyphosphate; BA, boric acid; MPP, pentaerythritol diphosphate.

sum of the effects of each of the individual components. In a project, LDH was used as a synergistic flame retardant agent for intumescent flame retardant (IFR) systems (Huang et al. 2012, Kalali et al. 2018). In another work, $NH_2SO_3^-$ intercalated Mg/Al-LDH (SA-LDH) was incorporated into ethylene vinyl acetate (EVA) in combination with IFR; the loading of only 1 wt% SA-LDH led to an increase in the LOI value of the EVA/IFR composite from 24.8% to 26.9% and a great reduction in the PHRR value (Li, Zhang, et al. 2018). The synergistic effect between other flame retardants and LDHs and their applications is summarized in Table 4.2.

It has been proven that some metal ions, such as Zn^{2+}, Fe^{3+} (Kalali, Wang, and Wang 2016), and Ni^{2+} (Hajibeygi, Shabanian, and Khonakdar 2015, Li, Zhang, et al. 2018), act as catalysts in polymer/LDH nanocomposites. Formation of a physical network in the presence of SDBS-modified Zn/Al-LDH was accelerated due to the catalytic action of Zn^{2+} metal cations as an adduct in combination with the lone pair of oxygen atom of epoxy, which results in an enhanced epoxy ring-opening process (Rastin et al. 2017). Carbonization of the original carbon layer in the intumescent flame retardant containing PP composite accelerated in the presence of Ni^{2+} and led to an enhanced cross-linking process and formation of more compact carbon layers (Kong et al. 2018).

4.6 CONCLUSIONS

Among the wide variety of nanomaterials discovered or synthesized to date, LDHs are known as two-dimensional nanostructured materials possessing layered structures and are widely used to construct functionalized additives. These materials have advantages such as friendliness, nontoxicity, biocompatibility, and low cost, and in addition, due to their unique properties, they have gained the interest of scientists regarding new applications. Moreover, LDH synthesis does not require harsh and highly complex conditions, which leads to an ease of preparation of multifunctional LDHs with controlled properties and applications. Currently, intercalation and modification of LDHs with adequate inorganic and organic compounds have enlarged their interlayer gallery distance, enhanced their physiochemical features, and also in some instances, improved their organo-phillicity, expanding their applications in advanced technologies including fire retardants for polymeric systems. Due to their tunable properties, high anion exchange capability, large surface area, and structural diversity, LDHs have been used alone or in combination with conventional flame retardants. Investigations showed that incorporation of low amounts of LDHs significantly enhanced the overall flame retardant properties of the polymer composites. Accordingly, with such findings, the author suggests employing the modified LDH at industrial scales, which is eco-friendly, cost-effective, highly efficient, and time-saving.

REFERENCES

Allada, Rama, Alexandra Navrotsky, Hillary Berbeco, and William Casey. 2002. "Thermochemistry and aqueous solubilities of hydrotalcite-like solids." *Science (New York, N.Y.)* no. 296:721–723. doi: 10.1126/science.1069797.

Anto Jeffery, A., C. Nethravathi, and Michael Rajamathi. 2014. "Two-dimensional nanosheets and layered hybrids of MoS2 and WS2 through exfoliation of ammoniated MS2 (M = Mo,W)." *The Journal of Physical Chemistry C* no. 118 (2):1386–1396. doi: 10.1021/jp410918c.

Bao, Yong-Zhong, Huang Zhi-Ming, Li Shen-Xing, and Weng Zhi-Xue. 2008. "Thermal stability, smoke emission and mechanical properties of poly(vinyl chloride)/hydrotalcite nanocomposites." *Polymer Degradation and Stability* no. 93 (2):448–455. doi: 10.1016/j.polymdegradstab.2007.11.014.

Becker, Cristiane M., Teo A. Dick, Fernando Wypych, Henri S. Schrekker, and Sandro C. Amico. 2012. "Synergetic effect of LDH and glass fiber on the properties of two- and three-component epoxy composites." *Polymer Testing* no. 31 (6):741–747. doi: 10.1016/j.polymertesting.2012.04.009.

Becker, Cristiane M., Aline D. Gabbardo, Fernando Wypych, and Sandro C. Amico. 2011. "Mechanical and flame-retardant properties of epoxy/Mg–Al LDH composites." *Composites Part A: Applied Science and Manufacturing* no. 42 (2):196–202. doi: 10.1016/j.compositesa.2010.11.005.

Bellotto, Maurizio, Bernadette Rebours, Olivier Clause, John Lynch, Dominique Bazin, and Eric Elkaïm. 1996. "Hydrotalcite decomposition mechanism: A clue to the structure and reactivity of spinel-like mixed oxides." *The Journal of Physical Chemistry* no. 100 (20):8535–8542. doi: 10.1021/jp960040i.

Benaddi, Hadja, Djafer Benachour, and Yves Grohens. 2016. "Preparation and characterization of polystyrene-MgAl layered double hydroxide nanocomposites using bulk polymerization." *Journal of Polymer Engineering* no. 36 (7):681–693. doi: 10.1515/polyeng-2015-0162.

Bertoldi, Massimo, Bice Fubini, Elio Giamello, Guido Busca, Ferruccio Trifirò, and Angelo Vaccari. 1988. "Structure and reactivity of zinc–chromium mixed oxides. Part 1.—The role of non-stoichiometry on bulk and surface properties." *Journal of the Chemical Society, Faraday Transactions 1: Physical Chemistry in Condensed Phases* no. 84 (5):1405–1421. doi: 10.1039/F19888401405.

Besserguenev, A. V., A. M. Fogg, R. J. Francis, S. J. Price, D. O'Hare, V. P. Isupov, and B. P. Tolochko. 1997. "Synthesis and structure of the gibbsite intercalation compounds [LiAl$_2$(OH)$_6$]X {X = Cl, Br, NO$_3$} and [LiAl$_2$(OH)$_6$]Cl·H$_2$O using synchrotron X-ray and neutron powder diffraction." *Chemistry of Materials* no. 9 (1):241–247. doi: 10.1021/cm960316z.

Bigey, L., C. Depège, A. de Roy, and J. Besse, P. 1997. "EXAFS and XANES study of layered double hydroxides." *J. Phys. IV France* no. 7 (C2):C2-949–C2-950.

Boclair, Joseph W., and Paul S. Braterman. 1998. "One-step formation and characterization of Zn(II)–Cr(III) layered double hydroxides, Zn$_2$Cr(OH)6X (X = Cl, 1/2SO$_4$)." *Chemistry of Materials* no. 10 (8):2050–2052. doi: 10.1021/cm980298g.

Boclair, Joseph W., and Paul S. Braterman. 1999. "Layered double hydroxide stability. 1. Relative stabilities of layered double hydroxides and their simple counterparts." *Chemistry of Materials* no. 11 (2):298–302. doi: 10.1021/cm980523u.

Boclair, Joseph W., Paul S. Braterman, Jianping Jiang, Shaowei Lou, and Faith Yarberry. 1999. "Layered double hydroxide stability. 2. Formation of Cr(III)-containing layered double hydroxides directly from solution." *Chemistry of Materials* no. 11 (2):303–307. doi: 10.1021/cm980524m.

Boehm, Hanns-Peter, Johann Steinle, and Carmen Vieweger. 1977. "[Zn$_2$Cr(OH)6]X·2H$_2$O, New layer compounds capable of anion exchange and intracrystalline swelling." *Angewandte Chemie International Edition in English* no. 16 (4):265–266. doi: 10.1002/anie.197702651.

Bonnet, S., C. Forano, A. de Roy, J. P. Besse, P. Maillard, and M. Momenteau. 1996. "Synthesis of hybrid organo–mineral materials: anionic tetraphenylporphyrins in layered double hydroxides." *Chemistry of Materials* no. 8 (8):1962–1968. doi: 10.1021/cm960020t.

Bookin, A. S., V. I. Cherkashin, and V. A. Drits. 1993. "Polytype diversity of the hydrotalcite-like minerals II. Determination of the polytypes of experimentally studied varieties." *Clays and Clay Minerals* no. 41 (5):558–564. doi: 10.1346/CCMN.1993 .0410505.

Camino, G., A. Maffezzoli, M. Braglia, M. De Lazzaro, and M. Zammarano. 2001. "Effect of hydroxides and hydroxycarbonate structure on fire retardant effectiveness and mechanical properties in ethylene-vinyl acetate copolymer." *Polymer Degradation and Stability* no. 74 (3):457–464. doi: 10.1016/S0141-3910(01)00167-7.

Chai, Hao, Yanjun Lin, David G. Evans, and Dianqing Li. 2008. "Synthesis and UV absorption properties of 2-naphthylamine-1,5-disulfonic acid intercalated Zn–Al layered double hydroxides." *Industrial & Engineering Chemistry Research* no. 47 (9):2855–2860. doi: 10.1021/ie071043s.

Chai, Hao, Xiangyu Xu, Yanjun Lin, David G. Evans, and Dianqing Li. 2009. "Synthesis and UV absorption properties of 2,3-dihydroxynaphthalene-6-sulfonate anion-intercalated Zn–Al layered double hydroxides." *Polymer Degradation and Stability* no. 94 (4):744–749. doi: 10.1016/j.polymdegradstab.2008.09.007.

Chen, Jir-Shyr, Mark D. Poliks, Christopher K. Ober, Yuanming Zhang, Ulrich Wiesner, and Emmanuel Giannelis. 2002. "Study of the interlayer expansion mechanism and thermal–mechanical properties of surface-initiated epoxy nanocomposites." *Polymer* no. 43 (18):4895–4904. doi: 10.1016/S0032-3861(02)00318-X.

Chen, Shengjiao, Juan Li, Yingke Zhu, Zibin Guo, and Shengpei Su. 2013. "Increasing the efficiency of intumescent flame retardant polypropylene catalyzed by polyoxometalate based ionic liquid." *Journal of Materials Chemistry A* no. 1 (48):15242–15246. doi: 10.1039/C3TA13538A.

Chiu, Shih-Hsuan, and Wun-Ku Wang. 1998. "Dynamic flame retardancy of polypropylene filled with ammonium polyphosphate, pentaerythritol and melamine additives." *Polymer* no. 39 (10):1951–1955. doi: 10.1016/S0032-3861(97)00492-8.

Clause, O., M. Gazzano, F. Trifiro', A. Vaccari, and L. Zatorski. 1991. "Preparation and thermal reactivity of nickel/chromium and nickel/aluminium hydrotalcite-type precursors." *Applied Catalysis* no. 73 (2):217–236. doi: 10.1016/0166-9834(91)85138-L.

Clause, O., B. Rebours, E. Merlen, F. Trifiró, and A. Vaccari. 1992. "Preparation and characterization of nickel aluminum mixed oxides obtained by thermal decomposition of hydrotalcite-type precursors." *Journal of Catalysis* no. 133 (1):231–246. doi: 10.101 6/0021-9517(92)90200-2.

Costa, Francis, Mahmoud Abdel-Goad, Udo Wagenknecht, and Gert Heinrich. 2005. "Nanocomposites based on polyethylene and Mg-Al layered double hydroxide. I. Synthesis and characterization." *Polymer* no. 46:4447–4453. doi: 10.1016/j.polymer.2 005.02.027.

Costa, Francis, Udo Wagenknecht, and Gert Heinrich. 2007. "LDPE/Mg–Al layered double hydroxide nanocomposite: Thermal and flammability properties." *Polymer Degradation and Stability* no. 92:1813–1823. doi: 10.1016/j.polymdegradstab.2 007.07.009.

Costache, Marius C., Matthew J. Heidecker, E. Manias, Giovanni Camino, Alberto Frache, Gunter Beyer, Rakesh K. Gupta, and Charles A. Wilkie. 2007. "The influence of carbon nanotubes, organically modified montmorillonites and layered double hydroxides on the thermal degradation and fire retardancy of polyethylene, ethylene–vinyl acetate copolymer and polystyrene." *Polymer* no. 48 (22):6532–6545. doi: 10.1016/ j.polymer.2007.08.059.

Costantino, Umberto, Natascia Coletti, Morena Nocchetti, Gian Gaetano Aloisi, and Fausto Elisei. 1999. "Anion exchange of methyl orange into Zn–Al synthetic hydrotalcite and photophysical characterization of the intercalates obtained." *Langmuir* no. 15 (13):4454–4460. doi: 10.1021/la981672u.

Costantino, Umberto, A. Gallipoli, M. Nocchetti, Giovanni Camino, Federica Bellucci, and Alberto Frache. 2005. "New nanocomposites constituted of polyethylene and organically modified ZnAl-hydrotalcites." *Polymer Degradation and Stability –POLYM DEGRAD STABIL* no. 90:586–590. doi: 10.1016/j.polymdegradstab.2005.05.019.

Depège, Corinne, Fatima-Zahrae El Metoui, Claude Forano, André de Roy, Jacques Dupuis, and Jean-Pierre Besse. 1996. "Polymerization of silicates in layered double hydroxides." *Chemistry of Materials* no. 8 (4):952–960. doi: 10.1021/cm950533k.

Ding, Peng, Bai Kang, Jin Zhang, Jingwen Yang, Na Song, Shengfu Tang, and Liyi Shi. 2015. "Phosphorus-containing flame retardant modified layered double hydroxides and their applications on polylactide film with good transparency." *Journal of Colloid and Interface Science* no. 440:46–52. doi: 10.1016/j.jcis.2014.10.048.

Ding, Peng, and Baojun Qu. 2006. "Synthesis of exfoliated PP/LDH nanocomposites via melt-intercalation: Structure, thermal properties, and photo-oxidative behavior in comparison With PP/MMT nanocomposites." *Polymer Engineering & Science* no. 46:1153–1159. doi: 10.1002/pen.20568.

Doğan, Mehmet, Sengul Dogan, Lemiye atabek savaş, Gulsah Ozcelik, and Umit Tayfun. 2021. "Flame retardant effect of boron compounds in polymeric materials." *Composites Part B: Engineering* no. 222:109088. doi: 10.1016/j.compositesb.2021.109088.

Dong, Liye, Chuangang Hu, Long Song, Xianke Huang, Nan Chen, and Liangti Qu. 2016. "A large-area, flexible, and flame-retardant graphene paper." *Advanced Functional Materials* no. 26 (9):1470–1476. doi: 10.1002/adfm.201504470.

Dong, Xiang, Ruxiang Qin, Shibin Nie, Jinian Yang, Chi Zhang, and Wei Wu. 2019. "Fire hazard suppression of intumescent flame retardant polypropylene based on a novel Ni-containing char-forming agent." *Polymers for Advanced Technologies* no. 30 (3):563–572. doi: 10.1002/pat.4492.

Edenharter, Andreas, Patrick Feicht, Bashar Diar-Bakerly, Günter Beyer, and Josef Breu. 2016. "Superior flame retardant by combining high aspect ratio layered double hydroxide and graphene oxide." *Polymer* no. 91:41–49. doi: 10.1016/j.polymer.2016.03.020.

Fogg, Andrew M., Jenifer S. Dunn, and Dermot O'Hare. 1998. "Formation of second-stage intermediates in anion-exchange intercalation reactions of the layered double hydroxide [LiAl$_2$(OH)$_6$]Cl·H$_2$O as observed by time-resolved, in situ X-ray diffraction." *Chemistry of Materials* no. 10 (1):356–360. doi: 10.1021/cm970553h.

Fogg, Andrew M., and Dermot O'Hare. 1999. "Study of the Intercalation of lithium salt in gibbsite using time-resolved in situ X-ray diffraction." *Chemistry of Materials* no. 11 (7):1771–1775. doi: 10.1021/cm981151s.

Ford, Robert G., Andreas C. Scheinost, Kirk G. Scheckel, and Donald L. Sparks. 1999. "The link between clay mineral weathering and the stabilization of Ni surface precipitates." *Environmental Science & Technology* no. 33 (18):3140–3144. doi: 10.1021/es990271d.

Gómez-Fernández, Sandra, Martin Günther, Bernhard Schartel, M. Angeles Corcuera, and Arantxa Eceiza. 2018. "Impact of the combined use of layered double hydroxides, lignin and phosphorous polyol on the fire behavior of flexible polyurethane foams." *Industrial Crops and Products* no. 125:346–359. doi: 10.1016/j.indcrop.2018.09.018.

Gao, Liping, Guangyao Zheng, Yonghong Zhou, Lihong Hu, Guodong Feng, and Meng Zhang. 2014. "Synergistic effect of expandable graphite, diethyl ethylphosphonate and organically-modified layered double hydroxide on flame retardancy and fire behavior of polyisocyanurate-polyurethane foam nanocomposite." *Polymer Degradation and Stability* no. 101:92–101.

Gilman, Jeffrey W. 1999. "Flammability and thermal stability studies of polymer layered-silicate (clay) nanocomposites. This work was carried out by the National Institute of

Standards and Technology (NIST), an agency of the U. S. government, and by statute is not subject to copyright in the United States.1." *Applied Clay Science* no. 15 (1):31–49. doi: 10.1016/S0169-1317(99)00019-8.

Guan, Weijiang, Wenjuan Zhou, Qianwen Huang, and Chao Lu. 2014. "Chemiluminescence as a novel indicator for interactions of surfactant–polymer mixtures at the surface of layered double hydroxides." *The Journal of Physical Chemistry C* no. 118 (5):2792–2798. doi: 10.1021/jp410030b.

Guo, Bingtuo, Yongzhuang Liu, Qi Zhang, Fengqiang Wang, Qingwen Wang, Yixing Liu, Jian Li, and Haipeng Yu. 2017. "Efficient flame-retardant and smoke-suppression properties of Mg–Al-layered double-hydroxide nanostructures on wood substrate." *ACS Applied Materials & Interfaces* no. 9 (27):23039–23047. doi: 10.1021/acsami.7b06803.

Guo, Xiaoxiao, Fazhi Zhang, David G. Evans, and Xue Duan. 2010. "Layered double hydroxide films: synthesis, properties and applications." *Chemical Communications* no. 46 (29):5197–5210. doi: 10.1039/C0CC00313A.

Hajibeygi, Mohsen, Meisam Shabanian, and Hossein Ali Khonakdar. 2015. "Zn–AL LDH reinforced nanocomposites based on new polyamide containing imide group: From synthesis to properties." *Applied Clay Science* no. 114:256–264. doi: 10.1016/j.clay.2015.06.008.

Haq, Muhammad. 2007. "Applications of unsaturated polyester resins." *Russian Journal of Applied Chemistry* no. 80:1256–1269. doi: 10.1134/S1070427207070464.

Hibino, T., and A. Tsunashima. 2000. "Calcination and rehydration behavior of Mg-Fe-CO3 hydrotalcite-like compounds." *Journal of Materials Science Letters* no. 19 (16):1403–1405. doi: 10.1023/A:1006754902156.

Hofmeister, W., and H. Von Platen. 1992. "Crystal chemistry and atomic order in brucite-related double-layer structures." *Crystallography Reviews* no. 3 (1):3–26. doi: 10.1080/08893119208032964.

Horrocks, A. R., M. Tune, and L. Cegielka. 1988. "The burning behaviour of textiles and its assessment by oxygen-index methods." *Textile Progress* no. 18 (1–3):1–186. doi: 10.1080/00405168908689004.

Hou, Yanbei, Shuilai Qiu, Yuan Hu, Chanchal Kumar Kundu, Zhou Gui, and Weizhao Hu. 2018. "Construction of bimetallic ZIF-derived Co–Ni LDHs on the surfaces of GO or CNTs with a recyclable method: Toward reduced toxicity of gaseous thermal decomposition products of unsaturated polyester resin." *ACS Applied Materials & Interfaces* no. 10 (21):18359–18371. doi: 10.1021/acsami.8b04340.

Huang, Guobo, Suqing Chen, Pingping Lu, Chenglin Wu, and Huading Liang. 2013. "Combination effects of graphene and layered double hydroxides on intumescent flame-retardant poly(methyl methacrylate) nanocomposites." *Applied Clay Science* no. 88–89. doi: 10.1016/j.clay.2013.11.002.

Huang, Guobo, Zhengdong Fei, Xiaoying Chen, Fangli Qiu, Xu Wang, and Jianrong Gao. 2012. "Functionalization of layered double hydroxides by intumescent flame retardant: Preparation, characterization, and application in ethylene vinyl acetate copolymer." *Applied Surface Science* no. 258 (24):10115–10122. doi: 10.1016/j.apsusc.2012.06.088.

Huang, Sheng-Chao, Cong Deng, Shui-Xiu Wang, Wen-Chao Wei, Hong Chen, and Yu-Zhong Wang. 2019. "Electrostatic action induced interfacial accumulation of layered double hydroxides towards highly efficient flame retardance and mechanical enhancement of thermoplastic polyurethane/ammonium polyphosphate." *Polymer Degradation and Stability* no. 165. doi: 10.1016/j.polymdegradstab.2019.05.006.

Huang, Sheng-Chao, Cong Deng, Ze-Yong Zhao, Hong Chen, Yu-Yang Gao, and Yu-Zhong Wang. 2020. "Phosphorus-containing organic-inorganic hybrid nanoparticles for the smoke suppression and flame retardancy of thermoplastic polyurethane."

Polymer Degradation and Stability no. 178:109179. doi: 10.1016/j.polymdegradstab.2020.109179.

Huang, Shu, Hongdan Peng, Weng Weei Tjiu, Zhe Yang, Hong Zhu, Tao Tang, and Tianxi Liu. 2010. "Assembling exfoliated layered double hydroxide (LDH) nanosheet/carbon nanotube (CNT) hybrids via electrostatic force and fabricating nylon nanocomposites." *The Journal of Physical Chemistry B* no. 114 (50):16766–16772. doi: 10.1021/jp1087256.

Ijdo, Wouter L., and Thomas J. Pinnavaia. 1998. "Staging of organic and inorganic gallery cations in layered silicate heterostructures." *Journal of Solid State Chemistry* no. 139 (2):281–289. doi: 10.1006/jssc.1998.7842.

Iyi, Nobuo, Keiji Kurashima, and Taketoshi Fujita. 2002. "Orientation of an organic anion and second-staging structure in layered double-hydroxide intercalates." *Chemistry of Materials* no. 14 (2):583–589. doi: 10.1021/cm0105211.

Jiang, Yi, Zhifeng Hao, Hongsheng Luo, Zehui Shao, Qian Yu, Ming Sun, Yong Ke, and Yilong Chen. 2018. "Synergistic effects of boron-doped silicone resin and a layered double hydroxide modified with sodium dodecyl benzenesulfonate for enhancing the flame retardancy of polycarbonate." *RSC Advances* no. 8 (20):11078–11086. doi: 10.1039/C8RA01086B.

Jiao, C. 2006. "Flame retardation of ethylene-vinyl acetate copolymer using nano magnesium hydroxide and nano hydrotalcite." *Journal of Fire Sciences* no. 24:47–64. doi: 10.1177/0734904106053160.

Jin, Xiaodong, Xiaoyu Gu, Chen Chen, Wufei Tang, Hongfei Li, Xiaodong Liu, Serge Bourbigot, Zongwen Zhang, Jun Sun, and Sheng Zhang. 2017. "The fire performance of polylactic acid containing a novel intumescent flame retardant and intercalated layered double hydroxides." *Journal of Materials Science* no. 52. doi: 10.1007/s10853-017-1354-5.

Kalali, Ehsan Naderi, Anabel Montes, Xin Wang, Lu Zhang, Marjan E. Shabestari, Zhi Li, and De-Yi Wang. 2018. "Effect of phytic acid–modified layered double hydroxide on flammability and mechanical properties of intumescent flame retardant polypropylene system." *Fire and Materials* no. 42 (2):213–220. doi: 10.1002/fam.2482.

Kalali, Ehsan Naderi, Xin Wang, and De-Yi Wang. 2016. "Synthesis of a Fe_3O_4 nanosphere@Mg–Al layered-double-hydroxide hybrid and application in the fabrication of multifunctional epoxy nanocomposites." *Industrial & Engineering Chemistry Research* no. 55 (23):6634–6642. doi: 10.1021/acs.iecr.5b04873.

Kaneyoshi, Masami, and William Jones. 1998. "Exchange of interlayer terephthalate anions from a Mg–Al layered double hydroxide: formation of intermediate interstratified phases." *Chemical Physics Letters* no. 296 (1):183–187. doi: 10.1016/S0009-2614(98)01012-4.

Kang, Nian-Jun, and De-Yi Wang. 2013. "A green functional nanohybrid: preparation, characterization and properties of a β-cyclodextrin based functional layered double hydroxide." *Journal of Materials Chemistry A* no. 1 (37):11376–11383. doi: 10.1039/C3TA12304A.

Kannan, S., and C. S. Swamy. 1992. "Synthesis and physicochemical characterization of cobalt aluminium hydrotalcite." *Journal of Materials Science Letters* no. 11 (23):1585–1587. doi: 10.1007/BF00740840.

Kannan, S., S. Velu, V. Ramkumar, and C. S. Swamy. 1995. "Synthesis and physicochemical properties of cobalt aluminium hydrotalcites." *Journal of Materials Science* no. 30 (6):1462–1468. doi: 10.1007/BF00375249.

Kashiwagi, Takashi, Jeffrey W. Gilman, Kathryn M. Butler, Richard H. Harris, John R. Shields, and Atsushi Asano. 2000. "Flame retardant mechanism of silica gel/silica." *Fire and Materials* no. 24 (6):277–289. doi: 10.1002/1099-1018(200011/12)24:6<277::AID-FAM746>3.0.CO;2-A.

Kaul, Pawan Kumar, A. Joel Samson, G. Tamil Selvan, Ivmv Enoch, and P. Mosae Selvakumar. 2017. "Synergistic effect of LDH in the presence of organophosphate on thermal and flammable properties of an epoxy nanocomposite." *Applied Clay Science* no. 135:234–243. doi: 10.1016/j.clay.2016.09.031.

Khaldi, M., A. De Roy, M. Chaouch, and J. P. Besse. 1997. "New varieties of zinc–chromium–sulfate lamellar double hydroxides." *Journal of Solid State Chemistry* no. 130 (1):66–73. doi: 10.1006/jssc.1997.7280.

Khan, Aamir I., and Dermot O'Hare. 2002. "Intercalation chemistry of layered double hydroxides: recent developments and applications." *Journal of Materials Chemistry* no. 12 (11):3191–3198. doi: 10.1039/B204076J.

Kiyoshi, Fuda, Kudo Noboru, Kawai Shinji, and Matsunaga Toshiaki. 1993. "Preparation of Zn/Ga-layered double hydroxide and its thermal decomposition behavior." *Chemistry Letters* no. 22 (5):777–780. doi: 10.1246/cl.1993.777.

Kloprogge, J. Theo, David Wharton, Leisel Hickey, and Ray L. Frost. 2002. "Infrared and Raman study of interlayer anions CO_3^{2-}, NO_3^-, SO_4^{2-} and ClO_4^- in Mg/Al-hydrotalcite." *American Mineralogist* no. 87 (5–6):623–629. doi: 10.2138/am-2002-5-604.

Kong, Q., T. Wu, J. Wang, H. Liu, and J. Zhang. 2018. "Improving the thermal stability and flame retardancy of PP/IFR composites by NiAl-layered double hydroxide." *Journal of Nanoscience and Nanotechnology* no. 18 (5):3660–3665. doi: 10.1166/jnn.201 8.14679.

Kopka, H., K. Beneke, and G. Lagaly. 1988. "Anionic surfactants between double metal hydroxide layers." *Journal of Colloid and Interface Science* no. 123 (2):427–436. doi: 10.1016/0021-9797(88)90263-9.

Kotal, Moumita, Suneel Kumar Srivastava, Saraswathi Kesavapillai Manu, Aravind Kumar Saxena, and Kailash Nath Pandey. 2013. "Preparation and properties of in-situ polymerized polyurethane/stearate intercalated layer double hydroxide nanocomposites." *Polymer International* no. 62 (5):728–735. doi: 10.1002/pi.4354.

Kumar, Manish, Samarshi Chakraborty, Pradeep Upadhyaya, and G. Pugazhenthi. 2018. "Morphological, mechanical, and thermal features of PMMA nanocomposites containing two-dimensional Co–Al layered double hydroxide." *Journal of Applied Polymer Science* no. 135 (5):45774. doi: 10.1002/app.45774.

Leroux, Fabrice, and Jean-Pierre Besse. 2001. "Polymer interleaved layered double hydroxide: a new emerging class of nanocomposites." *Chemistry of Materials* no. 13 (10):3507–3515. doi: 10.1021/cm0110268.

Li, Long, Kangjia Jiang, Yi Qian, Haoyue Han, Peng Qiao, and Haiming Zhang. 2020. "Effect of organically intercalation modified layered double hydroxides-graphene oxide hybrids on flame retardancy of thermoplastic polyurethane nanocomposites." *Journal of Thermal Analysis and Calorimetry* no. 142 (2):723–733. doi: 10.1007/s1 0973-020-09263-0.

Li, M., L. Pang, M. Chen, J. Xie, and Q. Liu. 2018. "Effects of aluminum hydroxide and layered double hydroxide on asphalt fire resistance." *Materials (Basel)* no. 11 (10). doi: 10.3390/ma11101939.

Li, Zhi, Junhao Zhang, François Dufosse, and De-Yi Wang. 2018. "Ultrafine nickel nanocatalyst-engineering of an organic layered double hydroxide towards a super-efficient fire-safe epoxy resin via interfacial catalysis." *Journal of Materials Chemistry A* no. 6 (18):8488–8498. doi: 10.1039/C8TA00910D.

Lim, Kien-Sin, Soo-Tueen Bee, Lee Tin Sin, Tiam-Ting Tee, C. T. Ratnam, David Hui, and A. R. Rahmat. 2016. "A review of application of ammonium polyphosphate as intumescent flame retardant in thermoplastic composites." *Composites Part B: Engineering* no. 84:155–174. doi: 10.1016/j.compositesb.2015.08.066.

Liu, Jiajia, Youji Tao, Keqing Zhou, Yongqian Shi, Xiaming Feng, Ganxin Jie, Richard K. K. Yuen, and Yuan Hu. 2017. "The influence of typical layered inorganic compounds on the improved thermal stability and fire resistance properties of polystyrene nano-composites." *Polymer Composites* no. 38 (S1):E320–E330. doi: 10.1002/pc.23792.

Liu, Lei, Wei Wang, and Yuan Hu. 2015. "Layered double hydroxide-decorated flexible polyurethane foam: significantly improved toxic effluent elimination." *RSC Advances* no. 5 (118):97458–97466. doi: 10.1039/C5RA19414H.

Liu, Shan, Zhengping Fang, Hongqiang Yan, Venkata S. Chevali, and Hao Wang. 2016. "Synergistic flame retardancy effect of graphene nanosheets and traditional retardants on epoxy resin." *Composites Part A: Applied Science and Manufacturing* no. 89:26–32. doi: 10.1016/j.compositesa.2016.03.012.

Liu, Y., Y. Gao, Z. Zhang, and Q. Wang. 2021. "Preparation of ammonium polyphosphate and dye co-intercalated LDH/polypropylene composites with enhanced flame retardant and UV resistance properties." *Chemosphere* no. 277:130370. doi: 10.1016/j.chemosphere.2021.130370.

Lv, Pin, Zhengzhou Wang, Keliang Hu, and Weicheng Fan. 2005. "Flammability and thermal degradation of flame retarded polypropylene composites containing melamine phosphate and pentaerythritol derivatives." *Polymer Degradation and Stability* no. 90 (3):523–534. doi: 10.1016/j.polymdegradstab.2005.04.003.

Ma, Renzhi, Zhaoping Liu, Liang Li, Nobuo Iyi, and Takayoshi Sasaki. 2006. "Exfoliating layered double hydroxides in formamide: a method to obtain positively charged nanosheets." *Journal of Materials Chemistry* no. 16 (39):3809–3813. doi: 10.1039/B605422F.

Mallakpour, Shadpour, and Masoud Hatami. 2017. "Condensation polymer/layered double hydroxide NCs: preparation, characterization, and utilizations." *European Polymer Journal* no. 90:273–300. doi: 10.1016/j.eurpolymj.2017.03.015.

Mallakpour, Shadpour, Masoud Hatami, and Chaudhery Mustansar Hussain. 2020. "Recent innovations in functionalized layered double hydroxides: fabrication, characterization, and industrial applications." *Advances in Colloid and Interface Science* no. 283:102216. doi: 10.1016/j.cis.2020.102216.

Manzi-Nshuti, Charles, Dongyan Wang, Jeanne M. Hossenlopp, and Charles A. Wilkie. 2008. "Aluminum-containing layered double hydroxides: the thermal, mechanical, and fire properties of (nano)composites of poly(methyl methacrylate)." *Journal of Materials Chemistry* no. 18 (26):3091–3102. doi: 10.1039/B802553C.

Manzi-Nshuti, Charles, Dongyan Wang, Jeanne M. Hossenlopp, and Charles A. Wilkie. 2009. "The role of the trivalent metal in an LDH: synthesis, characterization and fire properties of thermally stable PMMA/LDH systems." *Polymer Degradation and Stability* no. 94 (4):705–711. doi: 10.1016/j.polymdegradstab.2008.12.012.

Maria Fernández, José, Maria Angeles Ulibarri, Francisco M. Labajos, and Vicente Rives. 1998. "The effect of iron on the crystalline phases formed upon thermal decomposition of Mg-Al-Fe hydrotalcites." *Journal of Materials Chemistry* no. 8 (11):2507–2514. doi: 10.1039/A804867C.

Matusinovic, Zvonimir, Hongdian Lu, and Charles A. Wilkie. 2012. "The role of dispersion of LDH in fire retardancy: The effect of dispersion on fire retardant properties of polystyrene/Ca–Al layered double hydroxide nanocomposites." *Polymer Degradation and Stability* no. 97 (9):1563–1568. doi: 10.1016/j.polymdegradstab.2012.07.020.

Mendiboure, A., and Robert Schoellhorn. 1987. "Formation and anion exchange reactions of layered transition metal hydroxides [Ni$_{1-x}$Mx] (OH)$_2$(CO$_3$)x/2(H$_2$O)z (M=Fe,Co)."*ChemInform* no. 18 (23): 038. doi: 10.1002/chin.198723038.

Meyn, Martina, Klaus Beneke, and Gerhard Lagaly. 1990. "Anion-exchange reactions of layered double hydroxides." *Inorganic Chemistry* no. 29 (26):5201–5207. doi: 10.1021/ic00351a013.

Mishra, S., S. H. Sonawane, R. P. Singh, A. Bendale, and K. Patil. 2004. "Effect of nano-Mg $(OH)_2$ on the mechanical and flame-retarding properties of polypropylene composites." *Journal of Applied Polymer Science* no. 94 (1):116–122. doi: 10.1002/app.20750.

Miyata, Shigeo. 1980. "Physico-chemical properties of synthetic hydrotalcites in relation to composition." *Clays and Clay Minerals* no. 28 (1):50–56. doi: 10.1346/CCMN.1 980.0280107.

Mohammadi, Abbas, De-Yi Wang, Alireza Sheikh Hosseini, and Jimena De La Vega. 2019. "Effect of intercalation of layered double hydroxides with sulfonate-containing calix[4] arenes on the flame retardancy of castor oil-based flexible polyurethane foams." *Polymer Testing* no. 79:106055. doi: 10.1016/j.polymertesting.2019.106055.

Nevare, Mahendra R., Vikas V. Gite, Pramod P. Mahulikar, Abdul Ahamad, and Sandip D. Rajput. 2014. "Synergism between LDH and nano-zinc phosphate on the flammability and mechanical properties of polypropylene." *Polymer-Plastics Technology and Engineering* no. 53 (5):429–434. doi: 10.1080/03602559.2013.844244.

Nyambo, Calistor, Dan Chen, Shengpei Su, and Charles A. Wilkie. 2009a. "Does organic modification of layered double hydroxides improve the fire performance of PMMA?" *Polymer Degradation and Stability* no. 94 (8):1298–1306. doi: 10.1016/j.polymdegradstab.2009.03.023.

Nyambo, Calistor, Dan Chen, Shengpei Su, and Charles A. Wilkie. 2009b. "Variation of benzyl anions in MgAl-layered double hydroxides: Fire and thermal properties in PMMA." *Polymer Degradation and Stability* no. 94 (4):496–505. doi: 10.1016/j.polymdegradstab.2009.02.002.

Nyambo, Calistor, Everson Kandare, Dongyan Wang, and Charles Wilkie. 2008. "Flame-retarded polystyrene: investigating chemical interactions between ammonium poly-phosphate and MgAl layered double hydroxide." *Polymer Degradation and Stability* no. 93:1656–1663. doi: 10.1016/j.polymdegradstab.2008.05.029.

Nyambo, Calistor, Ponusa Songtipya, E. Manias, Maria M. Jimenez-Gasco, and Charles A. Wilkie. 2008. "Effect of MgAl-layered double hydroxide exchanged with linear alkyl carboxylates on fire-retardancy of PMMA and PS." *Journal of Materials Chemistry* no. 18 (40):4827–4838. doi: 10.1039/B806531D.

Nyambo, Calistor, and Charles A. Wilkie. 2009. "Layered double hydroxides intercalated with borate anions: fire and thermal properties in ethylene vinyl acetate copolymer." *Polymer Degradation and Stability* no. 94 (4):506–512. doi: 10.1016/j.polymdegradstab.2009.02.001.

Pérez-Ramírez, J., G. Mul, F. Kapteijn, and J. A. Moulijn. 2001. "A spectroscopic study of the effect of the trivalent cation on the thermal decomposition behaviour of Co-based hydrotalcites." *Journal of Materials Chemistry* no. 11 (10):2529–2536. doi: 10.1039/B104989P.

Pérez-Ramírez, Javier, Guido Mul, Freek Kapteijn, and Jacob A. Moulijn. 2001. "On the stability of the thermally decomposed Co-Al hydrotalcite against retrotopotactic transformation." *Materials Research Bulletin* no. 36 (10):1767–1775. doi: 10.1016/S0025-5408(01)00657-2.

P.Xu, Z., and H. C. Zeng. 1998. "Thermal evolution of cobalt hydroxides: a comparative study of their various structural phases." *Journal of Materials Chemistry* no. 8 (11):2499–2506. doi: 10.1039/A804767G.

Pang, Xiu-Yan, Yu Tian, and Xiu-Zhu Shi. 2017. "Synergism between hydrotalcite and silicate-modified expandable graphite on ethylene vinyl acetate copolymer combustion behavior." *Journal of Applied Polymer Science* no. 134 (12). doi: 10.1002/app.44634.

Pereira, Celeste M.C. 2009. "Flame retardancy of thermoset polymers based on nanoparticles and carbon nanotubes." *Solid State Phenomena* no. 151: 79–87. doi: 10.4028/www.scientific.net/SSP.151.79.

Pinnavaia, Thomas J. 1987. "Swelling clays and related complex layered oxides." In *Chemical Physics of Intercalation*, edited by A. P. Legrand and S. Flandrois, 233–252. Boston, MA: Springer US.

Prevot, V., C. Forano, J. P. Besse, and F. Abraham. 1998. "Syntheses and thermal and chemical behaviors of tartrate and succinate intercalated Zn_3Al and Zn_2Cr layered double hydroxides." *Inorganic Chemistry* no. 37 (17):4293–4301. doi: 10.1021/ic9801239.

Qian, Yi, Peng Qiao, Long Li, Haoyue Han, Haiming Zhang, and Guozhang Chang. 2020. "Hydrothermal synthesis of lanthanum-doped MgAl-layered double hydroxide/graphene oxide hybrid and its application as flame retardant for thermoplastic polyurethane." *Advances in Polymer Technology* no. 2020:1018093. doi: 10.1155/2020/1018093.

Qin, Huaili, Shimin Zhang, Chungui Zhao, Guangjun Hu, and Mingshu Yang. 2005. "Flame retardant mechanism of polymer/clay nanocomposites based on polypropylene." *Polymer* no. 46 (19):8386–8395. doi: 10.1016/j.polymer.2005.07.019.

Qiu, Lei, Yanshan Gao, Cheng Zhang, Qinghua Yan, Dermot O'Hare, and Qiang Wang. 2018. "Synthesis of highly efficient flame retardant polypropylene nanocomposites with surfactant intercalated layered double hydroxides." *Dalton Transactions* no. 47 (9):2965–2975. doi: 10.1039/C7DT03477F.

Rastin, Hadi, Mohammad Reza Saeb, Milad Nonahal, Meisam Shabanian, Henri Vahabi, Krzysztof Formela, Xavier Gabrion, Farzad Seidi, Payam Zarrintaj, Morteza Ganjaee Sari, and Pascal Laheurte. 2017. "Transparent nanocomposite coatings based on epoxy and layered double hydroxide: nonisothermal cure kinetics and viscoelastic behavior assessments." *Progress in Organic Coatings* no. 113:126–135. doi: 10.1016/j.porgcoat.2017.09.003.

Reshetnikov, Igor S., Marina Yu Yablokova, and Nikolay A. Khalturinskij. 1997. "Influence of surface structure on thermoprotection properties of intumescent systems." *Applied Surface Science* no. 115 (2):199–201. doi: 10.1016/S0169-4332(97)80205-X.

Sanz, J., and J. M. Serratosa. 1984. "Silicon-29 and aluminum-27 high-resolution MAS-NMR spectra of phyllosilicates." *Journal of the American Chemical Society* no. 106 (17):4790–4793. doi: 10.1021/ja00329a024.

Scheckel, Kirk, Andreas Scheinost, Robert Ford, and Donald Sparks. 2000. "Stability of layered Ni hydroxide surface precipitates — A dissolution kinetics study." *Geochimica et Cosmochimica Acta* no. 64:2727–2735. doi: 10.1016/S0016-7037(00)00385-9.

Schon, J. C., D. Adler, and G. Dresselhaus. 1988. "Theory of staging in intercalation compounds." *Journal of Physics C: Solid State Physics* no. 21 (33):5595. doi: 10.1088/0022-3719/21/33/005.

Schutz, A., and P. Biloen. 1987. "Interlamellar chemistry of hydrotalcites. I. Polymerization of silicate anions." *Journal of Solid State Chemistry France* no. 68:360–368. doi: 10.1016/0022-4596(87)90323-9.

Shan, Xueying, Lei Song, Weiyi Xing, Yuan Hu, and Siuming Lo. 2012. "Effect of nickel-containing layered double hydroxides and cyclophosphazene compound on the thermal stability and flame retardancy of poly(lactic acid)." *Industrial & Engineering Chemistry Research* no. 51 (40):13037–13045. doi: 10.1021/ie300589p.

Shaw, Brenda R., Yingping Deng, Francine E. Strillacci, Kathleen A. Carrado, and Mebrahtu G. Fessehaie. 1990. "Electrochemical surface analysis of nonconducting solids: ferricyanide and phenol as electrochemical probes of the surfaces of layered double hydroxide anion-exchanging clays." *Journal of The Electrochemical Society* no. 137 (10):3136. doi: 10.1149/1.2086173.

Shi, Yanqin, Feng Chen, Jingtao Yang, and Mingqiang Zhong. 2010. "Crystallinity and thermal stability of LDH/polypropylene nanocomposites." *Applied Clay Science* no. 50 (1):87–91. doi: 10.1016/j.clay.2010.07.007.

Soni, Daxit Bharatbhai, and Gnanu Bhatt. 2022. "A review on flame retardants used in polyurethane foam." *ECS Transactions* no. 107 (1):1153. doi: 10.1149/10701.1153ecst.

Starink, M. J. 2003. "The determination of activation energy from linear heating rate experiments: a comparison of the accuracy of isoconversion methods." *Thermochimica Acta* no. 404 (1):163–176. doi: 10.1016/S0040-6031(03)00144-8.

Thiel, J. P., C. K. Chiang, and K. R. Poeppelmeier. 1993. "Structure of lithium aluminum hydroxide dihydrate (LiAl2(OH)7.2H2O)." *Chemistry of Materials* no. 5 (3):297–304. doi: 10.1021/cm00027a011.

Vahabi, Henri, Elnaz Movahedifar, Maryam Jouyandeh, Mohammad Reza Saeb, and Sabu Thomas. 2021. "Improved flame retardancy in polyurethanes using layered double hydroxides." In *Materials and Chemistry of Flame-Retardant Polyurethanes Volume 2: Green Flame Retardants*, 137–160. American Chemical Society.

Valente, J. Sanchez, F. Figueras, M. Gravelle, P. Kumbhar, J. Lopez, and J. P. Besse. 2000. "Basic properties of the mixed oxides obtained by thermal decomposition of hydro-talcites containing different metallic compositions." *Journal of Catalysis* no. 189 (2):370–381. doi: 10.1006/jcat.1999.2706.

Visakh, P. M., A. O. Semkin, I. A. Rezaev, and A. V. Fateev. 2019. "Review on soft polyurethane flame retardant." *Construction and Building Materials* no. 227:116673. doi: 10.1016/j.conbuildmat.2019.116673.

Vucelic, M., W. Jones, and G. D. Moggridge. 1997. "Cation ordering in synthetic layered double hydroxides." *Clays and Clay Minerals* no. 45 (6):803–813. doi: 10.1346/CCMN.1997.0450604.

Vucelic, M., G. D. Moggridge, and W. Jones. 1995. "Thermal properties of terephthalate- and benzoate-intercalated LDH." *The Journal of Physical Chemistry* no. 99 (20):8328–8337. doi: 10.1021/j100020a068.

Wang, Biao, Keqing Zhou, Bibo Wang, Zhou Gui, and Yuan Hu. 2014. "Synthesis and characterization of CuMoO4/Zn–Al layered double hydroxide hybrids and their application as a reinforcement in polypropylene." *Industrial & Engineering Chemistry Research* no. 53 (31):12355–12362. doi: 10.1021/ie502232a.

Wang, De-Yi, Amit Das, Andreas Leuteritz, Regine Boldt, Liane Häußler, Udo Wagenknecht, and Gert Heinrich. 2011. "Thermal degradation behaviors of a novel nanocomposite based on polypropylene and Co–Al layered double hydroxide." *Polymer Degradation and Stability* no. 96 (3):285–290. doi: 10.1016/j.polymdegradstab.2010.03.003.

Wang, Hongchao, Shengnan Tan, and Rongjun Song. 2019. "Effect of trace chloride on the char formation and flame retardancy of the LLDPE filled with NiAl-layered double hydroxides." *Fire and Materials* no. 43 (1):110–120. doi: 10.1002/fam.2674.

Wang, Lili, Bin Li, Xiucheng Zhang, Chunxia Chen, and Feng Zhang. 2012. "Effect of intercalated anions on the performance of Ni–Al LDH nanofiller of ethylene vinyl acetate composites." *Applied Clay Science* no. 56:110–119. doi: 10.1016/j.clay.2011.12.004.

Wang, Lili, Bin Li, Xiaohong Zhao, Chunxia Chen, and Jingjing Cao. 2012. "Effect of rare earth ions on the properties of composites composed of ethylene vinyl acetate copolymer and layered double hydroxides." *PLoS One* no. 7:e37781. doi: 10.1371/journal.pone.0037781.

Wang, Lili, Milin Zhang, and Baibin Zhou. 2017. "Thermal Stability, Combustion Behavior, and Mechanical Property in a Flame-Retardant Polypropylene System." *Applied Sciences* no. 7 (1):55.

Wang, Linjiang, Xiangli Xie, Shengpei Su, Jiangxiang Feng, and Charles A. Wilkie. 2010. "A comparison of the fire retardancy of poly(methyl methacrylate) using montmoril-lonite, layered double hydroxide and kaolinite." *Polymer Degradation and Stability* no. 95 (4):572–578. doi: 10.1016/j.polymdegradstab.2009.12.012.

Wang, Peng-Ji, Xiao-Ping Hu, Dui-Jun Liao, Yi Wen, T. Richard Hull, Fei Miao, and Quan-Tong Zhang. 2017. "Dual fire retardant action: the combined gas and condensed phase effects of azo-modified NiZnAl layered double hydroxide on intumescent polypropylene." *Industrial & Engineering Chemistry Research* no. 56 (4):920–932. doi: 10.1021/acs.iecr.6b03953.

Wang, Xin, Ehsan Naderi Kalali, and De-Yi Wang. 2015. "Renewable cardanol-based surfactant modified layered double hydroxide as a flame retardant for epoxy resin." *ACS Sustainable Chemistry & Engineering* no. 3 (12):3281–3290. doi: 10.1021/ acssuschemeng.5b00871.

Wang, Z., X. Shen, T. Qian, K. Xu, Q. Sun, and C. Jin. 2018. "Fabrication of super-hydrophobic Mg/Al layered double hydroxide (LDH) coatings on medium density fiberboards (MDFs) with flame retardancy." *Materials (Basel)* no. 11 (7). doi: 10.3390/ ma11071113.

Wang, Zhenyu, En-hou Han, and Wei Ke. 2005. "Influence of nano-LDHs on char formation and fire-resistant properties of flame-retardant coating." *Progress in Organic Coatings* no. 53:29–37.

Wijitwongwan, Rattanawadee, Soontaree Intasa-Ard, and Makoto Ogawa. 2019. "Preparation of layered double hydroxides toward precisely designed hierarchical organization." *ChemEngineering* no. 3 (3):68.

Wu, Yifan, Mengqi Tang, Na Wang, Jun Qin, Xiaolang Chen, and Kun Zhang. 2018. "Preparation and investigation on morphology, thermal stability and crystallization behavior of HDPE/EVA/organo-modified layered double hydroxide nanocomposites." *Polymer Composites* no. 39 (S3):E1849–E1857. doi: 10.1002/pc.24834.

Xie, Xiang, Lin Wang, and Guo Zhang. 2011. "Preparation and flame retardancy of unsaturated polyester resin containing P-toluenesulfonate-pillared layered double hydroxide." *Advanced Materials Research* no. 399–401:1372–1375. doi: 10.4028/ www.scientific.net/AMR.399-401.1372.

Xu, Sailong, Lixia Zhang, Yanjun Lin, Rushi Li, and Fazhi Zhang. 2012. "Layered double hydroxides used as flame retardant for engineering plastic acrylonitrile–butadiene–styrene (ABS)." *Journal of Physics and Chemistry of Solids* no. 73 (12):1514–1517. doi: 10.101 6/j.jpcs.2012.04.011.

Xu, Wen-Zong, Shao-Qing Wang, Liang Liu, and Yuan Hu. 2016. "Synthesis of heptamolybdate-intercalated MgAl LDHs and its application in polyurethane elastomer." *Polymers for Advanced Technologies* no. 27 (2):250–257. doi: 10.1002/pat.3628.

Xu, Wenzong, Baoling Xu, Aijiao Li, Xiaoling Wang, and Guisong Wang. 2016. "Flame retardancy and smoke suppression of MgAl layered double hydroxides containing P and Si in polyurethane elastomer." *Industrial & Engineering Chemistry Research* no. 55 (42):11175–11185. doi: 10.1021/acs.iecr.6b02708.

Xu, Wenzong, Bingliang Zhang, Baoling Xu, and Aijiao Li. 2016. "The flame retardancy and smoke suppression effect of heptaheptamolybdate modified reduced graphene oxide/ layered double hydroxide hybrids on polyurethane elastomer." *Composites Part A: Applied Science and Manufacturing* no. 91:30–40. doi: 10.1016/j.compositesa.201 6.09.013.

Yadav, Anilkumar, Felipe M. de Souza, Tim Dawsey, and Ram K. Gupta. 2022. "Recent advancements in flame-retardant polyurethane foams: a review." *Industrial & Engineering Chemistry Research* no. 61 (41):15046–15065. doi: 10.1021/acs.iecr.2 c02670.

Yang, You-Hao, Yu-Chin Li, John Shields, and Rick D. Davis. 2015. "Layer double hydroxide and sodium montmorillonite multilayer coatings for the flammability reduction of flexible polyurethane foams." *Journal of Applied Polymer Science* no. 132 (14). doi: 10.1002/app.41767.

Ye, Lei, and Baojun Qu. 2008. "Flammability characteristics and flame retardant mechanism of phosphate-intercalated hydrotalcite in halogen-free flame retardant EVA blends."

Polymer Degradation and Stability no. 93:918–924. doi: 10.1016/j.polymdegradstab.2 008.02.002.

Yin, Huajie, and Zhiyong Tang. 2016. "Ultrathin two-dimensional layered metal hydroxides: an emerging platform for advanced catalysis, energy conversion and storage." *Chemical Society Reviews* no. 45 (18):4873–4891. doi: 10.1039/C6CS00343E.

Zammarano, Mauro, Massimiliano Franceschi, Severine Bellayer, Jeffrey Gilman, and Sergio Meriani. 2005. "Preparation and flame resistance properties of revolutionary self-extinguishing epoxy nanocomposites based on layered double hydroxides." *Polymer* no. 46:9314–9328. doi: 10.1016/j.polymer.2005.07.050.

Zhang, Linyan, Yanjun Lin, Zhenjun Tuo, David G. Evans, and Dianqing Li. 2007. "Synthesis and UV absorption properties of 5-sulfosalicylate-intercalated Zn–Al layered double hydroxides." *Journal of Solid State Chemistry* no. 180 (4):1230–1235. doi: 10.1016/j.jssc.2007.01.026.

Zhang, Ming, Peng Ding, and Baojun Qu. 2009. "Flammable, thermal, and mechanical properties of intumescent flame retardant PP/LDH nanocomposites with different divalent cations." *Polymer Composites* no. 30 (7):1000–1006. doi: 10.1002/pc.20648.

Zhang, Ming, Peng Ding, Baojun Qu, and Aiguo Guan. 2008. "A new method to prepare flame retardant polymer composites." *Journal of Materials Processing Technology* no. 208:342–347. doi: 10.1016/j.jmatprotec.2007.12.139.

Zhang, Sheng, and A. Richard Horrocks. 2003. "A review of flame retardant polypropylene fibres." *Progress in Polymer Science* no. 28 (11):1517–1538. doi: 10.1016/j.progpolymsci.2003.09.001.

Zhang, Sheng, Wufei Tang, Jun Sun, Yu Jiang, Xiangxing Bu, and Xiaoyu Gu. 2018. "Synergistic effects of modified hydrotalcite on improving the fire resistance of ethylene vinyl acetate containing intumescent flame retardants." *Polymer Composites* no. 39 (2):522–528. doi: 10.1002/pc.23964.

Zhang, Sheng, Yongxin Yan, Wenjia Wang, Xiaoyu Gu, Hongfei Li, Jianhua Li, and Jun Sun. 2017. "Intercalation of phosphotungstic acid into layered double hydroxides by reconstruction method and its application in intumescent flame retardant poly (lactic acid) composites." *Polymer Degradation and Stability* no. 147. doi: 10.1016/j.polymdegradstab.2017.12.004.

Zhang, Zejiang, Chenghua Xu, Fali Qiu, Xiujuan Mei, Bin Lan, and Shuosheng Zhang. 2004. "Study on fire-retardant nanocrystalline Mg-Al layered double hydroxides synthesized by microwave-crystallization method." *Science in China Series B: Chemistry* no. 47 (6):488–498. doi: 10.1360/04yb0006.

Zhao, Chun-Xia, Ya Liu, De-Yi Wang, De-Long Wang, and Yu-Zhong Wang. 2008. "Synergistic effect of ammonium polyphosphate and layered double hydroxide on flame retardant properties of poly(vinyl alcohol)." *Polymer Degradation and Stability* no. 93:1323–1331. doi: 10.1016/j.polymdegradstab.2008.04.002.

Zhao, Chunxia, Gang Peng, Bailing Liu, and Zhengwu Jiang. 2011. "Synergistic effect of organically modified layered double hydroxide on thermal and flame-retardant properties of poly(butyl acrylate–vinyl acetate)." *Journal of Polymer Research* no. 18 (6):1971–1981. doi: 10.1007/s10965-011-9604-8.

Zheng, Lu, Ting Wu, Qinghong Kong, Junhao Zhang, and Hong Liu. 2017. "Improving flame retardancy of PP/MH/RP composites through synergistic effect of organic CoAl-layered double hydroxide." *Journal of Thermal Analysis and Calorimetry* no. 129:1039–1046.

Zhou, Xia, Xiaowei Mu, Wei Cai, Junling Wang, Fukai Chu, Zhoumei Xu, Lei Song, Weiyi Xing, and Yuan Hu. 2019. "Design of hierarchical NiCo-LDH@PZS hollow dodecahedron architecture and application in high-performance epoxy resin with excellent fire safety." *ACS Applied Materials & Interfaces* no. 11 (44):41736–41749. doi: 10.1021/acsami.9b16482.

Zhu, Haifeng, Yongjun Feng, Pinggui Tang, Guojing Cui, David G. Evans, Dianqing Li, and Xue Duan. 2011. "Synthesis and UV absorption properties of aurintricarboxylic acid intercalated Zn–Al layered double hydroxides." *Industrial & Engineering Chemistry Research* no. 50 (23):13299–13303. doi: 10.1021/ie2016366.

Zhu, Jin, Fawn M. Uhl, Alexander B. Morgan, and Charles A. Wilkie. 2001. "Studies on the mechanism by which the formation of nanocomposites enhances thermal stability." *Chemistry of Materials* no. 13 (12):4649–4654. doi: 10.1021/cm010451y.

Zhu, Ping, Zhongji Gu, Shu Hong, and Hailan Lian. 2018. "Preparation and characterization of microencapsulated LDHs with melamine-formaldehyde resin and its flame retardant application in epoxy resin." *Polymers for Advanced Technologies* no. 29 (7):2147–2160. doi: 10.1002/pat.4323.

Zubair, Mukarram, Muhammad Daud, Gordon McKay, Farrukh Shehzad, and Mamdouh A. Al-Harthi. 2017. "Recent progress in layered double hydroxides (LDH)-containing hybrids as adsorbents for water remediation." *Applied Clay Science* no. 143:279–292. doi: 10.1016/j.clay.2017.04.002.

5 Polymer/MXene Composites with Enhanced Fire-Safe Characteristics

Wen-Wen Guo
Jiangnan University, Wuxi, Jiangsu, China

CONTENTS

DOI: 10.1201/9781003327158-5

5.1 INTRODUCTION

MXene, as a rising new star of the latest two-dimensional materials, has received growing interest since its first discovery in 2011 (Fu et al. 2021). It was the collective name for the new large family of two-dimensional transition metal carbides, nitrides, and carbonitrides, which were commonly prepared through the selective etching of the precursor MAX phases (Zhou, Zhao, et al. 2022). Due to the unique layered metal–non-metallic structure and surface terminal groups, MXenes possess outstanding physical and chemical properties, including high optical transparency, low density, good thermal and mechanical stability, hydrophilicity, and excellent thermal and electrical conductivities (He et al. 2021). These fantastic features make MXenes promising candidates to be applied in the fields of anticorrosion (Cai et al. 2021), separation (Huang, Ding, and Wang 2021), sensors (Li, Li, and Zeng 2021), biomedical (Lin et al. 2021), energy storage (Chaudhari et al. 2017), and photocatalysis (Zhong, Li, and Zhang 2021). In particular, the nanoscale structure, excellent characteristics, and good machinability make it suitable for application in polymer composites. This offers a favorable opportunity for the integration of the respective outstanding performances of MXene nanosheets and polymer matrices. Hence, MXene-based polymer composites have attracted considerable research interest for different applications.

To date, a variety of MXene/polymer composites with outstanding comprehensive performances have been developed by the strategies of in situ polymerization or ex situ blending (solution, emulsion or melt blending, etc.) ascribed to the rich active terminations on the MXene nanosheet surface (Gao et al. 2020). The fascinating properties of MXenes endow polymer composites with improved mechanical and thermal properties, enhanced electrical and thermal conductivities, excellent electromagnetic interference shielding performance, and so on. In particular, most polymeric materials are flammable and display high fire risk. Two-dimensional nanomaterials have been proven to be highly efficient flame retardants for polymeric materials ascribed to the so-called "tortuous path effect". In addition, nanoparticles with transition metal elements usually present significant smoke and toxic gas suppression. Hence, lamellar MXene nanosheets with transition metal elements can act as employable barriers and smoke suppression agents, which can remarkably improve the flame retardancy as well as smoke and toxic gas suppression of polymer composites (Yu et al. 2019). Moreover, the fire protection performance can be tuned by the types and composition of MXene and polymers. As reported in previous studies, MXene and its derivatives have been employed to fabricate various polymer composites such as epoxy resin (EP) (Zhou et al. 2021), thermoplastic polyurethane (TPU) (Yu et al. 2019), acrylonitrile-butadiene-styrene (ABS) resin (Zhu, Wang, et al. 2021), polystyrene (PS) (Si et al. 2019), polylactide (PLA) (Zhou et al. 2021), cellulose nanofibril (CNF) (Zhan et al. 2021), and poly (vinyl alcohol) (PVA) (Ning et al. 2021). Due to their optimized comprehensive performance, MXene/polymer composites can be an ideal material to meet industrial application requirements.

However, on account of the strong van der Waals force, MXene nanosheets tend to aggregate and restack in the polymer matrix, which leads to the deterioration of the overall performance of MXene/polymer composites. In addition, regarding the application of flame-retardant additives, MXenes used alone rarely endow polymer composites with satisfactory flame-retardant ratings. Hence, manipulating the surface of MXene nanosheets to obtain good processability and high flame-retardant efficiency is crucial for polymer processing and application. It is of great importance to develop a suitable strategy to functionalize $Ti_3C_2T_x$ nanosheets to prepare high-performance nanocomposites. The surface modification strategy of MXenes has been widely employed to prevent MXene nanosheets from aggregating and improve the interfacial compatibility with the polymer matrix, which can be simply divided into three aspects: organic functionalization, inorganic functionalization, and blending synergistic technology. Additionally, modification technology can be divided into covalent and non-covalent (van der Waals forces, electrostatic interactions, and hydrogen bonds) modifications. In this context, it is highly desirable to summarize the progress of MXenes and their derivatives in applications of flame-retardant polymer composites.

In this chapter, we review the recent research progress regarding the application of MXenes and their derivatives in the fabrication of flame-retardant polymer composites. In addition to the major focus on the flammability characteristics of MXene-based polymer nanocomposites, some other properties, including thermal stability, mechanical and electrical properties, electromagnetic interference shielding performance, antibacterial properties, fire-warning performance, etc., are also discussed. The flame-retardant mechanism of different MXene/polymer composites is also discussed. Ultimately, we provide a brief comment on challenges and opportunities for future application of this new class of polymer composites.

5.2   PREPARATION METHODS OF MXENES

Two-dimensional transition metal carbides, nitrides, or carbonitrides, known as MXenes, are generally fabricated via the selective etching of MAX phases. The so-called MAX phases are a family of ternary transition metal carbides, nitrides, and carbonitrides with layered hexagonal structures, which have a general chemical formula of $M_{n+1}AX_n$ (n = 1, 2, or 3), where "M" refers to the early transition metals, "A" belongs to the IIIA or IVA elements, and "X" represents carbon, nitrogen, or both (Khazaei et al. 2019, Dong et al. 2020, Gao et al. 2020), as depicted in Figure 5.1 (Chen, Yang, and Zhang 2021, Alhabeb et al. 2017), in the structure of layered hexagonal MAX precursor, the X layer is sandwiched between two M layers and they were linked by ionic/metallic/covalent bonds, while the layer of A atoms interpenetrated into every two $M_{n+1}X_n$ layers and interacted with M atom via pure metallic bonds. Hence, the M-A bond is relatively weaker than the M-X bond, which makes it possible to fabricate MXenes by selectively removing the A layer (Chen, Yang, and Zhang 2021, Gong et al. 2021).

More than 70 types of known MAX phases and novel members have been continually developed, which impart unusual designability and diversity to the structure of MXenes (Ronchi, Arantes, and Santos 2019, Chen et al. 2021). At present, the MXene family is composed of 30 synthesized members and dozens more theoretically predicted members. Generally, MXene fabrication methods can

(a)

(b)

MAX Phase   Multilayer MXene   Singel layer MXene

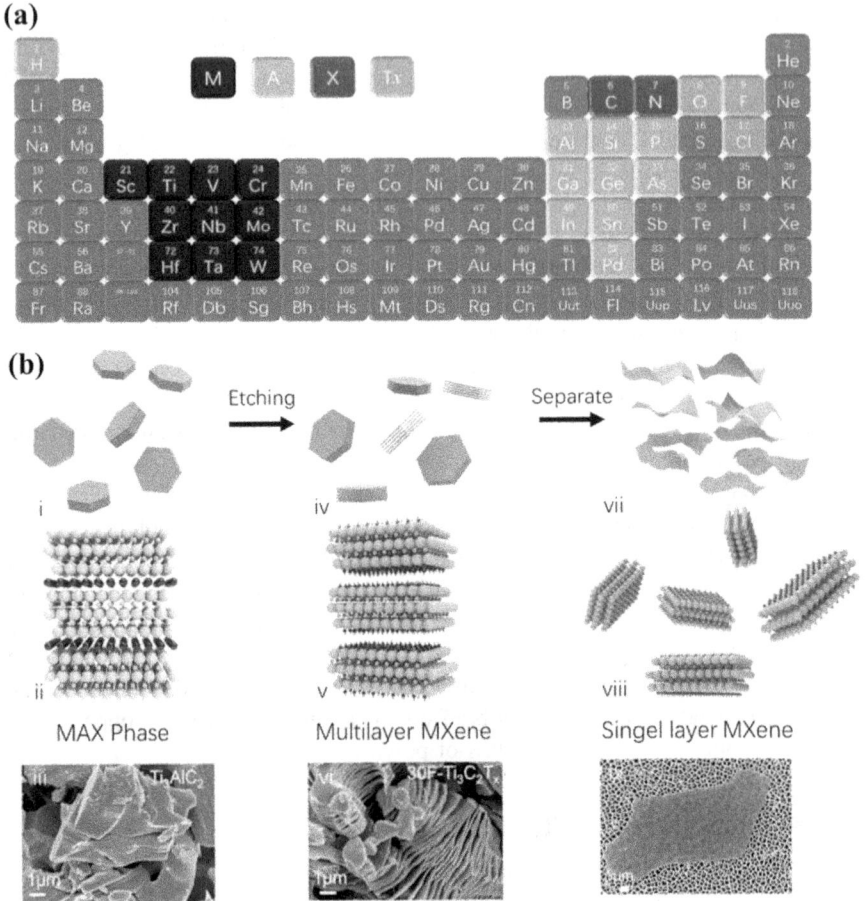

FIGURE 5.1 (a) The family of MXenes (and MAX phases) in Periodic table. (b) Consecutive steps of schematic steps for MXene synthesis from their MAX phase, taking Ti_3AlC_2 for example; (i–iii) schematic diagram, atomic structure diagram, and SEM images of the stepped MAX phase; (iv–vi) schematic diagram, atomic structure diagram, and SEM images of accordion $Ti_3C_2T_x$ MXene; (vii–ix) schematic diagram, atomic structure diagram, and TEM images of single-layer $Ti_3C_2T_x$ MXene. (Reprinted from Chen, N. et al., *Sci. Technol. Adv. Mater.*, 22, 917–930, 2021; Alhabeb, M. et al., *Chem. Mater.*, 29, 7633–7644, 2017. With permission.)

be divided into two strategies: top-down and bottom-up approaches (Gao et al. 2020). The former strategy refers to preparing MXenes by etching from MAX or non-MAX precursors, which is the most commonly used means to fabricate multilayered MXenes. The latter strategy involves methods such as chemical vapor deposition (CVD) and atomic layer deposition. MXenes obtained by bottom-up methods, including CVD, possess the virtues of large transverse sizes and fewer defects; however, high manufacturing complexity and costs have restricted the large-scale application of bottom-up synthesis methods. Hence, in this chapter, we focus on the development of MXenes from MAX or non-MAX precursors.

5.2.1 FLUORIDE ACID ETCHING METHOD

The fluoride acid etching approach is the most widely used and effective method for removing the A atomic layer from the bulky and rigid MAX phase to fabricate MXenes (Jiang et al. 2022). Lu et al. synthesized an organ-like $Ti_3C_2T_x$ MXene from Ti_3AlC_2 utilizing a 45 wt% HF solution as the etchant (Yang et al. 2019). Birkel et al. introduced a new member, $V_4C_3T_x$, to the MXene family by the chemical exfoliation of 413 MAX phase V_4AlC_3 using aqueous HF solution (40 wt%) (Tran et al. 2018). Moreover, many oxidants, such as H_2O_2, HNO_3, persulfate and permanganate salts, have been employed to assist the chemical etching of Si by HF or other etchants. Gogotsi *et al.* produced two-dimensional Ti_3C_2 from Ti_3SiC_2 through an oxidant-assisted selective etching method (Alhabeb et al. 2018). The resulting Ti_3C_2 displays the same structure as that of Ti_3AlC_2. However, hydrofluoric acid is a highly corrosive acid with acute toxicity, and the completion of etching often requires high concentrations of HF solution, which causes considerable safety issues for both the health of people and environmental protection. Therefore, many less hazardous methods were gradually developed.

5.2.2 FLUORIDE SALT ETCHING METHOD

As an alternative, fluoride acid solutions can also be formed in situ by bifluoride (e.g., $NaHF_2$, KHF_2, NH_4HF_2) or mixtures of fluoride salts (such as LiF, NaF, KF, and CsF) with other inorganic acids (for example HCl or H_2SO_4) (Feng, Yu, Wang, et al. 2017, Gong et al. 2021), which has been demonstrated as a safely effective method to produce two-dimensional MXene in lieu of HF.

5.2.2.1 Bifluoride

Several reports have found that bifluorides such as NH_4HF_2, KHF_2 and $NaHF_2$ can be used as etchants for fabricating MXenes. Yu et al. produced two-dimensional MXene Ti_3C_2 by etching Ti_3AlC_2 using NH_4HF_2, and the as-prepared Ti_3C_2 with a larger inter-planar spacing was more stable than the sample exfoliated by HF (Feng, Yu, Jiang, et al. 2017). In addition, Yu et al. employed three bifluorides, $NaHF_2$, KHF_2, and NH_4HF_2, as etchants to synthesize two-dimensional MXene Ti_3C_2, and the possible reaction equations between Ti_3AlC_2 and different bifluorides were generalized (Feng, Yu, Wang, et al. 2017). However, common etching preparations are mostly carried out in aqueous solutions, which obstructs the usage of MXenes in water-sensitive applications. It was proven that the MAX phase can be successfully etched to obtain fluorine-terminated MXenes in a number of NH_4HF_2-containing polar organic solvents without the use of water (Natu et al. 2020). This would open avenues for expanding the scope of MXene applications.

5.2.2.2 Fluoride Salts with Acid

Fluoride salts, including LiF, NaF, KF, NH_4F, etc., mixed with acids, including HCl and H_2SO_4, in the right proportions can form suitable etching agents. Among them, LiF and HCl are typical fluoride salts and inorganic acid partners (Figure 5.2a) (Zeng et al. 2022). Zhang et al. first compared the Ti_3C_2Tx MXenes obtained by

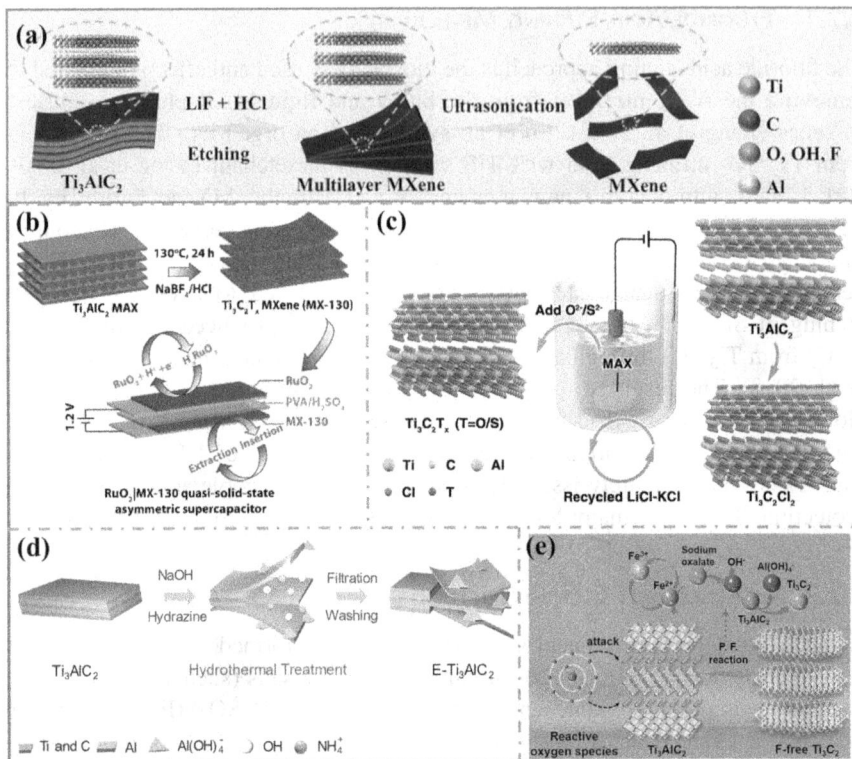

FIGURE 5.2 (a) Synthesis rout of MXene etching via LiF/HCl (Reprinted from Zeng, Q. et al., *Chem. Eng. J.*, 446, 136899, 2022. With permission.) (b) Single-step synthesis route of Ti3C2Tx MXene through low-cost NaBF4/HCl and its application as negative electrode material in a quasi-solid-state asymmetric supercapacitor device (Reprinted from Ghosh, M. et al., *Chemistryselect*, 7, e202201166, 2022. With permission.) (c) Synthesis of MXene from the MAX phase by molten-salt-assisted electrochemical etching and the in situ modification of surface terminations (Reprinted from Shen, M. et al., *Angew. Chem. Int. Ed.*, 60, 27013–27018, 2021. With permission.) (d) Schematic illustration of the one-step synthesis of Ti_3C_2-OH involving hydrazine decomposition during alkali treatment (Reprinted from Al Mayyahi, A. et al., *Mater. Today Commun.*, 32, 103835, 2022. With permission.) (e) Schematic illustration of photo-Fenton strategy for the fabrication of fluorine-free Ti_3C_2. (Reprinted from Liang, L. et al., *ACS Nano*, 16, 7971–7981, 2022. With permission.)

LiF/HCl and HF solutions, and the results revealed that the as-prepared $Ti_3C_2T_x$ MXenes from the combination of LiF and HCl (concentration: 9 M) solution exhibited better electrochemical performance than the HF etched sample (Zhang, Zhang, and Zhao 2022). The Li^+ can spontaneously intercalate into the layers of Ti_3C_2Tx during the etching process, which remarkably expands the lattice space of the obtained Ti_3C_2Tx. In previous works reported by Zhou's group (Liu et al. 2017, Wang et al. 2020, Guo et al. 2020), four different fluoride salts in hydrochloric acid (HCl), including LiF, NaF, KF and NH4F, have been proven to be suitable etchants for the preparation of Ti_3C_2 MXene, Ti_2C MXene, V_2C MXene, and Mo_2CT_x

MXene. In addition to the above fluoride salts, $NaBF_4$ can also be used as an etchant together with the HCl solution. Ghosh et al. reported that $Ti_3C_2T_x$ MXene can be successfully prepared under mild reaction conditions comprising an aqueous solution of $NaBF_4$ and HCl (Ghosh et al. 2022). The preparation process is shown in Figure 5.2b. The achieved specific capacitance (262 F/g) of the as-synthesized MXene is superior to that of $Ti_3C_2T_x$ obtained by HF treatment. The etching approach by combining HCl with fluoride salts offers a relatively milder way to fabricate MXenes and improve their potential for practical applications.

5.2.3 THE FLUORINE-FREE ETCHING METHOD

With the development of etching technology, harmless fluorine-free etchants have attracted increasing attention, which could overcome the disadvantages, including high toxicity and long etching course, of conventional HF-containing synthetic routes. To date, great efforts have been implemented to fabricate fluoride-free MXenes from MAX, and the methods mainly involve the electrochemical etching method, hydrothermal-assisted alkali etching method, Lewis acid molten salt treatment method, etc.

5.2.3.1 Electrochemical Etching Method

As a much safer and milder approach, the electrochemical etching (E-etching) process could be regarded as a practicable alternative approach for MXene fabrication. In the E-etching process, the bulk layered MAX phases were generally used as the working electrodes, which would delaminate through the interactions of ions, expansion of the layered electrodes and schism of the weak interlayer bonding (Jiang et al. 2022). The thermal-assisted electrochemical etching method has been used to synthesize MXenes including Ti_2CT_x (Pang et al. 2019), V_2CT_x and Cr_2CT_x, and the as-prepared Cr_2CT_x can reach 25 μm with a flower-like architecture. In addition, the novel molten-salt-assisted E-etching method was also employed to prepare fluorine-free $Ti_3C_2Cl_2$ (Shen et al. 2021), as depicted in Figure 5.2c. Electrons are utilized as reaction agents, and anode etching and cathode reduction can be spatially isolated. Therefore, metallic impurities do not appear in $Ti_3C_2Cl_2$. Furthermore, by adding various inorganic salts, including LiCl-KCl and Ag/AgCl, the surface can be terminated and modified by -Cl to -O and/or -S, which could shorten the modification processes and enrich the types of surface terminations. The E-etching method affords an effective solution to the long-standing problem following HF-involved methods and can be a promising universal method for MXene preparation.

5.2.3.2 Hydrothermal-Assisted Alkali Etching Method

To avoid the employment of HF-relevant methods, the alkali-based etching method was another effective fluorine-free method to prepare delaminated MXene, which was ascribed to its strong binding ability with A elements, especially for Ti_3AlC_2 (Wang, Chen, and Song 2020). In recent study by Amama et al. (Al Mayyahi et al. 2022), the delaminated, OH-terminated MXene (Ti_3C_2-OH) was synthesized via the one-step hydrothermal process that involves alkali (NaOH)-assisted aluminum etching in the presence of hydrazine as the delaminating agent (Figure 5.2d), and the prepared Ti_3C_2-OH sheets were directly produced without any post-synthesis treatment. Soon afterwards, the hydrothermal-assisted NaOH solution etching method was also applied to develop two-dimensional MoB MBene from MoAlB (Xiong et al. 2022), and the prepared MBene exhibited high electrochemical performance. Moreover, KOH was another promising alkali etchant for the preparation of MXenes. Li et al. prepared h-Ti_3C_2/Ti_3AlC_2 catalyst by a relatively mild alkaline etching process using KOH, and the influences of the KOH concentrations and different temperatures on the products were systematically studied (Zhang, He, et al. 2022). These works confirm that the alkaline etching method provides a new strategy for preparing MXene materials.

5.2.3.3 Lewis Acid Molten Salt Treatment Method

The etching process of MAX phases can be realized by the oxidation of the A elements using Lewis acid molten salts such as $ZnCl_2$, $FeCl_2$, $CuCl_2$, and AgCl, which commonly possess a higher electrochemical redox potential (Xiu, Wang, and Qiu 2020). Pang et al. fabricated Ti_3C_2-Cu/Co hybrids through the Lewis acidic molten salt etching method, and the existence of metallic Cu/Co atoms and their interactions with Ti_3C_2 lamellar structures were elucidated for the first time (Bai et al. 2021). The study provided deeper insights into the molten salt mechanism and new ideas for designing MXene-metal based active materials. $CuCl_2$, NaCl, and KCl can also be utilized as molten salt etchants; typically, Ti_3AlC_2 was mixed with molten salts with a molar ratio of 1:3:2:2 for Ti_3AlC_2:$CuCl_2$:NaCl:KCl. Subsequently, the mixture was heated to 680 °C under a heating rate of 4 °C/min with flowing argon in a tube furnace for 24 h (Liu, Orbay, et al. 2022). However, a high temperature and inert atmosphere are usually required for the synthesis.

5.2.3.4 Other Fluorine-Free Etching Methods

A low-temperature "soft chemistry" approach based on photoFenton (P.F.) reaction has been proven to be a proper method for the fabrication of MXenes (Figure 5.2e), and the prepared F-free Ti_3C_2 possessed a high purity of up to 95 wt% (Liang et al. 2022). It was demonstrated that the continuous generation of highly reactive HO$^\bullet$ and $O_2^\bullet$ radicals during the P.F. reaction could weaken the Ti–Al metallic bonds in the MAX precursor and promote the formation of high concentration OH$^-$ anions, which would give rise to the sequential topochemical deintercalation of Al layers. These explorations have opened up new avenues for fluorine-free preparation of MXenes.

5.3 PREPARATION METHODS OF MXENE/POLYMER COMPOSITES

For decades, polymer nanocomposite fabrication has been an attractive strategy to develop multifunctional composites. As a new star of the nanomaterial family, MXene, which unusually combines ceramic and metallic behaviors in its structure (Ronchi, Arantes, and Santos 2019), has collected a variety of attractive properties, including high thermal and mechanical stability, hydrophilicity and electrical conductivity (Xiu, Wang, and Qiu 2020). Moreover, MXenes also possess rich surface chemistry, and their surface can be easily modified with different functional groups to obtain high-quality MXenes (He et al. 2021). Hence, these distinctive features make MXenes promising candidates for applications in the fields of sensors, energy storage, electromagnetic shielding and so forth. Except for this, similar to graphene, one of the main advantages of MXenes is their hydrophilic performance with high conductivity, which offers good interaction with polymers and favors their application in polymeric composites. With deeper research on MXene-based composites, it was found that the composites formed from the combination of MXenes with polymers displayed attractive comprehensive performance. Therefore, the development of MXene/polymer composites has attracted increasing attention. The preparation methods for incorporating nanoparticles into polymer matrices can be mainly summarized into two strategies: in situ polymerization and blending approaches.

5.3.1 IN SITU POLYMERIZATION

Generally, the in situ polymerization method mainly refers to two steps. First, nanofillers such as MXenes are uniformly mixed with polymer monomers or pre-polymers and sometimes accompanied by catalysts or initiators by means of ultrasonic dispersion, mechanical blending, etc. Then, the MXene/polymer nanocomposites were manufactured by the following polymerization reaction (bulk polymerization or solution polymerization) process carried out under certain conditions. MXene sheets are usually modified with active groups, which is beneficial for covalently grafting MXenes onto polymer monomers or prepolymers. In this way, MXenes could be homogeneously dispersed in the polymer matrix; moreover, the composites can be formed in a one-time polymerization course without secondary thermal processing, which can avoid the comprehensive performance from the damage of thermal degradation. However, the presence of MXene nanoparticles will increase the viscosity of the system, which would complicate the polymerization reaction and compounding process.

5.3.2 BLENDING APPROACHES

Blending approaches are the simplest and most widely used method for preparing polymer nanocomposites. This method involves directly dispersing and mixing various inorganic nanoparticles into polymers to obtain nanocomposites. The typical blending methods were further divided into solution blending, emulsion or suspension blending and melt blending methods. Blending approaches have the advantages of being simple to operate, easy to realize industrialization, and suitable for nanoparticles with various shapes.

5.3.2.1 Solution Mixing

The solution mixing method is to first dissolve the polymer monomer or matrix into suitable solvents, and then nanoparticles are added under stirring or sonication to obtain a uniform dispersion. Finally, polymer nanocomposites were prepared by removing the solvent by casting, flocculation, electrospinning and filtration. The solution mixing method is conducive to the uniform dispersion of nanoparticles in the polymer matrix, which is commonly used for preparing polymer nanocomposites. However, this method requires a large amount of solvent, and problems such as solvent removal and poor solubility of many polymers have greatly limited the application of this method.

5.3.2.2 Emulsion Blending

The procedure of the emulsion blending method is very similar to that of the solution blending method, except that the polymer solution is replaced by a polymer emulsion or suspension, which is mainly suitable for situations where the polymer is difficult to dissolve.

5.3.2.3 Melt Blending

Melt compounding is the most attractive strategy because it presents high efficiency, low energy consumption and easy scaling up during processing. MXenes were directly dispersed into a polymer, which was heated to its melting point. After the plasticizing and dispersing processes, polymer nanocomposites were obtained. Compared with solution blending, the method avoids the use of organic reagents, so it is more environmentally friendly, economical, and suitable for industrial production. Nevertheless, the mixing temperature of lending compounding is normally higher than the melting temperature of the polymer; therefore, this method cannot be used for polymers whose decomposition temperature is lower than their melting point.

5.4 APPLICATION OF MXENES IN FLAME-RETARDANT POLYMERS

As mentioned above, owing to the metal and nonmetal components as well as the terminal surface groups, MXene presents many fascinating properties that enable it to be a promising two-dimensional layered nanofiller to be employed in the field of polymer nanocomposites. In particular, this section mainly focuses on the research of MXenes in improving the flame retardancy of polymer nanocomposites.

5.4.1 Utilization of Bare MXenes

In recent years, MXenes have been extensively used as flame-retardant additives for polymers (Chen et al. 2022, Gong et al. 2021). MXene can be easily synthesized via selectively removal the metal layers from the MAX precursor, and simultaneously some new terminating groups including hydroxyl, fluorine and oxygen groups would appear on the surface of MXene after the etching process, which are beneficial for the mixing procedure between polymer matrix and MXene. Owing to its fascinating thermal and flame-retardant properties, neat MXene has been directly employed to prepare fireproof polymer nanocomposites (Hai et al. 2020). Among a variety of MXene materials, $Ti_3C_2T_x$, developed from MAX-phase Ti_3AlC_2, is the most widely investigated two-dimensional MXene in many research fields (Liu, Yang, et al. 2020).

Taking advantage of the strong interaction between PI chains and MXene nanosheets, hydrophobic and flame-retardant PI/MXene aerogels with low density and high separation efficiency of oil and water were fabricated, and the hybrid aerogel also presented excellent reusability under an extreme environment (Wang et al. 2019). In addition, PI/MXene aerogels can be made into PI/MXene films on the basis of a secondary hot-pressing orientation strategy, and the resultant film exhibits good flexibility, an enhanced tensile strength of 102 MPa and a thermal conductivity of 5.12 W/(m·K) (Zhu, Zhao, et al. 2021). Neat MXene has also been applied to other aerogels. Superelastic and responsive anisotropic silica nanofiber/ polyvinylpyrrolidone/MXene hybrid aerogels have been developed via a "soft fiber-stiff sheet" synergistic strategy, which displayed an adjustable combination of insulating and conductive features (Li, Qin, et al. 2022).

In addition to MXenes containing a single transition metal, MXenes with double transition metal components have also attracted great attention. To investigate whether bimetallic MXenes possess higher flame-retardant efficiency, $Ti_3C_2T_x$ and bimetallic MXenes ($Mo_2Ti_2C_3T_x$) were separately introduced into the EP matrix for comparative investigation (Gong et al. 2022). The results show that the EP/ $Mo_2Ti_2C_3T_x$ composite has a notable downward trend in fire risk, which was superior to EP/Ti3C2Tx composites. The anti-dripping ability is an important combustion behavior for flame-retardant polymeric materials.

The potential of MXene in enhancing the anti-dripping performance of polymers was first explored (Li et al. 2019) based on three different types of PVA or PU composites (films, foams, and coatings). The results showed that the oxidization of MXene to C/TiO_2 in the combustion process could form a protective char layer and then improve the anti-dripping performance.

5.4.2 SURFACE MODIFICATION OF MXENE

As a new star of the two-dimensional materials family, MXene-based polymer nanocomposites have been widely reported, but there are still several issues that need to be settled. Self-restacking and agglomeration, poor interfacial compatibility with the polymer matrix and easy oxidation under humid environments have been unavoidable obstructions for its application in the field of polymer nanocomposites, which would inevitably have a negative effect on the flame-retardant performance (He et al. 2019, Si et al. 2019). Hence, manipulating the surface of MXene nanosheets to obtain better or novel functions is crucial for their polymer processing and application (Yu et al. 2019). In general, the surface modification strategies for MXene can be simply divided into three aspects: organic functionalization, inorganic functionalization and blending synergistic technology. In addition, the commonly used compounds for organic flame-retardant modification are mainly divided into three groups: The P series, N series, and Si series and B series. The mechanism is diverse for different series of flame retardants. P-containing flame retardants usually have catalytic char forming or free radical capture effects; Si- or B-containing flame retardants can generate a high-temperature char layer shielding barrier; and N-containing flame retardants can release incombustible inert gases to dilute the combustible gas and inhibit the combustion process. Consequently, a flame-retardant system composed of more than one flame-retardant element can exert an excellent synergistic flame-retardant effect.

5.4.2.1　Organic Functionalized MXene Nanomaterials

The surface organic modification of MXenes has been widely employed to prevent MXene nanosheets from aggregating and improve the interfacial compatibility with the polymer matrix.

5.4.2.1.1　Electrostatic Self-Assembly Approach

Cationic surfactants: The electrostatic self-assembly method has been proven to be an efficient and facile method to solve the self-restacking and aggregation problems of MXene nanosheets. Due to the electrostatic attraction effect, cationic surfactants are suitable organic modifiers for MXene nanosheets. Cationic surfactants such as decyltrimethylammonium bromide (DTAB), dihexadecyldimethylammonium bromide (DDAB), octadecyltrimethylammonium bromide (OTAB) (Si et al. 2019, Zhang, Cao, et al. 2022), cetyltrimethyl ammonium bromide (CTAB), tetrabutyl phosphine chloride (TBPC) (Yu et al. 2019) and poly(diallyldimethylammonium chloride) (PDDA) (He et al. 2019), etc., have been widely employed to alter MXene through electrostatic self-assembly technology. Taking PDDA as an example, the positively charged PDDA molecular chains will be attracted by the negatively charged MXene, and the electrostatic interaction between them prompts the formation of long-chain surfactant PDDA-modified MXene nanosheets, which would prevent the newly peeled-off MXene nanosheets from self-stacking and agglomerating. Moreover, the attached PDDA molecule may also serve as a protective shield for MXene nanosheets, contributing to the delay of oxidation on the surface of MXene nanosheets. Notably, TPU with 3 wt% PDDA-modified MXene exhibited a significantly decreased peak heat release rate (decreased by 50%) and total smoke production (decreased by 47%) (He et al. 2019).

　　Other cationic modifiers: In addition to cationic surfactants, negatively charged MXenes can also interact with positively charged flame-retardant modifiers. Boron dipyrromethene (BODIPY), a fluorochrome composed of multiple elements (i.e., F, B, N), is a good option for fire-retardant applications, and amino-BODIPY can be used to decorate MXene nanosheets based on charge attraction (Figure 5.3a) (Zhu, Wang, et al. 2021). The as-prepared BODIPY-MXene was introduced into ABS resin to prepare flame-retardant nanocomposites. Compared to pure ABS, the enhancement of mechanical properties was achieved by incorporating 0.5 wt% BODIPY-MXene into ABS. Furthermore, the cone calorimeter coupled with FTIR analysis showed that the PHRR (−24.5%), SPR (−18.4%), peak concentration of toxic gases including HCN (−33.5%), NO (−22.0%), N_2O (−46.6%), NH_3 (−76.0%) and CO (−28.8%) of ABS/BODIPY-MXene0.5 were significantly reduced. To overcome the difficulty of uniform dispersion of MXenes, surface manipulation of $Ti_3C_2T_x$ was achieved by 3-aminopropylheptaisobutyl-polyhedral oligomeric silsesquioxane (AP-POSS) via electrostatic interactions (Figure 5.3b) (Yu et al. 2021). The obtained POSS-$Ti_3C_2T_x$ can steadily disperse in many polar solvents. The PS composite modified with POSS-$Ti_3C_2T_x$ shows significantly improved thermal and thermoxidative stability, reflected by 22 °C and 39 °C increments in the temperature of 5 wt% mass loss under nitrogen and air atmospheres, respectively. Moreover, superior flame retardancy, including a 39.1% reduction in PHRR and

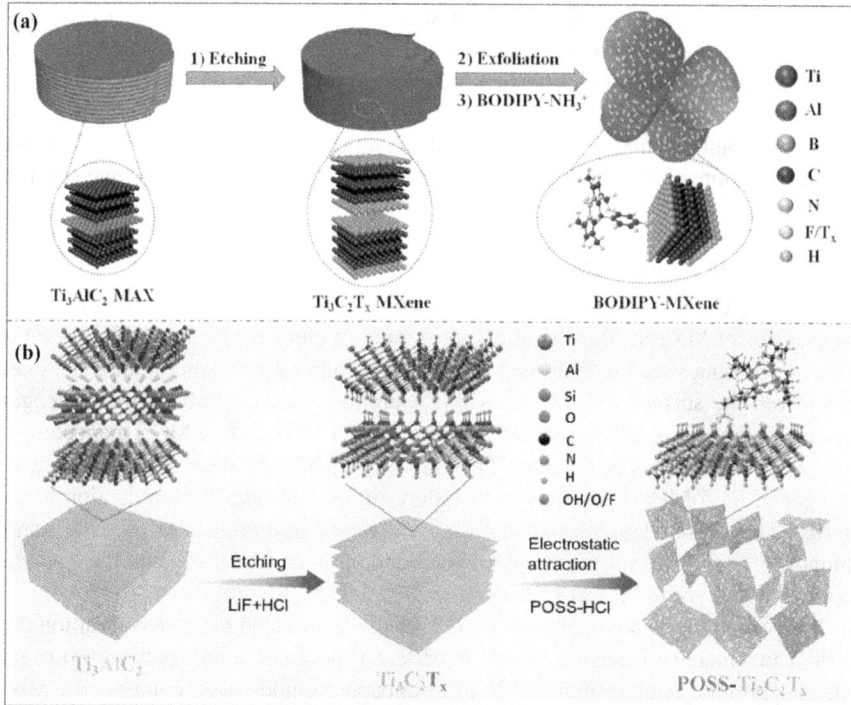

FIGURE 5.3 (a) Schematic preparation route of BODIPY-MXene nanosheets (Reprinted from Zhu, S. et al., *Compos. Part B: Eng.*, 223, 109130, 2021. With permission.) (b) Schematic illustration for the preparation of POSS-Ti$_3$C$_2$T$_x$ (Reprinted from Yu, B. et al., *J. Hazard. Mater.*, 401, 123342, 2021. With permission.)

54.4% and 35.6% reductions in peak CO and CO$_2$ production rates, respectively, was achieved. The excellent fire safety is ascribed to the combination of adsorption, catalytic and barrier effects of the modified Ti$_3$C$_2$T$_x$ nanosheets.

Layer-by-layer assembly: As a classical electrostatic assembly method, layer-by-layer self-assembly technology has been commonly utilized for flame-retardant modification of substrate surfaces. Chitosan (CS), acting as a natural positively charged polysaccharide, has been demonstrated as a synergistic flame-retardant in improving the fire resistance of polymeric materials and has been widely combined with negatively charged flame retardants such as phytic acid (PA), ammonium polyphosphate (APP), and MMT. Shi et al. first adopted Ti$_3$C$_2$T$_x$ and CS to functionalize APP through microencapsulation and layer-by-layer assembly (Liu, Yao, et al. 2021). The as-prepared core-shell-structured hybrid (APP@CS@Ti$_3$C$_2$T$_x$) was then employed to improve the interface compatibility, thermal stability and flame retardancy of TPU composites. The trapping effect of APP@CS combined with the "tortuous path" effect and thermal oxidization action of Ti$_3$C$_2$T$_x$ nanosheets led to the dramatically enhanced fire safety of TPU composites. Moreover, compared to those of pure TPU, the total smoke release (TSR), total CO yield and total CO$_2$ yield of TPU/9.5APP@CS@0.5Ti$_3$C$_2$Tx were significantly reduced by 77.3%, 75.3% and

75.3%, respectively, implying the great smoke and toxicity suppression effects of APP@CS@Ti$_3$C$_2$T$_x$ nanocomposites. In another work, phosphorylated chitosan (PCS) was employed to functionalize MXene to fabricate the PCS-MXene nanohybrid (Luo, Xie, Geng, Dai, et al. 2022), which was then introduced into the TPU matrix via solution mixing followed by the hot-pressing method. The fire risk and smoke emission of the as-prepared TPU/PCS-MXene nanocomposite were effectively suppressed.

5.4.2.1.2 *Functionalization via Hydrogen Bonding*

Considering the abundant active terminal groups such as -OH, =O, and -F left on the surfaces of MXene after the selective etching of elements from MAX, it is apt to afford anchoring sites for hydrogen bonding with other active compounds. Yu et al. engineered the surface of Ti$_3$C$_2$T$_x$ with melamine cyanurate (MCA) via hydrogen bonding interactions and subsequently prepared TPU/Ti$_3$C$_2$T$_x$@MCA nanocomposites (Shi et al. 2020). The resultant TPU/Ti$_3$C$_2$T$_x$@MCA-3.0 displayed a high tensile strength of 61.5 MPa, a high strain at failure of 588%, and a 40% reduction in the PHRR. The outstanding mechanical and fire retardancy properties of TPU nanocomposites are due to interfacial hydrogen bonding in combination with the catalytic action and "labyrinth" effect of two-dimensional Ti$_3$C$_2$T$_x$ nanosheets.

P-N-containing flame retardants (P-N FRs) have attracted extensive attention due to their prominent dispersion in and excellent synergistic effect between nitrogen and phosphorus. Among many P-N FRs, phosphoramide-based compounds, especially polyphosphoramides, have been regarded as flame retardants with high efficiency. Wang and Song et al. proved the synthesis of a novel MXene-phenyl phosphonic diamino-hexane (MXene-PPDA) nanohybrid through the intercalation of the PPDA molecular chain into the MXene interlayer, which enlarged the interlayer spacing of the MXene nanosheets (Figure 5.4a–e). MXene-PPDA can homogeneously disperse in the PLA matrix (Xue et al. 2020). The combustion test results show that incorporating only 1.0 wt% MXene-PPDA can enable PLA to reach a UL-94 V-0 rating, indicating significantly enhanced fire safety. Similarly, Shi et al. developed a novel titanium carbide-polyphenyl phosphate ester amide hybrid (Ti$_3$C$_2$T$_x$-PPPA) via esterification and hydrogen bonding-induced assembly for the subsequent fabrication of thermoplastic polyurethane (TPU) nanocomposites (Figure 5.4f–k) (Liu, Yang, et al. 2022). Compared to pristine TPU, the TPU nanocomposite with only 1.0 wt% Ti$_3$C$_2$T$_x$-PPPA displayed dramatically decreased THR, TSR and CO yields with reductions of 32.6%, 54.4% and 36.8%, respectively. The reduced fire risk of TPU nanocomposites was due to the chemical transformation, the catalytic charring effect and the barrier effect of the Ti$_3$C$_2$T$_x$ nanosheets and PPPA trapping free radicals.

5.4.2.1.3 *Functionalization via Covalent Bonds*

The covalent functionalization of nanomaterials can form a more stable and durable chemical structure, which is widely admired in flame-retardant modification applications. Benefiting from the unique lamellar structure with abundant terminating groups (such as -OH, =O, -F) on the surface, MXene nanosheets can be a versatile platform for covalent functionalization. 9,10-Dihydro-9-oxa-10-phosphaphenanthrene-10-oxide

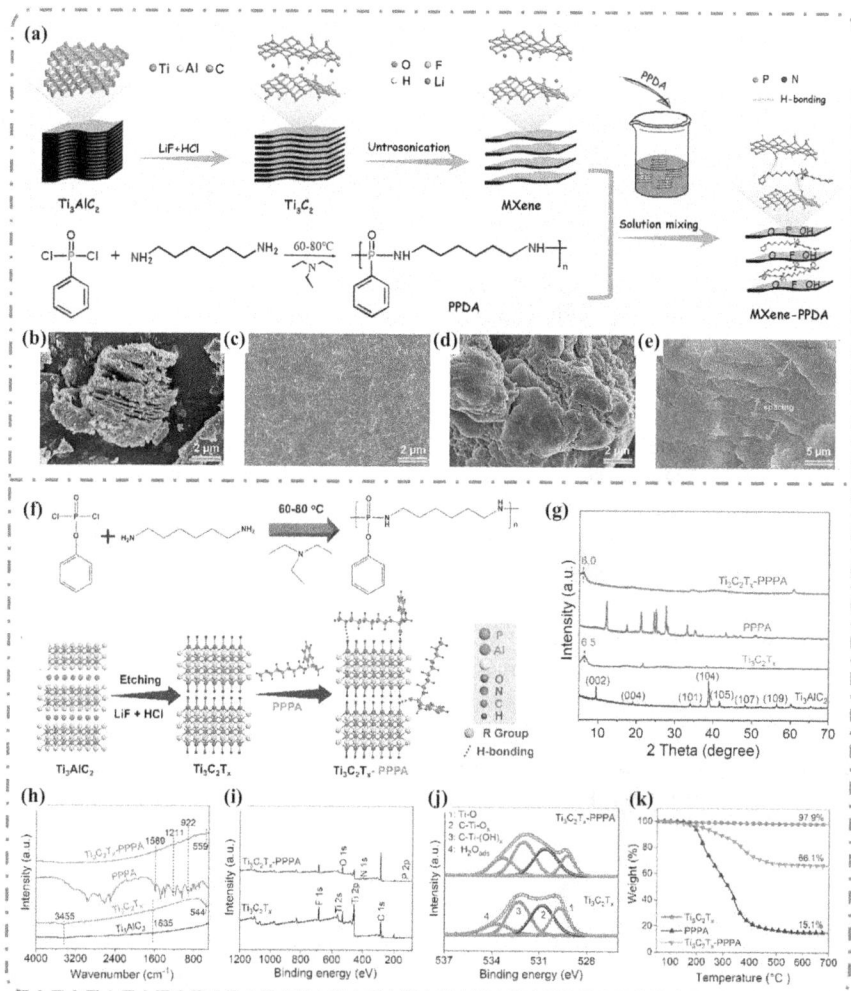

FIGURE 5.4 (a) Illustration for the preparation process of MXene-PPDA; SEM images of (b) Ti₃C₂, (c) MXene, (d, e) MXene-PPDA at different magnifications. (Reprinted from Xue, Y. et al., *Chem. Eng. J.*, 397, 125336, 2020. With permission.) (f) Schematic illustration for the preparation of Ti₃C₂Tₓ-PPPA hybrid; (g) XRD and (h) FTIR spectra of Ti₃AlC₂, PPPA, Ti₃C₂Tₓ and Ti₃C₂Tₓ-PPPA; (i) XPS survey spectra and (j) high-resolution O1s XPS spectra of Ti₃C₂Tₓ and Ti₃C₂Tₓ-PPPA; (k) TGA curves of Ti₃C₂Tₓ, PPPA and Ti₃C₂Tₓ-PPPA under nitrogen. (Reprinted from Liu, C. et al., *Compos. Commun.*, 29, 101055, 2022. With permission.)

(DOPO), a phosphorus-based FR, has been established to exhibit outstanding flame retardant performance in various polymers. Generally, DOPO displays high reactive activity with hydroxyl groups in many compounds based on the Atherton-Todd reaction. Thus, the two-dimensional Ti₃C₂ flakes can be modified by DOPO (Zhou et al. 2021). The combustion test results showed that PLA/Ti₃C₂-DOPO (3 wt %) reached a V-0 rating in the UL-94 test, and the PHRR, THR, CO, and TSP values were significantly reduced compared to those of pure PLA, implying the improved fire-safety performance of PLA composites. The enhanced fire safety of PLA/ Ti_3C_2-DOPO composites is due to the interplay of barrier, catalytic and condensed effects of Ti_3C_2-DOPO nanosheets. Moreover, an "Excellent" level (UPF 50+) in the UV-protection assessment can be observed for PLA composites ascribed to the UV-shielding function of MXene flakes. In addition, 10-(2,5-dihydroxyl phenyl)-9,10-dihydro-9-oxa-10-phosphaphenanthrene-10-oxide (DOPO-HQ) has also been applied to modify $Ti_3C_2T_x$ nanoflakes (Liu, Xu, et al. 2022), and the nanohybrid ($Ti_3C_2T_x$-D-H) can simultaneously improve the fire safety, smoke suppression and mechanical robustness of TPU composites.

5.4.2.1.4 Functionalization via Sol-Gel Technique

With the purpose of enhancing the dispersion of MXene nanosheets in EP and fulfilling the fabrication of multifunctional MXene-based materials, phytic acid (PA)-modified MXene (P-MXene) was prepared using (3-aminopropyl) trie-thoxysilane (APTES) and PA by an assembly technique (Meng et al. 2022). The prepared P-MXene was incorporated into EP to investigate the thermal and electrical conductivity and mechanical and flame-retardant properties. Inspired by the overlap architecture of the placoid scale, an in situ growth technique was employed to construct an MXene derivative (PEPA-IPTS@$Ti_3C_2T_x$) with a placoid scale-like biomimetic structure, which presented significant enhancement in the interfacial compatibility, mechanical properties, and flame retardancy of epoxy resin (EP) (Figure 5.5) (Yin, Wang, et al. 2022). In contrast to neat EP, the impact strength and tensile strength of EP nanocomposites increased by 43.5% and 30.1%, respectively. In addition, the THR and TSP of the EP composites with

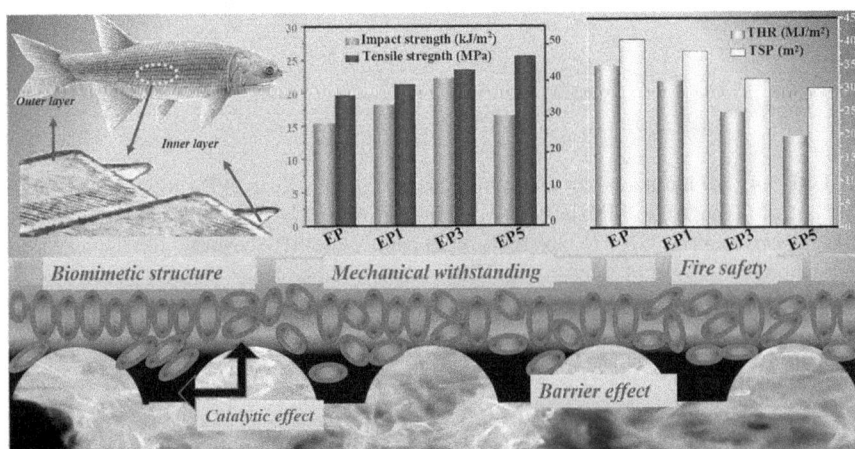

FIGURE 5.5 Fabrication of MXene derivative (PEPA-IPTS@$Ti_3C_2T_x$) with a placoid scale-like biomimetic structure and its improvement in mechanical property and fire safety of epoxy nanocomposites (Reprinted from Yin, Z. et al.,*Chem. Eng. J.*, 431, 133489, 2022. With permission.)

5 wt% PEPA-IPTS@$Ti_3C_2T_x$ were decreased by 44.1% and 26.8%, respectively, indicating a significantly reduced fire risk of the EP nanocomposites.

5.4.2.2 Inorganic Functionalized MXene Nanomaterials

5.4.2.2.1 Functionalization via Inorganic P/Si-Based Materials

At present, extensive research on MXene exfoliation has been carried out through ultrasonication and mechanical milling, and a gas-assisted exfoliation strategy has never been performed for MXenes. Pan et al. fabricated delaminated few-layered Ti_3C_2 nanomaterials through phosphorous vapor evolved from commercial red phosphorous (RP). The RP vapor deposits on the Ti_3C_2 nanosheet surface and simultaneously partially intercalates into the interlayers to develop a novel 2-D RP/Ti_3C_2 nanocomposite directly (Figure 5.6) (Yuan et al. 2021). The RP/Ti_3C_2 nanocomposite was then employed as the flame-retardant additive for EP and efficiently decreased its fire disaster.

SiO₂: To resolve the poor interfacial compatibility of MXene with the polymer matrix, a new type of MXene-based nanoflame retardant (m-MXene) was synthesized

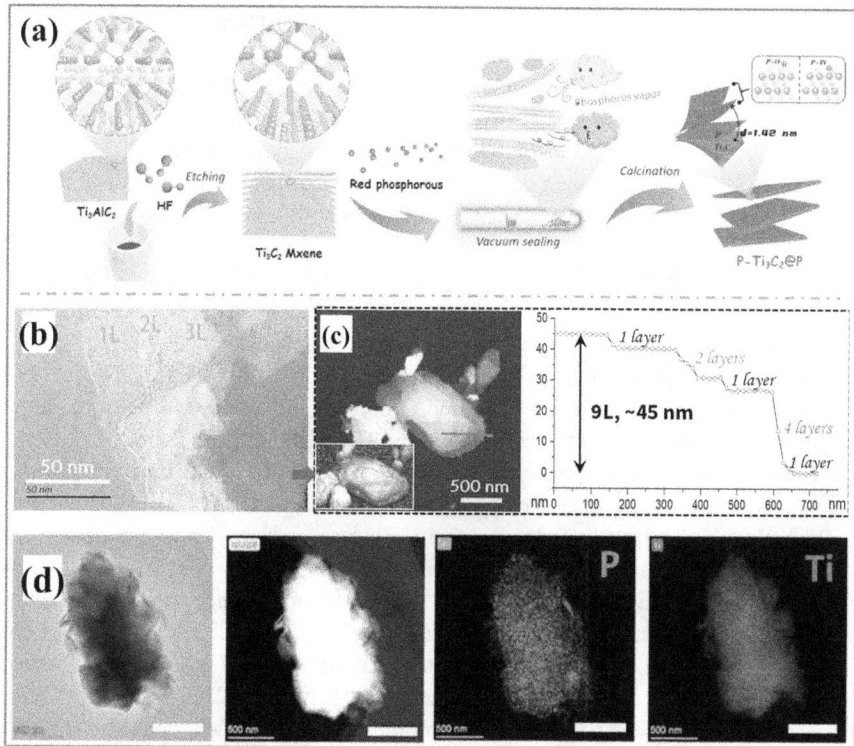

FIGURE 5.6 (a) Schematic illustration for the delamination and intercalation of Ti_3C_2 MXenes by phosphorus vapor; (b) TEM image of the terraced structure, (c) AFM image and height profile for P–Ti_3C_2@P and (d) elemental mapping in the scanning transmission electron microscope–high-angle annular dark field image (scale bar: 500 nm) (Reprinted from Yuan, Y. et al., *ACS Appl. Mater. Interfaces*, 13, 48196–48207, 2021. With permission.)

via in situ deposition of nanosilica on the MXene surface by sol–gel and surface treatment with a silane coupling agent. The PVA nanocomposite with 2% m-MXene presented a 34.9% reduction in PHRR and 32.9% and 97% increases in tensile strength and elongation at break, respectively, compared to neat PVA (Ning et al. 2021). In another work, based on the nacre-inspired thought, a function tunable three-dimensional MXene cross-linking network skeleton decorated by SiO_2 nanoparticles as a bridge linking was built in the CNF nanocomposites for fabrication the fireproof thermal conductive CNF nanocomposite films (Zhan et al. 2021).

Black phosphorus (BP): In light of the abundant phosphorus element, novel two-dimensional layered black phosphorus (BP) nanomaterials have exploited versatile categories to fabricate highly efficient flame retardants. An environmentally friendly two-electrode electrochemical exfoliation technique was carried out to construct the BP-Ti_3C_2 nanohybrids without the use of organic solvents (He et al. 2022). In another work, MXene nanosheets were combined with layered BP via π-π stacking and hydrogen bonding, and the obtained BP-MXene nanohybrids were then covered by polydopamine coating through the in situ polymerization of dopamine to prepare BP-MXene@PDA with better structural stability and dispersibility in the matrix (Luo, Xie, Geng, Chu, et al. 2022). The as-prepared BP-Ti_3C_2 and BP-MXene@PDA nanohybrids were both employed to promote the mechanical performance, thermal stability, and flame retardancy of TPU. Similarly, the well-known lamellar rGO has been utilized to be combined with $Ti_3C_2T_x$ via hydrogen bonding induced assembly (Liu, Wu, et al. 2020), the synthesized $Ti_3C_2T_x$-rGO hybrid displayed good compatibility and strong adhesion with TPU host. With incorporation of 2.0 wt% $Ti_3C_2T_x$-rGO, the TPU composites showed significantly reduced fire hazards.

5.4.2.2.2 Functionalization via Inorganic Metal Nanomaterials

Inorganic metal nanoparticles, such as metallic simple substances, layered double hydroxides (LDHs), metal organic frameworks (MOFs), metal hydroxides, metal sulfides and metal phosphides, which contain transition metals (Fe, Co, Ni, Cu, Mo, Sn, etc.), can greatly improve the suppression of the toxic smoke and gas of polymer composites during the combustion process (Zhou, Chu, et al. 2022). It was found that the multi-metal systems can play a synergistic effect on inhibiting the release of toxic gas and smoke of polymers (Yin, Lu, et al. 2022).

Recently, considering their large specific surface area, MXenes have been employed as templates to grow metal-containing compounds by in situ growth technology to construct highly efficient flame retardants and enhance their compatibility with polymer matrices. MOFs including ZIF-67 (Wan et al. 2022) and ZIF-8 (Yin, Lu, et al. 2022, Zhao et al. 2022), strontium hydroxystannate ($SrSn(OH)_6$) nanorods (Figure 5.7a) (Lu et al. 2022), LDH (Zhou, Gong, et al. 2022), metal phosphides (copper organophosphates (CuP) and zirconium amino-tris-(methylenephosphonate) (Zr-AMP)) and nano-Cu (Figure 5.7b) (Shi et al. 2021) have been anchored onto MXene nanosheets, and the prepared MXene@MOFs were used to reduce the toxic smoke. Zhou et al. synthesized MOF-derived LDH, which was then regularly arranged on the $Ti_3C_2T_x$ surface in a uniformly packaged state (Figure 5.7c) (Zhou, Chu, et al. 2022), which prominently increased the specific surface area and solved the problem of MXene agglomeration. Song et al. reported a novel surface

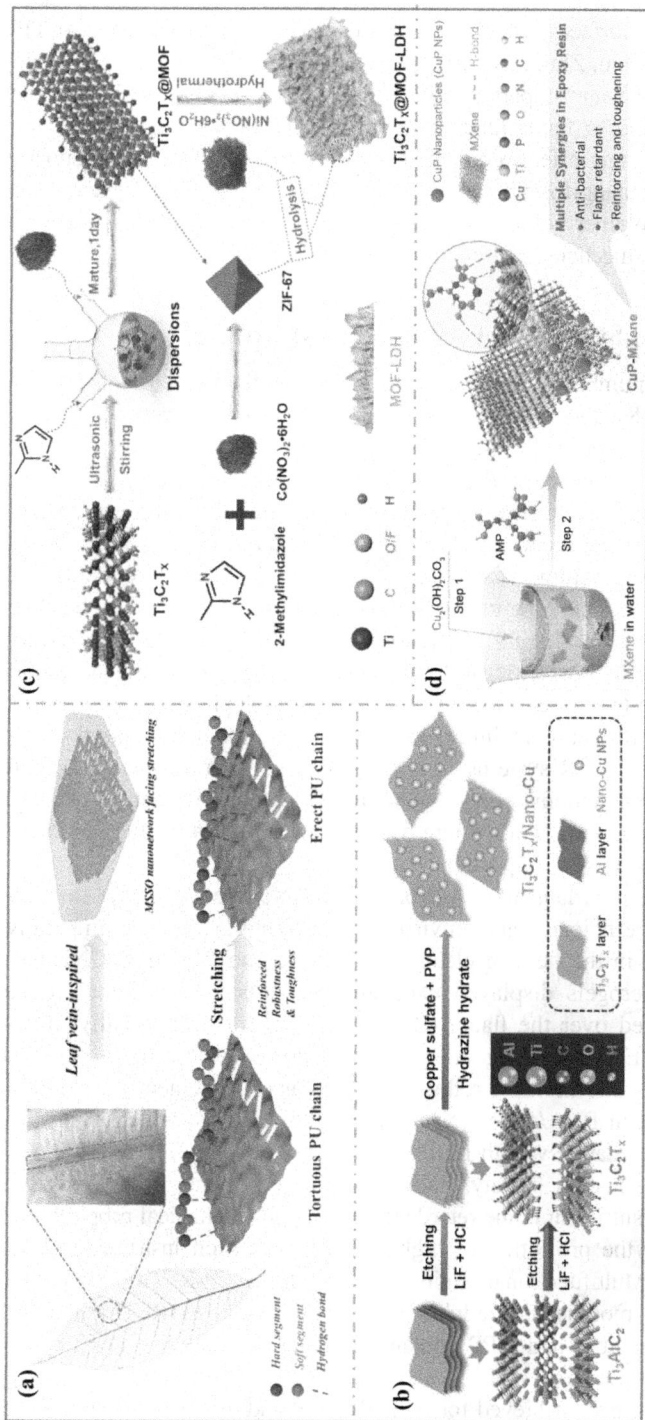

FIGURE 5.7 (a) The leaf vein-inspired strengthening and toughening mechanism of $Ti_3C_2T_x$@$SrSn(OH)_6$ for TPU composites (Lu, J. et al.,*Compos. Part B: Eng.* 228, 109425, 2022. With permission.) (b) Schematic diagrams for synthesis of $Ti_3C_2T_x$/Nano-Cu (Shi, Y. et al., *Compos. Part A: Appl. S.* 150, 106600, 2021. With permission.) (c) The preparation scheme of $Ti_3C_2T_x$@MOF-LDH (Zhou, Y. et al., Chemosphere, 297, 134134, 2022. With permission.) (d) The design and synthesis of CuP-MXene (Reprinted from Liu, L. et al., Chem. Eng. J., 430, 132712, 2022. With permission.)

modification strategy of in situ deposition of CuP (Figure 5.7d) or Zr-AMP nano-particles onto MXene surfaces, which were then introduced into the EP and TPU matrices, respectively (Liu, Zhu, et al. 2022, Liu, Zhu, et al. 2021). It is worth noting that 5.0 wt% CuP-MXene imparted EP with excellent antibacterial efficiency (above 99.9%), outstanding flame retardancy (passed UL-94 V-0 rating), and great reinforcing and toughening effects compared to virgin EP. Ascribed to the superior dual metal catalytic charring-forming effect and physical barrier effect, MXene-based hybrids can endow polymer composites with excellent flame retardant, smoke and toxic gas suppression, mechanical strength and toughness.

5.4.3   THE BLENDED SYNERGISTIC FLAME-RETARDANT SYSTEM

According to a large number of reports, two-dimensional flame retardants (such as GO, CNTs, LDH, MoS_2, and MXene) used alone rarely endow polymer composites with satisfactory flame-retardant ratings. Hence, synergy with other flame retardants is usually required to improve the flame safety of the polymer matrix. It is a very simple and easy way to construct a flame-retardant synergistic system utilizing MXene and other conventional flame retardants by the blending method. For environmental concerns, intumescent flame retardant (IFR), which generally con-tains three basic components of acid, carbonization and blowing sources, has been considered a promising and highly efficient halogen-free flame-retardant additive due to its advantages of low toxicity, low smoke, low corrosion, and no molten dripping during a fire (Wang et al. 2011). Ammonium polyphosphate (APP), as a typical phosphorus/nitrogen-containing FR, has been commonly used as an important component of IFR systems. It is also often used as a synergistic flame retardant for inorganic nanomaterials, including MXene nanosheets. The APP/MXene synergistic flame-retardant system has been widely used for the fabrication of EP, PLA, PVA and CNF composites.

Lightweight flame retardant and electromagnetic interference (EMI) shielding materials with high efficiency and environmental friendliness have imperative practical significance in the field of electronic devices. The fabricated CNF/APP/MXene composite aerogels displayed excellent fire resistance without obvious shrinkage when placed over the flame (Zhang et al. 2021). Additionally, it also showed outstanding EMI shielding effectiveness (55 dB) with a high electrical conductivity (12 S/m) and multiple reflections. Similar enhancement in fire safety can also be observed in PVA/APP/MXene composite aerogels, which passed the V-0 rating and had a relatively high LOI value of 42% (Sheng et al. 2019).

In the 5G communication society, high-performance polymer composites pos-sessing tunable EMI, sufficient flame retardancy and tough mechanical properties are strongly required for the protection of high-power or precision instruments under harsh environments. Multifunctional laminate composites (AEP-RFTn), which were composed of $Ti_3C_2T_x$-modified ramie fabric (RF) as the core layer (RFTn), untreated RF as the reinforcing fiber and APP containing EP as the adhesive (AEP), were fabricated by vacuum pressing (Wang et al. 2022). The UL-94 V0 rating and a high LOI value (40%) have been achieved for AEP-RFT1.2, and the PHRR and THR are decreased by 77.6% and 51.3%, respectively. In addition, its resultant SE can surpass

30 dB in the X band, which is sufficient for commercial usage. In another work, a traditional IFR system consisting of APP, dipentaerythritol, and melamine (P-C-N system) together with MXene was employed to fabricate an EP intumescent fire-resistant coating (TEIFC) (Huang et al. 2021). Except for the significantly improved fire-retardant performance, the integration of P-C-N and $Ti_3C_2T_x$ enhanced the thermal stability and Shore hardness of the coating. The influence of the IFR/MXene system on the flame retardancy of the PLA composite has also been studied (Huang et al. 2020). After blending with 0.5 wt% MXene and 11.5 wt% IFR, the PLA composites achieved the UL-94 V-0 level with an LOI value of 33.0%.

APP-derived flame retardants, such as silicon microencapsulated ammonium polyphosphate (SiAPP), have also been combined with cationic benzyldimethylhex-adecylammonium chloride (HDBAC)-modified $Ti_3C_2T_x$ (HDBAC-$Ti_3C_2T_x$), which was then incorporated into the PLA matrix (Figure 5.8a) (Shi et al. 2022). PLA composites containing 13.0 wt% SiAPP and 2.0 wt% HDBAC-$Ti_3C_2T_x$ can pass the UL-94 V-0 level with an LOI value of 33.3%, and the PHRR, THR, SPR and TSR are reduced by 49.8%, 31.9%, 60.3% and 52.7%, respectively. The formed phosphorus-containing species and TiO_2 contributed to producing intact expanded char layers, which were integrated with the barrier and catalytic effects of modified $Ti_3C_2T_x$ nanosheets to be responsible for significantly suppressing smoke and toxic gases and enhancing the fire resistance of PLA composites.

Some novel intumescent flame retardants are also used for the construction of synergistic flame-retardant systems with MXene nanosheets. Poly(vinylpho-sphonic acid) (PVPA) and polyethylenepolyamine (PEPA), halogen-free flame retardants with P and N elements, respectively, can be used together as intu-mescent flame-retardant systems. Yan et al. used PVPA and PEPA to improve the flame retardancy of $Ti_3C_2T_x$/PVA nanocomposites (Yan et al. 2022). The thermal stability, flame-retardant and mechanical properties of MXene/PVA nano-composites were investigated. Gelatin, derived from collagen, is a green natural biomaterial. It contains many nitrogen elements and can be used as a novel blowing source. The gelatin was integrated with phosphated lignocellulose nanofibrils (PLCNFs) and MXene to fabricate a hybrid composite aerogel (PGM) with high porosity and ultralight weight (Figure 5.8b) (Li, Chen, et al. 2022). Ascribed to the ternary synergistic flame-retardant effects, the PGM aerogel exhibited excellent fire retardancy with fire-warning performance. Moreover, great specific EMI shielding effectiveness (14230 $dB/(cm^2 \cdot g)$) was also achieved for the PGM aerogel. The prepared PGM aerogel can be a potential candidate in both fire protection and EMI shielding fields.

5.5 CONCLUSIONS AND PERSPECTIVES

Since its first discovery in 2011, MXene has attracted huge research interests in the fabrication of MXene-based polymer composites which have been extensively used in the fields of anti-corrosion, anti-friction, separation, biomedical, wearable devices, sensors, batteries and energy storage, electromagnetic wave absorption and interfer-ence shielding and so on. Among them, the application of MXene/polymer compo-sites in the flame-retardant field is an important branch and systematic summary is

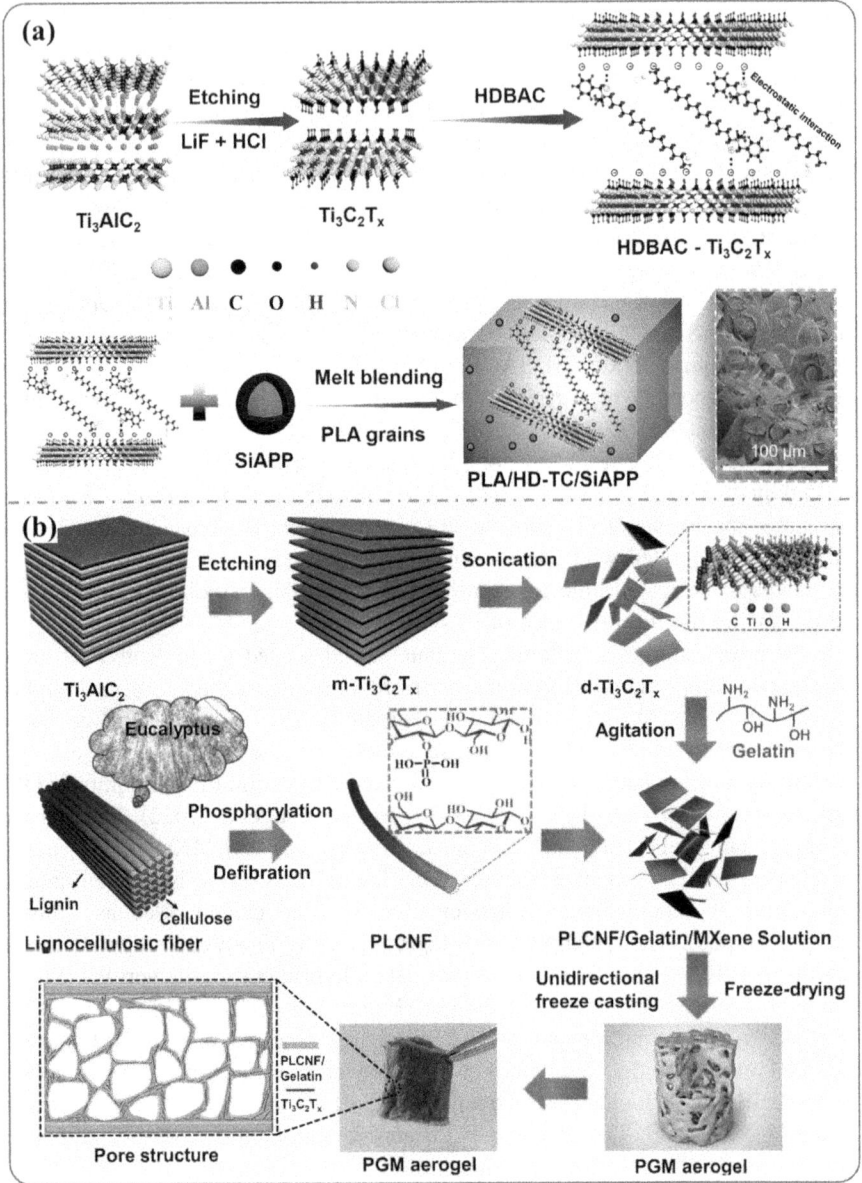

FIGURE 5.8 (a) Schematic diagrams for HDBAC-Ti$_3$C$_2$T$_x$ and PLA/HD-TC/SiAPP composites (Reprinted from Shi, Y. et al., Compos. Part B: Eng., 236, 109792, 2022. With permission.) (b) Fabrication of the phosphated lignocellulose nanofibrils/gelatin/MXene aerogel. (Reprinted from Li, Y. et al., Chem. Eng. J., 431, 133907, 2022. With permission.)

limited in this regard. The chapter focused on the synthetic routes and flame-retardant applications of the emerging two-dimensional MXene nanosheets in polymer composite fields.

For the fabrication of MXene nanosheets, the fluoride acid or fluoride salts etching method have widely used. Moreover, in order to avoid the employment of HF-relevant methods, several novel fluorine-free etching strategies including the electrochemical etching method, hydrothermal-assisted alkali etching method and Lewis acid molten salt treatment method have attracted increasing attention, which could overcome the disadvantages including high toxicity and long etching course of the conventional HF-containing synthetic routes. The fabricated MXene with tailorable surface chemistries, unique lamellar structure, and high thermal and mechanical stability has become a promising two-dimensional nanoflame retardant that has been applied to different polymer systems.

However, similar to graphene oxide nanosheets, the issues such as self-restacking and agglomeration, poor interfacial compatibility with polymer matrix and easy oxidation under humid environment has significantly blocked its application in the field of polymer nanocomposites, which would inevitably bring negative effect on flame-retardant performance. Hence, surface modification is crucial for MXene to obtain better or novel functions. A series of MXene-based flame-retardant hybrids emerge based on covalent and non-covalent functionalization with some traditional organic and inorganic flame-retardant components, which show high dispersibility and compatibility with polymer matrix. In addition, the constructed MXene-based hybrid systems with excellent catalytic charring and barrier effects could endow polymer composites with high-efficient flame retardancy and smoke suppression. Besides, due to the unique layered nature and electrical conductivity, MXene/polymer composites also display excellent electromagnetic interference/microwave absorption functions, antibacterial property as well as fire-warning performance, etc. With the rapid development of the economy, the demand for polymer nanocomposites with multiple functions is increasing. Therefore, the development of multifunctional and high-performance MXene/polymer composites with excellent flame retardancy, fire alarm, and electromagnetic shielding efficiency has important practical significance and broad application prospects.

REFERENCES

Al Mayyahi, A., S. Sarker, B. M. Everhart, X. Q. He, and P. B. Amama. 2022. "One-step fluorine-free synthesis of delaminated, OH-terminated Ti_3C_2: high photocatalytic NO_x storage selectivity enabled by coupling TiO_2 and Ti_3C_2-OH." *Materials Today Communications* no. 32. doi: Artn 103835 10.1016/J.Mtcomm.2022.103835.

Alhabeb, M., K. Maleski, T. S. Mathis, A. Sarycheva, C. B. Hatter, S. Uzun, A. Levitt, and Y. Gogotsi. 2018. "Selective etching of silicon from Ti_3SiC_2 (MAX) to obtain 2D titanium carbide (MXene)." *Angewandte Chemie-International Edition* no. 57 (19):5444–5448. doi: 10.1002/anie.201802232.

Alhabeb, M., K. Maleski, B. Anasori, P. Lelyukh, L. Clark, S. Sin, and Y. Gogotsi. 2017. "Guidelines for synthesis and processing of two-dimensional titanium carbide ($Ti_3C_2T_x$ MXene)." *Chemistry of Materials* no. 29 (18):7633–7644. doi: 10.1021/acs.chemmater.7b02847.

Bai, Y., C. Liu, T. Chen, W. Li, S. Zheng, Y. Pi, Y. Luo, and H. Pang. 2021. "MXene-copper/cobalt hybrids via Lewis acidic molten salts etching for high performance symmetric supercapacitors." *Angewandte Chemie International Edition* no. 60 (48):25318–25322. doi: 10.1002/anie.202112381.

Cai, M., X. Q. Fan, H. Yan, Y. T. Li, S. J. Song, W. Li, H. Li, Z. B. Lu, and M. H. Zhu. 2021. "In situ assemble $Ti_3C_2T_x$ MXene@MgAl-LDH heterostructure towards anticorrosion and antiwear application." *Chemical Engineering Journal* no. 419. doi: Artn 130050 10.1016/J.Cej.2021.130050.

Chaudhari, N. K., H. Jin, B. Kim, D. S. Baek, S. H. Joo, and K. Lee. 2017. "MXene: an emerging two-dimensional material for future energy conversion and storage applications." *Journal of Materials Chemistry A* no. 5 (47):24564–24579. doi: 10.1039/c7ta09094c.

Chen, N. J., W. Q. Yang, and C. F. Zhang. 2021. "Perspectives on preparation of two-dimensional MXenes." *Science and Technology of Advanced Materials* no. 22 (1):917–930. doi: 10.1080/14686996.2021.1972755.

Chen, W. H., P. J. Liu, Y. Liu, and Z. X. Liu. 2022. "Recent advances in two-dimensional $Ti_3C_2T_x$ MXene for flame retardant polymer materials." *Chemical Engineering Journal* no. 446:137239. doi: Artn 137239 10.1016/J.Cej.2022.137239.

Chen, X., Y. Zhao, L. Li, Y. Wang, J. Wang, J. Xiong, S. Du, P. Zhang, X. Shi, and J. Yu. 2021. "MXene/polymer nanocomposites: preparation, properties, and applications." *Polymer Reviews* no. 61 (1):80–115. doi: 10.1080/15583724.2020.1729179.

Dong, L. M., C. Ye, L. L. Zheng, Z. F. Gao, and F. Xia. 2020. "Two-dimensional metal carbides and nitrides (MXenes): preparation, property, and applications in cancer therapy." *Nanophotonics* no. 9 (8):2125–2145. doi: 10.1515/nanoph-2019-0550.

Feng, A. H., Y. Yu, F. Jiang, Y. Wang, L. Mi, Y. Yu, and L. X. Song. 2017. "Fabrication and thermal stability of NH_4HF_2-etched Ti_3C_2 MXene." *Ceramics International* no. 43 (8):6322–6328. doi: 10.1016/j.ceramint.2017.02.039.

Feng, A. H., Y. Yu, Y. Wang, F. Jiang, Y. Yu, L. Mi, and L. X. Song. 2017. "Two-dimensional MXene Ti_3C_2 produced by exfoliation of Ti_3AlC_2." *Materials & Design* no. 114:161–166. doi: 10.1016/j.matdes.2016.10.053.

Fu, B., J. X. Sun, C. Wang, C. Shang, L. J. Xu, J. B. Li, and H. Zhang. 2021. "MXenes: synthesis, optical properties, and applications in ultrafast photonics." *Small* no. 17 (11). doi: Artn 2006054 10.1002/Smll.202006054.

Gao, L. F., C. Li, W. C. Huang, S. Mei, H. Lin, Q. Ou, Y. Zhang, J. Guo, F. Zhang, S. X. Xu, and H. Zhang. 2020. "MXene/polymer membranes: synthesis, properties, and emerging applications." *Chemistry of Materials* no. 32 (5):1703–1747. doi: 10.1021/acs.chemmater.9b04408.

Ghosh, M., S. Szunerits, N. Cao, S. Kurungot, and R. Boukherroub. 2022. "Single-step synthesis of exfoliated $Ti_3C_2T_x$ MXene through NaBF4/HCl etching as electrode material for asymmetric supercapacitor." *Chemistryselect* no. 7 (19). doi: ARTN e202201166 10.1002/slct.202201166.

Gong, K., L. Yin, H. Pan, S. Mao, L. Liu, and K. Zhou. 2022. "Novel exploration of the flame retardant potential of bimetallic MXene in epoxy composites." *Composites Part B: Engineering* no. 237:109862. doi: 10.1016/j.compositesb.2022.109862.

Gong, K., K. Zhou, X. Qian, C. Shi, and B. Yu. 2021. "MXene as emerging nanofillers for high-performance polymer composites: a review." *Composites Part B: Engineering* no. 217:108867. doi: 10.1016/j.compositesb.2021.108867.

Guo, Y., S. Jin, L. Wang, P. He, Q. Hu, L.-Z. Fan, and A. Zhou. 2020. "Synthesis of two-dimensional carbide Mo_2CT_x MXene by hydrothermal etching with fluorides and its thermal stability." *Ceramics International* no. 46 (11, Part B):19550–19556. doi: 10.1016/j.ceramint.2020.05.008.

Hai, Y., S. Jiang, C. Zhou, P. Sun, Y. Huang, and S. Niu. 2020. "Fire-safe unsaturated polyester resin nanocomposites based on MAX and MXene: a comparative investigation of their properties and mechanism of fire retardancy." *Dalton Transactions* no. 49 (18):5803–5814. doi: 10.1039/D0DT00686F.

He, L. X., X. Jiang, X. Zhou, Z. X. Li, F. K. Chu, X. Wang, W. Cai, L. Song, and Y. Hu. 2022. "Integration of black phosphorene and MXene to improve fire safety and mechanical properties of waterborne polyurethane." *Applied Surface Science* no. 581:152386. doi: Artn 152386 10.1016/J.Apsusc.2021.152386.

He, Lingxin, Junling Wang, Bibo Wang, Xin Wang, Xia Zhou, Wei Cai, Xiaowei Mu, Yanbei Hou, Yuan Hu, and Lei Song. 2019. "Large-scale production of simultaneously exfoliated and Functionalized Mxenes as promising flame retardant for polyurethane." *Composites Part B: Engineering* no. 179:107486. doi: 10.1016/j.compositesb.2019.107486.

He, S. S., X. Sun, H. Zhang, C. D. Yuan, Y. P. Wei, and J. J. Li. 2021. "Preparation strategies and applications of MXene-polymer composites: a review." *Macromolecular Rapid Communications* no. 42 (19). doi: Artn 2100324 10.1002/Marc.202100324.

Huang, H., D. Dong, W. Li, X. Zhang, L. Zhang, Y. Chen, X. Sheng, and X. Lu. 2020. "Synergistic effect of MXene on the flame retardancy and thermal degradation of intumescent flame retardant biodegradable poly (lactic acid) composites." *Chinese Journal of Chemical Engineering* no. 28 (7):1981–1993. doi: 10.1016/j.cjche.2020.04.014.

Huang, L., L. Ding, and H. Wang. 2021. "MXene-based membranes for separation applications." *Small Science* no. 1 (7):2100013. doi: 10.1002/smsc.202100013.

Huang, S., L. Wang, Y. C. Li, C. B. Liang, and J. L. Zhang. 2021. "Novel $Ti_3C_2T_x$ MXene/ epoxy intumescent fire-retardant coatings for ancient wooden architectures." *Journal of Applied Polymer Science* no. 138 (27):50649. doi: ARTN e50649 10.1002/app.50649.

Jiang, L. B., D. Zhou, J. J. Yang, S. Y. Zhou, H. Wang, X. Z. Yuan, J. Liang, X. D. Li, Y. N. Chen, and H. Li. 2022. "2D single- and few-layered MXenes: synthesis, applications and perspectives." *Journal of Materials Chemistry A* no. 10 (26):13651–13672. doi: 10.1039/d2ta01572b.

Khazaei, M., A. Mishra, N. S. Venkataramanan, A. K. Singh, and S. Yunoki. 2019. "Recent advances in MXenes: from fundamentals to applications." *Current Opinion in Solid State & Materials Science* no. 23 (3):164–178. doi: 10.1016/j.cossms.2019.01.002.

Li, B.-X., L. Qin, D. Yang, Z. Luo, T. Zhao, and Z.-Z. Yu. 2022. "Superelastic and responsive anisotropic silica nanofiber/polyvinylpyrrolidone/MXene hybrid aerogels for efficient thermal insulation and overheating alarm applications." *Composites Science and Technology* no. 225:109484. doi: 10.1016/j.compscitech.2022.109484.

Li, L., X. Y. Liu, J. F. Wang, Y. Y. Yang, Y. X. Cao, and W. J. Wang. 2019. "New application of MXene in polymer composites toward remarkable anti-dripping performance for flame retardancy." *Composites Part A: Applied Science and Manufacturing* no. 127:105649. doi: ARTN 105649 10.1016/j.compositesa.2019.105649.

Li, Q. T., Y. Q. Li, and W. Zeng. 2021. "Preparation and application of 2D MXene-based gas sensors: a review." *Chemosensors* no. 9 (8). doi: ARTN 225 10.3390/ chemosensors9080225.

Li, Y., Y. Chen, X. He, Z. Xiang, T. Heinze, and H. Qi. 2022. "Lignocellulose nanofibril/ gelatin/MXene composite aerogel with fire-warning properties for enhanced electromagnetic interference shielding performance." *Chemical Engineering Journal* no. 431:133907. doi: 10.1016/j.cej.2021.133907.

Liang, L., L. Niu, T. Wu, D. Zhou, and Z. Xiao. 2022. "Fluorine-free fabrication of MXene via photo-Fenton approach for advanced lithium–sulfur batteries." *ACS Nano* no. 16 (5):7971–7981. doi: 10.1021/acsnano.2c00779.

Lin, X. P., Z. J. Li, J. M. Qiu, Q. Wang, J. X. Wang, H. Zhang, and T. K. Chen. 2021. "Fascinating MXene nanomaterials: emerging opportunities in the biomedical field." *Biomaterials Science* no. 9 (16):5397–5431. doi: 10.1039/d1bm00526j.

Liu, A., Q. Yang, X. Ren, F. Meng, L. Gao, M. Gao, Y. Yang, T. Ma, and G. Wu. 2020. "Energy- and cost-efficient NaCl-assisted synthesis of MAX-phase Ti_3AlC_2 at lower temperature." *Ceramics International* no. 46 (5):6934–6939. doi: 10.1016/j.ceramint.2 019.11.008.

Liu, C., W. Wu, Y. Q. Shi, F. Q. Yang, M. H. Liu, Z. X. Chen, B. Yu, and Y. Z. Feng. 2020. "Creating MXene/reduced graphene oxide hybrid towards highly fire safe thermoplastic polyurethane nanocomposites." *Composites Part B: Engineering* no. 203:108486. doi: ARTN 108486 10.1016/j.compositesb.2020.108486.

Liu, C., D. Yang, M. N. Sun, G. J. Deng, B. H. Jing, K. Wang, Y. Q. Shi, L. B. Fu, Y. Z. Feng, Y. C. Lv, and M. H. Liu. 2022. "Phosphorous-nitrogen flame retardants engineering MXene towards highly fire safe thermoplastic polyurethane." *Composites Communications* no. 29. doi: Artn 101055 10.1016/J.Coco.2021.101055.

Liu, C., K. Xu, Y. Shi, J. Wang, S. Ma, Y. Feng, Y. Lv, F. Yang, M. Liu, and P. Song. 2022. "Fire-safe, mechanically strong and tough thermoplastic polyurethane/MXene nanocomposites with exceptional smoke suppression." *Materials Today Physics* no. 22:100607. doi: 10.1016/j.mtphys.2022.100607.

Liu, C., A. Yao, K. Chen, Y. Shi, Y. Feng, P. Zhang, F. Yang, M. Liu, and Z. Chen. 2021. "MXene based core-shell flame retardant towards reducing fire hazards of thermoplastic polyurethane." *Composites Part B: Engineering* no. 226:109363. doi: 10.1016/j.compositesb.2021.109363.

Liu, F., A. Zhou, J. Chen, J. Jia, W. Zhou, L. Wang, and Q. Hu. 2017. "Preparation of Ti_3C_2 and Ti_2C MXenes by fluoride salts etching and methane adsorptive properties." *Applied Surface Science* no. 416:781–789. doi: 10.1016/j.apsusc.2017.04.239.

Liu, L., M. Zhu, Z. Ma, X. Xu, S. M. Seraji, B. Yu, Z. Sun, H. Wang, and P. Song. 2022. "A reactive copper-organophosphate-MXene heterostructure enabled antibacterial, self-extinguishing and mechanically robust polymer nanocomposites." *Chemical Engineering Journal* no. 430:132712. doi: 10.1016/j.cej.2021.132712.

Liu, L., M. Zhu, Y. Shi, X. Xu, Z. Ma, B. Yu, S. Fu, G. Huang, H. Wang, and P. Song. 2021. "Functionalizing MXene towards highly stretchable, ultratough, fatigue- and fire-resistant polymer nanocomposites." *Chemical Engineering Journal* no. 424:130338. doi: 10.1016/j.cej.2021.130338.

Liu, L., M. Orbay, S. Luo, S. Duluard, H. Shao, J. Harmel, P. Rozier, P.-L. Taberna, and P. Simon. 2022. "Exfoliation and delamination of $Ti_3C_2T_x$ MXene prepared via molten salt etching route." *ACS Nano* no. 16 (1):111–118. doi: 10.1021/acsnano.1c08498.

Lu, J., P. Jia, C. Liao, Z. Xu, F. Chu, M. Zhou, B. Yu, B. Wang, and L. Song. 2022. "Leaf vein-inspired engineering of MXene@SrSn(OH)6 nanorods towards super-tough elastomer nanocomposites with outstanding fire safety." *Composites Part B: Engineering* no. 228:109425. doi: 10.1016/j.compositesb.2021.109425.

Luo, Y., Y. H. Xie, W. Geng, J. H. Chu, H. Wu, D. L. Xie, X. X. Sheng, and Y. Mei. 2022. "Boosting fire safety and mechanical performance of thermoplastic polyurethane by the face-to-face two-dimensional phosphorene/MXene architecture." *Journal of Materials Science & Technology* no. 129:27–39. doi: 10.1016/j.jmst.2022.05.003.

Luo, Y., Y. H. Xie, W. Geng, G. F. Dai, X. X. Sheng, D. L. Xie, H. Wu, and Y. Mei. 2022. "Fabrication of thermoplastic polyurethane with functionalized MXene towards high mechanical strength, flame-retardant, and smoke suppression properties." *Journal of Colloid and Interface Science* no. 606:223–235. doi: 10.1016/j.jcis.2021.08.025.

Meng, W., Y. Song, K. Zhai, H. Di, H. Ma, X. Bi, J. Xu, and L. Fang. 2022. "Assembling MXene with bio-phytic acid: improving the fire safety and comprehensive properties of epoxy resin." *Polymer Testing* no. 110:107564. doi: 10.1016/j.polymertesting.2022.107564.

Natu, V., R. Pai, M. Sokol, M. Carey, V. Kalra, and M. W. Barsoum. 2020. "2D $Ti_3C_2T_z$ MXene synthesized by water-free etching of Ti_3AlC_2 in polar organic solvents." *Chem* no. 6 (3):616–630. doi: 10.1016/j.chempr.2020.01.019.

Ning, H. Z., Z. Y. Ma, Z. H. Zhang, D. Zhang, and Y. H. Wang. 2021. "A novel multi-functional flame retardant MXene/nanosilica hybrid for poly(vinyl alcohol) with simultaneously improved mechanical properties." *New Journal of Chemistry* no. 45 (9):4292–4302. doi: 10.1039/d0nj04897f.

Pang, S. Y., Y. T. Wong, S. G. Yuan, Y. Liu, M. K. Tsang, Z. B. Yang, H. T. Huang, W. T. Wong, and J. H. Hao. 2019. "Universal strategy for HF-free facile and rapid synthesis of two-dimensional MXenes as multifunctional energy materials." *Journal of the American Chemical Society* no. 141 (24):9610–9616. doi: 10.1021/jacs.9b02578.

Ronchi, R. M., J. T. Arantes, and S. F. Santos. 2019. "Synthesis, structure, properties and applications of MXenes: current status and perspectives." *Ceramics International* no. 45 (15):18167–18188. doi: 10.1016/j.ceramint.2019.06.114.

Shen, M., W. Jiang, K. Liang, S. Zhao, R. Tang, L. Zhang, and J.-Q. Wang. 2021. "One-pot green process to synthesize MXene with controllable surface terminations using molten salts." *Angewandte Chemie International Edition* no. 60 (52):27013–27018. doi: 10.1002/anie.202110640.

Sheng, X., S. Li, Y. Zhao, D. Zhai, L. Zhang, and X. Lu. 2019. "Synergistic effects of two-dimensional MXene and ammonium polyphosphate on enhancing the fire safety of polyvinyl alcohol composite aerogels." *Polymers* no. 11 (12):1964.

Shi, Y., C. Liu, Z. Duan, B. Yu, M. Liu, and P. Song. 2020. "Interface engineering of MXene towards super-tough and strong polymer nanocomposites with high ductility and excellent fire safety." *Chemical Engineering Journal* no. 399:125829. doi: 10.1016/j.cej.2020.125829.

Shi, Y., C. Liu, L. Fu, Y. Feng, Y. Lv, Z. Wang, M. Liu, and Z. Chen. 2021. "Highly efficient MXene/nano-Cu smoke suppressant towards reducing fire hazards of thermoplastic polyurethane." *Composites Part A: Applied Science and Manufacturing* no. 150:106600. doi: 10.1016/j.compositesa.2021.106600.

Shi, Y., Z. Wang, C. Liu, H. Wang, J. Guo, L. Fu, Y. Feng, L. Wang, F. Yang, and M. Liu. 2022. "Engineering titanium carbide ultra-thin nanosheets for enhanced fire safety of intumescent flame retardant polylactic acid." *Composites Part B: Engineering* no. 236:109792. doi: 10.1016/j.compositesb.2022.109792.

Si, J.-Y., B. Tawiah, W.-L. Sun, B. Lin, C. Wang, A. C. Y. Yuen, B. Yu, A. Li, W. Yang, H.-D. Lu, Q. N. Chan, and G. H. Yeoh. 2019. "Functionalization of MXene nanosheets for polystyrene towards high thermal stability and flame retardant properties." *Polymers* no. 11 (6):976.

Tran, M. H., T. Schafer, A. Shahraei, M. Durrschnabel, L. Molina-Luna, U. I. Kramm, and C. S. Birkel. 2018. "Adding a new member to the MXene family: synthesis, structure, and electrocatalytic activity for the hydrogen evolution reaction of $V_4C_3T_x$." *ACS Applied Energy Materials* no. 1 (8):3908–3914. doi: 10.1021/acsaem.8b00652.

Wan, M., C. Shi, X. Qian, Y. Qin, J. Jing, and H. Che. 2022. "Metal-organic framework ZIF-67 functionalized MXene for enhancing the fire safety of thermoplastic polyurethanes." *Nanomaterials* no. 12 (7):1142.

Wang, B., Z. Yin, Y. Zhang, P. Jia, R. He, F. Yu, and Y. Hu. 2022. "Robust flame-retardant, super mechanical laminate epoxy composites with tunable electromagnetic interference shielding." *Materials Today Physics* no. 26:100724. doi: 10.1016/j.mtphys.2022.100724.

Wang, C. D., S. M. Chen, and L. Song. 2020. "Tuning 2D MXenes by surface controlling and interlayer engineering: methods, properties, and synchrotron radiation characterizations." *Advanced Functional Materials* no. 30 (47). doi: Artn 2000869 10.1002/Adfm.202000869.

Wang, L., D. Liu, W. Lian, Q. Hu, X. Liu, and A. Zhou. 2020. "The preparation of V_2CT_x by facile hydrothermal-assisted etching processing and its performance in lithium-ion battery." *Journal of Materials Research and Technology* no. 9 (1):984–993. doi: 10.1016/j.jmrt.2019.11.038.

Wang, N.-N., H. Wang, Y.-Y. Wang, Y.-H. Wei, J.-Y. Si, A. C. Y. Yuen, J.-S. Xie, B. Yu, S.- E. Zhu, H.-D. Lu, W. Yang, Q. N. Chan, and G.-H. Yeoh. 2019. "Robust, light-weight, hydrophobic, and fire-retarded polyimide/MXene aerogels for effective oil/

water separation." *ACS Applied Materials & Interfaces* no. 11 (43):40512–40523. doi: 10.1021/acsami.9b14265.

Wang, X., L. Song, H. Yang, H. Lu, and Y. Hu. 2011. "Synergistic effect of graphene on antidripping and fire resistance of intumescent flame retardant poly(butylene succinate) composites." *Industrial & Engineering Chemistry Research* no. 50 (9):5376–5383. doi: 10.1021/ie102566y.

Xiong, W., X. Y. Feng, Y. Xiao, T. Huang, X. Y. Li, Z. C. Huang, S. H. Ye, Y. L. Li, X. Z. Ren, X. Z. Wang, X. P. Ouyang, Q. L. Zhang, and J. H. Liu. 2022. "Fluorine-free prepared two-dimensional molybdenum boride (MBene) as a promising anode for lithium-ion batteries with superior electrochemical performance." *Chemical Engineering Journal* no. 446. doi: Artn 137466 10.1016/J.Cej.2022.137466.

Xiu, L.-Y., Z.-Y. Wang, and J.-S. Qiu. 2020. "General synthesis of MXene by green etching chemistry of fluoride-free Lewis acidic melts." *Rare Metals* no. 39 (11):1237–1238. doi: 10.1007/s12598-020-01488-0.

Xue, Y., J. Feng, S. Huo, P. Song, B. Yu, L. Liu, and H. Wang. 2020. "Polyphosphoramide-intercalated MXene for simultaneously enhancing thermal stability, flame retardancy and mechanical properties of polylactide." *Chemical Engineering Journal* no. 397:125336. doi: 10.1016/j.cej.2020.125336.

Yan, X. F., J. Fang, J. J. Gu, C. K. Zhu, and D. M. Qi. 2022. "Flame retardancy, thermal and mechanical properties of novel intumescent flame retardant/MXene/poly(vinyl alcohol) nanocomposites." *Nanomaterials* no. 12 (3). doi: Artn 477 10.3390/Nano12030477.

Yang, Z. J., A. Liu, C. L. Wang, F. M. Liu, J. M. He, S. Q. Li, J. Wang, R. You, X. Yan, P. Sun, Y. Duan, and G. Y. Lu. 2019. "Improvement of gas and humidity sensing properties of organ-like MXene by alkaline treatment." *ACS Sensors* no. 4 (5):1261–1269. doi: 10.1021/acssensors.9b00127.

Yin, Z., J. Lu, N. Hong, W. Cheng, P. Jia, H. Wang, W. Hu, B. Wang, L. Song, and Y. Hu. 2022. "Functionalizing $Ti_3C_2T_x$ for enhancing fire resistance and reducing toxic gases of flexible polyurethane foam composites with reinforced mechanical properties." *Journal of Colloid and Interface Science* no. 607:1300–1312. doi: 10.1016/j.jcis.2021 .09.027.

Yin, Z., B. Wang, Q. Tang, J. Lu, C. Liao, P. Jia, L. Cheng, and L. Song. 2022. "Inspired by placoid scale to fabricate MXene derivative biomimetic structure on the improvement of interfacial compatibility, mechanical property, and fire safety of epoxy nano-composites." *Chemical Engineering Journal* no. 431:133489. doi: 10.1016/j.cej.2021 .133489.

Yu, B., A. C. Y. Yuen, X. D. Xu, Z. C. Zhang, W. Yang, H. D. Lu, B. Fei, G. H. Yeoh, P. A. Song, and H. Wang. 2021. "Engineering MXene surface with POSS for reducing fire hazards of polystyrene with enhanced thermal stability." *Journal of Hazardous Materials* no. 401:123342. doi: ARTN 123342 10.1016/j.jhazmat.2020.123342.

Yu, B., B. Tawiah, L.-Q. Wang, A. C. Y. Yuen, Z.-C. Zhang, L.-L. Shen, B. Lin, B. Fei, W. Yang, A. Li, S.-E. Zhu, E.-Z. Hu, H.-D. Lu, and G. H. Yeoh. 2019. "Interface decoration of exfoliated MXene ultra-thin nanosheets for fire and smoke suppressions of thermoplastic polyurethane elastomer." *Journal of Hazardous Materials* no. 374:110–119. doi: 10.1016/j.jhazmat.2019.04.026.

Yuan, Y. S., Y. T. Pan, W. C. Zhang, M. J. Feng, N. Wang, D. Y. Wang, and R. J. Yang. 2021. "Delamination and engineered interlayers of Ti_3C_2 MXenes using phosphorous vapor toward flame-retardant epoxy nanocomposites." *Acs Applied Materials & Interfaces* no. 13 (40):48196–48207. doi: 10.1021/acsami.1c11863.

Zeng, Q., Y. Zhao, X. Lai, C. Jiang, B. Wang, H. Li, X. Zeng, and Z. Chen. 2022. "Skin-inspired multifunctional MXene/cellulose nanocoating for smart and efficient fire protection." *Chemical Engineering Journal* no. 446:136899. doi: 10.1016/j.cej.2022 .136899.

Zhan, Y. J., B. F. Nan, Y. C. Liu, E. X. Jiao, J. Shi, M. E. Lu, and K. Wu. 2021. "Multifunctional cellulose-based fireproof thermal conductive nanocomposite films assembled by in-situ grown SiO$_2$ nanoparticle onto MXene." *Chemical Engineering Journal* no. 421:129733. doi: Artn 129733 10.1016/J.Cej.2021.129733.

Zhang, Q., J. He, X. Fu, S. Xie, R. Fan, H. Lai, W. Cheng, P. Ji, J. Sheng, Q. Liao, W. Zhu, and H. Li. 2022. "Fluorine-free strategy for hydroxylated Ti$_3$C$_2$/Ti$_3$AlC$_2$ catalysts with enhanced aerobic oxidative desulfurization and mechanism." *Chemical Engineering Journal* no. 430:132950. doi: 10.1016/j.cej.2021.132950.

Zhang, X., W. Zhang, and H. Zhao. 2022. "Comparative study on fabrication and energy storage performance of Ti$_3$C$_2$T$_x$ MXene by using hydrofluoric acid and in situ forming of hydrofluoric acid-based approaches." *International Journal of Energy Research* no. 46 (11):15559–15570. doi: 10.1002/er.8252.

Zhang, Y., J. Yu, J. Lu, C. Zhu, and D. Qi. 2021. "Facile construction of 2D MXene (Ti$_3$C$_2$T$_x$) based aerogels with effective fire-resistance and electromagnetic interference shielding performance." *Journal of Alloys and Compounds* no. 870:159442. doi: 10.1016/j.jallcom.2021.159442.

Zhang, Z. R., H. X. Cao, Y. F. Quan, R. Ma, E. B. Pentzer, M. J. Green, and Q. S. Wang. 2022. "Thermal stability and flammability studies of MXene-organic hybrid polystyrene nanocomposites." *Polymers* no. 14 (6). doi: Artn 1213 10.3390/Polym14 061213.

Zhao, X., M. Zhu, C. Tang, K. Quan, Q. Tong, H. Cao, J. Jiang, H. Yang, and J. Zhang. 2022. "ZIF-8@MXene-reinforced flame-retardant and highly conductive polymer composite electrolyte for dendrite-free lithium metal batteries." *Journal of Colloid and Interface Science* no. 620:478–485. doi: 10.1016/j.jcis.2022.04.018.

Zhong, Q., Y. Li, and G. K. Zhang. 2021. "Two-dimensional MXene-based and MXene-derived photocatalysts: recent developments and perspectives." *Chemical Engineering Journal* no. 409. doi: Artn 128099 10.1016/J.Cej.2020.128099.

Zhou, C., X. H. Zhao, Y. S. Xiong, Y. H. Tang, X. T. Ma, Q. Tao, C. M. Sun, and W. L. Xu. 2022. "A review of etching methods of MXene and applications of MXene conductive hydrogels." *European Polymer Journal* no. 167. doi: ARTN 111063 10.1016/j.eurpolymj.2022.111063.

Zhou, K., K. Gong, F. Gao, and L. Yin. 2022. "Facile strategy to synthesize MXene@LDH nanohybrids for boosting the flame retardancy and smoke suppression properties of epoxy." *Composites Part A: Applied Science and Manufacturing* no. 157:106912. doi: 10.1016/j.compositesa.2022.106912.

Zhou, Y. Y., Y. C. Lin, B. Tawiah, J. Sun, R. K. K. Yuen, and B. Fei. 2021. "DOPO-decorated two-dimensional MXene nanosheets for flame-retardant, ultraviolet-protective, and reinforced polylactide composites." *ACS Applied Materials & Interfaces* no. 13 (18):21876–21887. doi: 10.1021/acsami.1c05587.

Zhou, Y., F. Chu, L. Ding, W. Yang, S. Zhang, Z. Xu, S. Qiu, and W. Hu. 2022. "MOF-derived 3D petal-like CoNi-LDH array cooperates with MXene to effectively inhibit fire and toxic smoke hazards of FPUF." *Chemosphere* no. 297:134134. doi: 10.1016/j.chemosphere.2022.134134.

Zhu, S.-E., F.-D. Wang, J.-J. Liu, L.-L. Wang, C. Wang, A. C. Y. Yuen, T. B. Y. Chen, I. I. Kabir, G. H. Yeoh, H.-D. Lu, and W. Yang. 2021. "BODIPY coated on MXene nanosheets for improving mechanical and fire safety properties of ABS resin." *Composites Part B: Engineering* no. 223:109130. doi: 10.1016/j.compositesb.2021 .109130.

Zhu, Y., X. Zhao, Q. Peng, H. Zheng, F. Xue, P. Li, Z. Xu, and X. He. 2021. "Flame-retardant MXene/polyimide film with outstanding thermal and mechanical properties based on the secondary orientation strategy." *Nanoscale Advances* no. 3 (19):5683–5693. doi: 10.1039/D1NA00415H.

6 Graphene(oxide) and Its Hybrid Materials for Fire-Safe Polymers

Xiaming Feng and Hongyu Yang
Chongqing University, Chongqing, PR China

CONTENTS

6.1 INTRODUCTION

Polymer materials have been widely used in various applications, including transportation, construction, and electronics. Unfortunately, these fields are of high fire risk, which always leads to a great loss of life and property due to the high flammability of polymer materials. Making polymers flame retarded is highly desirable. Through decades of efforts, different kinds of flame-retardant additives and techniques have been developed to address the problem of heat release and smoke

DOI: 10.1201/9781003327158-6

155

production of polymers during combustion(Lazar, Kolibaba, and Grunlan 2020; Lu, Song, and Hu 2011). Two-dimensional (2D) flame retardants are at the center of an ever-growing research effort due to their unique properties, which are interesting for both fundamental science and applications. The universal physical barrier effect and specific catalytic action of 2D flame retardants have been well demonstrated for clay, layered double hydroxide, montmorillonite, etc. (Wang et al. 2017; Cai et al. 2021). However, a fancier and more specific platform is always desired. Graphene, a 2D carbon crystal structure with the repeating resemblance of a honeycomb in a hexagonal lattice, possesses exceptional physical properties, such as a high specific surface area, excellent mechanical modulus and strength, and excellent electronic and thermal conductivities. Some of these promising properties make graphene a high-performance flame retardant for polymer materials (Sang et al. 2016). In this chapter, various preparation methods of graphene(oxide) flame retardants will be summarized first. To enhance the flame retardant effect of graphene(oxide), different kinds of surface modification and functionalization will be discussed. Finally, some promising structures and applications of graphene(oxide)-based materials will be introduced.

6.2   PREPARATION OF GRAPHENE(OXIDE) FLAME RETARDANTS

Considering the large amount of usage as flame retardants, the first issue faced by graphene(oxide) is fabrication on an acceptable scale. Many methods have been developed to prepare graphene(oxide), for example, the original exfoliation by adhesive tape (Geim and Novoselov 2007; Geim 2009), which is certainly not suitable for fabricating graphene(oxide) flame retardants. Similarly, chemical vapor deposition (CVD) and epitaxy growth technology are also excluded because of their high cost and low yield (Chen, Zhang, and Chen 2015; Wang et al. 2011). Moreover, unlike graphene used in electronics, defect-free structures are not necessary for graphene flame retardants. Sometimes, certain and specific functional sites are preferable. To date, several methods have been demonstrated to prepare graphene(oxide) flame retardants, such as the widely used oxidation–reduction method (Marcano et al. 2010; Yu et al. 2016).

6.2.1   OXIDATION–REDUCTION METHOD

Since its discovery, Hummers' oxidation method has always been a popular technique to fabricate graphene oxide on an acceptable scale (Marcano et al. 2010; Yu et al. 2016). This makes graphene(oxide) able to be used as a nanofiller for polymer materials. Before being adopted as a flame retardant, the energetic nature and stability of graphene(oxide) must be clarified. Kim's report indicated that although the obtained graphene is very stable against ignition, its thermal stability is essentially decreased when potassium salt is present from the oxidation process (Kim et al. 2010). That is, contaminated graphene oxide is highly flammable and has a serious fire hazard. Generally, efficient purification is necessary before applying graphene oxide as a flame retardant. For the purpose of investigating the flame retardant effect of graphene oxide, Higginbotham et al. developed a series of graphite oxide flame retardant polymer nanocomposites by incorporating 1, 5, and 10 wt% graphene oxide in polycarbonate (PC), acrylonitrile butadiene styrene, and high-impact polystyrene, respectively (Higginbotham et al. 2009). It was found that the addition of graphene oxide obviously decreased the total heat release and peak heat release rates for all three polymers. The PC composites are able to self-extinguish in a very short time in the vertical open flame test. A condensed-phase effect ascribed to the formation of a thermal protection/mass loss barrier is believed to be the dominant mechanism of graphene oxide. This protective heat-shielding layer could slow down the escape of volatile products generated from the degrading polymers. Moreover, the graphene oxide can be directly used as flame retardants to fabricate the PDMS foam composites (Cao et al. 2020).

The oxidation degree plays a pivotal role in influencing the properties of graphene oxide, as well as the flame retardant effect. In Han's study (Han et al. 2013), graphite oxides with different oxidation degrees were synthesized by a modified Hummers' method. Additionally, as a control sample, graphene nanosheets were obtained by thermal reduction of graphene oxide. These graphene oxides and graphene were utilized to improve the flame retardancy of polystyrene (PS). The X-ray diffraction results and scanning electron microscope images suggest that graphene oxides and graphene were homogenously dispersed in the PS matrix without obvious aggregates. It was found that the thermal stability and flame retardancy (reduction in peak heat release rate) decreased with increasing oxygen groups in graphene oxide. Accordingly, the PS composite with 5 wt% graphene exhibited the highest reduction in peak heat release rate (~50%) compared to pure PS. Comparatively, the thermal stability and flame retardant effect of graphene are enhanced due to the elimination of oxygen-containing groups. That is, in general, graphene is better than graphene oxide in improving the flame retardancy of polymer materials. Furthermore, graphene is proven to possess a better flame retardant effect than conventional inorganic flame retardants (Huang, Gao, et al. 2012), Na-montmorillonite and multiwalled carbon nanotubes at the same addition content. Compared to neat poly(vinyl alcohol) (PVA), the incorporation of 3 wt% graphene leads to a reduction in the peak heat release rate by 49%. In contrast, there are only 30% and 35% for Na-montmorillonite and multiwalled carbon nanotubes, respectively. It is believed that the remarkable flame-retardant effect of graphene comes from the ability to form a compact, dense, and uniform char during combustion.

6.2.2 ELECTROCHEMICAL EXFOLIATION

Electrochemical exfoliation has been widely recognized as a very promising method to fabricate graphene (Liu et al. 2019), which integrates the merits of a simple apparatus, low cost, convenience for implementation, high efficiency, large production scale, and environmental protection. Feng and coworkers applied an electrochemical approach to prepare graphene flame retardants for the first time (Feng et al. 2016), as shown in Figure 6.1. A biobased phytic acid was used as the electrolyte and flame-retarding modifier simultaneously. It is worth noting that the addition of electrochemically exfoliated graphene leads to a great enhancement

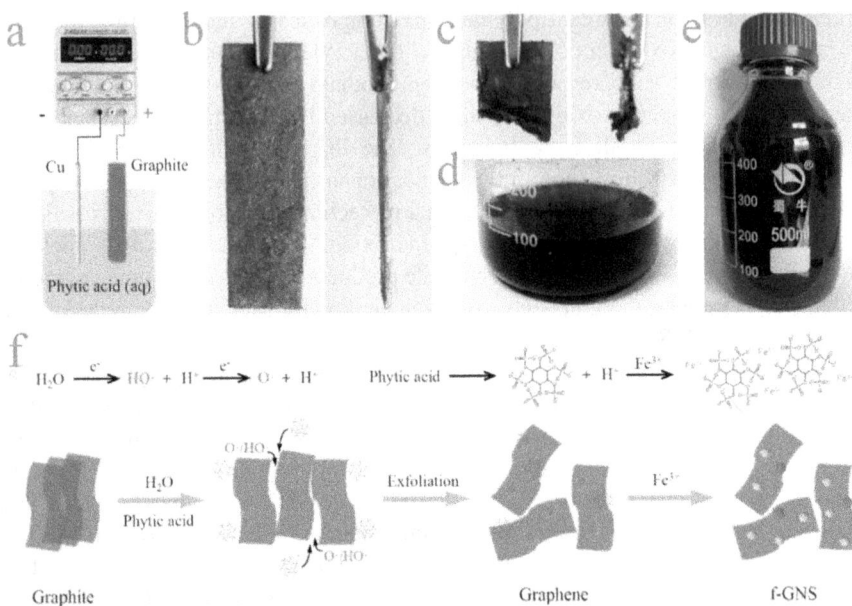

FIGURE 6.1   (a) Scheme for the preparation of electrochemically exfoliated graphene, photos of graphite foils (b) before and (c) after exfoliation, (d) exfoliated graphitic material suspended in phytic acid aqueous solution, (e) dispersed exfoliated in tetrahydrofuran, and (f) proposed mechanism for the functionalization of graphene (f-GNS).

Source: Reprinted from Feng, X.M., Wang, X., Cai, W., Qiu, S.L., Hu, Y. and Liew, K.M., *ACS Appl. Mater. Interfaces*, 8, 25552–25562, 2016. With permission.

in the fire safety of polylactic acid in terms of a reduced peak heat release rate, low carbon monoxide production, and formation of a high graphitized protective char layer. This work provides a new roadmap to develop high-performance graphene flame retardants in one green step, which is very facile yet efficient. Subsequently, various electrolytes with flame-retardant effects have been reported to develop graphene flame retardants through electrochemical exfoliation-modification strategies, such as sodium ligninsulfonate (Cai et al. 2016), sodium p-styrene sulfonate (Zhao et al. 2020), hex(4-carboxylphenoxy)cyclotriphosphazene (Cai et al. 2017), and allyloxy nonyl alcohol polyoxyethylene (10) ether ammonium sulfate (DNS86) (Qiu et al. 2020b).

Due to the high dispersibility in the aqueous phase, electrochemically exfoliated graphene can be easily incorporated as a flame retardant into water-soluble polymers. For example, Qiu et al. prepared graphene nanosheets by an electrochemical method assisted with the anionic surfactant DNS86 (Qiu et al. 2020b). Due to the good compatibility and strong interfacial interaction with PVA molecular chains, the addition of exfoliated graphene can improve not only the mechanical properties but also the flame retardancy of PVA. A 60.9% reduction in the peak heat release rate at a loading content of 2.0 wt% was achieved. To overcome the compatibility problem between electrochemically exfoliated graphene and water-insoluble polymers, a mussel-inspired modification was conducted by polymerizing dopamine on the surface of electrochemically exfoliated graphene (Cai et al. 2018). The abundant hydroxyl groups of polydopamine facilitate a forceful chain interaction with the polar thermoplastic polyurethane matrix (TPU). The TPU composite with only 2 wt% graphene flame retardants exhibited a 59.4% reduction in the peak heat release rate and a 27.1% reduction in the total heat release. A constituted tortuous path and hindered escape of pyrolysis products caused by functionalized graphene are proposed. A similar strategy has been utilized to reinforce polystyrene materials and polyacrylonitrile fibers (Qiu et al. 2020a), in which polystyrene sulfonate-functionalized graphene nanosheets and ferric diethylenetriaminepentakis (methyl-phosphonic acid)-functionalized graphene nanosheets were facilely synthesized by an electrochemical exfoliation method. All these reports suggest the great potential of the electrochemical exfoliation method in the eco-friendly and scalable production of functional graphene flame retardants.

6.2.3 Mechanical Exfoliation

Ball milling is the easiest way to achieve the exfoliation of graphite into graphene. There are two routes, dry ball milling and wet ball milling. Dry ball milling can lead to a high transient temperature, which favors some surface chemical reactions. For example, Kim et al. prepared graphene phosphonic acid (GPA) as an efficient flame retardant by the dry ball milling method (Kim et al. 2014). The graphite was directly ball milled with red phosphorus to achieve the simultaneous exfoliation and covalent attachment of phosphorus to the edges of graphene sheets. Hanji coated with solution-cast GPA shows excellent flame retardancy, which is attributed to the formation of a supporting protective char layer caused by GPA. Inspired by this work, heavily aluminated graphene nanoplatelets as efficient flame retardants have been reported by Jeon et al. (2017). They ball-milled graphite in the presence of solid aluminum beads at 500 rpm for 48 h followed by simple removal of the unreacted aluminum and other metallic impurities. The as-prepared aluminated graphene was used as a flame retardant to reinforce the PVA matrix. Both chemical (condensation) and physical (cooling and blocking) mechanisms were proposed. Additionally, to demonstrate a sustainable pathway, Zabihi et al. reported ball milling assisting scalable production of graphene flame retardants using waste fish deoxyribonucleic acid with flame retardant elements, phosphorus and nitrogen (Zabihi et al. 2020). The deoxyribonucleic acid-functionalized graphene nanoplatelets have a high production yield, high oxygen, nitrogen and phosphorus contents, and good water dispersibility. The flame-retardant effect of deoxyribonucleic acid-functionalized graphene was clarified for epoxy resin, polyvinyl alcohol, and polystyrene materials, as evidenced by the V-0 rating in the UL-94 vertical burning test. A multilayer char residue consisting of a compact layer and a porous layer was detected and believed to be the main flame retarding mechanism.

In addition, the wet ball milling method is proven to be an efficient and controllable way to exfoliate graphite. Very recently, Sn powder was applied to assist the exfoliation and modification of graphite during the wet ball milling process (Chen et al. 2022). The graphite was well exfoliated, as evidenced by atomic force microscopy and transmission electron microscope images. Subsequently, the flame retardant effect of the resulting graphene on epoxy resin was systematically investigated. It is believed that the formed Sn–O–P bonds could promote the compactness of the residual char in the condensed phase, thus protecting epoxy resin from further burning. Although ball milling is simple to exfoliate graphite, it is always time-consuming and inefficient. For the purpose of improving efficiency, a thermal shock combined with ball milling method was demonstrated to fabricate graphene from graphite (Tran et al. 2019). The much-decreased number of layers and the increase in specific surface area suggest successful exfoliation. The exfoliated graphene nanosheets exhibited a satisfactory combination effect with conventional flame retardants aluminum trihydroxide and zinc borate for improving the flame retardancy of polycarbonate and chlorine-sulfonated polyethylene rubber. All these results suggest that ball milling provides a promising option for a simple, cost-effective, yet eco-friendly preparation of graphene flame retardants.

6.2.4 LIQUID SONICATION EXFOLIATION

The liquid sonication exfoliation method has attracted great attention in preparing 2D nanomaterials, including graphene, molybdenum disulfide, boron nitride, and black phosphorus (Ciesielski and Samorì 2014; Hernandez et al. 2008). Because graphite is exfoliated in liquid solution, it is easy to achieve the simultaneous exfoliation and surface modification of graphene assisted by specific molecules, which makes it feasible to customize graphene as a flame retardant. Attia et al. reported a simple preparation of graphene flame retardants by a liquid sonication method and studied their flame-retardant effect on acrylonitrile-butadiene-styrene (Attia, Abd El-Aal, and Hassan 2016). Maleate diphosphate was applied as a dispersant and flame-retardant modifier. High-quality graphene layers without defects were obtained by using probe ultrasonication. The as-prepared modified graphene was decorated with TiO_2 nanoparticles to further improve the flame-retardant effect. Certainly, the inclusion of nanoparticle-decorated graphene flame retardants significantly enhanced the flame retardancy of acrylonitrile-butadiene-styrene. Specifically, an approximately 49% reduction in the peak heat release rate and total heat release, an approximately 50% reduction in the average mass loss rate, and a 37% reduction in the emission of carbon dioxide were observed.

Liquid sonication exfoliation is similar to the ball milling method. They are simple, facile and easy to achieve. However, both are low efficiency and time-consuming in fabricating graphene flame retardants. Therefore, improving the exfoliation efficiency is desirable. In Wang's work (Wang et al. 2019), ball milling coupled with the ultrasonication method was demonstrated to prepare graphene with a fully exfoliated structure from expanded graphite. The exfoliated structure and thin layer morphology of the as-prepared graphene were systematically characterized with transmission electron microscopy, scanning electron microscopy, X-ray diffraction, atomic force microscopy, and Raman spectroscopy. The resulting graphene exhibited a promising combination effect with alumina trihydrate (ATH) in enhancing the flame retardancy of polyethylene (PE). The cone calorimeter test results suggest a considerable reduction in the peak heat release rate of the PE/ATH composite at a loading content as low as 0.2 wt% graphene. A protective char layer originating from graphene is well characterized and accounts for the formation of an effective barrier, as well as the enhanced flame retardancy. Recently, to further improve the exfoliation efficiency, similar ball milling coupled with ultrasonication was conducted by using ionic liquid as a dispersant and modifier during the exfoliation process (Wang et al. 2021). Subsequently, the ionic liquid-modified graphene was compounded with conventional flame-retardant ammonium polyphosphate to reinforce the fire safety of bismaleimide resin. The addition of 10 wt% graphene-based flame retardants led to the limiting oxygen index increasing to ~43% and the peak heat release rate, total heat release and total smoke release decreasing by 41.8%, 47.8%, and 52.3%, respectively. A multifunctional effect of the graphene-based hybrid in modified bismaleimide resin was concluded. Specifically, the physical barrier and chemical char formation attributed to the coworking of graphene, ammonium polyphosphate, and ionic liquid are highlighted.

6.2.5 Carbonization

Carbonization of precursors is a bottom-up and efficient method to fabricate graphene products, especially graphene quantum dots (Beckert et al. 2015). Very recently, Ma introduced the pyrolysis and carbonization method in synthesizing graphene flame retardants (Ma et al. 2022). The amphiphilic graphene quantum dots (C-GQDs) were pyrolyzed by citric acid and dodecylamine at 200 °C in air. It can be easily generated in large-scale production. The generated C-GQDs were incorporated into polystyrene (PS) by the Pickering emulsion polymerization method to monitor their flame-retardant effect. Additionally, C-GQDs with controllable surface characteristics can be used as stabilizers in PS emulsion polymerization systems, which can wrap around the surface of PS microspheres through electrostatic interactions. The microscale combustion calorimeter test results indicated a 40% reduction in the peak heat release rate of the PS nanocomposite. The increased fire retardancy is believed to be caused by the promotion ability of graphene, which facilitates the formation of a physical protective barrier on the surface, thus impeding the permeation of heat and pyrolysis products. Nitrogen-doped graphene quantum dots (N-GQDs) and nitrogen and phosphorous codoped graphene quantum dots (NP-GQDs) were developed by a hydrothermal carbonization method with citric acid as the main reactant to effectively reduce the heat release and smoke production of polyacrylonitrile (PAN) (Rahimi-Aghdam et al. 2020). Both N-GQD and NP-GQD exhibited excellent flame-retardant effects on the PAN matrix, which was evidenced by the obviously reduced peak heat release rate, total heat release, CO production rate, and total smoke production, as well as the increased residual chars. Comparatively, the NP-GQDs have better performance than the N-GQDs in improving the flame retardancy of PAN composites, which is attributed to the synergistic effect of nitrogen and phosphorous.

In summary, the controllable production of graphene flame retardants has become a primary issue before their large-scale application. Various methods have been proposed and proven to be candidates. For example, the recently reported phosphorization of exfoliated graphite in an elevated temperature furnace

(Moradkhani et al. 2020). In comparison, oxidation–reduction method is widely used, controllable, and capable of mass production, but in hazardous and not environmental-friendly. The advantages of mechanical exfoliation and liquid sonication exfoliation are easy to operate and customizable. The disadvantage is that the size and thickness of as-prepared nanosheets are not precisely controllable, which always show quite a wide size distribution. The carbonization method generally leads to a defect-rich and small size graphene(oxide) nanosheets, which is favorable to catalysts, not for preparing polymer nanocomposites. Electrochemical exfoliation is a promising method that integrates the advantages of inexpensive and readily available raw materials, green and efficient exfoliation processes, and simple and facile posttreatments. However, it remains a challenge to prepare graphene flame retardants with controllable dispersibility and high flame-retardant efficiency.

6.3 MODIFIED GRAPHENE(OXIDE) AS FLAME RETARDANTS

There are two drawbacks persisting in pristine graphene(oxide) flame retardants: poor compatibility and fair flame-retardant efficiency. The dispersion state and interfacial interaction between nanofillers and the polymer matrix significantly affect the properties of polymer nanocomposites, including flame retardancy. For flame retardancy, the basic physical barrier effect and free radical trapping ability of pristine graphene are not sufficient to achieve satisfactory flame retardancy for polymer materials. Therefore, the proper surface modification and functionalization of graphene are essential. Generally, the modification and functionalization of graphene(oxide) can occur via a covalent grafting pathway, noncovalent interaction route, and heteroatom doping in plane.

6.3.1 Covalently Modified Graphene(oxide)

Surface modification through covalent bond grafting is one of the most popular methods to improve the properties of graphene(oxide), such as surface characteristics, dispersibility, compatibility with polymer matrices, and flame-retardant capability, depending on the kinds of modifiers. The organic modifiers can be small molecules, oligomers, and polymers. Furthermore, polymer modifiers can be long-chain polymers and hyperbranched polymers. For polymer modifiers, there are two strategies to conduct covalent bonding, "grafting from" and "grafting to".

First, 9,10-dihydro-9-oxa-10-phosphaphenanthrene-10-oxide (DOPO), a famous phosphorus-containing flame retardant, was grafted onto the surface of graphene oxide by reacting epoxy ring groups (Liao et al. 2012). The multifunctional action of DOPO as both a flame-retardant moiety and reducing agent has been proposed. The resulting DOPO-modified graphene was able to disperse uniformly in epoxy resin. The limiting oxygen index (LOI) test results confirmed the excellent flame retardancy of epoxy nanocomposites with 10 wt% DOPO-modified graphene. The synergistic effect between DOPO and reduced graphene oxide dominates the fascinating flame retardant effect. Subsequently, Guo and coworkers compared the flame-retardant effect of pristine graphene, graphene oxide, and organic phosphate-functionalized graphene oxides to illustrate the necessity of surface functionalization of graphene(oxide) (Guo et al. 2011). The representative highly flammable epoxy resin was selected as the polymer matrix. From the thermogravimetric analysis and micro combustion calorimeter results, one can find that pristine graphene and functionalized graphene oxide are of pros and cons, but both are better than pristine graphene oxide.

Naturally, one can predict that functionalized graphene would behave the best in enhancing the flame retardancy of polymers. Commercially available octa-aminophenyl polyhedral oligomeric silsesquioxanes (OapPOSS) were utilized as flame-retardant modifiers and reducing agents simultaneously (Wang, Song, et al. 2012), as shown in Figure 6.2. The resulting OapPOSS-functionalized graphene (OapPOSS-rGO) greatly influenced the thermal properties and flame retardancy of the epoxy resin. Specifically, only 2.0 wt% OapPOSS-rGO was able to improve the onset thermal decomposition temperature of the epoxy nanocomposite by 43°C. In addition, 49%, 37%, and 58% enhancements in the peak heat release rate, total heat release and CO production rate were observed, respectively. Bao et al. investigated the effect of

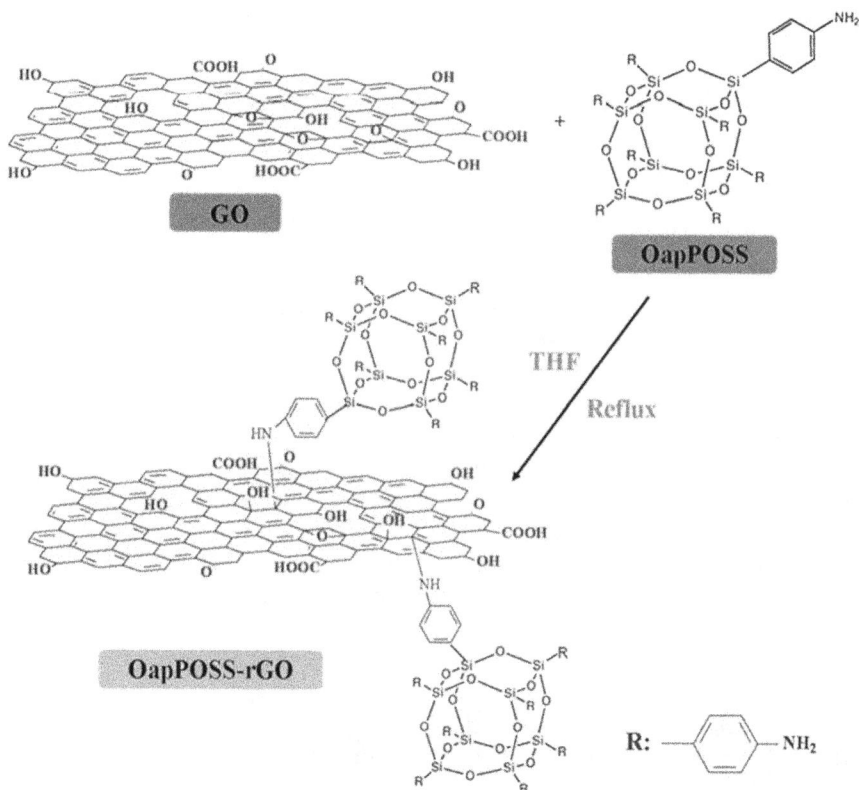

FIGURE 6.2 Schematic illustration of the synthetic route of functionalized and reduced graphene oxide (OapPOSS-rGO).

Source: Reprinted from Wang, X., Song, L., Yang, H.Y., Xing, W.Y., Kandola, B. and Hu, Y., *J. Mater. Chem.*, 22, 22037–22043, 2012. With permission.

functionalized graphene oxide on the flame retardancy of polystyrene and proposed a combination of condensed phase flame retardant strategies (Bao et al. 2012). The graphene oxide was first modified with hexachlorocyclotriphosphazene (HCCP) and hydroxyethyl acrylate (TEA) and then incorporated into polystyrene by an in situ polymerization method. The char-catalyzing action of HCCP and graphene oxide contributed to the formation of protective graphitic char from polystyrene in mutual cooperative mode. The combination of different chars facilitated the physical barrier effect and thus improved the flame retardancy of polystyrene nanocomposites.

To improve the flame retardant efficiency, it is better to increase the grafting content of the functional modifier. Therefore, oligomer and polymer flame-retardant modifiers are preferable. Generally, the "grafting to" strategy easily controls the molecular weight of the modifier and the grafting content. For example, Wang et al. reported the synthesis of polyphosphamide flame retardant chains and their grafting onto the surface of graphene nanosheets (Wang, Xing, Feng, Yu, Song, et al. 2014). The as-prepared graphene flame retardant (PPA-g-G) endowed epoxy resin with significantly increased flame retardancy, which is much better than that of bare PPA, as evidenced by the increased LOI values and reduced heat release and fire growth rate index. The higher char yield during combustion was caused by the phosphate species degraded from PPA and template action of graphene nanosheets, which is able to shield the mass and heat transfer and thus protect the underlying polymers against flame. Similarly, a polysilicone flame retardant (PMDA) was first copolymerized with four silicone monomers and then covalently grafted onto the surface of graphene oxide to obtain high-performance graphene oxide flame retardants (Wang 2021). The grafting of PMDA makes graphene oxide hydrophobic and uniformly distributed in organic chloroform and epoxy resin. The good dispersion state and synergistic effect between PMDA and graphene oxide nanosheets resulted in considerably improved thermal stability and flame retardancy of epoxy composites. The degraded polysilicone could form a thermally stable silica layer, heavily supporting and reinforcing the protective char layer of epoxy resin.

In addition, the hyperbranched flame retardant modifier is another option to increase the grafting content onto graphene(oxide). Various hyperbranched molecules have been reported to functionalize graphene(oxide). For example, Hu's group first prepared a hyperbranched flame retardant by the reaction of N-aminoethyl piperazine and di(acryloyloxyethyl)methylphosphonate (Hu et al. 2014). The obtained modifier was then grafted onto the surface of graphene oxide. The effect of the resulting graphene oxide flame retardant on cross-linked polyethylene (XLPE) was demonstrated.

All the thermal stability, flame retardancy, and anti-aging performance of the XLPE nanocomposites were dramatically improved. Inspired by the synergistic effect of phosphorus and silicon, a phosphorus-containing hyperbranched polysiloxane (P-HBPSi) modifier was fabricated by the sol-gel method (Huang et al. 2021). It was grafted onto graphene oxide by the reaction of epoxy and hydroxyl groups. Certainly, the graphene oxide flame retardant made the thermoplastic polyurethane nanocomposites nonflammable, as evidenced by the 63.5% reduction in the peak heat release rate and 20.9% reduction in the total heat release. It is believed that the free radical scavengers originating from phosphorus species in the gas phase, barrier effect of layered graphene oxide, catalytic charring and unique Si-O-Si framework in the condensed phase contribute to high-performance fire safety.

6.3.2 NONCOVALENTLY MODIFIED GRAPHENE(OXIDE)

Different from the covalent modification of graphene(oxide), the noncovalent strategy is much easier to achieve, expeditious, and entirely preserves the original quality and properties of graphene. The widely used mechanisms to achieve noncovalent functionalization of graphene(oxide) are π–π stacking interactions and the formation of hydrogen bonds. For example, a commercially available phosphorus-nitrogen compound, phenoxycycloposphazene, was utilized to noncovalently modify graphene oxide via a simple solution blending (Chen, Liu, Min, et al. 2018). The obtained functionalized graphene oxide was self-assembled with polyethyleneimine onto the surface of a poly(vinyl alcohol) (PVA) film by a layer-by-layer method and an electrostatic interaction mechanism, as displayed in Figure 6.3. As a result, PVA with multilayer graphene oxide-based coatings exhibited excellent flame retardancy, which is ascribed to the efficient transfer blocking of heat and mass during combustion.

Zhao et al. demonstrated a hydrogen bonding mechanism to achieve the noncovalent modification of graphene oxide (Zhao et al. 2019). A phosphorus/nitrogen (P/N)-containing flame retardant named diethyl-*N*,*N*-bis(2-hydroxyethyl) phosphoramide (DEPA) was prepared and then decorated onto graphene oxide through hydrogen bonding between the hydroxyl groups of DEPA and oxygen-containing species of graphene oxide. The hydrogen bonds were predicted by theoretical analysis and evidenced by experimental results. The introduction of 2 wt% graphene oxide

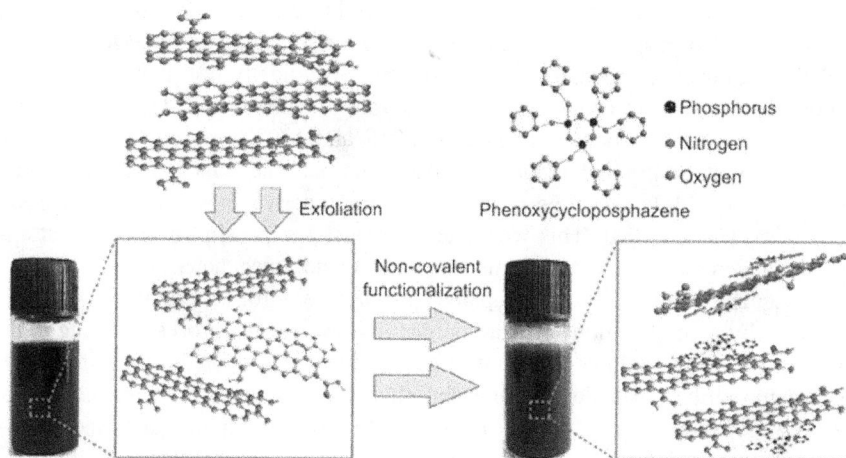

FIGURE 6.3 Schematic diagram showing overall processing of non-covalently FGO.

Source: Reprinted from Chen, W.H., Liu, P.J., Min, L.Z., Zhou, Y.M., Liu, Y., Wang, Q. and Duan, W.F., *Nanomicro Lett.*, 10, 39, 2018. With permission.

flame retardant substantially improved the flame retardancy of epoxy resin. A 29.3% reduction in the peak heat release rate, 73% reduction in total smoke production, and 65% reduction in smoke release rate were found. An efficient condensed phase flame-retardant mechanism was clarified for excellent fire safety. It is easy to imagine that the combination of different mechanisms could contribute to a stronger interaction between the modifier and graphene(oxide). Yuan and coworkers illustrated the noncovalent modification of graphene with commercial melamine via strong π-π interactions, hydrogen bonding, and electrostatic attraction (Yuan et al. 2015). The melamine-modified graphene oxides are uniformly distributed in the polypropylene matrix with intercalation and exfoliation microstructures. As a result, the poly-propylene composites with modified graphene oxide show better thermal stability and higher fire resistance than those of control graphene oxide. The space-confined degradation of melamine into graphitic carbon nitride (g-C3N4) was observed, which facilitates the in situ formation of protective char layers and restrains the escape of flammable volatile pyrolyzed products.

6.3.3 Doped Graphene(oxide)

Heteroatom doping is an important way to achieve high-performance graphene (oxide) catalysts for various application areas, including chemical conversion, photocatalysts, electrochemical sensors, fuel cells, and energy storage (Rao, Gopalakrishnan, and Govindaraj 2014). For flame retardant applications, the same strategy has been proven to be a promising method. First, Yuan and coworkers studied the effects of boron/phosphorus doping on the thermal oxidation stability of reduced graphene oxide (Yuan et al. 2016). The doped reduced graphene oxide was prepared by a high-temperature annealing method using boric acid and phosphoric acid as boron and phosphorus sources. Various techniques have been performed to confirm the in-plane doping structure of the resulting graphene. It was found that a very low level of doping (~1.1 at%) is able to significantly enhance the thermal oxidative stability of reduced graphene oxides. Specifically, the temperature cor-responding to the maximum weight loss rate of boron-doped graphene and phosphorus-doped graphene increased by 52°C and 130°C, respectively, in com-parison to that of pristine reduced graphene oxide. The conclusion that doped heteroatoms could decrease the reactivity of carbon active sites and inhibit carbon gasification was clarified. This work clearly shows the possibility of heteroatom-doped graphene as an efficient flame retardant for polymer materials.

Subsequently, Feng et al. prepared phosphorus/nitrogen codoped graphene (PN-rGO) by a combination of hydrothermal and microwave processes (Feng et al. 2018) and studied its flame retardant effect on epoxy resin for the first time. This result demonstrated that nitrogen and phosphorus atoms are doped into the lattice, adopting pyrrolic-N, pyridinic-N, quaternary-N, and pyrophosphate and meta-phosphate forms. Due to the improved thermal oxidization resistance of codoped graphene, the char residue of the epoxy composite was highly thermally stable. As a result, reductions of 30.9%, 29.3%, and 51.3% in the peak heat release rate, total heat release, and total smoke production for the epoxy composite (containing 5 wt% PN-rGO) were observed, respectively, compared to those of the original epoxy resin.

The work extended the application of heteroatom-doped graphene as a high-performance flame retardant. In Yuan's recent work (Zhan et al. 2021), phosphorus-doped graphene was applied as a synergist to conventional Co flame retardant compounds, cophenyl phosphonic acid, and Co-organic framework hybrids. The thermal stability and flame retardancy of various formulated polypropylene composites were evaluated. The increased char yield and decreased heat release and toxic gas products suggest the high fire safety of polypropylene composites. A condensed flame retardant mechanism catalyzing the formation of high-quality carbonaceous protection layers was proposed.

As stated above, the proper surface modification and functionalization of graphene(oxide) can not only improve the dispersibility and compatibility with polymer materials but also enhance the flame-retardant efficiency from different aspects, including improved thermal oxidation stability, enhanced charring ability, and synergistic effects with other elements. This is why the surface covalent and non-covalent modification or functionalization of graphene(oxide) for application as flame retardants are of great importance from academic and industry perspectives, and have been widely used. It is noted that for flame safety applications, these two modifications and functionalizations are different from the traditional strategies.

6.4   GRAPHENE(OXIDE)-BASED FLAME RETARDANT HYBRIDS

The flame-retardant efficiency of graphene(oxide) can be significantly increased by specific modifications and functionalization. However, the single usage of these surface functionalized graphene(oxide) flame retardants can always not pass the highest level (V-0 ranking) of UL-94 vertical burning tests. Limited by the grafting mechanism, the content of flame-retardant modifiers grafted onto graphene(oxide) is restricted to small scales. Hybridization of graphene with flame retardants provides a possibility to overcome this problem persisting over a long period.

For simplicity, the hybridization of graphene can be classified into organic-graphene(oxide) hybrids and inorganic-graphene(oxide) hybrids.

6.4.1   Organic–Inorganic Hybrids

The in situ sol-gel process has been widely used to fabricate organic–inorganic hybrid flame retardants containing exfoliated graphene. The organic sites in the organic–inorganic hybrids are not only conducive to the dispersion of inorganic substances in polymer materials but also endow them with high flame-retardant ability. Qian et al. introduced an organic–inorganic flame retardant (DOPO-VTS) containing both silicon and organophosphorus to graphene nanosheets via a simple sol-gel method (Qian et al. 2013). The novel hybrid flame retardants were homogeneously dispersed in the epoxy resin matrix. Remarkably, the epoxy composite with 5 wt% hybrid flame retardant was able to pass the UL-94 vertical burning test V-0 level. The LOI value reached 29.5. A combination of condensed phase and gas phase flame retardant mechanisms was proposed. Specifically, the nanocomposites technique, phosphorus–silicon synergism in the condensed phase and DOPO flame retardancy in the gas phase are crucial for the significantly improved flame retardancy of epoxy resin. A similar method was used to achieve the hybridization of graphene and phenyl-*bis*-(triethoxysilylpropyl) phosphamide (PBTP) (Zehao Wang et al. 2014). Die dramatic decreases (43.0% and 44.7%, respectively) in the peak heat release rate and total heat release were observed for epoxy resin with a loading of only 1 wt% graphene-based hybrid flame retardant. Yu et al. developed a high-performance graphene-based hybrid flame retardant by wrapping graphene nanosheets with phosphorus- and nitrogen (P-N)-containing flame retardants (Yu et al. 2015). The synergistic effect of graphene and phosphorus and nitrogen (P-N)-containing flame retardants is believed to be the main reason for the notably improved flame retardancy of epoxy resin.

6.4.2 INORGANIC–INORGANIC HYBRIDS

Different from organic–inorganic hybrids, graphene-based inorganic–inorganic hybrids are mainly developed to reduce not only the heat release but also the fire hazard of polymer materials, including toxic gaseous pyrolysis products and smoke particles. As we all know, the catalytic effect of pristine graphene(oxide) is quite fair. Therefore, the integrity of highly catalytic active transition metal compounds and graphene nanosheets with high specific surface areas is the most obvious method to prepare high-performance graphene-based inorganic–inorganic hybrid flame retardants. Initially, various transition metal oxides were decorated onto graphene nanosheets via wet chemical methods, including Co_3O_4, NiO, SnO_2 and others (Hong et al. 2013; Feng et al. 2015; Wang, Xing, Feng, Yu, Lu, et al. 2014). For example, as shown in Figure 6.4, Feng et al. developed a titanium dioxide/graphene hybrid flame retardant by a facile in situ solvothermal route (Feng et al. 2015). Transmission electron microscopy images clearly demonstrated that TiO_2 nanoparticles were homogeneously anchored on the surface of reduced graphene oxide nanosheets. Its effect on suppressing the fire hazard of poly(vinyl chloride) (PVC) materials was comprehensively investigated. Only 2.0 wt% incorporation of the graphene/TiO_2 hybrid leads to an obviously reduced peak heat release rate and HCl release rate and improved thermal stability.

Afterwards, element-doped transition metal oxides were developed to further modify the catalytic capability of graphene nanosheets. They are proven to behave better in enhancing the fire safety of polymer materials compared to these simple transition metal oxide/graphene hybrids. Wang et al. reported the effect of manganese-cobalt oxide/graphene hybrid flame retardants on improving the flame retardancy of poly(butylene terephthalate) (PBT) (Wang et al. 2014). Jiang and coworkers reported the preparation of Ce-doped MnO_2/graphene hybrid flame retardants and their effect on reducing the fire hazard of epoxy composites (Jiang, Bai, Tang, Song, et al. 2014). The PBT composite with the manganese-cobalt oxide/graphene hybrid exhibited a lower peak heat release rate and smoke production rate than the pure PBT and PBT composite with the same amount of cobalt oxide/graphene hybrid. Moreover, the suppressed pyrolysis products containing aromatic compounds, carbonyl compounds, carbon monoxide, and carbon dioxide clearly suggest the positive effect of element doping for transition metal oxides in graphene

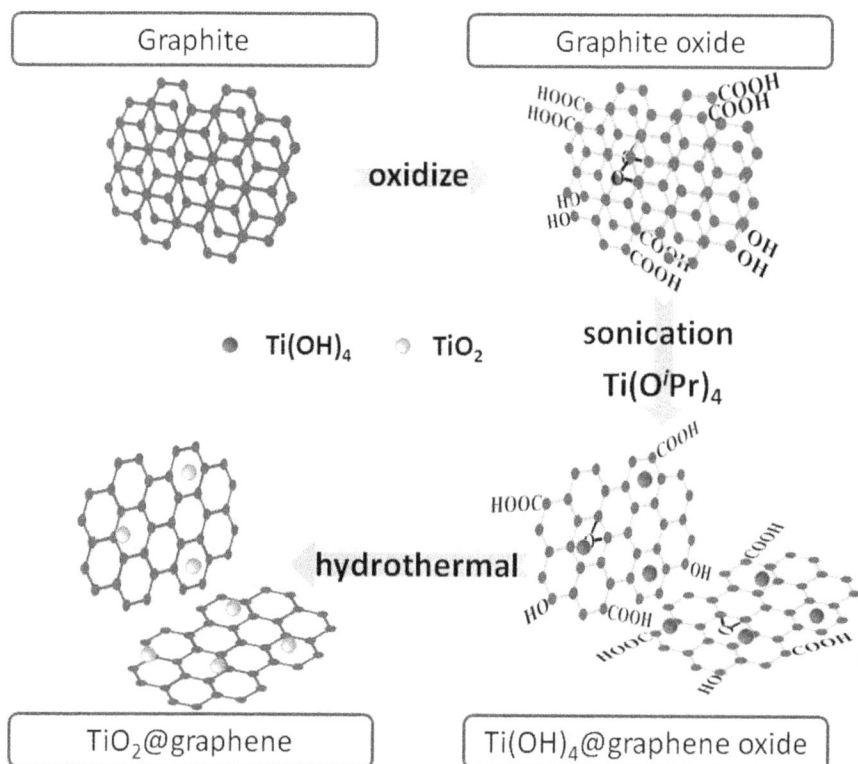

FIGURE 6.4 Schematic of the preparation of the graphene/TiO$_2$ hybrid.

Source: Reprinted from Feng, X.M., Xing, W.Y., Song, L., Hu, Y. and Liew, K.M., *Chem. Eng. J.*, **260, 524–531, 2015. With permission.**

hybrids. In the case of Ce-doped MnO$_2$/graphene hybrid flame retardants, similar conclusions were obtained.

Following the same idea, layered double hydroxides with various metal ions were decorated on graphene nanosheets, such as the Ni-Fe layered double hydroxide. As demonstrated by Wang et al. (2013), the synthesis of Ni-Fe layered double hydroxide (Ni-Fe LDH) and the reduction of graphene oxide can be achieved at the same time during the one-pot solvothermal process. The anchored lamellar Ni-Fe LDHs greatly improved the thermal stability of reduced graphene oxide. Importantly, the addition of 2.0 wt% Ni-Fe LDH/graphene hybrid endowed epoxy resin with excellent fire safety, including the increased onset thermal degradation temperature, obviously reduced peak heat release rate and total heat release values, and decreased production of toxic CO. The catalytic action of Ni-Fe LDHs and the adsorption and barrier effect of graphene are believed to promote the formation of graphitized carbons and improve the thermal oxidative resistance of the char layer.

In addition to transition metal oxides and hydroxides, the other transition metal compounds are also the emphases of research of graphene-based hybrid flame retardants, such as transition metal sulfides (Wang et al. 2013; Jiang, Bai, Tang,

Hu, et al. 2014), transition metal carbides (Liu et al. 2020), metal-organic frameworks (Zeng et al. 2022), etc. For example, the molybdenum disulfide (MoS_2)-modified graphene (MoS_2/GNS) hybrids reported by Wang et al. (D. Wang et al. 2013) and zinc sulfide/graphene hybrid flame retardants developed by Jiang, Bai, Tang, Hu, et al. (2014) both exhibited higher flame-retardant effects than pristine graphene and bare transition metal sulfides. Recently, following graphene, Mxene has drawn increasing attention in preparing high-performance flame retardants. The combination of graphene and MXene is certainly a promising candidate for fabricating flame-retardant polymer composites. Shi's group prepared an MXene/reduced graphene oxide hybrid to improve the flame retardancy of thermoplastic polyurethane (TPU) (Liu et al. 2020). Specifically, a titanium carbide-reduced graphene oxide ($Ti_3C_2T_x$-rGO) hybrid was synthesized by hydrogen bond-induced assembly of $Ti_3C_2T_x$ and rGO. TPU composites with the 2.0 wt% $Ti_3C_2T_x$-rGO hybrid exhibited greatly decreased (81.2% and 54.0%) peaks of the smoke production rate and total smoke release, respectively. Moreover, an approximately 54.4% reduction in the peak carbon monoxide production rate and a 46.2% reduction in the total carbon monoxide yield were obtained. The physical barrier effect and the catalytic charring capability of the $Ti_3C_2T_x$-rGO hybrid are believed to be responsible for the excellent fire safety of TPU composites.

6.4.3 Ternary Hybrids

It is easy to imagine that the synergistic effect is not solely a phenomenon between two components but also happens among three or more parties. Here, we only review some graphene-based ternary hybrid flame retardants as

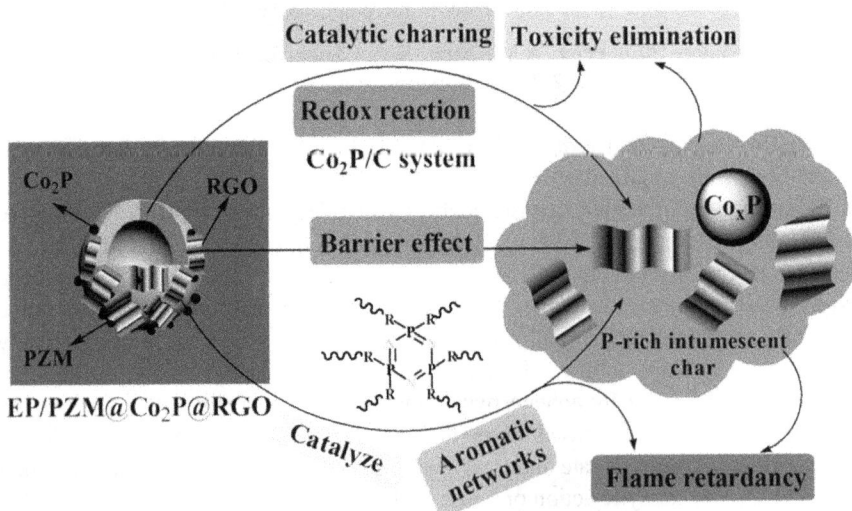

FIGURE 6.5 Schematic illustration of the mechanism for flame retardancy and toxicity elimination of PZM@Co_2P@RGO in flaming epoxy composites.

Source: Reprinted from Qiu, S.L., Xing, W.Y., Mu, X.W., Feng, X.M., Ma, C., Yuen, R.K.K. and Hu, Y., *ACS Appl. Mater. Interfaces*, 8, 32528–32540, 2016. With permission.

representatives. For example, Qiu and coworkers prepared a three-dimensional nanostructure based on cobalt phosphide nanoparticles and heteroatom-doped mesoporous carbon spheres interconnected with graphene (PZM@Co$_2$P@RGO) (Qiu et al. 2016). The morphology and structure of the ternary hybrid were comprehensively characterized and well confirmed. As illustrated in Figure 6.5, the resulting PZM@Co$_2$P@RGO ternary hybrid possessed fascinating catalytic activity regarding the combustion process of epoxy resin, including the barrier effect, redox reaction toxicity elimination, and catalytic charring of aromatic networks. As a result, an approximately 48% reduction in the peak heat release rate and a 30% reduction in the total heat release, lower toxic CO production, and the formation of a high-graphitized protective char layer were achieved. Following the same strategy, a polydopamine@KH560/carbon nitride/graphene ternary system was designed and synthesized to improve the fire retardancy of waterborne epoxy intumescent coatings (Chen et al. 2021). A much lower backside temperature, increased expansion height after the furnace test, and the lowest smoke density rating (SDR) were observed for the composite intumescent coating compared to the neat epoxy coating.

6.5 GRAPHENE(OXIDE) FLAME RETARDANT SYNERGISTS

Using graphene(oxide) as a synergist to conventional flame retardants is a more practical and feasible way to fabricate flame retardant polymer materials (Feng et al. 2013). Graphene(oxide) is only compounded or blended with conventional flame retardants into a polymer matrix. Initially, Hu's group reported the synergistic effect of graphene with an ammonium polyphosphate/melamine intumescent flame-retardant system on the anti-dripping and flame retardancy of poly(butylene succinate) (PBS) composites (X. Wang et al. 2011). The limiting oxygen index (LOI) values for pure PBS and PBS with a single intumescent flame retardant were 23.0 and 31.0, respectively. Upon replacing 2.0 wt% of intumescent flame retardant with the same amount of graphene, the limiting oxygen index (LOI) value of the PBS composite increased to 33.0. In addition, the intumescent flame-retardant PBS composite with 2.0 wt% graphene is able to reach the UL-94 V-0 level and exhibit excellent anti-dripping properties. The combined effect of melamine polyphosphate and graphene on improving the fire retardancy of poly(vinyl alcohol) (PVA) was demonstrated by Huang, Liang, et al. (2012). A simple solution blending method was utilized to prepare the flame-retardant PVA composites. Compared to pure PVA, the PVA composite reinforced with 10 wt% melamine polyphosphate and 1 wt% graphene exhibited an LOI value of 29.6 and UL-94 V-0 grade, as well as an approximately 60% reduction in the peak heat release rate. A compact and dense intumescent char jammed with graphene sheets was observed and accounted for the high flame resistance.

Subsequently, to show the advancement of graphene as a flame retardant synergist, Hu's group compared the synergistic effect of graphene and a commonly used synergist, polyhedral oligomeric silsesquioxanes (POSS), with melamine phosphate on the fire retardancy of poly(butylene succinate) (Wang, Hu, et al. 2012). The comparative study demonstrated that the PBS composite formulated with 18 wt% melamine phosphate and 2 wt% graphene reached the UL-94 V-0 ranking and exhibited an LOI value of 34, while the PBS formulated with 18 wt% melamine phosphate and 2 wt% POSS only passed the UL-94 V-1 level and showed an LOI value of 33. Moreover, the addition of graphene resulted in a lower peak heat release rate and total heat release, as well as better thermal-oxidative resistance of the char layer, compared to the POSS. Additionally, the synergistic effect of graphene with intumescent flame retardants on polypropylene was characterized (Yuan et al. 2017). The intumescent flame-retardant system is composed of ammonium polyphosphate and charring-foaming agent, a macromolecular triazine derivative containing hydroxyethylamino, triazine rings, and ethylenediamino groups. It is demonstrated that the loadings of graphene play an important role in improving the fire safety of polypropylene composites under different flame scenarios. Adding less graphene is able to enhance the swelling of char and lead to better insulation of the char. A further increase in graphene decreased the flame retardant properties instead of reinforcing them. The char structure, melt viscosity, and barrier effect are believed to be the main factors.

Moreover, functionalized graphene(oxide) has also been used as a synergist for conventional flame retardants. For example, Wang et al. reported the synergistic action of benzoic acid-functionalized graphene (BFG) (Wang et al. 2019). In this work, k-carrageenan (KC), ammonium polyphosphate (APP), and melamine (MEL) were used as the carbon source, acid source, and gas source, respectively. A "four-source flame-retardant system" (KC-IFR/BFG) was constructed to improve the fire retardancy of waterborne epoxy resin coatings. One can clearly find that the BFG is able to increase the flame retardancy of the epoxy composites. The LOI value was up to 29.8, while 18 wt% KC-IFR and 2 wt% BFG were incorporated. Meanwhile, the composite could pass the UL-94 test V-0 level. A huge reduction in peak heat release and smoke release rate were also achieved. Most recently, a zeolitic imidazolate framework-67 (ZIF-67)-functionalized graphene oxide (ZIF-67@GO) was designed and developed as a synergist for ammonium polyphosphate (APP) to reduce the fire risk of thermoplastic polyurethane (TPU) (Liu et al. 2022). The addition of 6.5 wt% APP and 0.5 wt% ZIF-67@GO decreased the peak heat release rate and peak smoke production rate by 81.2% and 48.7%, respectively. The combustion period of the flame-retardant TPU composite was obviously prolonged from 200 s of neat TPU to 600 s during the cone calorimetry test. The APP/ZIF-67@GO promoted charring and absorbed more smoke, toxic gas, and soot.

As a comparative study, Zhang et al. reported the preparation of a metal-organic framework and graphene oxide hybrid (MOF@GO) and then compared its synergistic effect with conventional inorganic synergists, layered double hydroxide and nano clay (Zhang et al. 2022). Ammonium polyphosphate was chosen as the intumescent flame retardant (IFR), and epoxy resin was selected as the polymer matrix. The mechanical strength and effectiveness of the char barrier that provides flame resistance were thoroughly investigated through dynamic mechanical analysis (DMA) and X-ray tomography (X-CT) analysis. The minimum strength and modulus of char reside for the epoxy/IFR composite reinforced with the MOF@GO hybrid tripled to 4930 Pa and 4368 Pa, respectively. Moreover, the multiscale porous structure within carbonaceous char was observed by the formation of step functions in the compressive stress–strain curve. This work provides an in-depth study to establish the relationship between the barrier effect and flame retardancy.

6.6 GRAPHENE(OXIDE)-ASSEMBLED FIRE-SAFE MATERIALS

Due to the unique geometrical characteristics and surface features, graphene(oxide) easily assembles into different structures assisted by polymer materials, which possess various properties and performances (Zhu et al. 2013; Ahmad, Fan, and Hui 2018), such as 2D nacre-like composites and 3D ultralight aerogels. The layer-by-layer assembled nacre-like structures have attracted great attention in fabricating mechanically strong and tough nanocomposites (Ming et al. 2015). Additionally, the long-range ordered high-content graphene platelets could act as an efficient protective layer for substrates from destruction, including fire attack.

6.6.1 2D LAYER-BY-LAYER ASSEMBLED STRUCTURES

In the early stage, nacre-inspired integrated nanocomposites were developed utilizing graphene oxide (GO) and montmorillonite (MMT) nanosheets with poly(vinyl alcohol) (PVA) via a vacuum-assisted filtration self-assembly process (Ming et al. 2015). The resulting composite membranes were proven to possess strong mechanical properties and excellent fire resistance. Afterwards, these nacre-like structures were fabricated in situ as fire-proof coatings and coated on various flammable materials through a layer-by-layer self-assembly method. For example, a multilayered coating on wood pulp paper (WPP) was fabricated with phenoxycyclophosphamide-functionalized graphene oxide (FGO) and chitosan-functionalized carbon nanotubes (CNTs) through a layer-by-layer assembly method (Chen, Liu, Liu, et al. 2018). The resulting WPP composite exhibited dramatically improved mechanical properties and fire resistance while maintaining intrinsic flexibility. Shi et al. constructed a multi-layered coating on polystyrene (PS) spheres by composing graphitic carbon nitride nanosheets (g-C$_3$N$_4$ NSs) and reduced graphene oxide (RGO) using a layer-by-layer assembly method (Shi et al. 2019). The thermal stability of the composites was improved in terms of the higher temperature corresponding to the maximum decomposition rate and the higher residual yield. Furthermore, the dramatically decreased release of total pyrolysis gaseous products and the peak heat release rate were observed as indicators of high flame retardancy.

The layer-by-layer self-assembly method has been widely used to improve the fire retardancy of flexible polyurethane foams. For instance, as shown in Figure 6.6, Carosio et al. constructed nacre-like coatings for flexible polyurethane foams by layer-by-layer assembly of graphene oxide nanoplatelets assisting with branched poly(ethylene imine) (positive charge) and polyacrylic acid (negative charge) (Carosio et al. 2018). The coated foams were able to restrain fire spread in the flammability test and prevent ignition when heat fluxes were applied. Pan and coworkers formulated assembled coatings for flexible polyurethane foams with sodium alginate (SA)/graphene oxide (GO) and polyethylenimine (PEI) (Pan et al. 2016), in which SA and GO were assembled by electrostatic attraction and

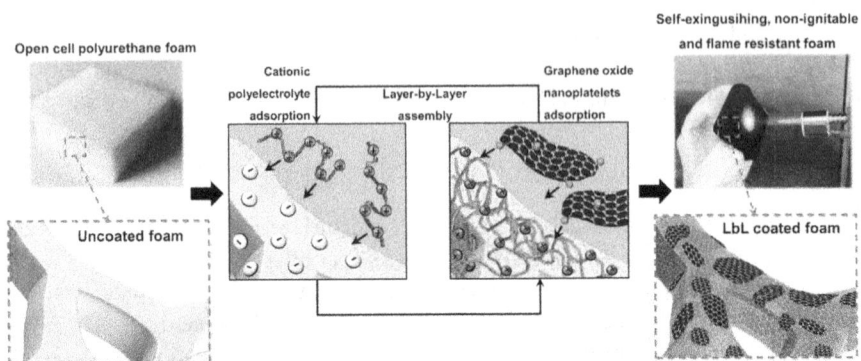

FIGURE 6.6 Schematic representation of the deposited LbL assembly. PU foams are alternatively dipped in cationic polyelectrolyte solution and negative graphene oxide nanoplatelet suspension. Each cycle deposits a bilayer (BL), and the process is repeated to deposit 3BL and 6BL.

Source: Reprinted from Carosio, F., Maddalena, L., Gomez, J., Saracco, G. and Fina, A., *Adv. Mater. Interfaces*, **5, 1801288, 2018. With permission.**

hydrogen bonding. It was found that the coated polyurethane foams are more thermally stable than the pure foam in the temperature range from 430 to 600 °C. The "delayed effect" caused by the physical barrier action of coatings contributed to the reduction in the heat release rate and smoke production rate. Maddalena et al. built fire resistance coatings for open cell polyurethane foams through the layer-by-layer self-assembly of chitosan and graphene oxide (GO) platelets (Maddalena et al. 2018). Similar to sodium alginate, chitosan, as a natural polysaccharide, is used as a polymer electrolyte and charring agent. Coupled with GO plantlets, it is able to slow the release of combustible volatiles, thus suppressing melt dripping and reducing the peak heat release rate and total smoke production. Subsequently, the same group reported the possibility of using DNA as a negatively charged and an all-in-one intumescent flame retardant with graphene oxide for constructing layer-by-layer assembled coatings (Maddalena, Carosio, and Fina 2021). The GO+DNA coatings were found to be efficient in restraining flame spread and decreasing the combustion rate and smoke production, as evidenced by the huge reduction in the peak heat release rate and total smoke release. Notably, the DNA facilitated the formation of a protective char layer and stood out of the reported flame-retardant coatings of PU foams.

In recent years, on the basis of the excellent fire resistance and flame/temperature-sensitive electrical resistance of graphene oxide assembled structures, an advanced fire alarm system application was proposed and demonstrated (Wu et al. 2018; Huang et al. 2019; Zhang et al. 2020; Guo et al. 2020; Yu et al. 2021). For example, a graphene oxide wide ribbon (GOWR)-wrapped melamine formaldehyde sponge was prepared by Tang's group (Xu et al. 2019). This composite foam integrated light-weight, hydrophobicity, reversible compressibility, acidic/alkaline tolerance, and flame resistance. The mechanism of the temperature-induced electrical resistance transition of graphene oxide was applied to achieve the fire warning function.

Specifically, once encountering a flame attack or abnormally high temperature, the GOWR sheets on the sponge skeleton are thermally reduced in situ, which results in a distinct transition in electrical resistance. Upon proper circuit connection, one can find the signal corresponding to the fire/high-temperature accident. To date, various efforts have been devoted to accelerating the response time of graphene oxide-based fire alarm systems. For example, Chen et al. reported a polyurethane sponge (PUS) deposited with flame retardant-modified graphene oxide (FGO) and chitosan-functionalized carbon nanotubes (CNTs) by layer-by-layer assembly (Chen, Liu, Liu, et al. 2018). A response time of ~1 s in fire and a superlong warning period of 2640 s were achieved due to the high thermal conductivity and ultrahigh structural stability. Very recently, Cao and coworkers reported a multiamino molecule, named HCPA, as a cross-linker, fire retardant, and reducing agent for decorating graphene oxide (GO) nanoplatelets (Cao et al. 2022). Excellent mechanical properties, flame retardancy, and ultrafast fire alarm performance were achieved. These advanced fire detection and alarming techniques exhibit great potential in building smart and safe constructions.

6.6.2 3D NETWORK STRUCTURES

Upon proper templates and techniques, graphene(oxide) nanosheets are able to assemble with polymers into 3D network structures (Wicklein et al. 2015; Wang et al. 2018), sometimes called aerogels (Guan et al. 2016; Zhang et al. 2022). Generally, graphene(oxide)-based aerogels are lightweight, thermally insulating, and fire retardant. One very promising application is thermal insulation materials for construction and transportation. For instance, Wicklein et al. fabricated super-insulating, fire-retardant and strong anisotropic foams by freeze-casting suspensions of cellulose nanofibers, graphene oxide and sepiolite nanorods (Wicklein et al. 2015). Compared to conventional polymer-based insulating foams, such as expanded polystyrene and polyurethane foams, the as-prepared graphene oxide-based foams are ultralight, fire resistant, and thermally insulating. Specifically, its thermal conductivity is only half of that of expanded polystyrene. This report indicates a promising research direction for developing high-performance flame-retardant graphene(oxide)-based polymer nanocomposites.

6.7 SUMMARY AND PERSPECTIVES

Here, we reviewed the progress of graphene(oxide) and its hybrid materials for fire-safe polymers. In the first section, we introduced the primary methods for developing graphene(oxide) flame retardants, which have always been neglected in previous review papers. As important as the flame-retardant efficiency, facile, sustainable, environmentally friendly techniques that can prepare graphene(oxide) flame retardants on a large scale are essential in achieving their whole potential in practical applications. Subsequently, the various surface modifications and hybridization of graphene(oxide) with organic and inorganic compounds were summarized. Generally, organic surface modification and hybridization with organic compounds can not only improve the flame retardant effect of graphene(oxide) but also facilitate homogeneous dispersion and compatibility between phases. Comparatively, inorganic hybridization is more effective in enhancing the catalytic ability of graphene(oxide) in reducing the fire hazard of polymer materials, including smoke production and toxic gas release. Additionally, we simply reviewed the synergistic effect of graphene(oxide) with conventional flame retardants, such as ammonium polyphosphate, melamine phosphate, melamine polyphosphate, and charring agents. Finally, we presented the latest research status of graphene(oxide)-assembled fire-safe materials, including 2D layer-by-layer assembled structures and 3D network structures. Some promising applications of these advanced fire-safe structures were introduced, such as fire alarm sensors and ultralight thermal insulating aerogels. The rising number of recently published papers means that this research direction is receiving increasing attention from the academic community. However, some works are required to address some points, such as a deep understanding of the molecular level of graphene(oxide) reinforcing the flame retardancy of polymers, the environmental concerns and impacts of nanosized graphene(oxide) flame retardants, and how to reuse and upcycle graphene(oxide) products. The authors believe that in the next decade, some high-performance graphene(oxide) flame retardants and their flame-retarded polymer composite products will be released from research labs and commercialized in industry.

REFERENCES

Ahmad, Hassan, Mizi Fan, and David Hui. 2018. "Graphene Oxide Incorporated Functional Materials: A Review." *Composites Part B: Engineering* 145: 270–280. 10.1016/j.compositesb.2018.02.006.

Attia, Nour F., N.S. Abd El-Aal, and M.A. Hassan. 2016. "Facile Synthesis of Graphene Sheets Decorated Nanoparticles and Flammability of Their Polymer Nanocomposites." *Polymer Degradation and Stability* 126: 65–74. 10.1016/j.polymdegradstab.2016.01.017.

Bao, Chenlu, Yuqiang Guo, Bihe Yuan, Yuan Hu, and Lei Song. 2012. "Functionalized Graphene Oxide for Fire Safety Applications of Polymers: A Combination of Condensed Phase Flame Retardant Strategies." *Journal of Materials Chemistry* 22 (43): 23057. 10.1039/c2jm35001g.

Beckert, M., M. Menzel, F. J. Tölle, B. Bruchmann, and R. Mülhaupt. 2015. "Nitrogenated Graphene and Carbon Nanomaterials by Carbonization of Polyfurfuryl Alcohol in the Presence of Urea and Dicyandiamide." *Green Chemistry* 17 (2): 1032–1037. 10.1039/C4GC01676A.

Cai, Wei, Xiaming Feng, Weizhao Hu, Ying Pan, Yuan Hu, and Xinglong Gong. 2016. "Functionalized Graphene from Electrochemical Exfoliation for Thermoplastic Polyurethane: Thermal Stability, Mechanical Properties, and Flame Retardancy." *Industrial & Engineering Chemistry Research* 55 (40): 10681–10689. 10.1021/acs.iecr.6b02579.

Cai, Wei, Xiaming Feng, Bibo Wang, Weizhao Hu, Bihe Yuan, Ningning Hong, and Yuan Hu. 2017. "A Novel Strategy to Simultaneously Electrochemically Prepare and Functionalize Graphene with a Multifunctional Flame Retardant." *Chemical Engineering Journal* 316: 514–524. 10.1016/j.cej.2017.01.017.

Cai, Wei, Bi-Bo Wang, Xin Wang, Yu-Lu Zhu, Zhao-Xin Li, Zhou-Mei Xu, Lei Song, Wei-Zhao Hu, and Yuan Hu. 2021. "Recent Progress in Two-Dimensional Nanomaterials Following Graphene for Improving Fire Safety of Polymer (Nano)Composites." *Chinese Journal of Polymer Science* 39 (8): 935–956. 10.1007/s10118-021-2575-2.

Cai, Wei, Junling Wang, Ying Pan, Wenwen Guo, Xiaowei Mu, Xiaming Feng, Bihe Yuan, Xin Wang, and Yuan Hu. 2018. "Mussel-Inspired Functionalization of Electrochemically Exfoliated Graphene: Based on Self-Polymerization of Dopamine and Its Suppression Effect on the Fire Hazards and Smoke Toxicity of Thermoplastic Polyurethane." *Journal of Hazardous Materials* 352: 57–69. 10.1016/j.jhazmat.2018.03.021.

Cao, Cheng-Fei, Peng-Huan Wang, Jian-Wang Zhang, Kun-Yu Guo, Yang Li, Qiao-Qi Xia, Guo-Dong Zhang, et al. 2020. "One-Step and Green Synthesis of Lightweight, Mechanically Flexible and Flame-Retardant Polydimethylsiloxane Foam Nanocomposites via Surface-Assembling Ultralow Content of Graphene Derivative." *Chemical Engineering Journal* 393: 124724. 10.1016/j.cej.2020.124724.

Cao, Cheng-Fei, Bin Yu, Zuan-Yu Chen, Yong-Xiang Qu, Yu-Tong Li, Yong-Qian Shi, Zhe-Wen Ma, et al. 2022. "Fire Intumescent, High-Temperature Resistant, Mechanically Flexible Graphene Oxide Network for Exceptional Fire Shielding and Ultra-Fast Fire Warning." *Nano-Micro Letters* 14 (1): 92. 10.1007/s40820-022-00837-1.

Carosio, Federico, Lorenza Maddalena, Julio Gomez, Guido Saracco, and Alberto Fina. 2018. "Graphene Oxide Exoskeleton to Produce Self-Extinguishing, Nonignitable, and Flame Resistant Flexible Foams: A Mechanically Tough Alternative to Inorganic Aerogels." *Advanced Materials Interfaces* 5 (23): 1801288. 10.1002/admi.201801288.

Chen, Chunlin, Shaotang Dong, Guoqing Xiao, Fei Zhong, Zhengwei Yang, Chunyan Chen, and Shuyi Zeng. 2021. "Flame Retardant Properties of Waterborne Epoxy Intumescent Coatings Reinforced by Polydopamine@KH560/Carbon Nitride/Graphene Ternary System." *Colloids and Surfaces A: Physicochemical and Engineering Aspects* 620: 126506. 10.1016/j.colsurfa.2021.126506.

Chen, Wenhua, Pengju Liu, Yuan Liu, Qi Wang, and Wenfeng Duan. 2018. "A Temperature-Induced Conductive Coating via Layer-by-Layer Assembly of Functionalized Graphene Oxide and Carbon Nanotubes for a Flexible, Adjustable Response Time Flame Sensor." *Chemical Engineering Journal* 353: 115–125. 10.1016/j.cej.2018.07.110.

Chen, Wenhua, Pengju Liu, Lizhen Min, Yiming Zhou, Yuan Liu, Qi Wang, and Wenfeng Duan. 2018. "Non-Covalently Functionalized Graphene Oxide-Based Coating to Enhance Thermal Stability and Flame Retardancy of PVA Film." *Nano-Micro Letters* 10 (3): 39. 10.1007/s40820-018-0190-8.

Chen, Xiangping, Lili Zhang, and Shanshan Chen. 2015. "Large Area CVD Growth of Graphene." *Synthetic Metals* 210: 95–108. 10.1016/j.synthmet.2015.07.005.

Chen, Yiyang, Hongjuan Wu, Rui Duan, Kailun Zhang, Weihua Meng, Yan Li, and Hongqiang Qu. 2022. "Graphene Doped Sn Flame Retardant Prepared by Ball Milling and Synergistic with Hexaphenoxy Cyclotriphosphazene for Epoxy Resin." *Journal of Materials Research and Technology* 17: 774–788. 10.1016/j.jmrt.2022.01.017.

Ciesielski, Artur, and Paolo Samorì. 2014. "Graphenevasonication Assisted Liquid-Phase Exfoliation." *Chemical Society Reviews* 43 (1): 381–398. 10.1039/C3CS60217F.

Feng, Xiaming, Xin Wang, Wei Cai, Shuilai Qiu, Yuan Hu, and Kim Meow Liew. 2016. "Studies on Synthesis of Electrochemically Exfoliated Functionalized Graphene and Polylactic Acid/Ferric Phytate Functionalized Graphene Nanocomposites as New Fire Hazard Suppression Materials." *ACS Applied Materials & Interfaces* 8 (38): 25552–25562. 10.1021/acsami.6b08373.

Feng, Xiaming, Xin Wang, Weiyi Xing, Bin Yu, Lei Song, and Yuan Hu. 2013. "Simultaneous Reduction and Surface Functionalization of Graphene Oxide by Chitosan and Their Synergistic Reinforcing Effects in PVA Films." *Industrial & Engineering Chemistry Research* 52 (36): 12906–12914. 10.1021/ie402073x.

Feng, Xiaming, Weiyi Xing, Lei Song, Yuan Hu, and Kim Meow Liew. 2015. "TiO$_2$ Loaded on Graphene Nanosheet as Reinforcer and Its Effect on the Thermal Behaviors of Poly

(Vinyl Chloride) Composites." *Chemical Engineering Journal* 260: 524–531. 10.1016/j.cej.2014.08.103.

Feng, Yuezhan, Chengen He, Yingfeng Wen, Yunsheng Ye, Xingping Zhou, Xiaolin Xie, and Yiu-Wing Mai. 2018. "Superior Flame Retardancy and Smoke Suppression of Epoxy-Based Composites with Phosphorus/Nitrogen Co-Doped Graphene." *Journal of Hazardous Materials* 346: 140–151. 10.1016/j.jhazmat.2017.12.019.

Geim, A. K. 2009. "Graphene: Status and Prospects." *Science* 324 (5934): 1530–1534. 10.1126/science.1158877.

Geim, A. K., and K. S. Novoselov. 2007. "The Rise of Graphene." *Nature Materials* 6 (3): 183–191. 10.1038/nmat1849.

Guan, Li-Zhi, Jie-Feng Gao, Yong-Bing Pei, Li Zhao, Li-Xiu Gong, Yan-Jun Wan, Helezi Zhou, et al. 2016. "Silane Bonded Graphene Aerogels with Tunable Functionality and Reversible Compressibility." *Carbon* 107: 573–582. 10.1016/j.carbon.2016.06.022.

Guo, Kun-Yu, Qian Wu, Min Mao, Heng Chen, Guo-Dong Zhang, Li Zhao, Jie-Feng Gao, Pingan Song, and Long-Cheng Tang. 2020. "Water-Based Hybrid Coatings toward Mechanically Flexible, Super-Hydrophobic and Flame-Retardant Polyurethane Foam Nanocomposites with High-Efficiency and Reliable Fire Alarm Response." *Composites Part B: Engineering* 193: 108017. 10.1016/j.compositesb.2020.108017.

Guo, Yuqiang, Chenlu Bao, Lei Song, Bihe Yuan, and Yuan Hu. 2011. "In Situ Polymerization of Graphene, Graphite Oxide, and Functionalized Graphite Oxide into Epoxy Resin and Comparison Study of On-the-Flame Behavior." *Industrial & Engineering Chemistry Research* 50 (13): 7772–7783. 10.1021/ie200152x.

Han, Yongqin, Ying Wu, Mingxia Shen, Xianli Huang, Jiajia Zhu, and Xiaogang Zhang. 2013. "Preparation and Properties of Polystyrene Nanocomposites with Graphite Oxide and Graphene as Flame Retardants." *Journal of Materials Science* 48 (12): 4214–4222. 10.1007/s10853-013-7234-8.

Hernandez, Yenny, Valeria Nicolosi, Mustafa Lotya, Fiona M. Blighe, Zhenyu Sun, Sukanta De, I. T. McGovern, et al. 2008. "High-Yield Production of Graphene by Liquid-Phase Exfoliation of Graphite." *Nature Nanotechnology* 3 (9): 563–568. 10.1038/nnano.2008.215.

Higginbotham, Amanda L., Jay R. Lomeda, Alexander B. Morgan, and James M. Tour. 2009. "Graphite Oxide Flame-Retardant Polymer Nanocomposites." *ACS Applied Materials & Interfaces* 1 (10): 2256–2261. 10.1021/am900419m.

Hong, Ningning, Lei Song, T. Richard Hull, Anna A. Stec, Bibo Wang, Ying Pan, and Yuan Hu. 2013. "Facile Preparation of Graphene Supported Co_3O_4 and NiO for Reducing Fire Hazards of Polyamide 6 Composites." *Materials Chemistry and Physics* 142 (2–3): 531–538. 10.1016/j.matchemphys.2013.07.048.

Hu, Weizhao, Jing Zhan, Xin Wang, Ningning Hong, Bibo Wang, Lei Song, Anna A Stec, T. Richard Hull, Jian Wang, and Yuan Hu. 2014. "Effect of Functionalized Graphene Oxide with Hyper-Branched Flame Retardant on Flammability and Thermal Stability of Cross-Linked Polyethylene." *Industrial & Engineering Chemistry Research* 53 (8): 3073–3083. 10.1021/ie4026743.

Huang, Guobo, Jianrong Gao, Xu Wang, Huading Liang, and Changhua Ge. 2012. "How Can Graphene Reduce the Flammability of Polymer Nanocomposites?" *Materials Letters* 66 (1): 187–189. 10.1016/j.matlet.2011.08.063.

Huang, Guobo, Huading Liang, Yong Wang, Xu Wang, Jianrong Gao, and Zhengdong Fei. 2012. "Combination Effect of Melamine Polyphosphate and Graphene on Flame Retardant Properties of Poly(Vinyl Alcohol)." *Materials Chemistry and Physics* 132 (2–3): 520–528. 10.1016/j.matchemphys.2011.11.064.

Huang, Neng-Jian, Cheng-Fei Cao, Yang Li, Li Zhao, Guo-Dong Zhang, Jie-Feng Gao, Li-Zhi Guan, Jian-Xiong Jiang, and Long-Cheng Tang. 2019. "Silane Grafted Graphene Oxide Papers for Improved Flame Resistance and Fast Fire Alarm

Response." *Composites Part B: Engineering* 168: 413–420. 10.1016/j.compositesb. 2019.03.053.

Huang, Wenjie, Jingshu Huang, Bin Yu, Yuan Meng, Xianwu Cao, Qunchao Zhang, Wei Wu, Dean Shi, Tao Jiang, and Robert K.Y. Li. 2021. "Facile Preparation of Phosphorus Containing Hyperbranched Polysiloxane Grafted Graphene Oxide Hybrid toward Simultaneously Enhanced Flame Retardancy and Smoke Suppression of Thermoplastic Polyurethane Nanocomposites." *Composites Part A: Applied Science and Manufacturing* 150: 106614. 10.1016/j.compositesa.2021.106614.

Jeon, In-Yup, Sun-Hee Shin, Hyun-Jung Choi, Soo-Young Yu, Sun-Min Jung, and Jong-Beom Baek. 2017. "Heavily Aluminated Graphene Nanoplatelets as an Efficient Flame-Retardant." *Carbon* 116: 77–83. 10.1016/j.carbon.2017.01.071.

Jiang, Shu-Dong, Zhi-Man Bai, Gang Tang, Yuan Hu, and Lei Song. 2014. "Synthesis of ZnS Decorated Graphene Sheets for Reducing Fire Hazards of Epoxy Composites." *Industrial & Engineering Chemistry Research* 53 (16): 6708–6717. 10.1021/ie500023w.

Jiang, Shu-Dong, Zhi-Man Bai, Gang Tang, Lei Song, Anna A. Stec, T. Richard Hull, Jing Zhan, and Yuan Hu. 2014. "Fabrication of Ce-Doped MnO_2 Decorated Graphene Sheets for Fire Safety Applications of Epoxy Composites: Flame Retardancy, Smoke Suppression and Mechanism." *J. Mater. Chem. A* 2 (41): 17341–17351. 10.1039/ C4TA02882A.

Kim, Franklin, Jiayan Luo, Rodolfo Cruz-Silva, Laura J. Cote, Kwonnam Sohn, and Jiaxing Huang. 2010. "Self-Propagating Domino-like Reactions in Oxidized Graphite." *Advanced Functional Materials* 20 (17): 2867–2873. 10.1002/adfm.201000736.

Kim, Min-Jung, In-Yup Jeon, Jeong-Min Seo, Liming Dai, and Jong-Beom Baek. 2014. "Graphene Phosphonic Acid as an Efficient Flame Retardant." *ACS Nano* 8 (3): 2820–2825. 10.1021/nn4066395.

Lazar, Simone T., Thomas J. Kolibaba, and Jaime C. Grunlan. 2020. "Flame-Retardant Surface Treatments." *Nature Reviews Materials* 5 (4): 259–275. 10.1038/s41578-019-0164-6.

Liao, Shu-Hang, Po-Lan Liu, Min-Chien Hsiao, Chih-Chun Teng, Chung-An Wang, Ming-Der Ger, and Chin-Lung Chiang. 2012. "One-Step Reduction and Functionalization of Graphene Oxide with Phosphorus-Based Compound to Produce Flame-Retardant Epoxy Nanocomposite." *Industrial & Engineering Chemistry Research* 51 (12): 4573–4581. 10.1021/ie2026647.

Liu, Chuan, Wei Wu, Yongqian Shi, Fuqiang Yang, Minghua Liu, Zhixin Chen, Bin Yu, and Yuezhan Feng. 2020. "Creating MXene/Reduced Graphene Oxide Hybrid towards Highly Fire Safe Thermoplastic Polyurethane Nanocomposites." *Composites Part B: Engineering* 203: 108486. 10.1016/j.compositesb.2020.108486.

Liu, Fei, Chaojun Wang, Xiao Sui, Muhammad Adil Riaz, Meiying Xu, Li Wei, and Yuan Chen. 2019. "Synthesis of Graphene Materials by Electrochemical Exfoliation: Recent Progress and Future Potential." *Carbon Energy* 1 (2): 173–199. 10.1002/cey2.14.

Liu, Qinyong, Huawei Wang, Hefu Li, Jun Sun, Xiaoyu Gu, and Sheng Zhang. 2022. "Constructing a Novel Synergistic Flame Retardant by Hybridization of Zeolitic Imidazolate Framework-67 and Graphene Oxide for Thermoplastic Polyurethane." *Polymers for Advanced Technologies* 33 (7): 2374–2385. 10.1002/pat.5694.

Lu, Hongdian, Lei Song, and Yuan Hu. 2011. "A Review on Flame Retardant Technology in China. Part II: Flame Retardant Polymeric Nanocomposites and Coatings: Flame Retardant Technology in China." *Polymers for Advanced Technologies* 22 (4): 379–394. 10.1002/pat.1891.

Ma, Rong, Ruiqing Shen, Yufeng Quan, and Qingsheng Wang. 2022. "Tunable Flammability Studies of Graphene Quantum Dots-Based Polystyrene Nanocomposites Using Microscale Combustion Calorimeter." *Journal of Thermal Analysis and Calorimetry* 147 (19): 10383–10390. 10.1007/s10973-022-11277-9.

Maddalena, Lorenza, Federico Carosio, and Alberto Fina. 2021. "In Situ Assembly of DNA/ Graphene Oxide Nanoplates to Reduce the Fire Threat of Flexible Foams." *Advanced Materials Interfaces* 8 (21): 2101083. 10.1002/admi.202101083.

Maddalena, Lorenza, Federico Carosio, Julio Gomez, Guido Saracco, and Alberto Fina. 2018. "Layer-by-Layer Assembly of Efficient Flame Retardant Coatings Based on High Aspect Ratio Graphene Oxide and Chitosan Capable of Preventing Ignition of PU Foam." *Polymer Degradation and Stability* 152: 1–9. 10.1016/j.polymdegradstab.2018. 03.013.

Marcano, Daniela C., Dmitry V. Kosynkin, Jacob M. Berlin, Alexander Sinitskii, Zhengzong Sun, Alexander Slesarev, Lawrence B. Alemany, Wei Lu, and James M. Tour. 2010. "Improved Synthesis of Graphene Oxide." *ACS Nano* 4 (8): 4806–4814. 10.1021/ nn1006368.

Ming, Peng, Zhaofei Song, Shanshan Gong, Yuanyuan Zhang, Jianli Duan, Qi Zhang, Lei Jiang, and Qunfeng Cheng. 2015. "Nacre-Inspired Integrated Nanocomposites with Fire Retardant Properties by Graphene Oxide and Montmorillonite." *Journal of Materials Chemistry A* 3 (42): 21194–21200. 10.1039/C5TA05742F.

Moradkhani, Ghane, Mohammad Fasihi, Thibault Parpaite, Loic Brison, Fouad Laoutid, Henri Vahabi, and Mohammad Reza Saeb. 2020. "Phosphorization of Exfoliated Graphite for Developing Flame Retardant Ethylene Vinyl Acetate Composites." *Journal of Materials Research and Technology* 9 (4): 7341–7353. 10.1016/j.jmrt.2020.04.085.

Pan, Haifeng, Bihao Yu, Wei Wang, Ying Pan, Lei Song, and Yuan Hu. 2016. "Comparative Study of Layer by Layer Assembled Multilayer Films Based on Graphene Oxide and Reduced Graphene Oxide on Flexible Polyurethane Foam: Flame Retardant and Smoke Suppression Properties." *RSC Advances* 6 (115): 114304–114312. 10.1039/C6RA15522G.

Qian, Xiaodong, Lei Song, Bin Yu, Bibo Wang, Bihe Yuan, Yongqian Shi, Yuan Hu, and Richard K. K. Yuen. 2013. "Novel Organic–Inorganic Flame Retardants Containing Exfoliated Graphene: Preparation and Their Performance on the Flame Retardancy of Epoxy Resins." *Journal of Materials Chemistry A* 1 (23): 6822. 10.1039/c3ta10416h.

Qiu, Minghui, Dong Wang, Liping Zhang, Min Li, Mingming Liu, and Shaohai Fu. 2020a. "Simultaneously Electrochemical Exfoliation and Functionalization of Graphene Nanosheets: Multifunctional Reinforcements in Thermal, Flame-retardant, and Mechanical Properties of Polyacrylonitrile Composite Fibers." *Polymer Composites* 41 (4): 1561–1573. 10.1002/pc.25478.

Qiu, Minghui, Dong Wang, Liping Zhang, Min Li, Mingming Liu, Shaohai Fu. 2020b. "Electrochemical Exfoliation of Water-Dispersible Graphene from Graphite towards Reinforcing the Mechanical and Flame-Retardant Properties of Poly (Vinyl Alcohol) Composites." *Materials Chemistry and Physics* 254: 123430. 10.1016/j.matchemphys. 2020.123430.

Qiu, Shuilai, Weiyi Xing, Xiaowei Mu, Xiaming Feng, Chao Ma, Richard K. K. Yuen, and Yuan Hu. 2016. "A 3D Nanostructure Based on Transition-Metal Phosphide Decorated Heteroatom-Doped Mesoporous Nanospheres Interconnected with Graphene: Synthesis and Applications." *ACS Applied Materials & Interfaces* 8 (47): 32528–32540. 10.1021/ acsami.6b11101.

Rahimi-Aghdam, Taher, Zahra Shariatinia, Minna Hakkarainen, and Vahid Haddadi-Asl. 2020. "Nitrogen and Phosphorous Doped Graphene Quantum Dots: Excellent Flame Retardants and Smoke Suppressants for Polyacrylonitrile Nanocomposites." *Journal of Hazardous Materials* 381: 121013. 10.1016/j.jhazmat.2019.121013.

Rao, C.N.R., K. Gopalakrishnan, and A. Govindaraj. 2014. "Synthesis, Properties and Applications of Graphene Doped with Boron, Nitrogen and Other Elements." *Nano Today* 9 (3): 324–343. 10.1016/j.nantod.2014.04.010.

Sang, Bin, Zhi-wei Li, Xiao-hong Li, Lai-gui Yu, and Zhi-jun Zhang. 2016. "Graphene-Based Flame Retardants: A Review." *Journal of Materials Science* 51 (18): 8271–8295. 10.1007/s10853-016-0124-0.

Shi, Yongqian, Chuan Liu, Libi Fu, Fuqiang Yang, Yuancai Lv, and Bin Yu. 2019. "Hierarchical Assembly of Polystyrene/Graphitic Carbon Nitride/Reduced Graphene Oxide Nanocomposites toward High Fire Safety." *Composites Part B: Engineering* 179: 107541. 10.1016/j.compositesb.2019.107541.

Tran, Vinh Q., Hai T. Doan, Nhiem T. Nguyen, and Cuong V. Do. 2019. "Preparation of Graphene Nanoplatelets by Thermal Shock Combined with Ball Milling Methods for Fabricating Flame-Retardant Polymers." *Journal of Chemistry* 2019: 1–6. 10.1155/2019/5284160.

Wang, Chunfeng, Jihua Wang, Zhenlong Men, Yongliang Wang, and Zhidong Han. 2019. "Thermal Degradation and Combustion Behaviors of Polyethylene/Alumina Trihydrate/Graphene Nanoplatelets." *Polymers* 11 (5): 772. 10.3390/polym11050772.

Wang, Dong, Qiangjun Zhang, Keqing Zhou, Wei Yang, Yuan Hu, and Xinglong Gong. 2014. "The Influence of Manganese–Cobalt Oxide/Graphene on Reducing Fire Hazards of Poly(Butylene Terephthalate)." *Journal of Hazardous Materials* 278: 391–400. 10.1016/j.jhazmat.2014.05.072.

Wang, Dong, Keqing Zhou, Wei Yang, Weiyi Xing, Yuan Hu, and Xinglong Gong. 2013. "Surface Modification of Graphene with Layered Molybdenum Disulfide and Their Synergistic Reinforcement on Reducing Fire Hazards of Epoxy Resins." *Industrial & Engineering Chemistry Research* 52 (50): 17882–17890. 10.1021/ie402441g.

Wang, Jiangbo. 2021. "Functionalization of Graphene Oxide with Polysilicone: Synthesis, Characterization, and Its Flame Retardancy in Epoxy Resin." *Polymers* 13 (21): 3857. 10.3390/polym13213857.

Wang, Lin, Lin-Hai Tian, Guo-Dong Wei, Feng-Mei Gao, Jin-Ju Zheng, and Wei-You Yang. 2011. "Epitaxial Growth of Graphene and Their Applications in Devices: Epitaxial Growth of Graphene and Their Applications in Devices." *Journal of Inorganic Materials* 26 (10): 1009–1019. 10.3724/SP.J.1077.2011.01009.

Wang, Teng, Yang You, Zhang, and Wang. 2019. "Synthesis of K-Carrageenan Flame-Retardant Microspheres and Its Application for Waterborne Epoxy Resin with Functionalized Graphene." *Polymers* 11 (10): 1708. 10.3390/polym11101708.

Wang, Xin, Yuan Hu, Lei Song, Hongyu Yang, Bin Yu, Baljinder Kandola, and Dario Deli. 2012. "Comparative Study on the Synergistic Effect of POSS and Graphene with Melamine Phosphate on the Flame Retardance of Poly(Butylene Succinate)." *Thermochimica Acta* 543: 156–164. 10.1016/j.tca.2012.05.017.

Wang, Xin, Ehsan Naderi Kalali, Jin-Tao Wan, and De-Yi Wang. 2017. "Carbon-Family Materials for Flame Retardant Polymeric Materials." *Progress in Polymer Science* 69: 22–46. 10.1016/j.progpolymsci.2017.02.001.

Wang, Xin, Lei Song, Hongyu Yang, Hongdian Lu, and Yuan Hu. 2011. "Synergistic Effect of Graphene on Antidripping and Fire Resistance of Intumescent Flame Retardant Poly(Butylene Succinate) Composites." *Industrial & Engineering Chemistry Research* 50 (9): 5376–5383. 10.1021/ie102566y.

Wang, Xin, Lei Song, Hongyu Yang, Weiyi Xing, Baljinder Kandola, and Yuan Hu. 2012. "Simultaneous Reduction and Surface Functionalization of Graphene Oxide with POSS for Reducing Fire Hazards in Epoxy Composites." *Journal of Materials Chemistry* 22 (41): 22037. 10.1039/c2jm35479a.

Wang, Xin, Weiyi Xing, Xiaming Feng, Bin Yu, Hongdian Lu, Lei Song, and Yuan Hu. 2014. "The Effect of Metal Oxide Decorated Graphene Hybrids on the Improved Thermal Stability and the Reduced Smoke Toxicity in Epoxy Resins." *Chemical Engineering Journal* 250: 214–221. 10.1016/j.cej.2014.01.106.

Wang, Xin, Weiyi Xing, Xiaming Feng, Bin Yu, Lei Song, and Yuan Hu. 2014. "Functionalization of Graphene with Grafted Polyphosphamide for Flame Retardant Epoxy Composites: Synthesis, Flammability and Mechanism." *Polymer Chemistry.* 5 (4): 1145–1154. 10.1039/C3PY00963G.

Wang, Xin, Shun Zhou, Weiyi Xing, Bin Yu, Xiaming Feng, Lei Song, and Yuan Hu. 2013. "Self-Assembly of Ni–Fe Layered Double Hydroxide/Graphene Hybrids for Reducing Fire Hazard in Epoxy Composites." *Journal of Materials Chemistry A* 1 (13): 4383. 10.1039/c3ta00035d.

Wang, Yan, Xining Jia, Hui Shi, Jianwei Hao, Hongqiang Qu, and Jingyu Wang. 2021. "Graphene Nanoplatelets Hybrid Flame Retardant Containing Ionic Liquid and Ammonium Polyphosphate for Modified Bismaleimide Resin: Excellent Flame Retardancy, Thermal Stability, Water Resistance and Unique Dielectric Properties." *Materials* 14 (21): 6406. 10.3390/ma14216406.

Wang, Zehao, Ping Wei, Yong Qian, and Jiping Liu. 2014. "The Synthesis of a Novel Graphene-Based Inorganic–Organic Hybrid Flame Retardant and Its Application in Epoxy Resin." *Composites Part B: Engineering* 60: 341–349. 10.1016/j.compositesb.2013.12.033.

Wang, Zicheng, Renbo Wei, Junwei Gu, Hu Liu, Chuntai Liu, Chunjia Luo, Jie Kong, et al. 2018. "Ultralight, Highly Compressible and Fire-Retardant Graphene Aerogel with Self-Adjustable Electromagnetic Wave Absorption." *Carbon* 139: 1126–1135. 10.1016/j.carbon.2018.08.014.

Wicklein, Bernd, Andraž Kocjan, German Salazar-Alvarez, Federico Carosio, Giovanni Camino, Markus Antonietti, and Lennart Bergström. 2015. "Thermally Insulating and Fire-Retardant Lightweight Anisotropic Foams Based on Nanocellulose and Graphene Oxide." *Nature Nanotechnology* 10 (3): 277–283. 10.1038/nnano.2014.248.

Wu, Qian, Li-Xiu Gong, Yang Li, Cheng-Fei Cao, Long-Cheng Tang, Lianbin Wu, Li Zhao, et al. 2018. "Efficient Flame Detection and Early Warning Sensors on Combustible Materials Using Hierarchical Graphene Oxide/Silicone Coatings." *ACS Nano* 12 (1): 416–424. 10.1021/acsnano.7b06590.

Xu, Hui, Yang Li, Neng-Jian Huang, Zhi-Ran Yu, Peng-Huan Wang, Zhao-Hui Zhang, Qiao-Qi Xia, et al. 2019. "Temperature-Triggered Sensitive Resistance Transition of Graphene Oxide Wide-Ribbons Wrapped Sponge for Fire Ultrafast Detecting and Early Warning." *Journal of Hazardous Materials* 363: 286–294. 10.1016/j.jhazmat.2018.09.082.

Yu, Bin, Yongqian Shi, Bihe Yuan, Shuilai Qiu, Weiyi Xing, Weizhao Hu, Lei Song, Siuming Lo, and Yuan Hu. 2015. "Enhanced Thermal and Flame Retardant Properties of Flame-Retardant-Wrapped Graphene/Epoxy Resin Nanocomposites." *Journal of Materials Chemistry A* 3 (15): 8034–8044. 10.1039/C4TA06613H.

Yu, Huitao, Bangwen Zhang, Chaoke Bulin, Ruihong Li, and Ruiguang Xing. 2016. "High-Efficient Synthesis of Graphene Oxide Based on Improved Hummers Method." *Scientific Reports* 6 (1): 36143. 10.1038/srep36143.

Yu, Zhi-Ran, Min Mao, Shi-Neng Li, Qiao-Qi Xia, Cheng-Fei Cao, Li Zhao, Guo-Dong Zhang, Zhan-Jiang Zheng, Jie-Feng Gao, and Long-Cheng Tang. 2021. "Facile and Green Synthesis of Mechanically Flexible and Flame-Retardant Clay/Graphene Oxide Nanoribbon Interconnected Networks for Fire Safety and Prevention." *Chemical Engineering Journal* 405: 126620. 10.1016/j.cej.2020.126620.

Yuan, Bihe, Ao Fan, Man Yang, Xianfeng Chen, Yuan Hu, Chenlu Bao, Saihua Jiang, et al. 2017. "The Effects of Graphene on the Flammability and Fire Behavior of Intumescent Flame Retardant Polypropylene Composites at Different Flame Scenarios." *Polymer Degradation and Stability* 143: 42–56. 10.1016/j.polymdegradstab.2017.06.015.

Yuan, Bihe, Haibo Sheng, Xiaowei Mu, Lei Song, Qilong Tai, Yongqian Shi, Kim Meow Liew, and Yuan Hu. 2015. "Enhanced Flame Retardancy of Polypropylene by

Melamine-Modified Graphene Oxide." *Journal of Materials Science* 50 (16): 5389–5401. 10.1007/s10853-015-9083-0.

Yuan, Bihe, Weiyi Xing, Yixin Hu, Xiaowei Mu, Junling Wang, Qilong Tai, Guojun Li, Lu Liu, Kim Meow Liew, and Yuan Hu. 2016. "Boron/Phosphorus Doping for Retarding the Oxidation of Reduced Graphene Oxide." *Carbon* 101: 152–158. 10.1016/j.carbon.2016.01.080.

Zabihi, Omid, Mojtaba Ahmadi, Quanxiang Li, Mahmoud Reza Ghandehari Ferdowsi, Roya Mahmoodi, Ehsan Naderi Kalali, De-Yi Wang, and Minoo Naebe. 2020. "A Sustainable Approach to Scalable Production of a Graphene Based Flame Retardant Using Waste Fish Deoxyribonucleic Acid." *Journal of Cleaner Production* 247: 119150. 10.1016/j.jclepro.2019.119150.

Zeng, Lingshuai, Danqi Cheng, Zhenxing Mao, Yikai Zhou, and Tao Jing. 2022. "ZIF-8/Nitrogen-Doped Reduced Graphene Oxide as Thin Film Microextraction Adsorbents for Simultaneous Determination of Novel Halogenated Flame Retardants in Crayfish-Aquaculture Water Systems." *Chemosphere* 287: 132408. 10.1016/j.chemosphere.2021.132408.

Zhan, Yuanyuan, Sheng Shang, Bihe Yuan, Shasha Wang, Xianfeng Chen, and Gongqing Chen. 2021. "Carbonization Mechanism of Polypropylene Catalyzed by Co Compounds Combined with Phosphorus-Doped Graphene to Improve Its Fire Safety Performance." *Materials Today Communications* 26: 101792. 10.1016/j.mtcomm.2020.101792.

Zhang, Jing, Juan P. Fernández-Blázquez, Xiao-Lu Li, Rui Wang, Xiuqin Zhang, and De-Yi Wang. 2022. "A Facile Technique to Investigate the Char Strength and Fire Retardant Performance towards Intumescent Epoxy Nanocomposites Containing Different Synergists." *Polymer Degradation and Stability* 202: 110000. 10.1016/j.polymdegradstab.2022.110000.

Zhang, Zhao-Hui, Zuan-Yu Chen, Yi-Hao Tang, Yu-Tong Li, Dequan Ma, Guo-Dong Zhang, Rabah Boukherroub, et al. 2022. "Silicone/Graphene Oxide Co-Cross-Linked Aerogels with Wide-Temperature Mechanical Flexibility, Super-Hydrophobicity and Flame Resistance for Exceptional Thermal Insulation and Oil/Water Separation." *Journal of Materials Science & Technology* 114: 131–142. 10.1016/j.jmst.2021.11.012.

Zhang, Zhao-Hui, Jian-Wang Zhang, Cheng-Fei Cao, Kun-Yu Guo, Li Zhao, Guo-Dong Zhang, Jie-Feng Gao, and Long-Cheng Tang. 2020. "Temperature-Responsive Resistance Sensitivity Controlled by L-Ascorbic Acid and Silane Co-Functionalization in Flame-Retardant GO Network for Efficient Fire Early-Warning Response." *Chemical Engineering Journal* 386: 123894. 10.1016/j.cej.2019.123894.

Zhao, Bin, Peng-Wei Liu, Dong-Yue Liu, Thomas J. Kolibaba, Cong-Yun Zhang, Yan-Ting Liu, and Ya-Qing Liu. 2019. "Functionalized Graphene Oxide Based on Hydrogen-Bonding Interaction in Water: Preparation and Flame-Retardation on Epoxy Resin." *Macromolecular Materials and Engineering* 304 (8): 1900164. 10.1002/mame.201900164.

Zhao, Zhixin, Wei Cai, Zhoumei Xu, Xiaowei Mu, Xiyun Ren, Bin Zou, Zhou Gui, and Yuan Hu. 2020. "Multi-Role p-Styrene Sulfonate Assisted Electrochemical Preparation of Functionalized Graphene Nanosheets for Improving Fire Safety and Mechanical Property of Polystyrene Composites." *Composites Part B: Engineering* 181: 107544. 10.1016/j.compositesb.2019.107544.

Zhu, Jiahua, Minjiao Chen, Qingliang He, Lu Shao, Suying Wei, and Zhanhu Guo. 2013. "An Overview of the Engineered Graphene Nanostructures and Nanocomposites." *RSC Advances* 3 (45): 22790. 10.1039/c3ra44621b.

7 Two-Dimensional Graphitic Carbon Nitride for Reducing Fire Hazards of Polymer Composites

Bin Yu and Hong-Liang Ding
State Key Laboratory of Fire Science, University of Science and Technology of China, Hefei, Anhui, PR China

Miao Liu and Yong-Qian Shi
Fuzhou University, Fuzhou, PR China

CONTENTS

DOI: 10.1201/9781003327158-7

7.1   INTRODUCTION

Various polymer materials have been commercialized for more than 50 years in modern applications due to their attractive characteristics, such as tunable properties, excellent chemical and corrosion resistance, superior mechanical properties, good electrical properties and easy processing (Anwar et al., 2015; Yong et al., 2012; Zhang, Garrison, et al., 2017). However, most polymer materials suffer from serious shortcomings, including inherent flammability and the generation of a large amount of toxic gas and smoke during combustion, which extremely restrict their applications in some industrial fields for safety considerations, such as aerospace, automotive industry and electrical or electronic devices (Rad et al., 2019; Xiao et al., 2021). Therefore, it is necessary to enhance the fire safety property of polymers. The addition of a very small amount of organic or inorganic additive is a simple and common method of dramatically improving the flame retardancy of polymeric materials (Qiu et al., 2020; Shi et al., 2020; Yao et al., 2021). Moreover, among various kinds of additives, the nanofillers have been omnipresently investigated in the field of composite materials, and have been widely applied in a variety of manufacturing industries (Vaithylingam et al., 2017; Xu et al., 2008).

Carbon-based materials comprise an attractive and charming family. The discovery and universal application of a series of novel carbon-based materials, including fullerene, graphene/graphene oxide (GO), carbon nanotubes (CNTs), carbon black, graphite and carbon fiber (CF) will make a great revolutionary development both in the academic and industrial communities (Cui et al., 2016; Gaddam et al., 2019; Ma et al., 2019). Due to their excellent comprehensive properties, excellent dispersibility and stability in polymer matrices, carbon-based materials have been widely used as reinforcing materials to enhance the mechanical properties and fire safety performance of polymers (Zhao et al., 2017; Zhu et al., 2017).

Graphitic carbon nitride (g-C$_3$N$_4$) is a typical two-dimensional (2D) layered material that is traditionally formed from the regular arrangement of s-triazine (poly-s-triazine type) or s-heptazine (poly-s-heptazine type) units (Cao, BarisKumru, et al., 2020; Yang et al., 2012). In the past decade, g-C$_3$N$_4$ has attracted increasing attention due to its outstanding photoelectrochemical properties (Liu et al., 2016), thermal stability, low toxicity and low cost (Schwarzer et al., 2013) when used as a heterogeneous photocatalyst for organic transformations (Datta et al., 2010; Kurpil et al., 2018), hydrogen evolution (Katharina. Schwinghammer, 2013; Zhang, Li, et al., 2017) or pollution degradation (Cheng et al., 2013). Moreover, the typical 2D layered structure of g-C$_3$N$_4$ can effectively block heat propagation during combustion (Shi, Yu, et al., 2017; Wang, Wang, et al., 2012). Compared with other layered materials, such as graphene, GO, black phosphorene or montmorillonite (MMT), the g-C_3N_4 has the outstanding advantages in the easy, simple and fast synthesis with low cost (Xiao et al., 2021). In recent years, g-C_3N_4 and its modifications have been widely investigated as flame-retardant nanoadditives to prepare polymer composites.

Many research and review articles have focused on the application of g-C_3N_4 in catalysis and the catalytic mechanism. However, unlike its catalyst applications, few review articles have reported the flame-retardant enhancement effect of the g-C_3N_4 or its derivatives when they are used in polymeric materials. This chapter aims to briefly summarize the recent investigations on the flame-retardant application of g-C_3N_4 in polymeric materials, which is mainly organized into five sections. The Section 7.1 refers to the research background of g-C_3N_4. Then, the preparation and functionalization method of g-C_3N_4 is briefly introduced. Moreover, how to fabricate the polymer/g-C_3N_4 nanocomposites is also discussed, including solvent-blending, melt-blending, in situ polymerization and some other methods. Special emphasis is placed on the Section 7.4, referring to the flame-retardant polymer/graphitic carbon nitride composites. It is necessary to overview the effect of g-C_3N_4-based nanomaterials on fire safety performance and other important properties, such as thermal stability and mechanical behaviors. Ultimately, a concise summary of the prospective opportunities for g-C_3N_4-based flame retardants as an emerging flame-retardant strategy is given to expect better development in the future.

7.2 PREPARATION AND FUNCTIONALIZATION OF g-C_3N_4

7.2.1 PREPARATION OF g-C_3N_4

Since g-C_3N_4 shows promising applications in various fields, many synthesis methods and functionalization techniques, including solvothermal, thermal condensation, hydrothermal, electrodeposition and magnetron sputtering methods have been developed to prepare g-C_3N_4 nanomaterials with different microstructures and morphologies. In this section, numerous synthesis methods for the preparation of g-C_3N_4 nanomaterials are presented as follows.

7.2.1.1 Solvothermal Method

The solvothermal method is mainly a heterogeneous synthesis method, where the original mixture is reacted with each other at a certain temperature and under an autogenic pressure of organic solvent in a sealed system. The solvothermal method has been widely used in the fabrication of various nanomaterials due to its advantages, such as excellent uniformity, mild reaction conditions, suitable flowability, relatively simple process and easy control. Recently, g-C_3N_4 nanomaterials with different micromorphologies have been obtained by changing the experimental conditions, including the reaction temperature, pressure, time, solvents and precursors (Cui et al., 2012; Montigaud et al., 2000).

Melamine is one of the most popular and commonly used precursors among many precursors, while the most common solvents selected for the solvothermal treatment of precursors involve benzene, trimethylamine, acetonitrile, carbon tetrachloride and so on. However, most solvents used in the solvothermal method are highly toxic and dangerous. Andreyev et al. (2003) used hexachlorobenzene and sodium azide as reactants and obtained a fine-grained homogeneous carbon nitride powder, while the synthesis process was conducted in an enclosed system under high pressure 7.7 GPa at 400–500 °C for 50–70 h. In Bai's study, carbon tetrachloride and ammonium chloride were used to synthesize the g-C$_3$N$_4$ nanocrystals at 400 °C for 20 h without any catalyst (Bai et al., 2003). The as-prepared g-C$_3$N$_4$ showed superior crystallinity, and the average size was approximately 11 nm. The synthesis process significantly reduced the reaction temperature of g-C$_3$N$_4$ and broadened the selection range of precursors. Lv et al. (2003) successfully obtained crystalline carbon nitride powder through the liquid-solid reaction between lithium nitride and cyanuric chloride at 355 °C for 12 h under 5–6 MPa. Meanwhile, other crystalline carbon nitrides named α-C$_3$N$_4$ and β-C$_3$N$_4$ can also be synthesized.

To further reduce the reaction temperature, the well-crystallized g-C$_3$N$_4$ nanocrystallite was prepared via the reaction between melamine and sodium amino using

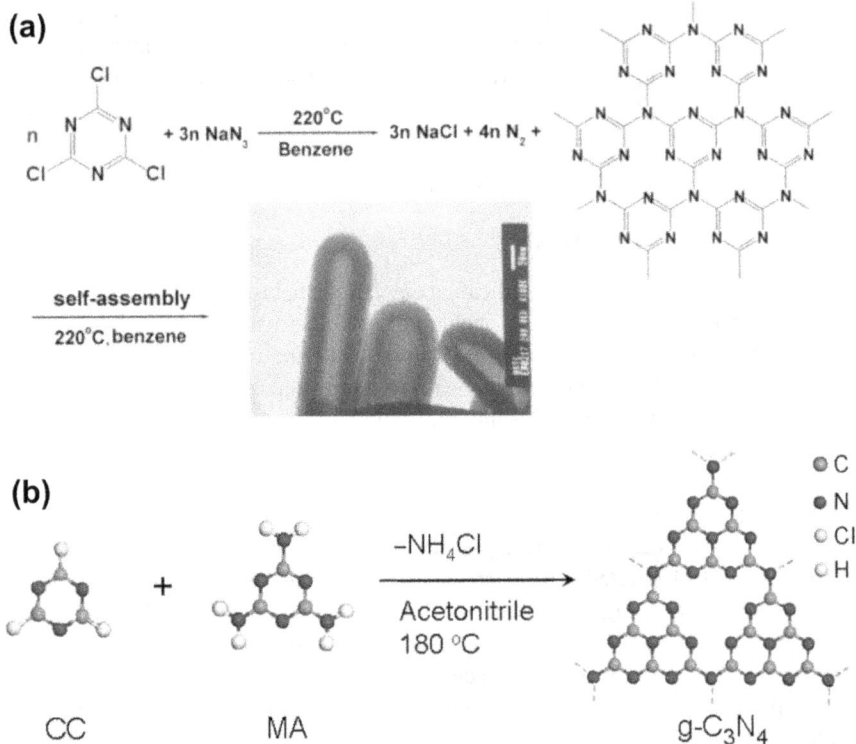

FIGURE 7.1 (a) The synthesis diagram of g-C$_3$N$_4$ from sodium azide and cyanuric chloride. (Reprinted from Guo, Q., Xie, Y., Wang, X., Zhang, S., Hou, T. and Lv, S., *Chem. Commun.*, 1, 26–27, 2004. With permission.) (b) The Polymerization processes of cyanuric chloride (CC) and melamine (MA). (Reprinted from Cui, Y., Ding, Z., Fu, X. and Wang, X., *Angew. Chem. Int. Ed.*, 51, 11814–11818, 2012. With permission.)

benzene as the solvent at 180–220 °C for 8–12 h (Guo et al., 2003). TEM images indicated that the obtained product showed different morphologies, including particles and tiny hollow balls, as well as square frameworks. Sodium azide can replace sodium amino groups and can also be used as a raw material to synthesize high-quality g-C_3N_4 nanotubes by a simple benzene-thermal process at 220 °C for 15 h (Figure 7.1a) (Guo, Xie, et al., 2004). Moreover, Guo, Yang, et al. (2004) selected ammonium chloride as the nitrogen source and iron as the catalyst to prepare g-C_3N_4 nanotubes with uniform size at 220 °C for 15 h. Cui et al. (2012) prepared g-C_3N_4 at a relatively low temperature (150 °C) with sodium amino and melamine as the nitrogen source in subcritical acetonitrile (Figure 7.1b).

7.2.1.2 Thermal Condensation Method

Thermal treatment of organic compounds containing a decent amount of nitrogen as a precursor is one of the most popular and simple methods to prepare g-C_3N_4 (Figure 7.2). Taking melamine as an example, g-C_3N_4 with lamellar micromorphology can be obtained through the thermal condensation process in the temperature range of 500–650 °C. When the temperature rises over 660 °C, carbon nitride rapidly decomposes and cannot stably exist, which may decompose and generate many byproducts during this reaction.

The reaction conditions and types of precursors have a significant effect on the physical and chemical properties of the g-C_3N_4 nanomaterial. Yan et al. (2009) prepared g-C_3N_4 photocatalyst powder by directly heating melamine in a semi-closed system via a two-step heat process. In the second step, melamine was heated to 520, 550 and 580 °C for 2 h for further deamination treatment. The product obtained after thermal treatment at 520 °C for 2 h exhibited better photodegradation activity than the other two products. The higher heat treatment temperature can reduce the content of hydrogen in the products and increase the degrees of condensation (higher C/N ratio), which leads to the lower structural integrality and worse thermal stability of the final g-C_3N_4 product. Similarly, some pretreatment processes for precursors also have a great influence on the thermal condensation process. Yan et al. (2012) demonstrated that the sublimation of melamine during the heating process could be effectively restrained by using sulfuric acid to pretreat melamine, resulting in a different thermal condensation process. The specific surface area of the obtained g-C_3N_4 was slightly larger than that of the directly prepared products.

Some other nitrogenous compounds including urea, thiourea and ammonium thiocyanate can also be heated to fabricate the g-C_3N_4 material. Qin et al. (2015) synthesized g-C_3N_4 nanosheets (CNU-BA) derived from urea and barbituric acid (BA) via a facile one-pot method. Figure 7.3 shows the scanning electron microscope (SEM) and transmission electron microscope (TEM) images of CNU and CNU-BA$_{0.03}$. The obtained material exhibited superior photocatalytic performance for the reduction of CO_2. Komatsu (Komatsu et al., 2001) obtained g-C_3N_4 through the pyrolysis of ammonium thiocyanate at 550 °C for 24 h, which could be stable up to more than 600 °C under an inert condition.

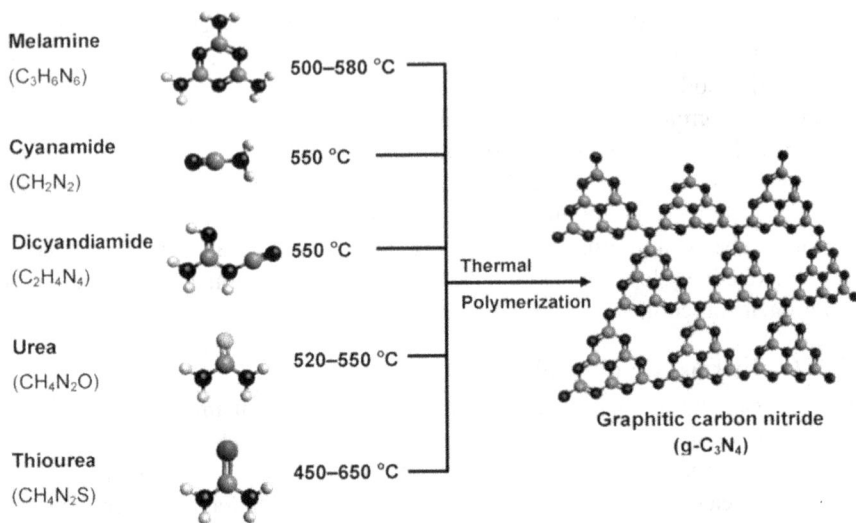

Melamine ($C_3H_6N_6$) 500–580 °C

Cyanamide (CH_2N_2) 550 °C

Dicyandiamide ($C_2H_4N_4$) 550 °C

Thermal Polymerization

Urea (CH_4N_2O) 520–550 °C

Thiourea (CH_4N_2S) 450–650 °C

Graphitic carbon nitride (g-C_3N_4)

FIGURE 7.2 The synthesis diagram of g-C_3N_4 by thermal condensation of precursors containing C and N. The black, blue, white, red and yellow spheres represent C, N, H, O and S atoms, respectively. (Reprinted from Ong, W., Tan, L., Ng, Y., Yong, S. and Chai, S., *Chem. Rev.* 116, 7159–7329, 2016. With permission.)

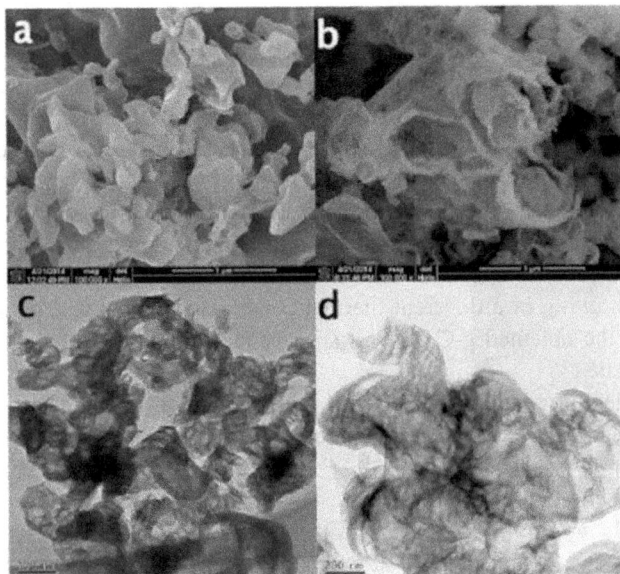

FIGURE 7.3 SEM and TEM images of CNU (a, c) and CNU–$BA_{0.03}$ (b, d). (Reprinted from Qin, J., Wang, S., Ren, H., Hou, Y. and Wang, X., *Appl. Catal. B-Environ.*, 179, 1–8, 2015. With permission.)

7.2.1.3 Electrodeposition Method

Organic compounds containing a decent amount of nitrogen used as precursors can significantly reduce the temperature of the C-N bonding reaction and provide enough C-N bonds for the electrodeposition reaction. Therefore, g-C_3N_4 can be prepared by an electrochemical deposition process at room temperature and pressure. In Li's study, g-C_3N_4 film was deposited on a Si (100) substrate at room temperature by the electrodeposition method (Li, Cao, & Zhu, 2004; Li, Cao, Zhu, et al., 2004). A saturated acetonitrile solution of melamine and melamine chloride was chosen as the deposition solution. In addition, the g-C_3N_4 film with different chemical compositions and crystal structures can be obtained by adjusting the molar ratio of the precursors. Bai et al. (2010) synthesized hollow carbon nitride microspheres via the liquid phase electrodeposition technique by choosing silica nanospheres as a template attached to the electrode surface, where dicyandia-mide was used as a precursor in acetone solution.

7.2.1.4 Magnetron Sputtering Method

The magnetron sputtering method is mainly used in the preparation of g-C_3N_4 films due to its lower working pressure and faster deposition rate. The nitrogen fraction and substrate temperature showed a great influence on the deposition rate and morphology of the g-C_3N_4 film (Hellgren et al., 2001). When the volume fraction of nitrogen was 50%, the formed film was denser and smoother. Broitman et al. (2001) compared the hardness, elasticity, wear rate and friction coefficient of different g-C_3N_4 films deposited under various conditions in N_2/Ar plasma.

The advantages and disadvantages of several synthesis methods of g-C_3N_4 are listed in Table 7.1. Considering the great advantage of solvothermal method, it is inevitable to use large amounts of organic solvents with high toxicity to prepare g-C_3N_4. The g-C_3N_4 prepared by the electrodeposition method has poor purity, and its crystal forms are complex. Similarly, it is difficult to prepare g-C_3N_4 with high purity by the magnetron sputtering method. Thermal condensation has become a common and popular method to synthesize g-C_3N_4 due to its simplicity and

---

TABLE 7.1

The Comparison of Serval Synthesis Methods for g-C_3N_4

Method	Advantage	Disadvantage
Solvothermal	Mild reaction condition; Simplicity; Easy to control.	High toxicity of solvent; Low yield
Thermal condensation	Abundant raw materials; Simple to prepare; By adding templates to adjust the microstructure	Small specific surface area; Difficult to remove all templates; Inevitable to use toxic solvents
Electrodeposition	Room temperature and pressure.	Complex crystal forms; Poor purity.
Magnetron sputtering	Suitable for preparation of film.	Difficult to prepare g-C_3N_4 with high purity

abundance of raw materials. Various kinds of templates can be added to adjust the microstructure of g-C$_3$N$_4$ to improve its comprehensive properties. Though fully removing the templates is always a great challenge, i.e., harmful to human health and the environment of the solvents used to remove the templates. On many counts, thermal condensation has become the most widely used method for the preparation of g-C$_3$N$_4$ in recent research, compared with three other methods.

7.2.2　Functionalization of g-C$_3$N$_4$

Although g-C$_3$N$_4$ materials possess many excellent properties and show great application potential in various fields, they still have some shortcomings, such as weak interactions with other materials and poor dispersion. Thus, the chemical functionalization of g-C$_3$N$_4$ is another effective method to improve its physical and chemical properties for different applications. Similar to many other carbon-based nanomaterials, such as CNTs, graphene/GO or fullerene, the modification process is mainly related to covalent and noncovalent routes (Figure 7.4). The main purpose of the covalent strategy is to establish a new chemical bond between the g-C$_3$N$_4$

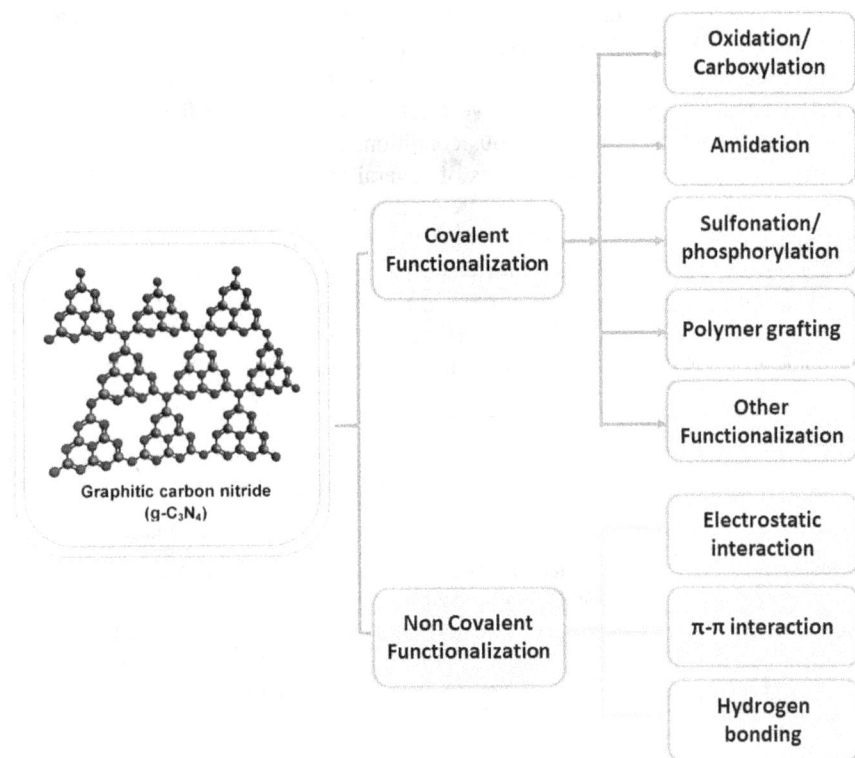

Graphitic carbon nitride
(g-C$_3$N$_4$)

Covalent Functionalization
- Oxidation/Carboxylation
- Amidation
- Sulfonation/phosphorylation
- Polymer grafting
- Other Functionalization

Non Covalent Functionalization
- Electrostatic interaction
- π-π interaction
- Hydrogen bonding

FIGURE 7.4 Two main methods of the functionalization of g-C$_3$N$_4$. (Reprinted from Majdoub, M., Anfar, Z. and Amedlous, A., *ACS Nano.*, 14, 12390–12469, 2020. With permission.)

structure and chemical modifiers, which involves oxidation, amidation, sulfonation and polymer grafting to modify the g-C_3N_4 surface. Different from the covalent approach, the noncovalent strategy relies heavily on intermolecular interactions and physical bonds, such as van der Waals forces, hydrogen bonding, π–π interactions and electrostatic interactions.

7.2.2.1   Covalent Method

7.2.2.1.1   Oxidation/Carboxylation

Oxidation or carboxylation reactions mainly introduce carboxyl or hydroxyl functional groups as well as doping oxygen atoms into the framework of g-C_3N_4 with the existence of a catalytic amount of an oxidant, including K_2CrO_7, H_2SO_4, HNO_3, NaOH and oxalic acid. The introduction of oxygen-containing groups into the g-C_3N_4 surface can significantly enhance its hydrophilicity and other properties. Li et al. used $K_2Cr_2O_7$ and H_2SO_4 as chemical oxidants to synthesize porous g-C_3N_4 with excellent water dispersibility (Li et al., 2015). The whole synthesis process was divided into two steps (Figure 7.5a). Firstly, dicyandiamide was heated

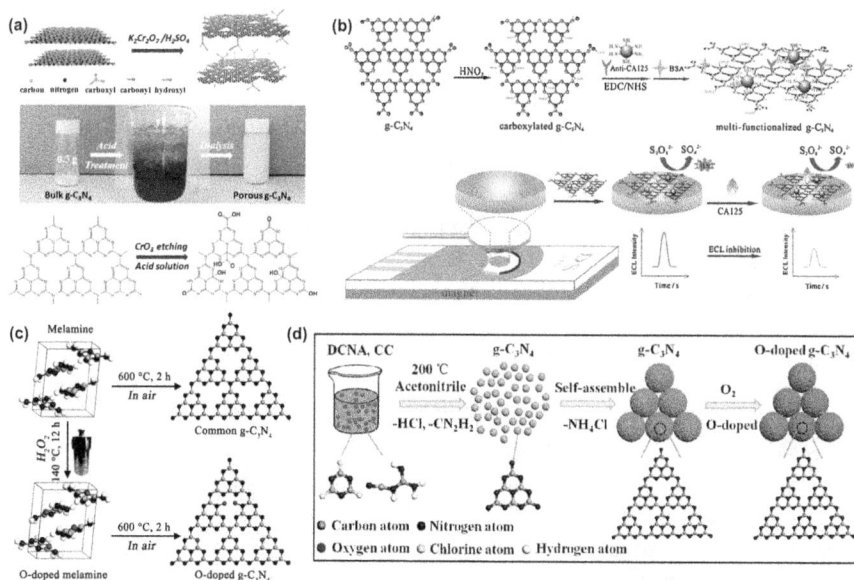

FIGURE 7.5   (a) Synthesis Procedure of porous g-C_3N_4 by chemical oxidation process. (Reprinted from Li, H., Sun, B., Sui, L., Qian, D. and Sun, M., *Phys. Chem. Chem. Phys.*, 17, 3309–3315, 2015. With permission.) (b) Preparation procedure of multi-functionalized g-C_3N_4. (Reprinted from Wu, L., Sha, Y., Li, W., Wang, S., Guo, Z., Zhou, J., et al., *Sensor. Actuat. B-Chem.*, 226, 62–68, 2016. With permission.) (c) Synthesis process of O-doped g-C_3N_4 nanosheets in Wu's study. (Reprinted from Sun, S., Li, J., Cui, J., Gou, X., Yang, Q. and Liang, S. et al., *Inorg Chem Front.*, 5, 1721–1727, 2018. With permission.) (d) Synthesis route of the formation process of oxygen self-doped g-C_3N_4 nanospheres in Wei's study. (Reprinted from Wei, F., Liu, Y., Zhao, H., Ren, X., Liu, J. and Hasan, T., et al. *Nanoscale*, 10, 4515–4522, 2018. With permission.)

at 550 °C for 4 h to prepare bulk g-C_3N_4. Then, the bulk g-C_3N_4 was oxidized using a mixed solution of 10 g of $K_2Cr_2O_7$ and 50 mL of H_2SO_4 (98 wt%) for 2 h at room temperature. Hydroxyl and carboxyl groups were produced on the surface, and simultaneously the prepared g-C_3N_4 showed a porous structure.

HNO_3 is another common oxidant with high efficiency for the oxidation of g-C_3N_4. In Wu's study, the multifunctionalized g-C_3N_4 material was synthesized after oxidation treatment with HNO_3, which could be used for the detection of the tumor marker carbohydrate antigen 125 (CA125) (Wu et al., 2016). The synthesis process was also very simple and easy to operate. The mixture of g-C_3N_4 powder and 100 mL of 5 mol/L HNO_3 was refluxed for 24 h at 125 °C (Figure 7.5b). The fabricated immunosensor possessed many superior properties, such as high sensitivity, superior specificity, excellent reproducibility and long-term stability. Unlike using acid as an oxidant, Wang et al. (2020) chose an alkali solution to modify g-C_3N_4 and prepare multifunctional g-C_3N_4 via a hydrothermal reaction. The OH/Na co-functionalized g-C_3N_4 exhibited enhanced photocatalytic efficiency due to the promotion of the directional transfer of charge carriers and an increase in the redox potential.

In addition, oxygen doping on the framework of g-C_3N_4 is another important method to enhance its properties. Sun et al. (2018) developed O-doped g-C_3N_4 nanosheets via a hydrothermal method using hydrogen peroxide (Figure 7.5c). The obtained O-doped g-C_3N_4 nanosheets exhibited significantly improved photocatalytic performance for the hydrogen evolution reaction, which was approximately 10.7 times higher than that of the original g-C_3N_4 nanosheets. Similarly, Wei et al. (2018) prepared an O-doped g-C_3N_4 nanomaterial with better photocatalytic activity via a copolymerization of dicyandiamide and 1,3,5-trichlorotriazine (Figure 7.5d). In summary, oxidation is one of the most valuable and effective methods to improve the comprehensive performances of g-C_3N_4, such as photocatalytic activity, photoelectrochemical activity and ionic conductivity. Moreover, the generation of oxygen-containing groups (-COOH or -OH) is conducive to further chemical covalent functionalization of g-C_3N_4.

7.2.2.1.2 Amidation

The main purpose of the amidation reaction is to establish a new amide linkage (–NH–C=O) on the surface of g-C_3N_4 via the reaction between -NH_2 and -COOH. Thus, there are two strategies to complete the amidation reaction. The first strategy principally relies on the direct chemical reaction between the amino group on the surface of structurally deficient g-C_3N_4 and the chemical modifier containing a carboxyl group. Another method involves the reaction of the amine-containing chemical modifier with the carboxylated g-C_3N_4 after carboxylation treatment to form the amide product. Guo et al. (2010) successfully prepared a novel polymeric photocatalyst (PMCNP) via the reaction between anhydride and the NH_2 group in g-C_3N_4 (Figure 7.6a). The synthesized photocatalyst exhibited excellent photocatalytic activity and performance due to the existence of coupling anhydride groups promoting the separation of the photoinduced electrons and holes. In Ye's study, a novel ferrocene-modified g-C_3N_4 material (Fc-CN) was developed by the amidation reaction between NH_2 groups on the mesoporous g-C_3N_4 (MCN) surface

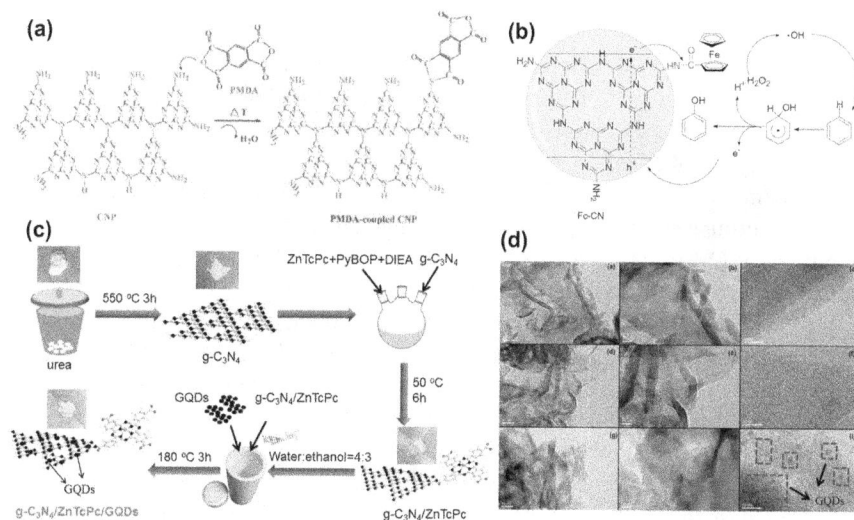

FIGURE 7.6 (a) The synthesis route of PMCNP in Guo's study. (Reprinted from Guo, Y., Chu, S., Yan, S., Wang, Y., Zou, Z., *Chem. Commun.*, 46, 7325–7327, 2010. With permission.) (b) Proposed reaction mechanism for the oxidation reaction of benzene catalyzed by Fc-CN. (Reprinted from Ye X., Zheng Y., Wang, X., *Chin. J. Chem. Rev.*, 32, 498–506, 2014. With permission.) (c) Synthesis process of g-C_3N_4/ZnTcPc/GQDs. (d) TEM images of g-C_3N_4 and g-C_3N_4/ZnTcPc/GQDs. (Reprinted from Xu, T., Wang, D., Dong, L., Shen, H., Lu, W., Chen, W., *Appl. Catal. B-Environ.*, 244, 96–106, 2019. With permission.)

and the carboxyl groups of ferrocenecarboxylic acid (Fc-COOH) (Ye et al., 2014). The obtained functionalized g-C_3N_4 served as heterogeneous catalysts to oxidize benzene to produce phenol using hydrogen peroxide as an oxidant under visible light irradiation. The mechanism for the catalytic reaction of benzene to generate phenol was proposed, as shown in Figure 7.6b. Due to the synergistic donor-acceptor interaction between the MCN matrix and Fc motifs, phenol can be directly achieved from benzene with the existence of H_2O_2 under visible light irradiation.

Moreover, some metal ions can be introduced into the framework of g-C_3N_4 via covalent immobilization in the amidation process. Lu et al. (2016) reported the covalent immobilization of zinc tetracarboxyphthalocyanine (ZnTcPc) onto the network of g-C_3N_4 to prepare a novel photocatalyst (g-C_3N_4/ZnTcPc) by combining g-C_3N_4 with zinc phthalocyanine. Compared with virgin g-C_3N_4 or pure ZnTcPc as a catalytic agent, g-C_3N_4/ZnTcPc showed much higher photocatalytic activity and maintained its photocatalytic property with high levels over a wide range of pH values. In another work, graphene quantum dots (GQDs) were used to further modify the obtained g-C_3N_4/ZnTcPc via the hydrothermal method (Figures 7.6c and d) (Xu et al., 2019). When rhodamine B, sulfaquinoxaline sodium and carbamazepine were used as model pollutants, the as-prepared g-C_3N_4/ZnTcPc/GQDs showed higher photocatalytic activity in the process of photodegradation under solar light irradiation.

7.2.2.1.3 Sulfonation/Phosphorylation

To introduce some acid groups (sulfonic or phosphoric) into the g-C$_3$N$_4$ surface, sulfonation and phosphorylation which are common and powerful approaches are adopted to enhance its comprehensive properties and to expand its applications. Researching on the sulfonation chemistry of g-C$_3$N$_4$ is gradually increasing. Velayutham (Velayutham et al., 2018) successfully prepared g-C$_3$N$_4$ with amino and imino groups by one-step calcination using melamine as the raw material. Then, sulfonated g-C$_3$N$_4$ (s-GCN) was obtained by surface functionalization treatment using chlorosulfonic acid as a modifier, which can serve as an effective bifunctional filler to modify the Nafion membrane for direct methanol fuel cells. Lu et al. (2020) reported similar work using sulfonic acid-functionalized g-C$_3$N$_4$ (SCN) nanosheets as nanofillers. Different from Velayutham's work, sulfonated g-C$_3$N$_4$ was added to the fluorinated poly(arylene ether) copolymers with disulfonated naphthyl pendants (SPFAE) (Figure 7.7a). The SPFAE/SCN membrane with only 0.5 wt% SCN exhibited a higher proton conductivity, which increased by 26%, as compared to the original membrane. The mechanical property and dimensional stability of the SPFAE/SCN membrane were also significantly reinforced due to interfacial interactions.

Ye et al. (2016) developed a novel mesoporous phosphorylated g-C$_3$N$_4$ (MPCN) by the phosphorylation reaction of concentrated phosphoric acid with bulk g-C$_3$N$_4$ (BCN). MPCN showed much higher activity in the photocatalytic CO$_2$ reduction process than unmodified BCN, which was mainly attributed to the synergistic cooperation of the phosphorylation-induced mesoporous structure and the PO$_4$$^{3-}$ groups on the surface. Moreover, MPCN displayed a higher photoinduced carrier separation efficiency than BCN, resulting in enhanced photocatalytic activity. Fang et al. (2019) developed phosphorylated polymeric carbon nitride (PCN) photoanodes via the phosphorylation process and immersion into a concentrated H$_3$PO$_4$ solution. The modified PCN films showed higher photocatalytic activity and efficiency of solar-driven water splitting due to obviously increased the valence electron density on the surface in this system.

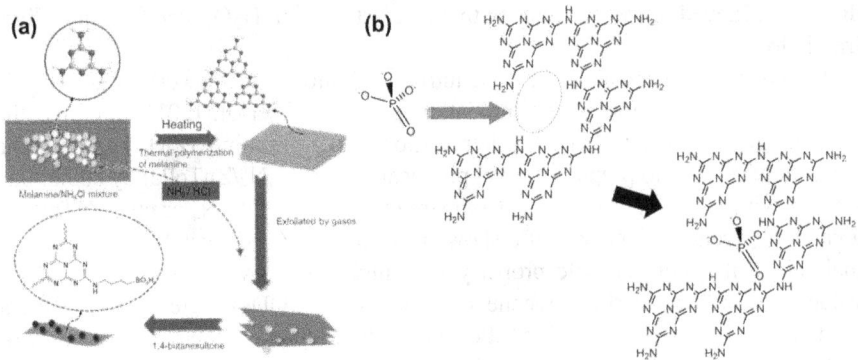

FIGURE 7.7 (a) Synthesis route of CN and SCN nanosheets. (b) Illustration of the modified g-C$_3$N$_4$ structure after phosphorylation (Reprinted from Lu, Y., Liu, Y., Li, N., Hu, Z., Chen, S., *J. Membr. Sci.*, 601, 117908, 2020. With permission.)

7.2.2.1.4 Polymer Grafting

Various polymer materials have been widely used in daily life due to their unique properties, including outstanding electrical conductivity, excellent corrosion resistance, good flame retardancy and self-healing performance. Therefore, grafting various polymer materials on the g-C_3N_4 surface seems to be a valuable modification approach, which can endow g-C_3N_4 materials with superior properties and expand their industrial application by integrating characteristics of various individual components. The polymer grafting process to functionalize g-C_3N_4 can be simply divided into two major categories. One category involves the in situ polymerization of monomers directly on the surface of g-C_3N_4. Another consists of the direct reaction of the presynthesized polymer with the active groups on the framework of g-C_3N_4.

Jiang et al. (2016) developed polyaniline (PANI)/carbon nitride nanosheet (CNNS) composite hydrogels via in situ polymerization. The PANI/CNNS composite hydrogel exhibited high contaminant photodegradation activity, which was mainly due to the cooperation of adsorptive preconcentration and subsequent photocatalytic oxidation. Furthermore, the as-synthesized g-C_3N_4-based composite hydrogel was easy to recycle due to its robust and stable 3D hierarchical nanostructure. Cao et al. (2019) chose a "grafting to" strategy to graft defined polymers onto g-C_3N_4 surfaces under visible-light irradiation. Ene-functional polymers were grafted onto the surface of -g-C_3N_4 under visible light due to the formation of surface radicals. The functionalized polymer/g-C_3N_4 hybrid materials showed several superior properties, such as improved dispersibility and hydrophobicity.

7.2.2.1.5 Other Covalent Functionalization

Similar to other carbon-based materials, g-C_3N_4 can also be functionalized with some metal atoms, metal complexes or nanoparticles. There have been many investigations about the functionalization of g-C_3N_4 with some 2D or 3D materials, including GO, MoS_2, polyoxometalates (POMs) and MOFs. Coulson et al. (2020) successfully realized the direct coordination of [Cp*IrCl] to the g-C_3N_4 surface. The obtained g-C_3N_4-based composite material could act as a catalyst for the hydrogenation of unsaturated molecules with high selectivity.

7.2.2.2 Noncovalent Method

As mentioned above, the covalent functionalization method is conducted by establishing a new chemical bond between the g-C_3N_4 surface and chemical modifier, whereas the noncovalent functionalization method is mainly related to intermolecular interactions and physical bonds, including van der Waals forces, hydrogen bonding, π–π interactions and electrostatic interactions. Compared with the formed covalent bond, the intermolecular interactions between g-C_3N_4 and other organic or inorganic compounds are very weak; thus, the g-C_3N_4 hybrid material modified by the noncovalent method exhibits inferior sustainable stability. However, noncovalent methods can combine the unique features of each compound and preserve their intrinsic properties.

Da Silva et al. (2018) synthesized novel hybrid photocatalysts based on a g-C$_3$N$_4$ material sensitized with free-base porphyrins via simple impregnation through noncovalent interactions. The hybrid photocatalysts were used to generate hydrogen (H$_2$) from water splitting. The porphyrin was assembled over the surface of g-C$_3$N$_4$ via π–π interactions. Undoubtedly, the hybrid catalysts showed higher photocatalytic activity than pure g-C$_3$N$_4$. The electron transfer from the excited state of the porphyrin to the conduction band of g-C$_3$N$_4$ promoted the separation and transport of the photogenerated charge carriers, resulting in higher photo-catalytic activity. Liu et al (Liu et al., 2017) successfully modified g-C$_3$N$_4$ nanosheets using cobalt meso-tetra-phenylporphyrin (CoTPP/g-C$_3$N$_4$) by self-assembly based on π–π supramolecular interactions. The prepared CoTPP/g-C$_3$N$_4$ displayed superior CO$_2$ catalytic activity, and thus showed excellent CO$_2$ reduction performance. Similarly, Chen, Wang, et al. (2015) chose Cu(II) meso-tetra(4-carboxyphenyl) porphyrin (CuTCPP) to hybridize with g-C$_3$N$_4$ by π–π stacking interactions. The CuTCPP/g-C$_3$N$_4$ hybrid exhibited much higher pho-tocatalytic efficiency than unmodified g-C$_3$N$_4$ for the degradation of phenol under visible light irradiation.

7.3 APPROACHES TO THE FABRICATION OF POLYMER/g-C$_3$N$_4$ NANOCOMPOSITES

Polymers have been widely investigated due to their important role in daily life and industry. Currently, polymer materials with special properties, such as photo-catalysis, biodegradability, and electrical conductivity, also attract increasing attention. The combination of novel g-C$_3$N$_4$ with traditional polymers has been proven to be a valuable approach to prepare useful polymers with excellent com-prehensive properties, which can combine the advantages of each compound. There are several methods to fabricate polymer/g-C$_3$N$_4$ nanocomposites, including solvent-blending, melt-blending, in situ polymerization and some other methods.

7.3.1 SOLVENT BLENDING

As a typical inorganic material, how to well disperse g-C$_3$N$_4$ in organic polymer monomers is always a great challenge. Solvent blending is probably the simplest and most commonly used processing technology to prepare polymer/g-C$_3$N$_4$ nanocomposites, which can ensure the excellent dispersion of g-C$_3$N$_4$ in polymer monomers through the strong interaction between each compound. The polymer matrix is mixed in organic solvents or aqueous solvents such as water, N,N-dimethyl formamide (DMF), toluene, acetone and so on. Then, g-C$_3$N$_4$ was added to the polymer solution using mechanical, magnetic or ultrasonication mixing for better dispersion. Ultimately, the g-C$_3$N$_4$ and polymer are evenly dispersed in the solvent after the vigorous stirring. The solvent needs to be removed by evaporation or distillation before pouring into the mold for fixation. The solvent-blending method is useful and effective for the combination of inorganic carbon-based hybrids and polymers, including epoxy resin (EP), polyacrylamide, polyimide (PI), polycarbonate and polystyrene (PS).

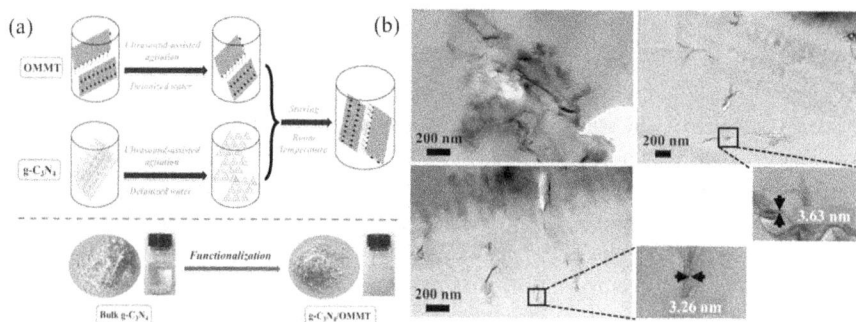

FIGURE 7.8 (a) Synthesis route of g-C$_3$N$_4$/OMMT nanohybrids. (b) TEM images of PS/g-C$_3$N$_4$ composites. (Reprinted from Shi, Y., Wang B., Duan, L., Zhu, Y., Gui, Z., Yuen, R. et al., *Ind. Eng. Chem. Res.*, 55, 7646–7654, 2016. With permission.)

Shi, Wang, et al. (2016) used organic-modified MMT (OMMT) to functionalize the g-C$_3$N$_4$ nanomaterial by electrostatic interactions (Figure 7.8a). Then, the obtained g-C$_3$N$_4$/OMMT was added to a PS matrix via the simple solvent blending method using DMF as the solvent (Figure 7.8b). The PS composites exhibited improved flame retardancy. The peak heat release rate (PHRR) value of the CBACM50 sample decreased from 1120 w/g for pure PS to 830 w/g, and the total heat release (THR) value decreased from 47.7 kJ/g to 46.1 kJ/g.

Sometimes the polymer matrix itself is liquid, and the g-C$_3$N$_4$-based hybrids can be added into the polymer matrix without any solvent for uniform dispersion. Xu et al. (2021) used zeolite imidazole frameworks (ZIFs) to modify g-C$_3$N$_4$ by electrostatic action. SiO$_2$ was introduced to the surface of ZIF-67/g-C$_3$N$_4$ (ZC) via the hydrolyzation of tetraethyl orthosilicate. Finally, the obtained ZIF-67/g-C$_3$N$_4$@SiO$_2$ (ZCS) nanometer hybrid material was added to EP with stirring during the curing. The EP composites with 3 wt% ZCS exhibited better fire resistance than neat EP. The PHRR, THR and total smoke production (TSP) values were significantly reduced by 46.8%, 21.7% and 19.3%, respectively.

7.3.2 MELTING AND BENDING

Melt blending is also a useful technique to fabricate polymer/g-C$_3$N$_4$ nanocomposites. Due to its low cost, this method is suitable for large-scale manufacturing in most industrial sectors. The polymer monomers are heated to a molten state at high temperature, whereas the g-C$_3$N$_4$ hybrids are directly mixed with the polymer matrix by a twin-screw extruder to prepare the polymer/g-C$_3$N$_4$ composites. Therefore, the melt blending method tends to require higher temperatures and specific experimental equipment. However, the melt-blending method can avoid using solvents, which need to be fully removed in the solvent-blending method.

Shi, Yu, et al. (2015) synthesized a series of CuCo$_2$O$_4$/g-C$_3$N$_4$ nanohybrids (C-CuCo$_2$O$_4$) with different weight ratios of each compound via a facile hydrothermal method. Then, the C-CuCo$_2$O$_4$ nanohybrid was added to thermoplastic polyurethane (TPU) monomers to fabricate the TPU composite. The SEM and TEM images of TPU composites indicated that C-CuCo$_2$O$_4$ exhibited excellent dispersion in the TPU matrix, while the CuCo$_2$O$_4$ or g-C$_3$N$_4$ nanosheets alone showed severe aggregation in TPU composites. The thermal stability and fire resistance performance of the TPU/C-CuCo$_2$O$_4$ composite were obviously enhanced. Compared with neat TPU, the PHRR and THR of TPU/C-CuCo$_2$O$_4$-7 samples declined by 37% and 31.3%, respectively. Zhu et al. (2022) confirmed that the g-C$_3$N$_4$ was incorporated with poly diaminodiphenyl phosphonic methane (PDMPD) to improve the flamer retardancy as well as the thermal stability of TPU. Zheng et al. (2020) chose polyamide 6 (PA 6) as a polymer matrix and fabricated flame-retardant PA 6 by the incorporation of an inorganic–organic hybrid (MgO/g-C$_3$N$_4$). Unlike unmodified g-C$_3$N$_4$, MgO/g-C$_3$N$_4$ (MCN) possessed a laminate structure with more holes on the surface and a larger specific surface area. Owing to its superior compatibility and dispersion in the PA 6 matrix, the fire resistance and mechanical properties of the PA 6 composites were effectively improved. With the addition of 20 wt% MCN, the PA 6/MCN samples reached a V-0 rating in the UL-94 test, while the LOI value also increased to 32.1%.

7.3.3 IN SITU POLYMERIZATION

In the manufacturing process, the g-C$_3$N$_4$-based hybrids can be directly mixed with the polymer matrix and polymerized with the assistance of heat or radiation to fabricate polymer/g-C$_3$N$_4$ composites. The polymer matrix and g-C$_3$N$_4$-based nanofillers are combined through the formed covalent cross-linking.

Yu et al. (2021) successfully fabricated a series of g-C$_3$N$_4$/polyaniline composites (g-C$_3$N$_4$/PANI) via a facile in situ polymerization in an ice water bath. When using oxytetracycline (OTC) as a model pollutant, the as-prepared g-C$_3$N$_4$/PANI material exhibited excellent photocatalytic activity for the degradation of wastewater under simulated sunlight irradiation. The optimal g-C$_3$N$_4$/PANI composite exhibited an obvious improvement in the degradation rate, which was 5-fold that of pure g-C$_3$N$_4$. The main reason was that such g-C$_3$N$_4$/PANI hybrid nanocomposites could improve photogenerated electron-hole separation and enhance extensive absorption, resulting in better photocatalytic activity. Sheng et al. (2015) used an in situ polymerization process via thermal imidization to fabricate g-C$_3$N$_4$/PI composite films. The SEM and TEM images of g-C$_3$N$_4$ are shown in Figure 7.9a, whereas the TEM images of g-C$_3$N$_4$/PI composite films are presented in Figure 7.9b. When the incorporation amount of g-C$_3$N$_4$ was only 2 wt%, the mechanical properties of g-C$_3$N$_4$/PI composites were significantly enhanced due to the strong interaction between each compound. The tensile strength and elongation at break of the PI material increased by 23% and 21%, respectively. The g-C$_3$N$_4$/PI composites also showed better dielectric properties with higher dielectric constants and loss tangents of approximately 5.42 and 0.00657, respectively. Moreover, the lower the content of g-C$_3$N$_4$, the better the optical transparency was.

In general, in situ polymerization technology is mainly used for the functionalization of g-C$_3$N$_4$, which involves surface modification of the g-C$_3$N$_4$. Thus, this method is not comparable to melt blending method for large-scale manufacturing,

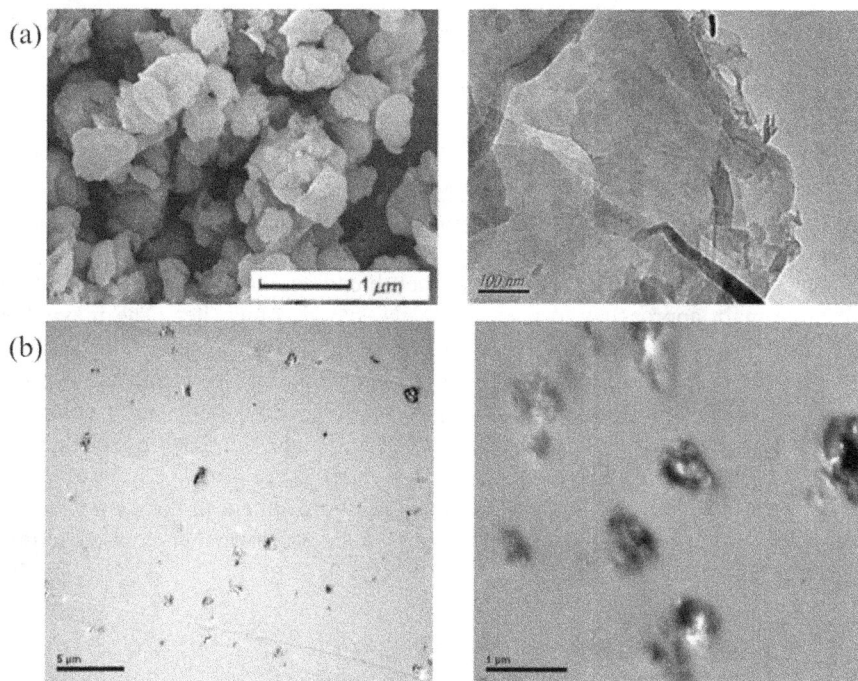

FIGURE 7.9 (a) SEM and TEM images of g-C_3N_4 and (b) TEM images of g-C_3N_4/PI composite films. (Reprinted from Sheng, W., and Chen Q., Yang, P., Chen, C., *High Perform. Polym.*, 27, 950–960, 2015. With permission.)

and the g-C_3N_4 hybrid material prepared in this way is usually used for photocatalysis, photodegradation or electrochemical sensors.

7.3.4 OTHER APPROACHES

As mentioned above, solvent blending, melt blending and in situ polymerization are currently the three main strategies used to fabricate polymer/g-C_3N_4 nanocomposites. With the development of science and technology, some other methods have also been used to prepare polymer/g-C_3N_4 nanocomposites.

Wang et al. (2017) successfully synthesized novel carbonized polyvinylpyrrolidone (PVP)/g-C_3N_4 (CPVP/g-C_3N_4) films via the electrospinning method. The photoelectrochemical (PEC) performance of CPVP/g-C_3N_4 films could be substantially controlled by varying the electrospinning time to adjust the thickness of the films. The as-prepared films exhibited excellent coverage, superior PEC properties and good stability. Shi, Liu, et al. (2019) fabricated multilayered PS nanocomposites hybridized with g-C_3N_4 nanosheets and reduced graphene oxide (RGO) via the layer-by-layer assembly method. It was confirmed that g-C_3N_4 and RGO were alternatively assembled to the surface of PS spheres by electrostatic interactions. The obtained PS nanocomposites exhibited excellent thermal stability and fire resistance performance.

Compared with mere the PS/g-C$_3$N$_4$ or PS/RGO samples, the PS/g-C$_3$N$_4$/RGO composites exhibited the lowest PHRR value of 640 W/g in the microscale combustion calorimetry (MCC) test, which was mainly attributed to the lamellar blocking effect of the g-C$_3$N$_4$ and RGO sheets.

In summary, the fabrication of polymer/g-C$_3$N$_4$ nanocomposites has been considered a valuable route to create novel high-performance materials that combine the advantages of polymers and g-C$_3$N$_4$. Compared with neat polymers, the polymer composites exhibit enhanced thermal stability and fire safety performance. At present, solvent blending, melt blending and in situ polymerization are three main approaches to prepare g-C$_3$N$_4$-based composites. The inevitable use of chemical solvents and low yield are the main problems of the solvent blending method. Thus, the melt blending technique has gained increasing interest because it is solvent-free, environmentally friendly and suitable for large-scale manufacturing in most industrial sectors. The in situ polymerization technique is more suitable for the functionalization of g-C$_3$N$_4$. Regardless of how they are prepared, all the polymer composites show an improvement in their comprehensive properties, which inspires an increasing number of researchers to work in this field fruitfully.

7.4 FLAME-RETARDANT POLYMER/GRAPHITIC CARBON NITRIDE COMPOSITES

The tri-s-triazine ring structure allows g-C$_3$N$_4$ to be directly used as a multifunctional heterogeneous metal-free catalyst in the oxidation of hydrocarbons and water splitting (Su et al., 2010; Wang et al., 2010; Wang et al., 2011). It is interesting that organic and inorganic compounds or metals can bind into the matrix to tune the functionality of g-C$_3$N$_4$. Many protocols, including protonation, doping, loading (metals and inorganic compounds) and chemical modification, are performed to control the performance of carbon nitride(Chen, Lin, et al., 2015; Liu et al., 2010; Sridharan et al., 2013; Wang, Yu, et al., 2013; Zhang et al., 2012; Zhang et al., 2010). The introduction of such functional atoms, groups and compounds into carbon nitride or its surface greatly improves the performances of the matrix. To date, g-C$_3$N$_4$ and its various derivatives have been widely applied in the fields of catalysts, hydrogen devices, lithium ion storage, optoelectronic equipment and bioimaging (Fang et al., 2013; Lin et al., 2013; Shi et al., 2014; Wang, Blechert, et al., 2012; Wang et al., 2009; Wang, Hong, et al., 2013; Zhang et al., 2013; Zhang et al., 2010). Recently, g-C$_3$N$_4$-based polymer composites have been investigated. Yan et al. (2011) revealed that g-C$_3$N$_4$-poly(3-hexylthiophene) composites contributed to enhancing hydrogen production from water under visible light. Myllymaa et al. (2009) demonstrated that the surface hydrophilicity of polypropylene (PP) was improved by the addition of a silicon-doped carbon nitride coating. Shi' group reported an investigation into the thermal stability, flame retardancy and mechanical properties of polymer nanocomposites containing g-C$_3$N$_4$ or its modifications (Shi et al., 2013; Shi, Long, et al., 2015; Shi, Yu, et al., 2015). The g-C$_3$N$_4$/spinel ZnCo$_2$O$_4$ nanohybrids were successfully synthesized for enhanced thermal stability of PS (Shi, Zhu, et al., 2015).

7.4.1 Thermal Stability and Flame-Retardant Potential of Neat g-C$_3$N$_4$

The g-C$_3$N$_4$ is the most stable carbon nitride family and begins to decompose when the temperature exceeds 600 °C. As a 2D nanomaterial, g-C$_3$N$_4$ is considered to be a promising flame retardant that contributes to enhancing the flame retardancy of polymers due to its excellent thermal stability, good chemical resistance and other properties (Cao et al., 2020; Lou et al., 2019; Yue et al., 2019). However, pure g-C$_3$N$_4$ is poorly dispersed in the polymer matrix, and chemical corrosion or thermal oxidation lead to partial scission of g-C$_3$N$_4$ (Chen et al., 2022; Zheng et al., 2020). These results indicate that the addition of g-C$_3$N$_4$ alone cannot achieve satisfactory flame retardancy. Therefore, it is particularly important to improve the flame retardancy of polymeric materials by functionalizing g-C$_3$N$_4$ using organic and inorganic components or combining g-C$_3$N$_4$ with traditional flame retardants (Shi, Wang, et al., 2019) (Figure 7.10).

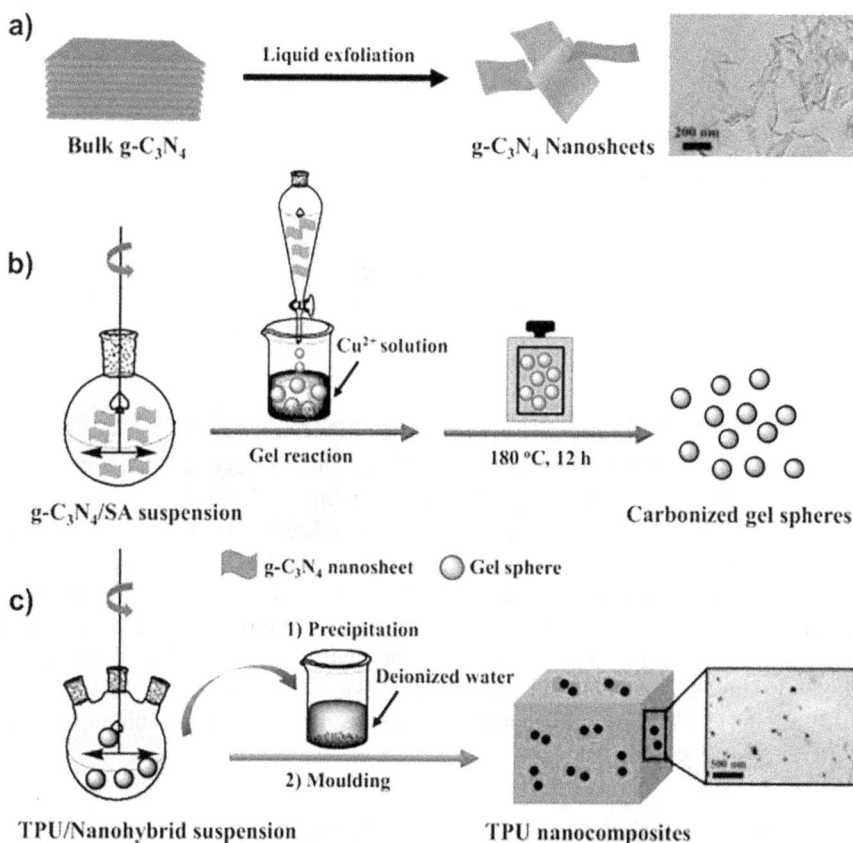

FIGURE 7.10 Schematic diagrams for (a) liquid exfoliation of bulk g-C$_3$N$_4$, (b) preparation of nanohybrids and (c) fabrication of TPU nanocomposites. (Reprinted from Shi, Y., Wang, L., Fu, L., Liu, C., Yu, B. and Yang, F., et al., *J. Colloid Interf. Sci.*, 539, 1–10, 2019. With permission.)

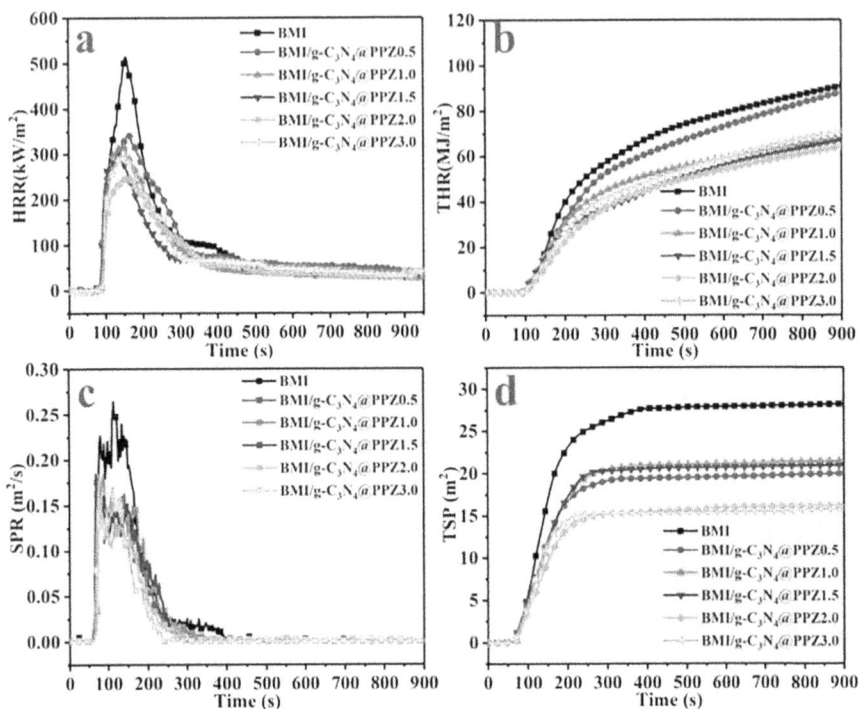

FIGURE 7.11 The HRR, smoke production rate (SPR), THR and TSP vs time curves of BMI and its nanocomposites (a–d). (Reprinted from Zhou, X., Qiu, S., Liu, J., Zhou, M., Cai, W., Wang, J., et al., *Chem. Eng. J.*, 401, 126094–126106, 2020. With permission.)

7.4.2 ORGANIC FLAME-RETARDANT-FUNCTIONALIZED g-C₃N₄

Organic flame retardant-functionalized g-C$_3$N$_4$ with a very low addition level can achieve high flame retardancy without sacrificing mechanical properties. When 2 wt% g-C$_3$N$_4$ modified by polyphosphazene (g-C$_3$N$_4$@PPZ) was incorporated into bismaleimide resin (BMI), the THR and TSP of the BMI nanocomposite were reduced by 29.4% and 42.9%, respectively, compared with those of pure BMI (Figure 7.11) (Zhou et al., 2020). Hua et al. loaded titanium dioxide in palygorskite (PGS) and modified g-C$_3$N$_4$ to prepare a ternary hybrid flame retardant (P-g-C$_3$N$_4$@PGS-Ti) (Hua et al., 2021), and the obtained results showed that the PHRR of the EP/P-g-C$_3$N$_4$@PGS-Ti nanocomposite was reduced by 36% at a loading of 5 wt% P-g-C$_3$N$_4$@PGS-Ti.

7.4.3 INORGANIC/g-C₃N₄ HYBRID FLAME RETARDANT

g-C$_3$N$_4$-based inorganic hybrids have received considerable attention from researchers. Shi et al. synthesized a series of g-C$_3$N$_4$-based hybrids by employing a coupling effect between −NH$_2$ or −NH groups present in g-C$_3$N$_4$ and functional groups such as −SH, −OH, −COOH or other metal cations. Novel highly effective

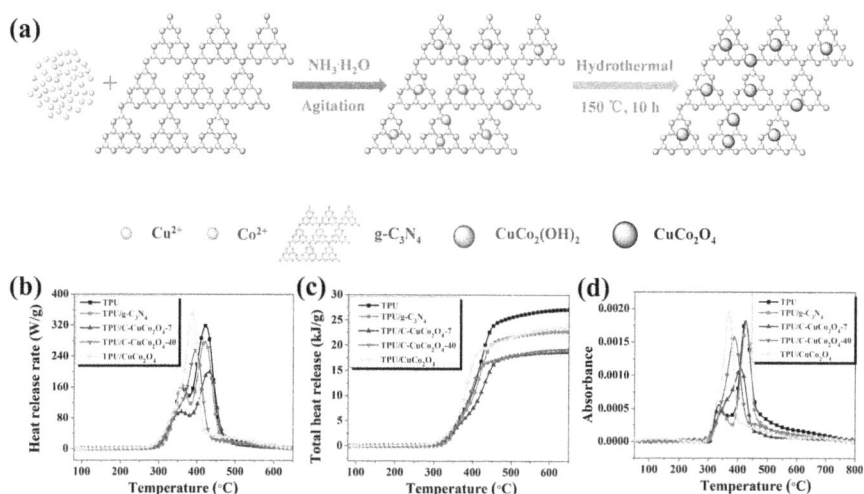

FIGURE 7.12 (a) Schematic illustration for solvothermal synthesis of C-CuCo$_2$O$_4$ nanohybrids; (b) HRR, (c) THR and (d) Gram–Schmidt curves of TPU and its nanocomposites. (Reprinted from Shi, Y., Yu, B., Zhou, K., Yuen, R., Gui, Z., Hu, Y., et al, *J. Hazard. Mater.*, 293, 87–96, 2015. With permission.)

CuCo$_2$O$_4$/g-C$_3$N$_4$ (C-CuCo$_2$O$_4$) nanohybrids were successfully synthesized by a simple hydrothermal method and then incorporated into TPU (Figure 7.12a) (Shi, Yu, et al., 2015).The obtained results indicated that C-CuCo$_2$O$_4$ nanohybrids, especially C-CuCo$_2$O$_4$-7 (proportion of g-C$_3$N$_4$ to CuCo$_2$O$_4$ of 93/7), significantly reduced the CO generation and fire hazards of TPU (Figures 7.12b–d). The g-C$_3$N$_4$ nanosheets and acidized multiwall CNTs (aMWCNTs) were alternately deposited on PS spheres by a layer-by-layer self-assembly technique (Figure 7.13a) (Shi, Long, et al., 2015). The PS sphere allowed the intimate assembly of the g-C$_3$N$_4$/ aMWCNT bilayer on its surface through electrostatic interactions (Figure 7.13b). Enhanced thermal stability and flame retardancy (e.g., a decrease of ca. 45% and 47% in PHRR and THR, respectively) were obtained for the ternary assembled systems instead of the binary materials (Figures 7.13c and d). The direct dispersion of hydrophobic g-C$_3$N$_4$ nanosheets in water or organic solvents without the assistance of dispersing agents is considered to be a great challenge. To address this issue, OMMT was used to functionalize g-C$_3$N$_4$ through electrostatic interactions and thereafter introduced into a PS matrix to fabricate nanocomposites by a simple solvent blending-precipitation method (Shi, Wang, et al., 2016). These nanohybrids were evenly dispersed in PS and showed strong interfacial interactions with the polymer matrix (Figures 7.14a and b). In addition, the fire hazards and total gaseous product release were dramatically inhibited by combining g-C$_3$N$_4$ with OMMT (Figures 7.14c and d). A series of g-C$_3$N$_4$/ammonium polyphosphate (APP) (CNAPP) hybrids were prepared to enhance the thermal and flame-retardant properties of PS. Cone calorimeter test results showed that the PHRR and THR of PS/CNAPP were reduced by 39% and 16%, respectively, for PS/CNAPP20 (proportion of g-C$_3$N$_4$ to APP of 1/4) (Shi, Xing, et al., 2016). The phosphorus-

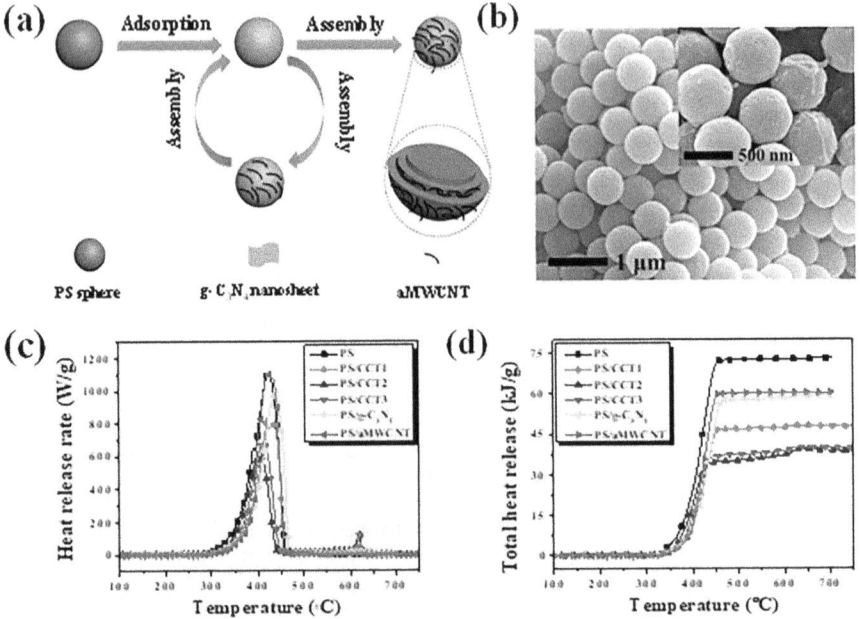

FIGURE 7.13 (a) Schematic illustration of the LBL assembly process for construction of PS/g-C$_3$N$_4$/MWCNT systems; (b) SEM images of assembled systems with 2 layers of g-C$_3$N$_4$/MWCNT; (c) HRR and (d) THR curves of PS and its assembled systems. (Reprinted from Shi, Y., Long, Z., Yu, B., Zhou, K., Gui, Z., Yuen, R., et al, J. Mater. Chem. A, 3, 17064–17073, 2015. With permission.)

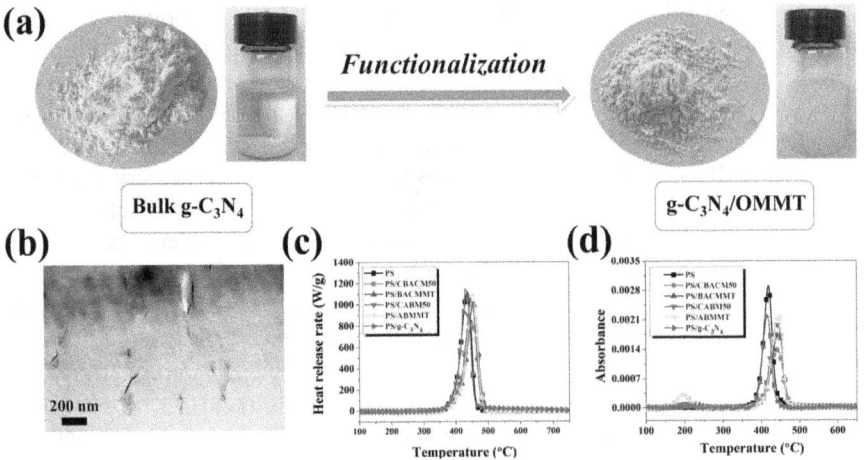

FIGURE 7.14 (a) Schematic illustration for facile preparation of g-C$_3$N$_4$/OMMT nanohybrids; (b) Ultrathin TEM images of PS/CBACM50; (c) HRR and (d) total pyrolysis gaseous products release curves of PS and its nanocomposites. (Reprinted from Shi, Y., Wang, B., Duan, L., Zhu, Y., Gui, Z., Yuen, R.,*Ind. Eng. Chem. Res.*, 55, 7646–7654, 2016. With permission.)

TABLE 7.2

Comparison in Different Work on Flame-Retardant Effect of g-C_3N_4-Based Hybrids

Sample No.	Additives Content	PHRR Reduction	THR Reduction	Gaseous Productions Reduction	Ref
TPU/C-CuCo$_2$O$_4$-7	2 wt%	37%	31.3%	38.9%	Shi, Yu, et al. (2015)
PS/gC$_3$N$_4$/aMWCNT	2 bilayers	45%	47%	67.8%	Shi, Long, et al., 2015
PS/CBACM50 [a]	2 wt%	27.3%	–	51.7%	Shi, Wang, et al. (2016)
PS/CNAPP	20 wt%	39%	16%	–	Shi, Xing, et al. (2016)
TPU/CAHPi20	10 wt%	40%	50%	–	Shi, Fu, et al. (2017)

Notes

[a] CBACM50 means that the g-C_3N_4/OMMT nanohybrid, where weight ratio of g-C_3N_4 to benzyldimethylhexadecyl-ammonium chloride modified MMT equals to 50/50 (wt%/wt%).

containing compounds acted as highly effective flame retardants. The combination of aluminum hypophosphite (AHPi) and g-C_3N_4 led to PHRR and TSP decreasing by 40% and 50%, respectively, for TPU composites containing CAHPi20 (weight ratio of g-C_3N_4 to AHPi of 20/80) (Shi, Fu, et al., 2017). All these investigations are summarized in Table 7.2.

7.4.4 Organic/g-C_3N_4 Hybrid Flame Retardants

Recently, organic/g-C_3N_4 hybrids have received increasing interest because they cannot only prevent the aggregation of g-C_3N_4 but also enhance its thermal stability and flame-retardant efficiency. Guo et al. (2020) synthesized novel organic aluminum hypophosphite (ALCPA)/g-C_3N_4 (CNALCPA) hybrids through salification and hybridization and then added them to polyamide 6. The results from combustion tests revealed that CNALCPA showed outstanding flame-retardant efficiency in PA6, and especially, the PA6 composite could attain an LOI of 38.3% and achieve a UL-94 V-0 rating when 20 wt% CNALCPA was added. The g-C_3N_4/phosphorus-rich aluminum phosphinates (OAHPi) hybrids were reported as smoke suppressants and flame retardants for polystyrene (PS) (Shi, Yu, et al., 2017). Combining g-C_3N_4 with OAHPi contributed to reductions in the PHRR, THR and peak of the smoke production rate (PSPR) of the PS matrix. Traditional flame retardants could further enhance flame-retardant efficiency of the g-C_3N_4. The g-C_3N_4 was coated by PAZn via a simple self-assembly, followed by incorporation into epoxy resin (EP) (Figure 7.15) (Zhang et al., 2020). The nanohybrid reduced the PHRR and PSPR of the EP composite by 71.38% and 25%, respectively. In addition, Dong et al. (2022) prepared polyaniline-coupled g-C_3N_4 doped with

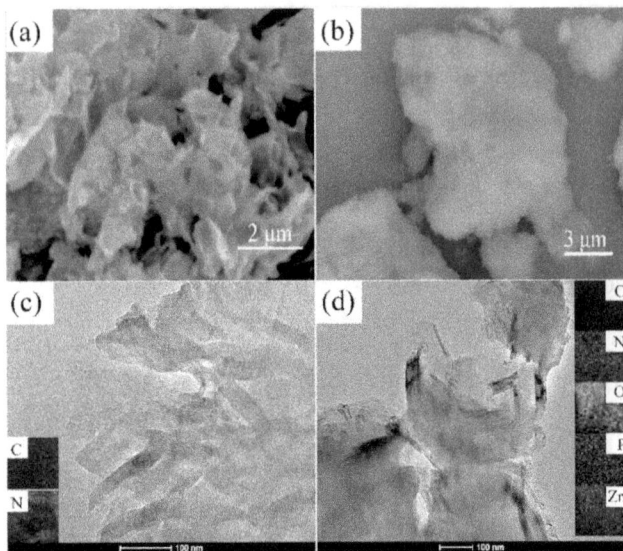

FIGURE 7.15 SEM images of (a) g-C₃N₄, (b) g-C₃N₄/PAZn; TEM images of (c) g-C₃N₄ and (d) g-C₃N₄/PAZn. (Reprinted from Zhang, W., Wu, W., Meng, W., Xie, W., Gui, Y., Xu, J., et al., *Polymers*, 12, 212–223, 2020. With permission.)

phytic acid (CP@PA) hybrids and found that the combination of CP@PA and APP in EP promoted fire safety of EP composite due to its self-carbonization and good compatibility.

7.4.5 SYNERGISTIC COMBINATION OF g-C₃N₄ AND TRADITIONAL FLAME RETARDANTS

When g-C₃N₄ alone is used as a flame retardant, its flame retardant efficiency is often negligible, and the synergistic combination of g-C₃N₄ and traditional flame retardants is an effective method to improve the flame-retardant efficiency of g-C₃N₄. Cao, Chi, Deng, Liu, et al. (2020) found that g-C₃N₄ combined with commercially available flame-retardant melamine pyrophosphate (MPP) and 9,10-dihydro-9-oxa-10-phosphaphenanthrene 10-oxide could remarkably enhance the flame retardancy of TPU. In particular, the PHRR and THR of PLA composites containing 10 wt% DOPO were reduced by 25.2% and 23.6%, respectively. As an important traditional inorganic flame retardant, APP is combined with g-C₃N₄ to increase the fire safety of polymers. Shi, Xing, et al. (2016) revealed that the thermal stability of the wrapping of APP by g-C₃N₄ (CNAPP) was higher than that of pure APP. In addition, the char layer of PS/CNAPP was more compact and denser than that of PS with an equal amount of APP. The combination of phosphorus and silicon codoped g-C₃N₄ (PSiCN) and APP was reported to be a high-efficiency flame-retardant protocol (Zhang et al., 2021). The results demonstrated that the EP composite achieved an LOI of 29% and reached a UL-94 V-0 rating after the addition of 2 wt% PSiCN and 8 wt% APP, and the PHRR, THR, PSPR and

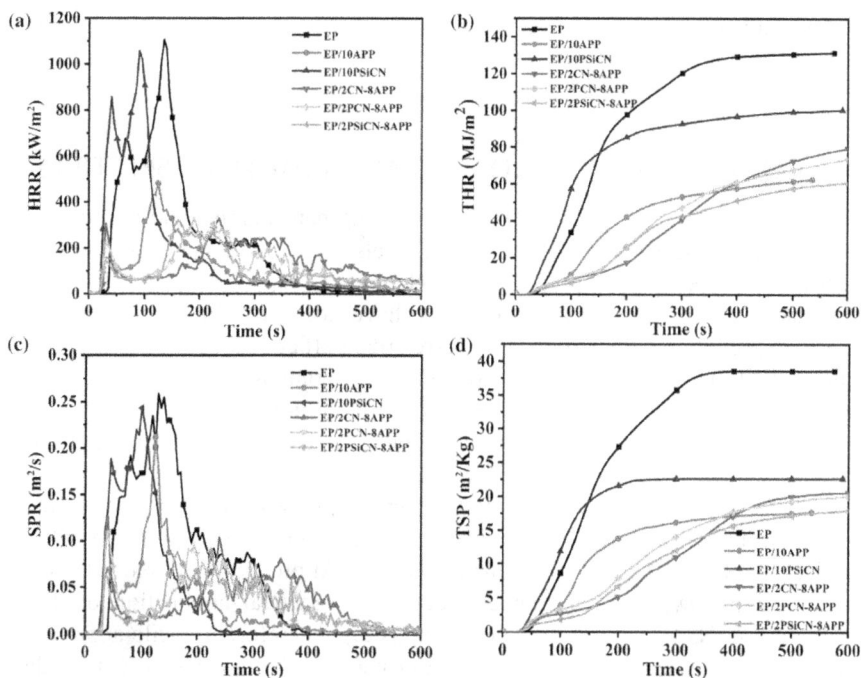

FIGURE 7.16 HRR, SPR, THR and TSRP vs time curves of BMI and BMI composites (a–d). (Reprinted from Zhang, W., Chen, Z., Yu, Y., Chen, T., Zhang, Q. and Li, C., *J. Appl. Polym. Sci.*, 139, 51614–51626, 2021. With permission.)

TSP of the EP/2PSiCN-8APP composite declined by 71.1%, 53.5%, 65.4% and 53.5%, respectively, in parallel with those of neat EP (Figure 7.16).

7.4.6 FLAME-RETARDANT ACTION OF g-C$_3$N$_4$

As reported by Pozdnjakov et al., pyrolysis gaseous products of a powdery g-C$_3$N$_4$ were detected by a mass spectrometric technique (Pozdnyakov et al., 2005). The thermolysis products of the 2D nanomaterial consisted of mainly CN species, such as C$_3$N$_2^+$ (m/z 64), C$_2$N$_3^+$ (m/z 66), CN$_4^+$ (m/z 68), C$_3$N$_3^+$ (m/z 78) and C$_3$N$_4^+$ (m/z 92). The 2D nanomaterial retards the permeation of heat and the escape of volatile degradation products, resulting from the physical barrier effect (Shi, Gui, et al., 2015; Shi et al., 2013). Additionally, acting as a polymeric semiconductor, g-C$_3$N$_4$ exhibited a catalytic effect in the decomposition of gaseous products (Shi, Yu, et al., 2015). The other components in the as-synthesized (nano)hybrids show highly catalytic action and free radical capture effects. The addition of spinel CuCo$_2$O$_4$ leads to a change in the decomposition route and increased char residues of the TPU matrix, while g-C$_3$N$_4$/aMWCNT bilayers induce the construction of the "tortuous path", which impedes the permeation of heat and the escape of pyrolysis volatile products. as shown in Figure 7.4 (Shi, Long, et al., 2015; Shi, Yu, et al., 2015). The phosphorus-containing compounds not only thermally decompose into the formation of P–O–C and P–N–C structures, which can remarkably improve the stability of the char layer but also effectively capture high-energy free radicals evolved from polymeric materials (Shi, Xing, et al., 2016; Shi, Yu, et al., 2017).

7.5   CONCLUSIONS, CHALLENGES AND FUTURE PERSPECTIVES

2D g-C_3N_4 has been demonstrated as a promising nanoadditive and reinforcer for reducing fire hazards of polymeric materials and simultaneously improving mechanical properties. Numerous flame retardants, e.g., APP, organic P-N flame retardants, and inorganic flame retardants, have been combined with g-C_3N_4 to further improve the dispersion and flame-retardant efficiency of g-C_3N_4 in order to achieve high-performance fire-safe polymer nanocomposites. This chapter summarizes recent advances in the utilization of g-C_3N_4-based flame-retardant systems in polymeric materials with significantly reduced fire hazards, and the related flame-retardant mechanisms have been proposed.

Despite some advances in preparation and flame-retardant application of 2D g-C_3N_4 or its derivatives, there remain many major challenges associated with the application of 2D g-C_3N_4 in fire-safety fields: (1) At present, the yields of g-C_3N_4 are relatively low, which is not conducive for practical application of flame-retardant g-C_3N_4-based polymer nanocomposites. Therefore, a facile approach to the scalable production of g-C_3N_4 urgently needs to be developed. (2) To date, liquid exfoliation has become the main method to prepare exfoliated g-C_3N_4 nanosheets. Only grams of g-C_3N_4 nanosheets can be obtained, which cannot meet industrial applications. It is imperative to explore a highly efficient exfoliation method for preparing 2D g-C_3N_4 nanosheets, e.g., ball-milling method. (3) There are very few active groups on the surface of 2D g-C_3N_4 because the preparation process under high temperature leads to the removal of most active groups. This will be difficult to perform the surface functionalization and to enhance the interfacial interaction of g-C_3N_4 with polymer matrices. How to realize surface functionalization of g-C_3N_4 nanosheets with massive modifiers and strong interfacial interactions within the polymer matrix still remains a huge challenge. (4) The flame retardancy mechanisms of g-C_3N_4, especially the catalytic action in the condensed phase, are still not clear and should be further investigated.

Overall, to develop high-performance g-C_3N_4-based flame-retardant polymer nanocomposites for practical applications, more efforts should be made in terms of massive production and exfoliation methods and scalable functionalization methods of g-C_3N_4. Finally, advanced characterization technologies should be developed to obtain an in-depth understanding of the catalytic and flame-retardant mechanisms of g-C_3N_4.

ACKNOWLEDGMENTS

The research work was funded by the National Natural Science Foundation of China (Nos. 52106187 and 52173070) and the National Key Research and Development Program of China (No. 2021YFB3700202).

REFERENCES

Andreyev, A., Akaishi, M., and Golberg, D. 2003. "Sodium flux-assisted low-temperature high-pressure synthesis of carbon nitride with high nitrogen content." *Chemical Physics Letters* 372(5–6): 635–639. doi:10.1016/s0009-2614(03)00471-8.

Anwar, Z., Kausar, A., Rafique, I., and Muhammad, B. 2015. "Advances in epoxy/graphene nanoplatelet composite with enhanced physical properties: a review." *Polymer-Plastics Technology and Engineering* 55(6): 643–662. doi:10.1080/03602559.2015.1098695.

Bai, X., Li, J., and Cao, C. 2010. "Synthesis of hollow carbon nitride microspheres by an electrodeposition method." *Applied Surface Science* 256(8): 2327–2331. doi:10.1016/j.apsusc.2009.10.061.

Bai, Y. J., Lue, B., Liu, Z. G., Li, L., Cui, D. L., Xu, X. G., and Wang, Q. L. 2003. "Solvothermal preparation of graphite-like C 3 N 4 nanocrystals." *Journal of Crystal Growth* 247(3/4): 505–508. doi:10.1016/s0022-0248(02)01981-4.

Broitman, E., Hellgren, N., Wänstrand, O., Johansson, M., Berlind, T., Sjöström, H., Sundgren, J.-E., Larsson, M., and Hultman, L. 2001. "Mechanical and tribological properties of CNx films deposited by reactive magnetron sputtering." *Wear* 248(1–2): 55–64. doi:10.1016/s0043-1648(00)00519-6.

Cao, Q., Baris K., Markus A., and Schmidt, B. 2020. "Graphitic carbon nitride and polymers: a mutual combination for advanced properties." *Materials Horizons* 7(3): 762–786. doi:10.1039/c9mh01497g.

Cao, Q., Kumru, B., Antonietti, M., and Schmidt, B. 2019. "Grafting polymers onto carbon nitride via visible-light-induced photofunctionalization." *Macromolecules* 52(13): 4989–4996. doi:10.1021/acs.macromol.9b00894.

Cao, X., Chi, X., Deng, X., Liu, T., Yu, B., Wang, B., Yuen, A. C. Y., Wu, W., and Li, R. K. Y. 2020. "Synergistic effect of flame retardants and graphitic carbon nitride on flame retardancy of polylactide composites." *Polymers for Advanced Technologies* 31(7): 1661–1670. doi:10.1002/pat.4894.

Chen, D., Wang, K., Hong, W., Zong, R., Yao, W., and Zhu, Y. 2015. "Visible light photoactivity enhancement via CuTCPP hybridized g-C3N4 nanocomposite." *Applied Catalysis B: Environmental* 166–167: 366–373. doi:10.1016/j.apcatb.2014.11.050.

Chen, Y., Lin, B., Yu, W., Yang, Y., Bashir, S. M., Wang, H., Takanabe, K., Idriss, H., and Basset, J. M. 2015. "Surface functionalization of g-C3N4: molecular-level design of noble-metal-free hydrogen evolution photocatalysts." *Chemistry–A European Journal* 21(29): 10290–10295. doi:10.1002/chem.201501742.

Chen, Z., Suo, Y., Yu, Y., Chen, T., Li, C., Zhang, Q., Jiang, J., and Chen, T. 2022. "Polymerization of hydroxylated graphitic carbon nitride as an efficient flame retardant for epoxy resins." *Composites Communications* 29. doi:10.1016/j.coco.2021.101018.

Cheng, N., Tian, J., Liu, Q., Ge, C., Qusti, A. H., Asiri, A. M., Al-Youbi, A. O., and Sun, X. 2013. "Au-nanoparticle-loaded graphitic carbon nitride nanosheets: green photocatalytic synthesis and application toward the degradation of organic pollutants." *ACS Applied Materials & Interfaces* 5(15): 6815–6819.

Coulson, B., Lari, L., Isaacs, M., Raines, D. J., Douthwaite, R. E., and Duhme-Klair, A. K. 2020. "Carbon nitride as a ligand: selective hydrogenation of terminal alkenes using [(eta(5) -C5 Me5)IrCl(g-C3 N4 -kappa(2) N,N')]Cl." *Chemistry* 26(30): 6862–6868. doi:10.1002/chem.201905749.

Cui, Y., Ding, Z., Fu, X., and Wang, X. 2012. "Construction of conjugated carbon nitride nanoarchitectures in solution at low temperatures for photoredox catalysis." *Angew Chem Int Ed Engl* 51(47): 11814–11818. doi:10.1002/anie.201206534.

Cui, Y., Kundalwal, S. I., and Kumar, S. 2016. "Gas barrier performance of graphene/polymer nanocomposites." *Carbon* 98: 313–333. doi:10.1016/j.carbon.2015.11.018.

Da Silva, E. S., Moura, N. M. M., Neves, M. G. P. M. S., Coutinho, A., Prieto, M., Silva, C. G., and Faria, J. L. 2018. "Novel hybrids of graphitic carbon nitride sensitized with free-base meso-tetrakis(carboxyphenyl) porphyrins for efficient visible light photocatalytic hydrogen production." *Applied Catalysis B: Environmental* 221: 56–69. doi:10.1016/j.apcatb.2017. 08.079.

Datta, K. K. R., Reddy, B. V. S., Ariga, K., and Vinu, A. 2010. "Gold nanoparticles embedded in a mesoporous carbon nitride stabilizer for highly efficient three-component coupling reaction." *Angewandte Chemie International Edition* 49(34): 5961–5965. doi:10.1002/ anie.201001699.

Dong, H., Wang, Y., Feng, T., Piao, J., Ren, J., Wang, Y., Liu, W., Chen, W., Li, S., Chen, X., and Jiao, C. 2022. "Phytic acid doped polyaniline-coupled g-C3N4 nanosheets for synergizing with APP promoting fire safety and waterproof performance of epoxy composites." *Polymer Degradation and Stability* 198. doi:10.1016/j.polymdegradstab.2022.109879.

Fang, Y., Li, X., and Wang, X. 2019. "Phosphorylation of polymeric carbon nitride photo-anodes with increased surface valence electrons for solar water splitting." *ChemSusChem* 12(12): 2605–2608. doi:10.1002/cssc.201900291.

Fang, Y., Lv, Y., Che, R., Wu, H., Zhang, X., Gu, D., Zheng, G., and Zhao, D. 2013. "Two-dimensional mesoporous carbon nanosheets and their derived graphene nanosheets: synthesis and efficient lithium ion storage." *Journal of the American Chemical Society* 135(4): 1524–1530. doi:10.1021/ja310849c.

Gaddam, S. K., Pothu, R., and Boddula, R. 2019. "Graphitic carbon nitride (g-C3N 4) reinforced polymer nanocomposite systems—A review." *Polymer Composites* 41(2): 430–442. doi:10.1002/pc.25410.

Guo, C., Zhao, Y., Ji, G., Wang, C., and Peng, Z. 2020. "Organic aluminum hypophosphite/ graphitic carbon nitride hybrids as halogen-free flame retardants for polyamide 6." *Polymers (Basel)* 12(10). doi:10.3390/polym12102323.

Guo, Q., Xie, Y., Wang, X., Lv, S., Hou, T., and Liu, X. 2003. "Characterization of well-crystallized graphitic carbon nitride nanocrystallites via a benzene-thermal route at low temperatures." *Chemical Physics Letters* 380(1–2): 84–87. doi:10.1016/j.cplett.2003. 09.009.

Guo, Q., Xie, Y., Wang, X., Zhang, S., Hou, T., and Lv, S. 2004. "Synthesis of carbon nitride nanotubes with the C(3)N(4) stoichiometry via a benzene-thermal process at low temperatures." *Chem Commun (Camb)* (1): 26–27. doi:10.1039/b311390f.

Guo, Q., Yang, Q., Zhu, L., Yi, C., Zhang, S., and Xie, Y. 2004. "A facile one-pot solvothermal route to tubular forms of luminescent polymeric networks [(C3N3)2(NH)3]." *Solid State Communications* 132(6): 369–374. doi:10.1016/j.ssc.2004.08.014.

Guo, Y., Chu, S., Yan, S., Wang, Y., and Zou, Z. 2010. "Developing a polymeric semi-conductor photocatalyst with visible light response." *Chem Commun (Camb)* 46(39): 7325–7327. doi:10.1039/c0cc02355h.

Hellgren, N., Johansson, M. P., Broitman, E., Sandström, P., Hultman, L., and Sundgren, J.-E. 2001. "Effect of chemical sputtering on the growth and structural evolution of magnetron sputtered CNx thin films." *Thin Solid Films* 382(1–2): 146–152. doi:10.101 6/s0040-6090(00)01690-4.

Hua, F., Wei, H., Ren, F., Wang, M., Li, L., Lv, B., Wang, H., Yang, Z., and Lei, Z. 2021. "Preparation of a ternary hybrid of P-g-C3N4@ PGS-Ti and its enhancement of the flame retardancy of epoxy resins." *Fire and Materials* 46(1): 95–106. doi:10.1002/fam.2950.

Jiang, W., Luo, W., Zong, R., Yao, W., Li, Z., and Zhu, Y. 2016. "Polyaniline/carbon nitride nanosheets composite hydrogel: a separation-free and high-efficient photocatalyst with 3D Hierarchical Structure." *Small* 12(32): 4370–4378. doi:10.1002/smll.201601546.

Kurpil, B., Kumru, B., Heil, T., Antonietti, M., and Savateev, A. 2018. "Carbon nitride creates thioamides in high yields by the photocatalytic Kindler reaction." *Green Chemistry* 20(4): 838–842. doi:10.1039.C7GC03734A

Li, C., Cao, C.-B., and Zhu, H.-S. 2004. "Graphitic carbon nitride thin films deposited by electrodeposition." *Materials Letters* 58(12–13): 1903–1906. doi:10.1016/j.matlet.2 003.11.024.

Li, C., Cao, C.-B., Zhu, H.-S., Lv, Q., Zhang, J.-T., and Xiang, X. 2004. "Electrodeposition route to prepare graphite-like carbon nitride." *Materials Science and Engineering: B* 106(3): 308–312. doi:10.1016/j.mseb.2003.10.006.

Li, H. J., Sun, B. W., Sui, L., Qian, D. J., and Chen, M. 2015. "Preparation of water-dispersible porous g-C3N4 with improved photocatalytic activity by chemical oxidation." *Phys Chem Chem Phys* 17(5): 3309–3315. doi:10.1039/c4cp05020g.

Liu, G., Niu, P., Sun, C., Smith, S. C., Chen, Z., Lu, G. Q., and Cheng, H.-M. 2010. "Unique electronic structure induced high photoreactivity of sulfur-doped graphitic C3N4." *Journal of the American Chemical Society* 132(33): 11642–11648. doi:10.1021/ja103798k.

Liu, J., Shi, H., Shen, Q., Guo, C., and Zhao, G. 2017. "A biomimetic photoelectrocatalyst of Co–porphyrin combined with a g-C3N4 nanosheet based on π–π supramolecular interaction for high-efficiency CO2 reduction in water medium." *Green Chemistry* 19(24): 5900–5910. doi:10.1039/c7gc02657a.

Liu, J., Wang, H., and Antonietti, M. 2016. "Graphitic carbon nitride "reloaded": emerging applications beyond (photo)catalysis." *Chemical Society Reviews* 45(8): 2308–2326. doi:10.1039/c5cs00767d.

Lou, Y. J., Zhao, Z. X., Chen, Z. W., Dai, Z. H., Fu, F. Y., Zhang, Y. Y., Zhang, L., and Liu, X. D. 2019. "Processability improvement of a 4-vinlyguiacol derived benzoxazine using reactive diluents." *Polymer* 160 316–324. doi:10.1016/j.polymer.2018.11.056.

Lu, W., Xu, T., Wang, Y., Hu, H., Li, N., Jiang, X., and Chen, W. 2016. "Synergistic photocatalytic properties and mechanism of g-C3N4 coupled with zinc phthalocyanine catalyst under visible light irradiation." *Applied Catalysis B: Environmental* 180: 20–28. doi:10.1016/j.apcatb.2015.06.009.

Lu, Y., Liu, Y., Li, N., Hu, Z., and Chen, S. 2020. "Sulfonated graphitic carbon nitride nanosheets as proton conductor for constructing long-range ionic channels proton exchange membrane." *Journal of Membrane Science* 601. doi:10.1016/j.memsci.202 0.117908.

Lv, Q., Cao, C., Li, C., Zhang, J., Zhu, H., Kong, X., and Duan, X. 2003. "Formation of crystalline carbon nitride powder by a mild solvothermal method." *Journal of Materials Chemistry* 13(6). doi:10.1039/b303210h.

Ma, L., Zhu, Y., Feng, P., Song, G., Huang, Y., Liu, H., Zhang, J., Fan, J., Hou, H., and Guo, Z. 2019. "Reinforcing carbon fiber epoxy composites with triazine derivatives functionalized graphene oxide modified sizing agent." *Composites Part B: Engineering* 176. doi:10.1016/j.compositesb.2019.107078.

Montigaud, H., Tanguy, B., Demazeau, G., Alves, I., and Courjault, S. 2000. "C3N4: dream or reality? Solvothermal synthesis as macroscopic samples of the C3N4 graphitic form." *Journal of Materials Science* 35(10): 2547–2552. doi:10.1023/A:10047985 09417.

Myllymaa, K., Myllymaa, S., Korhonen, H., Lammi, M. J., Saarenpää, H., Suvanto, M., Pakkanen, T. A., Tiitu, V., and Lappalainen, R. 2009. "Improved adherence and spreading of Saos-2 cells on polypropylene surfaces achieved by surface texturing and carbon nitride coating." *Journal of Materials Science: Materials in Medicine* 20(11): 2337–2347. doi:10.1007/s10856-009-3792-3.

Pozdnyakov, O., Blinov, L., Arif, M., Pozdnyakov, A., Filippov, S., and Semencha, A. 2005. "Mass spectrometry of carbon nitride C3N4." *Technical Physics Letters* 31(12): 1001–1003.

Qin, J., Wang, S., Ren, H., Hou, Y., and Wang, X. 2015. "Photocatalytic reduction of CO2 by graphitic carbon nitride polymers derived from urea and barbituric acid." *Applied Catalysis B: Environmental* 179: 1–8. doi:10.1016/j.apcatb.2015.05.005.

Qiu, S., Zou, B., Zhang, T., Ren, X., Yu, B., Zhou, Y., Kan, Y., and Hu, Y. 2020. "Integrated effect of NH2-functionalized/triazine based covalent organic framework black phosphorus on reducing fire hazards of epoxy nanocomposites." *Chemical Engineering Journal* 401. doi:10.1016/j.cej.2020.126058.

Rad, E. R., Vahabi, H., de Anda, A. R., Saeb, M. R., and Thomas, S. 2019. "Bio-epoxy resins with inherent flame retardancy." *Progress in Organic Coatings* 135: 608–612. doi:10.1016/j.porgcoat.2019.05.046.

Schwarzer, A., Saplinova, T., and Kroke, E. 2013. "Tri-s-triazines (s-heptazines)—from a "mystery molecule" to industrially relevant carbon nitride materials." *Coordination Chemistry Reviews* 257(13–14): 2032–2062. doi:10.1016/j.ccr.2012.12.006.

Sheng, W., Chen, Q., Yang, P., and Chen, C. 2015. "Synthesis, characterization, and enhanced properties of novel graphite-like carbon nitride/polyimide composite films." *High Performance Polymers* 27(8): 950–960. doi:10.1177/0954008314566894.

Shi, Y., Fu, L., Chen, X., Guo, J., Yang, F., Wang, J., Zheng, Y., and Hu, Y. 2017. "Hypophosphite/graphitic carbon nitride hybrids: preparation and flame-retardant application in thermoplastic polyurethane." *Nanomaterials* 7(9): 259. doi:10.3390/nano7090259.

Shi, Y., Gui, Z., Yu, B., Yuen, R. K., Wang, B., and Hu, Y. 2015. "Graphite-like carbon nitride and functionalized layered double hydroxide filled polypropylene-grafted maleic anhydride nanocomposites: comparison in flame retardancy, and thermal, mechanical and UV-shielding properties." *Composites Part B: Engineering* 79: 277–284. doi:10.1016/j.compositesb.2015.04.046.

Shi, Y., Jiang, S., Zhou, K., Bao, C., Yu, B., Qian, X., Wang, B., Hong, N., Wen, P., and Gui, Z. 2013. "Influence of g-C3N4 nanosheets on thermal stability and mechanical properties of biopolymer electrolyte nanocomposite films: a novel investigation." *ACS Applied Materials & Interfaces* 6(1): 429–437. doi:10.1021/am4044932.

Shi, Y., Jiang, S., Zhou, K., Wang, B., Gui, Z., Hu, Y., and Yuen, R. K. 2014. "Facile preparation of ZnS/gC 3 N 4 nanohybrids for enhanced optical properties." *RSC Advances* 4(6): 2609–2613. doi:10.1039/c3ra44256j.

Shi, Y., Liu, C., Duan, Z., Yu, B., Liu, M., and Song, P. 2020. "Interface engineering of MXene towards super-tough and strong polymer nanocomposites with high ductility and excellent fire safety." *Chemical Engineering Journal* 399. doi:10.1016/j.cej.2020.125829.

Shi, Y., Liu, C., Fu, L., Yang, F., Lv, Y., and Yu, B. 2019. "Hierarchical assembly of polystyrene/graphitic carbon nitride/reduced graphene oxide nanocomposites toward high fire safety." *Composites Part B: Engineering* 179. doi:10.1016/j.compositesb.2019.107541.

Shi, Y., Long, Z., Yu, B., Zhou, K., Gui, Z., Yuen, R. K. K., and Hu, Y. 2015. "Tunable thermal, flame retardant and toxic effluent suppression properties of polystyrene based on alternating graphitic carbon nitride and multi-walled carbon nanotubes." *Journal of Materials Chemistry A* 3(33): 17064–17073. doi:10.1039/c5ta04349b.

Shi, Y., Wang, B., Duan, L., Zhu, Y., Gui, Z., Yuen, R. K. K., and Hu, Y. 2016. "Processable Dispersions of Graphitic Carbon Nitride Based Nanohybrids and Application in Polymer Nanocomposites." *Industrial & Engineering Chemistry Research* 55(28): 7646–7654. doi:10.1021/acs.iecr.6b01237.

Shi, Y., Wang, L., Fu, L., Liu, C., Yu, B., Yang, F., and Hu, Y. 2019. "Sodium alginate-templated synthesis of g-C3N4/carbon spheres/Cu ternary nanohybrids for fire safety application." *Journal of Colloid Interface Science* 539: 1–10. doi:10.1016/j.jcis.2018.12.051.

Shi, Y., Xing, W., Wang, B., Hong, N., Zhu, Y., Wang, C., Gui, Z., Yuen, R. K. K., and Hu, Y. 2016. "Synergistic effect of graphitic carbon nitride and ammonium polyphosphate for enhanced thermal and flame retardant properties of polystyrene." *Materials Chemistry and Physics* 177: 283–292. doi:10.1016/j.matchemphys.2016.04.029.

Shi, Y., Yu, B., Duan, L., Gui, Z., Wang, B., Hu, Y., and Yuen, R. K. K. 2017. "Graphitic carbon nitride/phosphorus-rich aluminum phosphinates hybrids as smoke suppressants and flame retardants for polystyrene." *Journal of Hazardous materials* 332: 87–96. doi: 10.1016/j.jhazmat.2017.03.006.

Shi, Y., Yu, B., Zhou, K., Yuen, R. K., Gui, Z., Hu, Y., and Jiang, S. 2015. "Novel CuCo2O4/graphitic carbon nitride nanohybrids: highly effective catalysts for reducing CO generation and fire hazards of thermoplastic polyurethane nanocomposites." *Journal of Hazardous Materials* 293: 87–96.

Shi, Y., Zhu, Y., Yu, B., Gui, Z., She, S., Yuen, R. K., Liu, H., and Hu, Y. 2015. "Enhanced thermal stability of polystyrene by graphitic carbon nitride/spinel ZnCo 2 O 4 nano-hybrids and the catalytic mechanism investigation." *RSC Advances* 5(52): 41835–41838. doi: 10.1039/c5ra02787j.

Sridharan, K., Jang, E., and Park, T. J. 2013. "Novel visible light active graphitic C3N4–TiO2 composite photocatalyst: synergistic synthesis, growth and photocatalytic treatment of hazardous pollutants." *Applied Catalysis B: Environmental* 142: 718–728. doi: 10.1016/j.apcatb.2013.05.077.

Su, F., Mathew, S. C., Lipner, G., Fu, X., Antonietti, M., Blechert, S., and Wang, X. 2010. "mpg-C3N4-catalyzed selective oxidation of alcohols using O2 and visible light." *Journal of the American Chemical Society* 132(46): 16299–16301. doi: 10.1021/ja102866p.

Sun, S., Li, J., Cui, J., Gou, X., Yang, Q., Liang, S., Yang, Z., and Zhang, J. 2018. "Constructing oxygen-doped g-C3N4 nanosheets with an enlarged conductive band edge for enhanced visible-light-driven hydrogen evolution." *Inorganic Chemistry Frontiers* 5(7): 1721–1727. doi: 10.1039/c8qi00242h.

Vaithylingam, R., Ansari, M. N. M., and Shanks, R. A. 2017. "Recent advances in polyurethane-based nanocomposites: a review." *Polymer Plastics Technology & Engineering* 1–14. doi: 10.1080/03602559.2017.1280683.

Wang, J., Cui, W., Chen, R., He, Y., Yuan, C., Sheng, J., Li, J., Zhang, Y., Dong, F., and Sun, Y. 2020. "OH/Na co-functionalized carbon nitride: directional charge transfer and enhanced photocatalytic oxidation ability." *Catalysis Science & Technology* 10(2): 529–535. doi: 10.1039/c9cy02048a.

Wang, W., Yu, J. C., Xia, D., Wong, P. K., and Li, Y. 2013. "Graphene and g-C3N4 nanosheets cowrapped elemental α-sulfur as a novel metal-free heterojunction photo-catalyst for bacterial inactivation under visible-light." *Environmental Science & Technology* 47(15): 8724–8732. doi: 10.1021/es4013504.

Wang, X., Blechert, S., and Antonietti, M. 2012. "Polymeric graphitic carbon nitride for heterogeneous photocatalysis." *ACS Catalysis* 2(8): 1596–1606. doi: 10.1021/cs300240x.

Wang, X., Maeda, K., Thomas, A., Takanabe, K., Xin, G., Carlsson, J. M., Domen, K., and Antonietti, M. 2009. "A metal-free polymeric photocatalyst for hydrogen production from water under visible light." *Nature Materials* 8(1): 76–80. doi: 10.1038/nmat2317.

Wang, Y., Di, Y., Antonietti, M., Li, H., Chen, X., and Wang, X. 2010. "Excellent visible-light photocatalysis of fluorinated polymeric carbon nitride solids." *Chemistry of Materials* 22(18): 5119–5121. doi: 10.1021/cm1019102.

Wang, Y., Hong, J., Zhang, W., and Xu, R. 2013. "Carbon nitride nanosheets for photocatalytic hydrogen evolution: remarkably enhanced activity by dye sensitization." *Catalysis Science & Technology* 3(7): 1703–1711. doi: 10.1016/j.mtchem.2022.101084.

Wang, Y., Li, H., Yao, J., Wang, X., and Antonietti, M. 2011. "Synthesis of boron doped polymeric carbon nitride solids and their use as metal-free catalysts for aliphatic C–H bond oxidation." *Chemical Science* 2(3): 446–450. doi: 10.1039/c0sc00475h.

Wang, Y., Wang, X., and Antonietti, M. 2012. "Polymeric graphitic carbon nitride as a heterogeneous orgaeocatalyst: from photochemistry to multipurpose catalysis to sustainable chemistry." *Angewandte Chemie.* doi: 10.1002/anie.201101182.

Wang, Y., Zhao, X., Tian, Y., Wang, Y., Jan, A. K., and Chen, Y. 2017. "Facile electrospinning synthesis of carbonized polyvinylpyrrolidone (PVP)/g-C3 N4 hybrid films for photo-electrochemical applications." *Chemistry* 23(2): 419–426. doi:10.1002/chem.201604468.

Wei, F., Liu, Y., Zhao, H., Ren, X., Liu, J., Hasan, T., Chen, L., Li, Y., and Su, B. L. 2018. "Oxygen self-doped g-C3N4 with tunable electronic band structure for unprecedent-edly enhanced photocatalytic performance." *Nanoscale* 10(9): 4515–4522. doi:10.103 9/c7nr09660g.

Wu, L., Sha, Y., Li, W., Wang, S., Guo, Z., Zhou, J., Su, X., and Jiang, X. 2016. "One-step preparation of disposable multi-functionalized g-C3N4 based electrochemiluminescence immunosensor for the detection of CA125." *Sensors and Actuators B: Chemical* 226: 62–68. doi:10.1016/j.snb.2015.11.133.

Xiao, G., Yang, Z., Chen, C., Chen, C., Zhong, F., Wang, M., and Zou, R. 2021. "Novel carbon nitride@polydopamine/molybdenum disulfide nanoflame retardant improves fire performance of composite coatings." *Colloids and Surfaces A: Physicochemical and Engineering Aspects* 630. doi:10.1016/j.colsurfa.2021.127575.

Xu, T., Wang, D., Dong, L., Shen, H., Lu, W., and Chen, W. 2019. "Graphitic carbon nitride co-modified by zinc phthalocyanine and graphene quantum dots for the efficient photocatalytic degradation of refractory contaminants." *Applied Catalysis B: Environmental* 244: 96–106. doi:10.1016/j.apcatb.2018.11.049.

Xu, W., Yan, H., Wang, G., Qin, Z., Fan, L., and Yang, Y. 2021. "A silica-coated metal-organic framework/graphite-carbon nitride hybrid for improved fire safety of epoxy resins." *Materials Chemistry and Physics* 258. doi:10.1016/j.matchemphys.2020.123 810.

Yan, H., Chen, Y., and Xu, S. 2012. "Synthesis of graphitic carbon nitride by directly heating sulfuric acid treated melamine for enhanced photocatalytic H2 production from water under visible light." *International Journal of Hydrogen Energy* 37(1): 125–133. doi:1 0.1016/j.ijhydene.2011.09.072.

Yan, S. C., Li, Z. S., and Zou, Z. G. 2009. "Photodegradation performance of g-C3N4 fabricated by directly heating melamine." *Langmuir* 25(17): 10397–10401. doi:10.1 021/la900923z.

Yang, F., Lublow, M., Orthmann, S., Merschjann, C., Tyborski, T., Rusu, M., Kanis, M., Thomas, A., Arrigo, R., and Haevecker, M. 2012. "Preparation and characterization of metal-free graphitic carbon nitride film photocathodes for light-induced hydrogen evolution." *Physics.*

Yao, M., Wu, H., Liu, H., Zhou, Z., Wang, T., Jiao, Y., and Qu, H. 2021. "In-situ growth of boron nitride for the effect of layer-by-layer assembly modified magnesium hydroxide on flame retardancy, smoke suppression, toxicity and char formation in EVA." *Polymer Degradation and Stability* 183. doi:10.1016/j.polymdegradstab.2020.109417.

Ye, L., Wu, D., Chu, K. H., Wang, B., Xie, H., Yip, H. Y., and Wong, P. K. 2016. "Phosphorylation of g-C3N4 for enhanced photocatalytic CO2 reduction." *Chemical Engineering Journal* 304: 376–383. doi:10.1016/j.cej.2016.06.059.

Ye, X., Zheng, Y., and Wang, X. 2014. "Synthesis of ferrocene-modified carbon nitride photocatalysts by surface amidation reaction for phenol synthesis." *Chinese Journal of Chemistry* 32(6): 498–506. doi:10.1002/cjoc.201400229.

Yong, D., Shen, S. Z., Cai, K., and Casey, P. S. 2012. "Research progress on polymer–inorganic thermoelectric nanocomposite materials." *Progress in Polymer Science* 37(6): 820–841. doi:10.1016/j.progpolymsci.2011.11.003.

Yu, C., Tan, L., Shen, S., Fang, M., Yang, L., Fu, X., Dong, S., and Sun, J. 2021. "In situ preparation of g-C3N4/polyaniline hybrid composites with enhanced visible-light photocatalytic performance." *J Environ Sci (China)* 104: 317–325. doi:10.1016/j.jes.2 020.08.024.

Yue, X., Li, C., Ni, Y., Xu, Y., and Wang, J. 2019. "Flame retardant nanocomposites based on 2D layered nanomaterials: a review." *Journal of Materials Science* 54(20): 13070–13105. doi:10.1007/s10853-019-03841-w.

Zhang, C., Garrison, T. F., Madbouly, S. A., and Kessler, M. R. 2017. "Recent advances in vegetable oil-based polymers and their composites." *Progress in Polymer Science* S0079670016301198. doi:10.1016/j.progpolymsci.2016.12.009.

Zhang, G., Li, G., Lan, Z. A., Lin, L., Savateev, A., Heil, T., Zafeiratos, S., Wang, X., and Antonietti, M. 2017. "Optimizing optical absorption, exciton dissociation, and charge transfer of a polymeric carbon nitride with ultrahigh solar hydrogen production activity." *Angew Chem Int Ed Engl* 56(43): 13445–13449. doi:10.1002/anie.201706870.

Zhang, J., Grzelczak, M., Hou, Y., Maeda, K., Domen, K., Fu, X., Antonietti, M., and Wang, X. 2012. "Photocatalytic oxidation of water by polymeric carbon nitride nanohybrids made of sustainable elements." *Chemical Science* 3(2): 443–446. doi:10.1039/c1sc00644d.

Zhang, W., Chen, Z., Yu, Y., Chen, T., Zhang, Q., Li, C., Chen, Z., Gao, W., and Jiang, J. 2021. "Synthesis of phosphorus and silicon co-doped graphitic carbon nitride and its combination with ammonium polyphosphate to enhance the flame retardancy of epoxy resin." *Journal of Applied Polymer Science* 139(6). doi:10.1002/app.51614.

Zhang, W., Wu, W., Meng, W., Xie, W., Cui, Y., Xu, J., and Qu, H. 2020. "Core-shell graphitic carbon nitride/zinc phytate as a novel efficient flame retardant for fire safety and smoke suppression in epoxy resin." *Polymers (Basel)* 12(1). doi:10.3390/polym12010212.

Zhang, X., Xie, X., Wang, H., Zhang, J., Pan, B., and Xie, Y. 2013. "Enhanced photo-responsive ultrathin graphitic-phase C3N4 nanosheets for bioimaging." *Journal of the American Chemical Society* 135(1): 18–21. doi:10.1021/ja308249k.

Zhang, Y., Mori, T., Ye, J., and Antonietti, M. 2010. "Phosphorus-doped carbon nitride solid: enhanced electrical conductivity and photocurrent generation." *Journal of the American Chemical Society* 132(18): 6294–6295. doi:10.1021/ja101749y.

Zhao, Q., Wu, W., Wei, X., Jiang, S., Zhou, T., Li, Q., and Lu, Q. 2017. "Graphitic carbon nitride as electrode sensing material for tetrabromobisphenol-A determination." *Sensors and Actuators B: Chemical* 248: 673–681. doi:10.1016/j.snb.2017.04.002.

Zheng, T., Xia, W., Guo, J., Wang, K., Zeng, M., Wu, Q., and Liu, Y. 2020. "Preparation of flame-retardant polyamide 6 by incorporating MgO combined with g-C3N4." *Polymers for Advanced Technologies* 31(9): 1963–1971. doi:10.1002/pat.4920.

Zhou, X., Qiu, S., Liu, J., Zhou, M., Cai, W., Wang, J., Chu, F., Xing, W., Song, L., and Hu, Y. 2020. "Construction of porous g-C3N4@PPZ tubes for high performance BMI resin with enhanced fire safety and toughness." *Chemical Engineering Journal* 401. doi:10.1016/j.cej.2020.126094.

Zhu, Y., Cai, W., Zhao, Y., Mu, X., Zhou, X., Chu, F., Wang, B., and Hu, Y. 2022. "Graphite-like carbon nitride/polyphosphoramide nanohybrids for enhancement on thermal stability and flame retardancy of thermoplastic polyurethane elastomers." *ACS Applied Polymer Materials* 4(1): 121–128. doi:10.1021/acsapm.1c01060.

Zhu, Y., Shi, Y., Huang, Z.-Q., Duan, L., Tai, Q., and Hu, Y. 2017. "Novel graphite-like carbon nitride/organic aluminum diethylhypophosphites nanohybrid: Preparation and enhancement on thermal stability and flame retardancy of polystyrene." *Composites Part A: Applied Science and Manufacturing* 99: 149–156. doi:10.1016/j.compositesa.2017.03.023.

8 Layered Boron Nitride Derived Flame Retardants for Fire-Safe Polymeric Materials

Bin Fei

School of Fashion and Textiles, Hong Kong Polytechnic University, Kowloon, Hong Kong, China

Yuchun Li

College of Materials Science and Engineering, Beijing University of Chemical Technology, Chaoyang, PR China

Yang Ming

School of Fashion and Textiles, Hong Kong Polytechnic University, Kowloon, Hong Kong, China

CONTENTS

8.1 INTRODUCTION

Hexagonal boron nitride (h-BN) is a layered material with a graphite-like structure, where planar networks of BN hexagons are regularly stacked. For h-BN, boron and

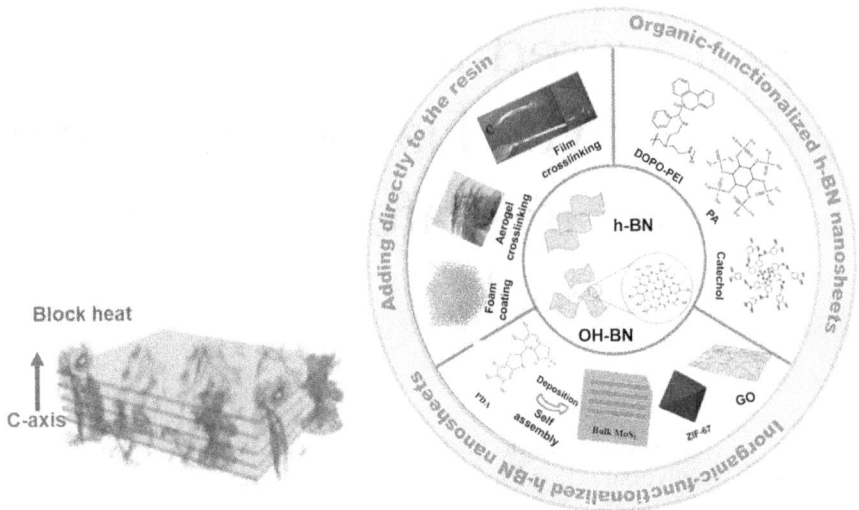

FIGURE 8.1 Left: The anisotropic thermal conductivity of h-BN nanosheets allows heat block on c-axis direction. Right: The application strategies of h-BN nanosheets in various matrixes to endow flame retardancy.

nitrogen atoms are covalently bonded to each other within individual layers, with van der Waals forces existing between adjacent layers. Similar to graphite, h-BN can tolerate the high-temperature damage due to its highly stable B-N bonds (Cai et al. 2021). However, h-BN nanosheets have low thermal expansion, high thermal conductivity, and high electric resistance compared with graphite (Gilbert et al. 2019). In addition, the layered structure of h-BN nanosheets is capable of presenting a barrier effect to slow the mass transfer. These fascinating physical properties make h-BN nanosheets an ideal candidate for fabricating fire safety polymer composites. Many researchers have tried to improve the flame retardancy of polymers with h-BN, where different approaches have been explored. Here we categorized them into three sections: directly adding h-BN nanosheets into various types of matrixes, organic functionalization of h-BN nanosheets, and inorganic functionalization of h-BN nanosheets, as summarized in Figure 8.1 (Yue et al. 2019).

8.2   ADDING DIRECTLY TO THE RESIN

The h-BN nanosheet possesses excellent flame retardancy by preventing flame spreading and blocking heat transfer due to its high thermal-chemical stability. It is suggested that the h-BN nanosheet mainly works in the condensed phase and can promote the formation of stable char residue during combustion (Lu, Hong, and Chen 2019). Therefore, h-BN nanosheets are used as synergists in resins, such as epoxy resin (EP) and thermoplastic polyurethane (TPU). Yuan (Guan et al. 2017) introduced h-BN nanosheets to the EP/modified SiO_2 system as a flame retardant and thermal conductivity additive. After adding 9%h-BN and 6%m-SiO_2, the pHRR and THR of EP significantly decreased by 45.3% and 33.5%, respectively, because the h-BN nanosheets possess a higher thermal conductivity and decomposition temperature as well as a lower coefficient of thermal expansion. Similarly, h-BN nanosheets were also introduced to the EP/R-GO/Ni(OH)$_2$ system to further improve the thermal conductivity of EP. It was proposed that the good dispersion and interfacial interaction of the RGO@Ni(OH)$_2$ hybrid in the matrix can inhibit the stacking aggregation behavior of h-BN sheets, contributing to bridging the adjacent h-BN sheets and hence further improving the thermal conductivity (Feng et al. 2020). Apart from, the h-BN nanosheets were also introduced to water-based polyurethane (WPU) to improve its flame retardancy: the pHRR and THR of WPU decreased by 31.9% and 14.2% by 4% h-BN addition (Yin et al. 2020).

8.2.1   BN-Based Coating

The extremely high resistance to thermal oxidation makes h-BN nanosheets an ideal nanomaterial for fabricating protective coatings. It is suggested that the low thermal conductivity along the c axis of h-BN nanosheets can be used to weaken heat spreading from the flame zone to the internal polymer surface, such as wood, cotton fabric, and PU foam, and hence can protect the matrix from burning fast.

Li (Qiu et al. 2019) deposited branched polyethylenimine (BPEI) and h-BN nanosheets–sodium alginate (SA) on the surface of FPUFs by LbL assembly. By coating 14.4 wt% BN/PEI/SA on the surface of FPUF, the released heat and smoke significantly decreased, especially for pHRR, which decreased by 50.1%. It was proposed that the physical barrier and tortuous path effects result from the anisotropic thermal conductivity of the h-BN nanosheets. The BN/PEI/SA system was also applied on the surface of unsaturated polyester resin (UPR), and after adding 20wt% flame retardant, the release of heat was significantly decreased (Chu et al. 2018). Ma also investigated a "sandwich-like" coating on a PU surface constructed successively by rod coating of the PVH/PA layer, spray coating of the GO/CNTs layer, and finally brush-coating of the PVH/PA/BN layer. Importantly, the pHRR and pSPR significantly decreased by 49.0% and 41.0%, respectively, and the coating can give rise to an obvious delay of ignition (about 8 s to burn) and a satisfactory early warning to the temperature changes. During combustion, PVH works in the gaseous phase by releasing nonflammable SO$_2$ to dilute the combustible gases and oxygen, while PA works in the condensed phase by catalyzing carbonization of the matrix. Additionally, the GO/CNTs layer and the thermally stable BN sheets can serve as physical barriers to retard heat and mass transfer (Ma et al. 2022).

In addition, a coating including h-BN nanosheets was also used on the surface of cotton to improve its flame retardancy by taking advantage of the strong bonding between diethylenetriamine and h-BN on the cotton surface. It was reported that h-BN hinders combustion by reducing the rate at which oxygen molecules reach the cotton surface (Ambekar et al. 2020). Hu developed a strong, fire-resistant wood by combining a densification treatment with an anisotropic thermally conductive flame-retardant coating of h-BN nanosheets to produce BN-densified wood, which provided a fast, in-plane thermal diffusion, slowing the conduction of heat. Therefore, the ignition time and ignition temperature of wood were enhanced, and the pHRR decreased by 25% (Gan et al. 2020).

8.2.2 BN Nanosheets Mixed with PVA Films

As a commercially biocompatible hydrophilic polymer, poly(vinyl alcohol) (PVA) membranes have been widely used in biomaterial areas. In order to improve the flame retardancy and mechanical properties of PVA films, h-BN nanosheets were introduced into aqueous PVA solutions. The layered structure of OH-BN presented a barrier function for hindering the transfer of degradation products and heat, thus enhancing the thermal stability and fire safety of PVA nanocomposite films. In addition, the strong hydrogen-bond interaction between h-BN nanosheets and PVA is regarded as the source of high performance. h-BN can effectively improve the tensile strength, toughness, and ductility as well as the flame retardant of PVA films while preserving its high visible light transmittance.

Cai, Zhang, et al. (2018) found that the well-dispersed state and layered structure allowed the incorporated OH-BN to present a barrier function to suppress the delivery of thermal degradation products of the PVA matrix, thus enhancing thermal stability and fire safety. By adding only 0.2wt% h-BN nanosheets, the flame retardancy of PVA was improved, with the pHRR decreasing by 25.1%. Meanwhile, the elongation at break, tensile strength, and Young's modulus of the PVA nanocomposite film were simultaneously increased by 109.3%, 73.6%, and 144.4%, respectively. Based on this, Wang, Ji, et al. (2019) introduced lignin nanoparticles to PVA/h-BN films to further improve the thermal conductivity, stability, flame retardancy, and flexibility, which were fabricated through vacuum filtration in conjunction with chemical crosslinking. The BN-OH/PVA/LNP composite film exhibited a higher through-plane thermal conductivity after crosslinking with glutaraldehyde (GA). The ignition time was delayed while the degradation temperature was increased. However, adding only h-BN nanosheets or other nanoparticles cannot realize satisfactory flame retardancy. Hence, the phosphorus based flame retardant APP was added to the PVA/h-BN film to further decrease the heat release during combustion (Zhang, Wang, et al. 2019). Moreover, h-BN nanosheets can combine with one-dimensional hydroxyapatite nanowires to form layer-structured composite films by liquid phase exfoliation, in which HAPNWs provide flexibility and BNNSs offer orientational thermal conductive pathways. The resultant films can burn under continuous fire for five minutes to maintain their original appearance (Cheng and Feng 2020).

8.2.3 BN Nanosheets Mixed with Aerogels

Aerogels have been well recognized as prospective materials utilized for pollutant treatment, catalytic supports, thermal insulation, energy storage devices, sensors, and biomedical applications, due to their highly porous structure, ultra-low density, high specific surface area, low thermal conductivity, and strong adsorption. Nonetheless, the thermal stability and mechanical strength of organic aerogels are relatively poor. Additionally, they are highly flammable due to their organic nature. The introduction of h-BN nanosheets can significantly enhance the mechanical, thermal stability, and flame retardance properties of polymeric materials due to their unique structure and extraordinary physicochemical performance. However, the interfacial affinity between h-BN and polymer matrix is usually weak; thus, h-BN nanosheets aggregate easily, deteriorating the barrier function and decreasing the thermal stability. The integration of individual BNNSs (hexagonal boron nitride nanosheets) into aerogels with a three-dimensional (3D) porous structure is considered to be one of the most effective strategies to resolve the agglomeration issue. Furthermore, this strategy may result in hierarchical systems possessing synergistic effects from individual components. In order to further improve the dispersions, BNNSs were first prepared via the exfoliation of hexagonal BN in alcohol solution and then introduced into the preparation of aerogels.

Zhang prepared APP/BNNS/ZF composite aerogels by combining h-BN nanosheets with APP and zinc ferrite (ZF) nanoparticles via a freeze-drying method, which was introduced to EP to improve its fire resistance. The APP/BNNS/ZF-EP composite containing 24.0 wt% of the APP/BNNS/ZF aerogel exhibits superior flame-retardant performance, in which the pHRR and THR are reduced by 86.2 and 86.5%, respectively, compared with those of pristine EP. It is suggested that the ternary flame-retardant additive is able to dilute flammable gases, cut off the supply of oxygen and prevent combustion, thereby reducing smoke release (Zhang et al. 2020). Li reported rGO-BN-NR composite films, in which an rGO -BN aerogel was formed by a freeze-drying method, and light aerogels were dispersed in the matrix (Li et al. 2020). In addition, the h-BN nanosheets were also directly dispersed in aqueous pectin suspensions to prepare the bio-composite aerogels. The aerogel with a mass ratio of pectin to BNNSs of 10/1 possessed improved thermal stability, enhanced compressive strength, reduced pHRR (by 45%), and peak CO_2 production (by 53%) during combustion (Yang et al. 2019).

8.3 FUNCTIONALIZED h-BN NANOSHEETS

8.3.1 ORGANIC-FUNCTIONALIZED h-BN NANOSHEETS

The dispersion state is the first factor restricting the function of 2D nanomaterials. The interfacial compatibility between h-BN and polymer matrix is usually weak, especially for nonpolar polymers, and the h-BN nanosheets aggregate in the matrix, deteriorating the barrier function to heat. Therefore, it is necessary to modify and functionalize the surface of h-BN nanosheets to solve their dispersion, such as by organic functionalization, thus improving their flame retardancy and thermal conductivity effect in polymer composites.

8.3.1.1 Modification by Phosphorous Flame Retardants in Solution

Currently, the organic functionalization of h-BN nanosheets is mainly conducted by covalent bonds by using organic modifiers with reactive functional groups, which can be chemically linked with h-BN nanosheets. In addition, due to the affinity of the nitrogen atoms, the synthesized nucleophilic flame retardant (Lewis bases) bonds with boron atoms of h-BN nanosheets (Lewis acids) and facilitates the direct exfoliation of bulk h-BN. Many recent studies have focused on the functionalization of h-BN nanosheets by organic phosphorous flame retardants.

SiO_2 inorganic coating was constructed on the surface of h-BN nanosheets, which offered an opportunity for introducing phytic acid (PA) for the h-BN nanosheets to prepare h-BN@SiO_2@PA nanosheets. By adding 2% h-BN@SiO_2@PA, the pHRR and THR of TPU decreased by 23.5% and 22.1%, respectively. Importantly, it was found that toxic gases (CO, CH_4, C_2H_6, TOC) were turned into CO_2 by the catalytic effect of h-BN@SiO_2@PA (Cai et al. 2019). Moreover, PA was doped with a polypyrrole shell and then combined with copper ions to wrap on the surface of the h-BN nanosheets to form CPBN nanoparticles. By adding 3% CPBN, the flame retardancy of TPU was improved, and the release of heat and smoke significantly decreased. In addition, a tensile test demonstrates that adding CPBN favors reinforcement of the mechanical properties of TPU (Wang, Zhang, et al. 2019). Similarly, a nucleophilic flame retardant (P-PEI) composed of DOPO and PEI was synthesized and applied to functionalize h-BN nanosheets, and then it was added to the TPU sample to simultaneously improve its flame retardancy and reinforcement (Figure 8.2a). By adding 5.0 wt% P-PEI-BN nanosheets, the maximum mass loss rate, pHRR, and pSPR were decreased by 20.8%, 68.0%, and 53.6%, respectively. Moreover, the mechanical properties of TPU composites were enhanced, and the storage modulus of TPU/fBN-5.0 exhibited a 13.8% increment (Cai et al. 2017).

Similar to the same modification method for h-BN nanosheets by Lewis acid catalysts, the flame retardant DOPO was also introduced on the surface of h-BN nanosheets by medium poly(glycidyl methacrylate) (PGMA) to form h-BN-DOPO nanosheets. It has been reported that by adding 20wt% BNNS-DOPO to EP, the pHRR and pSPR decreased by 47.6% and 46.5%, respectively (Yang et al. 2022). In order to efficiently improve the flame retardancy of EP, ZIF-8 nanoparticle in-situ decorated boron nitride nanosheets (BN-OH/ZIF-8) were fabricated via a self-assembly method and then ternary integrated BN-OH/ ZIF-8/ PA hybrids were prepared through the chemical etching effect of PA. By adding only 2% h-BN /ZIF-8/PA hybrids, the pHRR and pSPR were reduced by 23.9% and 25.8%, respectively (Yin et al. 2022). Based on the above results, the organic phosphorous functionalized h-BN nanosheets showed a better effect on flame retardancy, especially for suppressing toxic gases in TPU and EP (Table 8.1), especially for TPU, because electron holes were produced on the surface of h-BN and capable of oxidizing these toxic and reducing gases, including CO, CH_4, C_2H_6, and so on.

Moreover, the organic functionalized h-BN nanosheets modified by phosphorus flame retardants were also applied to other matrices, such as PLLA and UPR (Rosely et al. 2020). Wang prepared aminated BN nanosheets by using (3-aminopropyl) triethoxysilane (APTES), which were subsequently reacted with ene-terminated hyperbranched polyphosphate acrylate (HPPA) containing phosphorus and nitrogen elements through a Michael addition reaction (Wang et al. 2018). The release of heat and smoke of unsaturated polyester resin (UPR) significantly decreased by adding only 3%BN-HPPA nanosheets. The flame retardancy of epoxy acrylate (EA) was also improved in the presence of DPCS@OBN, which was modified on the surface of BN

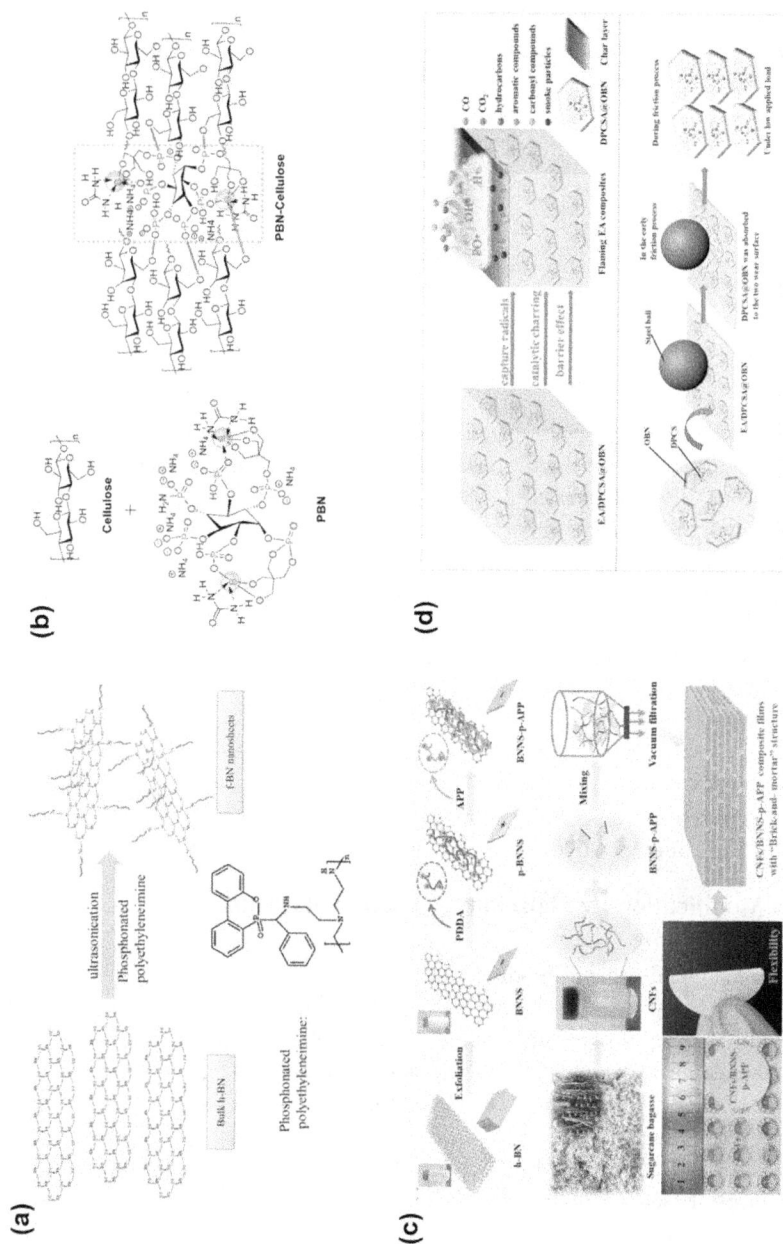

FIGURE 8.2 The structure of h-BN-DOPO (a) (Cai et al. 2017) and PBN (b) (Zhu et al. 2020); the scheme of prepared process of BNNS-p-APP nanosheets and CNF/BNNS-p-APP films (c) (Hu et al. 2021); the flame retardant mechanism of DPCS@OBN in CE resin (d) (Qiu et al. 2018).

TABLE 8.1

The Decrease Rate of pHRR and pSPR of TPU and EP Composites Containing h-BN Nanosheets Functionalized by Different Phosphorous Flame Retardants

Sample	ΔpHRR (%)	ΔTHR (%)	ΔpSPR (%)	References
TPU/h-BN@SiO$_2$@PA-2	23.5	22.1	29.2	Cai et al. (2019)
TPU/2.0CPBN	27.1	14.2	45.4	Wang, Zhang, et al.
TPU/3.0CPBN	35.7	19.8	33.3	(2019)
TPU/f-BN-2	34.4	21.3	27.3	Cai et al. (2017)
TPU/f-BN-5	60.1	25.4	39.1	
EP/BNNS-20	16.2	20.1	45.5	Yang et al. (2022)
EP/BNNS-DOPO-20	30.9	44.5	54.5	
EP/BN-OH	9.3	7.3	12.4	Yin et al. (2022)
EP/BN-OH/ZIF-8/PA	23.8	5.4	25.7	

nanosheets by embedding supermolecular aggregate (DPCS) via ionic bonding and electrostatic interaction between dicyandiamide and multivalent anions PA connected with cobalt ions. It was suggested that well-characterized DPCS@OBN nanohybrids significantly reduced the friction heat, and reduced the pHRR and THR by 42.7% and 48.3%, respectively. The introduction of DPCS@OBN into EA promoted the formation of a lubricating transfer layer on the wear track, as shown in Figure 8.2d (Qiu et al. 2018).

In addition, the functionalized h-BN nanosheets by flame retardant APP were also mixed with cellulose nanofibers to prepare CNF/BNNS-p-APP films; the thermal conductivity and flame resistance were simultaneously improved; the scheme of the preparationprocess is shown in Figure 8.2c (Hu et al. 2021). The functionalized h-BN nanosheets (PBN) were also applied on the waterborne coating, as shown in Figure 8.2b, which exhibited unique anti-corrosion and flame retardant performance (Li et al. 2022, Liu et al. 2020).

8.3.1.2 Modification by Phosphorous Flame Retardants via Ball Milling

Apart from chemical organic modification, the ball milling technique has recently become a new vitality in simultaneously exfoliating and functionalizing h-BN nanosheets by strong mechanochemical reactions. Although the lateral size is often reduced significantly during violent mechanical impact, the advantages of high exfoliation efficiency, low cost, and mass-production make ball milling the most promising approach to prepare BNNSs with functionalization. Han prepared exfoliated and flame-retardant functionalized h-BN nanosheets via a one-step ball milling process based on the synergetic effect of mechanical shear and chemical peeling of APP and sodium hydroxide. Then, it was used to improve the flame retardancy of EP by solution blending and program-controlled curing. By adding 5% functionalized h-BN nanosheets, the pHRR and pSPR decreased by 60.9% and 38.7%, respectively. Importantly, the thermal conductivity of EP/h-BN was significantly improved, with an enhancement (312.4% and 397.0%) in the thermal conductivity of EP/h-BN composites (Han et al. 2021). Ionic liquid functionalized h-BN nanosheets obtained by the ball milling process can endow them with a large specific surface area and rich flame-retardant functional groups. Han found that high-performance boron nitride nanosheets wrapped by the ionic liquid (h-BN@IL) 1-butyl-3-methyl-1H-imidazol-3-ium hexafluorophosphate [bmim][PF6] are workable to EP, which were first simultaneously exfoliated and functionalized by flame-retardant via a one-step ball milling process through a strong mechanochemical reaction. The EP/h-BN@IL sample showed the best flame retardancy with a dramatic decrease in pHRR (72.9%) and THR (75.7%). The char residue of the matrix in combustion was increased by the catalytic carbonization function of BNNS@IL. The barrier effect during thermal degradation was attributed to the BNNS@IL flakes with an in-plane oriented direction (Han et al. 2022). In addition, BN@APP was obtained by a one-step ball milling method and then introduced to PBS. After incorporating 1.0% BN@APP into PBS, the thermal conductivity of the PBS composite is increased by 62.8%. The heat and toxic smoke are reduced significantly, and the pHRR and pSPR of PBS composites were dramatically reduced by 28.2% and 31.9%, respectively, compared to pure PBS (Xu et al. 2022). Apart from the mixing procedure, ionic liquid functionalized h-BN nanosheets obtained by the ball milling process can also form a stable coating preserved on the surface of sponges to improve their fire safety and mechanical performance, such as plant fiber sponges (Chen et al. 2022).

It is suggested that the organic functionalized h-BN nanosheets by phosphorus flame retardants can block the release of heat and smoke during combustion, and the initial degradation temperature was delayed due to the presence of h-BN nanosheets, which improved the thermal stability of the matrix. Meanwhile, the layered structure of h-BN nanosheets acts as the first barrier to prevent the delivery of flammable pyrolysis gas to the inner matrix. In addition, the organic phosphorus flame retardants on the surface of h-BN nanosheets can capture free radicals produced by the cleavage of main chains and promote char formation during combustion. Therefore, phosphorus-modified h-BN nanosheets can be used as an integrated flame retardant to play an important role in blocking and catalyzing during combustion.

8.3.2 Inorganic-Functionalized h-BN Nanosheets

As a wide bandgap semiconductor, it is difficult to generate electron-hole pairs on the surface of h-BN nanosheets, which are usually regarded as undesirable catalysts to absorb the release of CO and CO_2. Therefore, by only adding h-BN nanosheets, the release of smoke was not significantly improved. Many research studies have been devoted to integrating metal-based nanostructures into h-BN nanosheets to prepare multifunctional nanoparticles, which are then introduced into polymers to further broaden their wide applications, such as flame retardants and super-hydrophobicity, in which h-BN nanosheets are used as supports.

However, the direct growth of inorganic nanoparticles on boron nitride nanosheets is difficult owing to the lack of essential active sites. Fortunately, dopamine chemistry presents strong bio-adhesion toward inert surfaces with polydopamine (PDA) as the

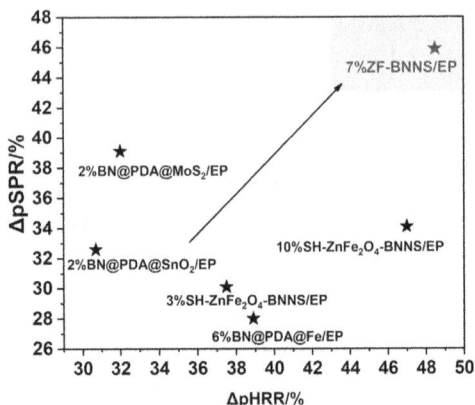

FIGURE 8.3 The decrease rate of pHRR and pSPR of EP composites containing different inorganic nanomaterials functionalized h-BN nanosheets (Cai, Guo, et al. 2018, Sui et al. 2021, Wang et al. 2020, Li et al. 2018).

bridging nano-layer. In parallel, the presence of PDA on BN nanosheets facilitated the interfacial growth of inorganic nanomaterials due to the intensive coordination interaction. Many studies have focused on inorganic functionalized h-BN nanosheets relying on the adhesion of PDA. Li prepared a nanohybrid (BN@PDA@Fe) by PDA nano-coating and in-situ interfacial growth of an iron-derived nanocatalyst (Fe) and then introduced it to EP (Li et al. 2018). By adding 6% BN@PDA@Fe, the pHRR of EP was reduced by 38.9% with notably suppressed CO and smoke production (Figure 8.3). It was suggested that dopamine was employed toward exfoliating bulk h-BN with hydrogen bond action and imparting a highly active surface. SnO_2 nanoparticles were synthesized in-situ on the surface of PDA@h-BN nanosheets with the assistance of PDA to prepare BN@PDA@SnO_2, and the preparation process of ternary h-BN@PDA@SnO_2 sheets is shown in Figure 8.4a. Because of the presence of well-dispersed h-BN@PDA@SnO_2, the pHRR, THR, and TSR of EP decreased 41.1%, 30.1%, and 21.6%, respectively (Cai, Guo, et al. 2018). In addition, PDA can be used as a green bridge to prepare the h-BN/MoS_2 assembly via electrostatic force (Sui et al. 2021). Compared with pure EP, the addition of 2.0 % h-BN@PDA@MoS_2 resulted in 32% and 17.8% decreases in the pHRR and THR of the EP composites, respectively (Figure 8.3). It is suggested that the h-BN@PDA@MoS2 assembly could form strong interactions with the EP matrix and increase the graphitization degree of char residues. Yang also used PDA as a medium and grafted Ti-O compounds on the surface of h-BN@PDA, which can generate TiO_2 and TiP_2O_7 at high temperatures (Figure 8.4b). The presence of h-BN@PDA@Ti effectively enhanced the strength and oxidation resistance of the char and improved the barrier effect of the char to heat and oxygen during the combustion of EP (Yang et al. 2021).

The aromatic structure of PDA allows strong non-covalent bonds through the interaction of π-π and van der Waals forces with the hexagonal structure of h-BN, which can be used to decorate the h-BN nanosheets and further improve the dispersion of h-BN nanosheets in the matrix.

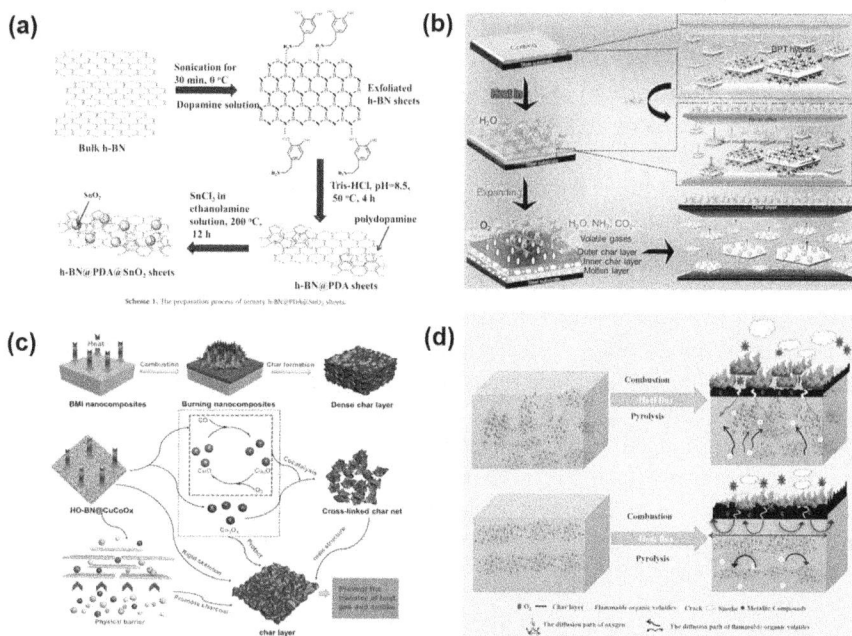

FIGURE 8.4 The preparation process of ternary h-BN@PDA@SnO$_2$ sheets (a) (Cai, Guo, et al. 2018); the flame retardant mechanism of the water-based expansive fireproof coating (b) (Yang et al. 2021); schematic diagram of flame retardant mechanism (c) (Zhou, Chu, et al. 2022); Schematic illustration of the flame retardation mechanisms of different dispersion ZF-BNNS in EP (d) (Zhang, Li, et al. 2019).

Apart from relying on PDA as a mediator between inorganic nanomaterials and h-BN nanosheets, some metal oxides can also grow in situ on the surface of h-BN nanosheets. Zhou functionalized h-BN nanosheets (h-BN@CuCoOx) with copper-cobalt metal oxide (CuCoOx) generated in situ on the surface of h-BN. The flame retardancy of bismaleimide resin (BMI) was significantly improved. The pHRR and THR were reduced by 37.2% and 37.5 due to the combined effect of the h-BN nanosheets and metal oxides on the catalyst and barrier, respectively, as shown in Figure 8.4c. In addition, BMI/0.5%h-BN@CuCoOx achieved an 80% increase in impact strength, revealing excellent toughness due to the strong interface interaction between the matrix and h-BN@CuCoOx (Zhou, Chu, et al. 2022). As an effective catalytic metal coordinate, the MOF can be combined with h-BN nanosheets, simultaneously improving the ability of heat/smoke suppression. In order to further reduce the release of smoke and heat, a double metal hydroxide (Co/Mg-LDH) with a lamellar structure was introduced into the surface of MOF@h-BN nanosheets to prepare BN@MOF-LDH. In addition, 3-aminopropyltriethoxysilane (APTES) was selected to modify the BN@MOF-LDH nanosheets to obtain hydrophobic BN@MOF-LDH@APTES, which exhibited extremely high filler efficiency in reducing the pHRR and TSP (Zhou, Qiu, et al. 2022).

Zinc ferrite ($ZnFe_2O_4$) was also used to functionalize h-BN nanosheets. Shi first prepared $ZnFe_2O_4$ decorated h-BN nanosheets, followed by stearic acid modification to obtain magnetic and hydrophobic properties, and then introduced them into EP to improve its flame retardancy (Shi et al. 2021). Wang also prepared hBN@ $ZnFe_2O_4$ nanosheets by the same procedure and then introduced them to the PVA film by solution forming. By the addition of 5% hBN@$ZnFe_2O_4$, the pHRR and THR decreased by 46.3% and 15.6%, respectively, which was mainly due to the barrier effect of the two-dimensional nanosheet structure of h-BN and the formation of a carbon layer catalyzed by $ZnFe_2O_4$ to enhance the shielding effect (Wang et al. 2020). Similar to $ZnFe_2O_4$, cobalt ferrite ($CoFe_2O_4$) and copper molybdate ($CuMoO_4$) were also used to decorate h-BN nanosheets by the hydrothermal method, and improved the flame retardancy of silicone and polyurethane elastomer (PUE), respectively (Xu et al. 2018, Lu et al. 2021).

It can be concluded that the metal compounds on the surface of h-BN nanosheets exhibited excellent catalytic effects, inhibiting the release of toxic gases, and promoting char formation. Moreover, high-temperature resistant metal oxides produced during combustion covered the surface of the char layer to enhance the structural stability of the char layer.

8.4 CONCLUSION

Based on the above results, it can be concluded that the h-BN nanosheets played an important role in the barrier for heat and/or smoke, and thermal conductivity. In addition, the characteristics of high-temperature resistance endowed h-BN nanosheets with the possibility of participating in char formation and eventually being wrapped by char residues. In the future, the multi-functions of h-BN nanosheets, including thermal conductivity, electrical insulation, anisotropic barrier effect, and flame retardancy, can be combined with other additives to simultaneously improve the comprehensive performance of the matrix. Additionally, the detailed surface group conversion of h BN in various environments of high temperatures should be clarified, to explore its contribution at the atomic scale (Gautam and Chelliah 2021), where micro spectrometers will be employed to obtain direct evidence. These precise maps of chemical conversions will provide a reliable direction for the further development of efficient flame retardants.

REFERENCES

Ambekar, Rushikesh S., Abhishek Deshmukh, Martha Y. Suarez-Villagran, Rakesh Das, Varinder Pal, Satyahari Dey, John H. Miller, Jr., Leonardo D. Machado, Partha Kumbhakar, and Chandra S. Tiwary. 2020. "2D hexagonal boron nitride-coated cotton fabric with self-extinguishing property." *Acs Applied Materials & Interfaces* 12 (40):45274–45280. doi: 10.1021/acsami.0c12647.

Cai, Wei, Wenwen Guo, Ying Pan, Junling Wang, Xiaowei Mu, Xiaming Feng, Bihe Yuan, Bibo Wang, and Yuan Hu. 2018. "Polydopamine-bridged synthesis of ternary h-BN@ PDA@SnO_2 as nanoenhancers for flame retardant and smoke suppression of epoxy composites." *Composites Part A-Applied Science and Manufacturing* 111:94–105. doi: 10.1016/j.compositesa.2018.05.015.

Cai, Wei, Ningning Hong, Xiaming Feng, Wenru Zeng, Yongqian Shi, Yi Zhang, Bibo Wang, and Yuan Hu. 2017. "A facile strategy to simultaneously exfoliate and functionalize boron nitride nanosheets via Lewis acid-base interaction." *Chemical Engineering Journal* 330:309–321. doi: 10.1016/j.cej.2017.07.162.

Cai, Wei, Bi-Bo Wang, Xin Wang, Yu-Lu Zhu, Zhao-Xin Li, Zhou-Mei Xu, Lei Song, Wei-Zhao Hu, and Yuan Hu. 2021. "Recent Progress in Two-dimensional Nanomaterials Following Graphene for Improving Fire Safety of Polymer (Nano)composites." *Chinese Journal of Polymer Science* 39 (8):935–956. doi: 10.1007/s10118-021-2575-2.

Cai, Wei, Bibo Wang, Longxiang Liu, Xia Zhou, Fukai Chu, Jing Zhan, Yuan Hu, Yongchun Kan, and Xin Wang. 2019. "An operable platform towards functionalization of chemically inert boron nitride nanosheets for flame retardancy and toxic gas suppression of thermoplastic polyurethane." *Composites Part B – Engineering* 178. doi: 10.1016/j.compositesb.2019.107462.

Cai, Wei, Dichang Zhang, Bibo Wang, Yongqian Shi, Ying Pan, Junling Wang, Weizhao Hu, and Yuan Hu. 2018. "Scalable one-step synthesis of hydroxylated boron nitride nanosheets for obtaining multifunctional polyvinyl alcohol nanocomposite films: multi-azimuth properties improvement." *Composites Science and Technology* 168:74–80. doi: 10.1016/j.compscitech.2018.09.004.

Chen, Tingjie, Zhiyong Liu, Xiaokang Hu, Gang Zhao, Zipeng Qin, John Tosin Aladejana, Xiangfang Peng, Yongqun Xie, and Binghui Wu. 2022. "Fire-resistant plant fiber sponge enabled by highly thermo-conductive hexagonal boron nitride ink." *Chemical Engineering Journal* 429. doi: 10.1016/j.cej.2021.132135.

Cheng, Le, and Jiachun Feng. 2020. "Flexible and fire-resistant all-inorganic composite film with high in-plane thermal conductivity." *Chemical Engineering Journal* 398. doi: 10.1016/j.cej.2020.125633.

Chu, Fukai, Dichang Zhang, Yanbei Hou, Shuilai Qiu, Junling Wang, Weizhao Hu, and Lei Song. 2018. "Construction of hierarchical natural fabric surface structure based on two-dimensional boron nitride nanosheets and its application for preparing biobased toughened unsaturated polyester resin composites." *ACS Applied Materials & Interfaces* 10 (46):40168–40179. doi: 10.1021/acsami.8b15355.

Feng, Yuezhan, Gaojie Han, Bo Wang, Xingping Zhou, Jianmin Ma, Yunsheng Ye, Chuntai Liu, and Xiaolin Xie. 2020. "Multiple synergistic effects of graphene-based hybrid and hexagonal boron nitride in enhancing thermal conductivity and flame retardancy of epoxy." *Chemical Engineering Journal* 379. doi: 10.1016/j.cej.2019.122402.

Gan, Wentao, Chaoji Chen, Zhengyang Wang, Yong Pei, Weiwei Ping, Shaoliang Xiao, Jiaqi Dai, Yonggang Yao, Shuaiming He, Beihan Zhao, Siddhartha Das, Bao Yang, Peter B. Sunderland, and Liangbing Hu. 2020. "Fire-resistant structural material enabled by an anisotropic thermally conductive hexagonal boron nitride coating." *Advanced Functional Materials* 30 (10). doi: 10.1002/adfm.201909196.

Gautam, C., and S. Chelliah. 2021. "Methods of hexagonal boron nitride exfoliation and its functionalization: covalent and non-covalent approaches." *RSC Advances* 11 (50):31284–31327. doi: 10.1039/d1ra05727h.

Gilbert, S. Matt, Thang Pham, Mehmet Dogan, Sehoon Oh, Brian Shevitski, Gabe Schumm, Stanley Liu, Peter Ercius, Shaul Aloni, Marvin L. Cohen, and Alex Zettl. 2019. "Alternative stacking sequences in hexagonal boron nitride." *2D Materials* 6 (2):021006. doi: 10.1088/2053-1583/ab0e24.

Guan, Qingbao, Li Yuan, Shenmei Wu, Aijuan Gu, and Guozheng Liang. 2017. "Enhanced thermal and dielectric properties of hybrid organic/inorganic shell microcapsule/thermosetting resin nanocomposites." *Polymer International* 66 (12):1940–1948. doi: 10.1002/pi.5481.

Han, Gaojie, Di Zhang, Chuiming Kong, Bing Zhou, Yongqian Shi, Yuezhan Feng, Chuntai Liu, and De-Yi Wang. 2022. "Flexible, thermostable and flame-resistant epoxy-based thermally conductive layered films with aligned ionic liquid-wrapped boron nitride nanosheets via cyclic layer-by-layer blade-casting." *Chemical Engineering Journal* 437. doi: 10.1016/j.cej.2022.135482.

Han, Gaojie, Xiaoyu Zhao, Yuezhan Feng, Jianmin Ma, Keqing Zhou, Yongqian Shi, Chuntai Liu, and Xiaolin Xie. 2021. "Highly flame-retardant epoxy-based thermal conductive composites with functionalized boron nitride nanosheets exfoliated by one-step ball milling." *Chemical Engineering Journal* 407. doi: 10.1016/j.cej.2020.127099.

Hu, Dechao, Huaqing Liu, Yong Ding, and Wenshi Ma. 2021. "Synergetic integration of thermal conductivity and flame resistance in nacre-like nanocellulose composites." *Carbohydrate Polymers* 264. doi: 10.1016/j.carbpol.2021.118058.

Li, Jingchao, Xiuying Zhao, Wenjie Wu, Zhaoxu Zhang, Yue Xian, Yutao Lin, Yonglai Lu, and Liqun Zhang. 2020. "Advanced flexible rGO-BN natural rubber films with high thermal conductivity for improved thermal management capability." *Carbon* 162:46–55. doi: 10.1016/j.carbon.2020.02.012.

Li, Yu-Tong, Wen-Jun Liu, Fei-Xiang Shen, Guo-Dong Zhang, Li-Xiu Gong, Li Zhao, Pingan Song, Jie-Feng Gao, and Long-Cheng Tang. 2022. "Processing, thermal conductivity and flame retardant properties of silicone rubber filled with different geometries of thermally conductive fillers: A comparative study." *Composites Part B-Engineering* 238. doi: 10.1016/j.compositesb.2022.109907.

Li, Zhi, Sara Isabel Montero Lira, Lu Zhang, Daniel Fernandez Exposito, Vignesh Babu Heeralal, and De-Yi Wang. 2018. "Bio-inspired engineering of boron nitride with iron-derived nanocatalyst toward enhanced fire retardancy of epoxy resin." *Polymer Degradation and Stability* 157:119–130. doi: 10.1016/j.polymdegradstab.2018.10.005.

Liu, Hong, Yang Du, Shaohua Lei, and Zhuoqun Liu. 2020. "Flame-retardant activity of modified boron nitride nanosheets to cotton." *Textile Research Journal* 90 (5–6):512–522. doi: 10.1177/0040517519871260.

Lu, Haitao, Huili Shi, Lei Sun, Keli Wang, and Xia Zhang. 2021. "Flame retardant and superhydrophobic composites via oriented arrangement of boron nitride nanosheets." *Journal of Materials Science* 56 (36):19955–19968. doi: 10.1007/s10853-021-06519-4.

Lu, Shaolin, Wei Hong, and Xudong Chen. 2019. "Nanoreinforcements of two-dimensional nanomaterials for flame retardant polymeric composites: an overview." *Advances in Polymer Technology* 2019. doi: 10.1155/2019/4273253.

Ma, Zhewen, Jianzhong Zhang, Lei Liu, Hua Zheng, Jinfeng Dai, Long-Cheng Tang, and Pingan Song. 2022. "A highly fire-retardant rigid polyurethane foam capable of fire-warning." *Composites Communications* 29. doi: 10.1016/j.coco.2021.101046.

Qiu, Shuilai, Yanbei Hou, Weiyi Xing, Chao Ma, Xia Zhou, Longxiang Liu, Yongchun Kan, Richard K. K. Yuen, and Yuan Hu. 2018. "Self-assembled supermolecular aggregate supported on boron nitride nanoplatelets for flame retardant and friction application." *Chemical Engineering Journal* 349:223–234. doi: 10.1016/j.cej.2018.05.053.

Qiu, Xiaoqing, Zhiwei Li, Xiaohong Li, Laigui Yu, and Zhijun Zhang. 2019. "Construction and flame-retardant performance of layer-by-layer assembled hexagonal boron nitride coatings on flexible polyurethane foams." *Journal of Applied Polymer Science* 136 (29). doi: 10.1002/app.47839.

Rosely, C. V. Sijla, Angel Mary Joseph, Andreas Leuteritz, and E. Bhoje Gowd. 2020. "Phytic acid modified boron nitride nanosheets as sustainable multifunctional nano-fillers for enhanced properties of poly(L-lactide)." *ACS Sustainable Chemistry & Engineering* 8 (4):1868–1878. doi: 10.1021/acssuschemeng.9b06158.

Shi, Huili, Huanhuan Wang, Haitao Lu, Xia Zhang, and Zhijun Zhang. 2021. "Magnetic field-induced orientation of modified boron nitride nanosheets in epoxy resin with improved flame and wear resistance." *Langmuir* 37 (27):8222–8231. doi: 10.1021/acs.langmuir.1c00927.

Sui, Yanlong, Peihong Li, Xueyan Dai, and Chunling Zhang. 2021. "Green self-assembly of h-BN@PDA@MoS2 nanosheets by polydopamine as fire hazard suppression materials." *Reactive & Functional Polymers* 165. doi: 10.1016/j.reactfunctpolym.2021.104965.

Wang, Dong, Xiaowei Mu, Wei Cai, Lei Song, Chao Ma, and Yuan Hu. 2018. "Constructing phosphorus, nitrogen, silicon-co-contained boron nitride nanosheets to reinforce flame retardant properties of unsaturated polyester resin." *Composites Part a-Applied Science and Manufacturing* 109:546–554. doi: 10.1016/j.compositesa.2018.04.003.

Wang, Junling, Dichang Zhang, Yan Zhang, Wei Cai, Congxue Yao, Yuan Hu, and Weizhao Hu. 2019. "Construction of multifunctional boron nitride nanosheet towards reducing toxic volatiles (CO and HCN) generation and fire hazard of thermoplastic polyurethane." *Journal of Hazardous Materials* 362:482–494. doi: 10.1016/j.jhazmat.2018.09.009.

Wang, Xiaodong, Yanjun Yin, Mingling Li, and Yuan Hu. 2020. "Hexagonal boron Nitride@ZnFe$_2$O$_4$ hybrid nanosheet: an ecofriendly flame retardant for polyvinyl alcohol." *Journal of Solid State Chemistry* 287. doi: 10.1016/j.jssc.2020.121366.

Wang, Xiu, Song-Lin Ji, Xin-Qi Wang, Hui-Yang Bian, Ling-Rui Lin, Hong-Qi Dai, and Huining Xiao. 2019. "Thermally conductive, super flexible and flame-retardant BN-OH/PVA composite film reinforced by lignin nanoparticles." *Journal of Materials Chemistry C* 7 (45):14159–14169. doi: 10.1039/c9tc04961d.

Xu, Wenzong, Aijiao Li, Yucheng Liu, Rui Chen, and Wu Li. 2018. "CuMoO$_4$@hexagonal boron nitride hybrid: an ecofriendly flame retardant for polyurethane elastomer." *Journal of Materials Science* 53 (16):11265–11279. doi: 10.1007/s10853-018-2390-5.

Xu, Xiaotong, Zhenlin Jiang, Keyu Zhu, Yun Zhang, Min Zhu, Chaosheng Wang, Huaping Wang, and Alex Ren. 2022. "Highly flame-retardant and low toxic polybutylene succinate composites with functionalized BN@APP exfoliated by ball milling." *Journal of Applied Polymer Science* 139 (21). doi: 10.1002/app.52217.

Yang, Liu, Jiachen Guo, Ling Zhang, and Chunzhong Li. 2022. "Significant improvement in the flame retardancy and thermal conductivity of the epoxy resin via constructing a branched flame retardant based on SI-ATRP initiated by dopamine-modified boron nitride." *Industrial & Engineering Chemistry Research* 61 (23):8031–8042. doi: 10.1021/acs.iecr.2c01044.

Yang, Wei, Anthony Chun Yin Yuen, Peng Ping, Rui-Chao Wei, Lei Hua, Zheng Zhu, Ao Li, San- E. Zhu, Li-Li Wang, Jing Liang, Timothy Bo Yuan Chen, Bin Yu, Jing-Yu Si, Hong-Dian Lu, Qing Nian Chan, and Guan Heng Yeoh. 2019. "Pectin-assisted dispersion of exfoliated boron nitride nanosheets for assembled bio-composite aerogels." *Composites Part A: Applied Science and Manufacturing* 119:196–205. doi: 10.1016/j.compositesa.2019.02.003.

Yang, Zhengwei, Guoqing Xiao, Chunlin Chen, Chunyan Chen, Mingtan Wang, Fei Zhong, Shuyi Zeng, and Lanxin Lin. 2021. "Synergistic decoration of organic titanium and polydopamine on boron nitride to enhance fire resistance of intumescent waterborne epoxy coating." *Colloids and Surfaces a-Physicochemical and Engineering Aspects* 621. doi: 10.1016/j.colsurfa.2021.126561.

Yin, Lian, Kaili Gong, Keqing Zhou, Xiaodong Qian, Congling Shi, Zhou Gui, and Lijun Qian. 2022. "Flame-retardant activity of ternary integrated modified boron nitride nanosheets to epoxy resin." *Journal of Colloid and Interface Science* 608:853–863. doi: 10.1016/j.jcis.2021.10.056.

Yin, Sihao, Xinlin Ren, Peichao Lian, Yuanzhi Zhu, and Yi Mei. 2020. "Synergistic effects of black phosphorus/boron nitride nanosheets on enhancing the flame-retardant properties of waterborne polyurethane and its flame-retardant mechanism." *Polymers* 12 (7). doi: 10.3390/polym12071487.

Yue, Xiaopeng, Chaofan Li, Yonghao Ni, Yongjian Xu, and Jian Wang. 2019. "Flame retardant nanocomposites based on 2D layered nanomaterials: a review." *Journal of Materials Science* 54 (20):13070–13105. doi: 10.1007/s10853-019-03841-w.

Zhang, Qiaoran, Zhiwei Li, Xiaohong Li, Laigui Yu, Zhijun Zhang, and Zhishen Wu. 2019. "Zinc ferrite nanoparticle decorated boron nitride nanosheet: preparation, magnetic field arrangement, and flame retardancy." *Chemical Engineering Journal* 356:680–692. doi: 10.1016/j.cej.2018.09.053.

Zhang, Qiaoran, Xiaoyang Wang, Xiaojun Tao, Zhiwei Li, Xiaohong Li, and Zhijun Zhang. 2019. "Polyvinyl alcohol composite aerogel with remarkable flame retardancy, chemical durability and self-cleaning property." *Composites Communications* 15:96–102. doi: 10.1016/j.coco.2019.07.003.

Zhang, Qiaoran, Mengmeng Zhang, Huili Shi, Zhiwei Li, Laigui Yu, Xiaohong Li, Zhijun Zhang, and Zhishen Wu. 2020. "Enhancing flame retardance of epoxy resin by incorporation into ammonium polyphosphate/boron nitride nanosheets/zinc ferrite three-dimensional porous aerogel via vacuum-assisted infiltration." *Journal of Applied Polymer Science* 137 (17). doi: 10.1002/app.48609.

Zhou, Yifan, Fukai Chu, Wenhao Yang, Shuilai Qiu, and Yuan Hu. 2022. "MOF-derived strategy to obtain CuCoOx functionalized HO-BN: a novel design to enhance the toughness, fire safety and heat resistance of bismaleimide resin." *Chemical Engineering Journal* 431. doi: 10.1016/j.cej.2021.134013.

Zhou, Yifan, Shuilai Qiu, Fukai Chu, Wenhao Yang, Yong Qiu, Lijun Qian, Weizhao Hu, and Lei Song. 2022. "High-performance flexible polyurethane foam based on hierarchical BN@MOF-LDH@APTES structure: enhanced adsorption, mechanical and fire safety properties." *Journal of Colloid and Interface Science* 609:794–806. doi: 10.1016/j.jcis.2021.11.089.

Zhu, Wenju, Mingyang Yang, Hao Huang, Zhao Dai, Bowen Cheng, and Shuaishuai Hao. 2020. "A phytic acid-based chelating coordination embedding structure of phosphorus-boron-nitride synergistic flame retardant to enhance durability and flame retardancy of cotton." *Cellulose* 27 (8):4817–4829. doi: 10.1007/s10570-020-03063-3.

9 Molybdenum Disulphide/Polymer Composites for Fire Safety Applications

Anthony Chun Yin Yuen

Department of Building Environment and Energy Engineering, The Hong Kong Polytechnic University, Hung Hom, Kowloon, Hong Kong SAR, PR China

Wei Wang and Ao Li

School of Mechanical and Manufacturing Engineering, University of New South Wales, Sydney, Australia

CONTENTS

9.1 INTRODUCTION

In the past decade, with the continuous development of two-dimensional materials such as graphene in various fields such as energy storage materials (Kuila, Mishra et al. 2013, Zhang, Hou et al. 2016), biomedicine (Han, Sun et al. 2019, Wang, Qiu et al.

2019), foldable and wearable devices (Qiao, Li et al. 2019, Pang, Yang et al. 2020), and solar cells (Mahmoudi, Wang et al. 2018) due to its excellent properties, including high specific surface area, transparency, high Young's modulus (0.3 TPa), and high electrical conductivities, researchers have discovered the magical properties of two-dimensional materials and gradually applied them in composite materials for better performance (Phiri, Gane et al. 2017, Xu, Zhang et al. 2020). Among the two-dimensional materials, transition metal dichalcogenides (TMDs), as promising semiconducting materials, also exhibit outstanding features owing to their atomic scale thickness, direct bandgap, strong spin-orbit coupling, and excellent mechanical and electronic characteristics (Choi, Choudhary et al. 2017, Manzeli, Ovchinnikov et al. 2017). Generally, TMDs have a general representation of MX_2, where M indicates the transition metal atom, such as Mo, Ta, etc., and X means the element in the chalcogenide family, including S, Se, etc. (Manzeli, Ovchinnikov et al. 2017). Although there are several kinds of TMDs, molybdenum disulphide (MoS_2) is the most popular one that presents fantastic characteristics, such as a high direct band gap (~1.9 eV for monolayer), feasible synthesis, and outstanding on/off ratio of 10 (Radisavljevic, Radenovic et al. 2011, Perkins, Friedman et al. 2013, Ganatra and Zhang 2014). In terms of MoS_2 nanosheets, it is interesting that the properties are determined by the polytype formats and the layer numbers. For example, by tuning the number of MoS_2 layers, it is possible to adjust the bandgap of MoS_2 from indirect to direct by exfoliating the bulk MoS_2 layer (~1.2 eV) to single-layered MoS_2 nanosheets, which thus could be employed in electronics such as sensors (Huang, Chu et al. 2019, Kumar, Zheng et al. 2020) and energy storage (Jiao, Hafez et al. 2018, Li, Li et al. 2020).

Generally, the structures of materials play a critical role in determining their properties. For the structure of MoS_2, the crystal has three polytype formats, namely 1T, 2H, and 3R. Here, the first number indicates the number of layers of MoS_2, and the second letter represents the crystallographic structure (T, H, and R depict trigonal, hexagonal, and rhombohedral, respectively). By controlling the synthesis process, rational conditions and modification could lead to the formation of various morphologies and structures, such as zero-dimensional nanoplatelets, one-dimensional nanorods and nanowires, and two-dimensional nanosheets. These polytype MoS_2 with various structures exhibit different properties. Among them, as shown in Figure 9.1 (Sethulekshmi, Jayan et al. 2021), in which six sulphur atoms are organized in a symmetric hexagonal frame, the 2H format of MoS_2 is the well-known mineral molybdenite that shows the most stable form that has a sandwich structure with one molybdenum layer and two sulphur layers connected by stable covalent bonds. The Van der Waal forces existed between the layers to hold the layers together (Gurarslan, Jiao et al. 2016, Biby, Ali et al. 2021). Nevertheless, the weak Van der Waals forces could not resist the stronger external mechanical forces which could be used to exfoliate the MoS_2 bulk layer and obtain two-dimensional nanosheets with a monolayer thickness of 6.5 Å.

As aforementioned, the monolayer or few-layered MoS_2 nanosheets show more fantastic properties than unexfoliated raw MoS_2. Therefore, achieving exfoliated MoS_2 nanosheets with single- or few-layered structures is critically important for further applications. Generally, single- and few-layer MoS_2 can be grown from precursors or exfoliated from bulk crystals, and these employed methods can be summarized as three general routes: micromechanical exfoliation, ion-intercalation, and liquid-phase

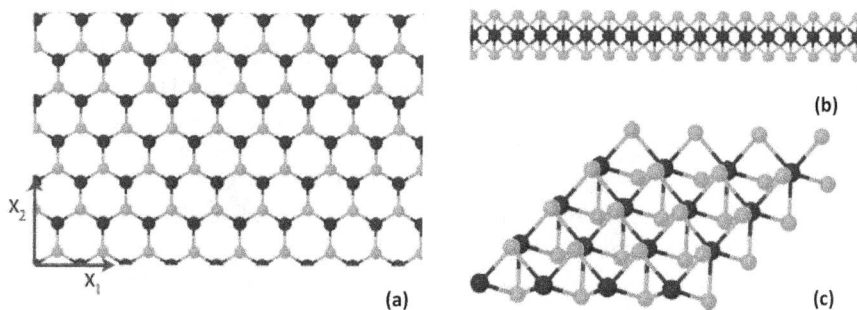

FIGURE 9.1 Atomic structure of monolayer MoS_2.

Source: Reprinted from Sethulekshmi, A. S., Jayan, J. S., Appukuttan, S., and Joseph, K., *Physica E Low Dimens. Syst. Nanostruct.*, 132, 114716, 2021.

exfoliation. The vertically stacked layers of MoS_2 are connected by van der Waals forces that can be overcome by stronger outside physical forces. Extensive approaches such as mechanical exfoliation (Lee, Yan et al. 2010, Radisavljevic, Radenovic et al. 2011, Li, Wu et al. 2014) and shear exfoliation(Varrla, Backes et al. 2015, Xu, Li et al. 2018) have been widely employed to fabricate single or few-layer MoS_2 nanosheets. Similar to graphene prepared by using the "Scotch tape method", single-layer MoS_2 could also be obtained by the same method (Li, Zhang et al. 2012). However, this method shows a low yield of high-quality monolayers. Therefore, the production of MoS_2 nanosheets in large quantities is still challenging. Alternatively, chemical exfoliation is also regarded as an efficient strategy to separate stacked layers through ion intercalation to disrupt interlayer attraction. For example, lithium can be used to exfoliate MoS_2 by inserting lithium atoms into the MoS_2 layer, and then the expansion of the hydrogen generated from the reaction between the intercalated lithium and water would separate the layers (Joensen, Frindt et al. 1986). During the chemical reaction process, n-butyl-lithium reduces the bulk MoS_2 by reducing almost all the Mo^{4+} cations to Mo^{3+}, and the lithium ions are inserted into the layers. Once Li_xMoS_2 recontacts water, the Mo^{3+} cations are re-oxidized, accompanied by the reduction of water and the release of hydrogen between the layers. Additionally, the sonication technique can be used to assist the chemical reaction to further separate the intercalated MoS_2 nanosheets. This strategy has been employed widely in preparing single- and few-layer MoS_2 nanosheets. However, hazardous chemical agents such as n-butyl lithium, inert atmosphere, long reaction times (>48 h), and post-exfoliation procedures to restore the structure bring challenges for safe and practical scale-up.

In addition to the methods mentioned above, there are also other approaches to realizing a few monolayer MoS_2 nanosheets, such as plasma-based exfoliation (Sharma, Verheijen et al. 2018), liquid-phase exfoliations (Huang, Deng et al. 2018), etc. Notably, Liu et al. introduced a layer-by-layer thinning method to fabricate single-layer and bilayer MoS_2 nanosheets. The plasma thinning method is very reliable with an almost 100% success rate and can be easily scaled up (Liu, Nan et al. 2013). Moreover, suitable chemical modification and incorporation of targeted components have also been approved as effective approaches to enhance

MoS$_2$ and broaden the application fields (Cheng, Dong et al. 2016, Wu, Gong et al. 2018, Fan, Fan et al. 2021, Luo, Li et al. 2022). Nevertheless, in terms of the 2H polytype, due to the presence of a direct band gap in the monolayer induced by quantum confinement caused by thickness, the modification methods can significantly change the properties of MoS$_2$ nanosheets by creating defects through the introduction of a negative charge into the lattice (Presolski and Pumera 2016). Therefore, rational structural design and reasonable modification methods play critical roles in improving the properties of MoS$_2$ nanosheets and their performance in applications. In addition to the properties stated above, such as outstanding Young's modulus, high specific surface area, low thermal conductivity, and high elasticity, MoS$_2$ nanosheets with a 2H polytype are regarded as the ideal platform for fabricating high-performance polymer composites. To date, extensive research on MoS$_2$ nanosheet-based polymer composites has been explored to realize improved thermal, mechanical, electrical, and fire-resistant performance.

In this chapter, we aim to highlight the development of MoS$_2$ nanosheets in the field of fire safety with a particular focus on polymer composites, coatings, and other applications associated with flame retardant materials. Generally, two-dimensional materials, due to their structural advantages, can be employed as nanofillers and coating components to enhance the flame retardancy of the flammable polymer matrix. The flame retardancy mechanism has been explored widely, such as the physical barrier effect, catalytic effect, and elimination of free radicals. However, the discussion of the mechanism should be combined with particular modification methods and specific applications. As indicated above, MoS$_2$ nanosheets are capable of flame retardant potential, particularly due to their intrinsic properties, two-dimensional structure, and active sites available for further modification. Accordingly, the upcoming section will systematically conclude different modification methods of MoS$_2$ nanosheets, alongside their applications, and the effect of structure and methods on flame retardancy. Furthermore, through the rational analysis and conclusions of the relation among modified MoS$_2$ structures, properties, and synthesis methods, reasonable relationships will be revealed and anticipated to provide professional solutions and strategic advice for the future research development of 2D MoS$_2$-based flame retardant composites applied in the fire safety field.

9.2   MoS$_2$ NANOSHEETS FOR FIRE-SAFE POLYMERS

9.2.1   MoS$_2$-BASED FIRE-RESISTANT POLYMER COMPOSITES

Despite the great effort and difficulties of chemical exfoliation, MoS$_2$ nanosheets commonly synthesized by intercalation of n-butyl lithium with a solvothermal reaction have been generally used in fire safety enhancement of polymer materials, such as PS, PE, TPU, EP, PBS and so on. The specific structure with layers of MoS$_2$ nanosheets is highly stable within a nitrogen ambient environment, with weight balance. Given the scenario of nanofillers inside a polymer matrix, MoS$_2$ nanosheets are primarily in a nitrogen ambient environment during polymer combustion. As a result, a specific structure with layers of MoS$_2$ nanosheets can demonstrate the barrier effect. With the growth of fire development, MoS$_2$ nanosheets will be disclosed from the inner polymer

TABLE 9.1A
Abbreviations Applied in Section 9.2.1

Acronyms

CS	Chitosan	*PZS*	Polyphosphazene
CTAB	Cetyl trimethyl ammonium bromide	*PEO*	Poly(ethylene oxide)
TPP	(4-carboxybutyl) Triphenylphosphonium bromide	*MH*	Magnesium hydroxide
EP	Epoxy resin	*PPN*	Polyphosphazene nanoparticle
TPEE	Thermoplastic polyester-ether elastomer	*PMMA*	Poly(methyl methacrylate)
EVA	Ethylene-vinyl acetate	*MCA*	Melamine cyanurate
ATH	Aluminium hydroxide	*PA6*	Polyamide 6
GNPs	Graphene nanoplatelets	*PP*	Polypropylene
AHP	Aluminium hypophosphite	*MWCNTs*	Multi-walled carbon nanotubes
PS	Polystyrene	*TNT*	Titanium dioxide nanotube
BD	Bismaleimide/diallyl bisphenol A	*LDH*	Layered double hydroxide
CNT	Carbon nanotubes	*UPR*	Unsaturated polyester resin
PLA	Poly(lactic acid)	*RGO*	Reduced graphene oxide
PU	Polyurethane	*GNS*	Graphene nanosheets
TPU	Thermoplastic polyurethane	*MTS*	3-mercaptopropyl trimethoxy silane
PEI	Polyethylenimine		
DOPO	9,10-dihydro-9-oxa-10-phosphaphenanthrene-10-oxide	*PVA*	Poly(vinyl alcohol)
		RPUF	Rigid polyurethane foam
PDA	Polydopamine	*MAP*	Melamine phosphate
BN	Boron nitride	*PBS*	Poly(butylene succinate)
PANI	Polyaniline		

to the outer side and thus transformed into MoOx nanoparticles. It can also be concluded that the generation procedure of MoOx nanoparticles is unstable. The process will generate a mass of lattice oxygen that will convert CO into CO_2. Therefore, both the barrier function and catalysis effect illustrate that MoS_2 nanosheets are desirable nano enhancers for improving the fire safety of polymer composites.

This section summarizes MoS_2 nanosheet-based fire-resistant polymer composites in Tables 9.1a and b. Due to the synthesis methods of the polymer composites, four categories are formed, including pristine MoS_2 nanosheets, organic-functionalized MoS_2 nanosheets, inorganic-functionalized MoS_2 nanosheets, bio-based and hybrid MoS_2 nanosheets.

9.2.1.1   Pristine MoS$_2$ Nanosheet Applications

At the early stage of the recent decade, pristine MoS_2 nanosheets were directly applied to suppress the fire hazards of polymer composites. Matusinovic et al. synthesized PS-MoS$_2$ and PMMA-MoS$_2$ nanocomposites using a solution mixing

TABLE 9.1B

Summary of Papers on MoS₂-Based Fire-Resistant Polymer Composites in the Past Decade

Synthesis Methods	Additives	Matrix	TGA		LOI	UL94	MCC/FTIR				Year	References
			$T_{50\%}$	Char Residue	LOI%	Rank	TSP	TSR	THR	PHRR		
Pristine	10% PS- MoS₂		28°C +						19.8% -	36.4% -	2012	(Matusinovic, Shukla et al. 2012)
	10% PMMA- MoS₂		21°C +						7.3% -	24.2% -		
	3 wt% PS-MoS₂		52°C +	2%					7.4% -	12.8% -	2013	(Zhou, Yang et al. 2013)
	0.9 wt% PEO-MoS₂		62.1°C +								2015	(Feng, Xing et al. 2015)
	PP/1.6 wt% MoS₂			43.3%					20.5% -	28.1% -	2016	(Feng, Wang et al. 2016)
Organic	1% CS-MoS₂	CS	26°C +						3% -	39% -	2014	(Zhou, Liu et al. 2014)
	3 wt% CTAB-MoS₂	PS	60°C +							19% -		
	5 wt% Melamine-MoS₂	PU									2019	(Malkappa, Ray et al. 2019)
	2 wt% f-MoS₂ (PEI, DOPO)	TPU	34.3°C +	3.8%					25.3% -	45.4% -	2017	(Cai, Zhan et al. 2017)
	2 wt% h-BN/PDA/MoS₂	EP	51.6°C +	3.65%	27.8%	V-1	26.6% -		17.8% -	32% -	2021	(Sui, Li et al. 2021)
	2 wt% h-BN/PANI/MoS₂	EP		11.1%				29.9% -	9.1% -	23.7% -	2021	(Cai, Zhu et al. 2021)
	5 wt% PANI-MoS₂	EP	0.8°C +	23%					16.8% -	38.4% -	2022	(Feng, Cai et al. 2022)
	3 wt% PZS-MoS₂	EP	14.3°C +	20.8%					30.3% -	41.3% -	2017	(Zhou, Qiu et al. 2017)
	2.0 wt% PPN-MoS₂	EP	41°C +					43.0% -	23.6% -	30.7% -	2018	(Qiu, Hu et al. 2018)
	5 wt% CTAB-MoS₂	PS	39°C +						27% -	40% -	2015	(Zhou, Liu et al. 2015)
	1 wt% CTAB-MoS₂	PMMA							20% -	25% -		
	3 wt% MoS₂-MCA	PLA		4.17%	26.4%	V-1					2019	(Xu, Tan et al. 2019)
	4 wt% MCA- MoS₂	PA6		1.3%					25% -	40% -	2016	(Feng, Wang et al. 2016)
	2.0 wt% MoS₂-CNTs	EP		22.2%					31% -	27% -	2015	(Zhou, Liu et al. 2015)

Group	Sample	Matrix	Temp			UL-94					Year	Reference
	3 wt% CTAB- MoS₂	PS		19%							2014	(Zhou, Jiang et al. 2014)
	2 wt% TPP- MoS₂	EP	32°C -				24% -	21% -	17% -	20% -	2019	(Wang, Yu et al. 2019)
	1 wt% MWCNTs-MoS₂	EP							13.9% -	26% -	2019	(Wang, Fu et al. 2019)
	2 wt% PANI- MoS₂	EP							22% -	43.3% -	2018	(Zhou, Liu et al. 2018)
	MoS₂-RGO	UPR								49% -	2016	(Wang1, Xing1 et al. 2016)
	2 wt% MoS₂-GNS	EP					30.5% -		25.3% -	45.8% -	2013	(Wang, Zhou et al. 2013)
	5 wt% MoS₂-HPA	UPR							39.6% -	43.2% -	2017	(Wang, Wen et al. 2017)
	1.0 wt% MoS₂-MTS	PVA	118°C +							52.5% -	2017	(Zhou, Gao et al. 2017)
	1% MAP- MoS₂	PBS									2016	(Zhou, Liu et al. 2016)
Inorganic	2 wt% MoS₂-TiO₂	BD		36.5%					15% -	32.5% -	2019	(Zhou, Qiu et al. 2019)
	2 wt% MH-MoS₂	EP	8°C -	20.3%						32% -	2019	(Zhao, Yin et al. 2019)
	2 wt% MoS₂-TiO₂	EP		19.7%					20% -	32.5% -	2018	(Zhou, Tang et al. 2018)
	2 wt% MoS₂-CoOOH	EP									2014	(Feng, Xing et al. 2014)
	2 wt% MoS₂-TNTs	EP		3.4%	26.8%		36.9% -		32.2% -	39% -	2018	(Cai, Wang et al. 2018)
	2 wt% CoFe-LDH- MoS₂	EP							24.8% -	62% -	2017	(Zhou, Gao et al. 2017)
	2 wt% NiFe-LDH- MoS₂	EP							33.9% -	65.5% -		
	3 wt% Fe-MoS₂	PS	47°C +	3.25%			22% -		21.8% -	21% -	2014	(Zhou, Zhang et al. 2014)
	1 wt% MoS₂-Bi₂Se₃	EP							18.2% -	22% -	2019	(Hou, Hu et al. 2019)
	1 wt% CuO₂-MoS₂	RPUF	2°C +	7.85%			35.8% -			26.3% -	2020	(Yuan, Wang et al. 2020)
	3 wt% NiFeTb-LDH-MoS₂	TPU	33°C +	13.74%					18.2% -	50.9% -	2022	(Qian, Su et al. 2022)
Bio	1 wt% CS-MoS₂			41.4%							2014	(Feng, Wang et al. 2014)
	2 wt% CS-MoS₂	EP							14.6% -	43.3% -	2015	(Wang, Song et al. 2015)
Hybrid	1 wt% TPEE/P-N/CNTs/MoS₂			14.6%	31.5%	V-0					2015	(Zhong, Li et al. 2015)
	2 wt% PNMoS₂/Co-Bi	EP	2.75°C +	22.9%	27.5%	V-1	41.8% -		27.9% -	28% -	2021	(Zou, Qiu et al. 2021)
	EVA/ATH/2 wt% MoS₂			27.6%	27%	V-2			7.5% -	64.5% -	2017	(Guo, Xue et al. 2017)

(Continued)

TABLE 9.1B (Continued)
Summary of Papers on MoS$_2$-Based Fire-Resistant Polymer Composites in the Past Decade

Synthesis Methods	Additives	Matrix	TGA T$_{50\%}$	TGA Char Residue	LOI LOI%	UL94 Rank	TSP	TSR	MCC/FTIR THR	MCC/FTIR PHRR	Year	References
	EVA/ATH/GNPs/2 wt% MoS$_2$			29.3%	29.5%	V-0			8.3% -	79.2% -		
	EVA/AHP/MCA/2 wt% MoS$_2$		21°C +	6.2%	38.5%	V-0			29.3% -	55.4% -	2016	(Zhou, Tang et al. 2016)
	1.0% PS/CS/MoS$_2$	PS	52°C +	2.7 %					5.4% -	23.2% -	2018	(Zhou, Tang et al. 2018)
	0.5 wt% MoS$_2$/Co$_2$O$_3$/CNT	PLA	0°C						11.3% -	13.4% -	2020	(Homa, Wenelska et al. 2020)
	0.5 wt% MoS$_2$/Fe$_2$O$_3$/CNT	PLA	17°C -						4.5% -	32.8% -		
	0.5 wt% MoS$_2$/Ni$_2$O$_3$/CNT	PLA	3°C -						10.4% -	40.7% -		

technique (Matusinovic, Shukla et al. 2012). The improvement in thermal stability of the PS-MoS$_2$ and PMMA-MoS$_2$ nanocomposites was observable, where the temperature at which 50% degradation occurs was 28°C higher and 21°C higher compared to the neat PS and PMMA. Also, a significant reduction in PHRR and HRR was achieved. Similar results were also demonstrated by Zhou et al. (2013). Meanwhile, pristine MoS$_2$ nanosheets were added to PEO (Feng, Xing et al. 2015) and PP (Feng, Wang et al. 2016) to enhance the functions of nanofillers in the thermal decomposition process.

9.2.1.2   Organic-Functionalized MoS$_2$ Nanosheet Applications

The cramped bandgap and the reaction product of molybdenum oxides impart a catalytic effect to MoS$_2$ during the fire development of the polymer. Nonetheless, the fire performance utilization of MoS$_2$ is still restricted by its relatively low interfacial compatibility. In regard to being exfoliated by the lithium-intercalation and hydrolysis approach, sufficient defects existing within MoS$_2$ nanosheets can conjugate with thiol-based compounds. Simultaneously, the mentioned exfoliated MoS$_2$ nanocomponents have negative charges, achieving electrostatic attraction to the oppositely charged compounds. Regarding the aforementioned characteristics, covalent functionalization is successfully promoted to provide highly efficient functionalized MoS$_2$ nanocomponents to improve polymer composite fire safety. Wang et al. synthesized layered MoS$_2$/GNS hybrids by the hydrothermal method and prepared MoS$_2$/GNS/EP composites by a simple blend under ultrasonic agitation (Wang, Zhou et al. 2013). The synergistic effect between MoS$_2$ and GNS on thermal stability and fire hazards was investigated by thermogravimetric analysis (TGA), microcombustion calorimeter (MCC), and cone calorimeter. Compared to pure EP, 2 wt% MoS$_2$/GNS/EP demonstrated a remarkable increment in T$_{onset}$ under either an air or nitrogen atmosphere. Additionally, the PHRR, THR, and TSR values of the MoS$_2$/GNS/EP composites were obviously decreased by 45.8%, 25.3%, and 30.5%, respectively. Zhou et al. developed an innovative way to decorate MoS$_2$ on the outer faces of CNTs to generate a new interface between epoxy and CNTs. The deposited MoS$_2$ nanolayers can restrain the accumulation of CNTs, improving the dispersibility of CNTs in EP (Zhou, Liu et al. 2015). The prepared composites were characterized by Raman spectroscopy, XRD, SEM, and TEM. Furthermore, the nanocomposites reduced the PHRR (27%) and THR (31%) and inhibited the combustion development of EP matched with the extension of CNTs or MoS$_2$ alone in the EP. The flame retardant mechanism was explained based on pyrolysis product and char residue analyses. In the gas part, the existence of MoS$_2$/CNT hybrids serves as a physical barrier that could set back and reduce the generated volume of combustible fumes, especially hydrocarbons and aromatic compounds; in the condensed phase, the extension of MoS$_2$/CNT hybrids led to the generation of a compact and insulating char layer to prevent the inner polymer matrix from further combustion.

9.2.1.3   Inorganic-Functionalized MoS$_2$ Nanosheet Applications

In addition to organic-functionalized MoS$_2$ nanosheet applications, inorganic modification has been applied in combination with MoS$_2$ to better utilize their

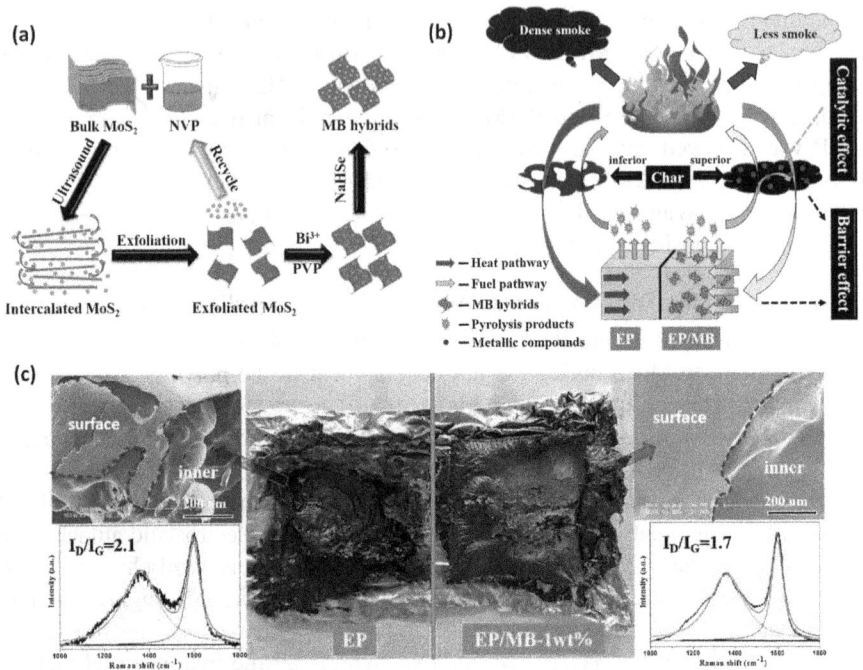

FIGURE 9.2 (a) Schematic demonstration of preparation process of MB hybrids; (b) Schematic figure for the mechanism of flame retardancy and smoke suppression; (c) Digital and SEM images and Raman spectra for both neat EP and EP/MB-1 wt%.

Source: Reprinted from Hou, Y., Hu, Y., Qiu, S., Liu, L., Xing, W., and Hu, W., *J. Hazard. Mater.*, 364, 720–732, 2019.

properties to enhance the fire safety of polymer composites. Hou et al. conducted a recyclable method to exfoliate MoS$_2$ nanosheets with *N*-vinyl pyrrolidone, as shown in Figure 9.2a (Hou, Hu et al. 2019). In-situ grown Bi$_2$Se$_3$ on the outer faces of MoS$_2$ endows the hybrid with a new composition and expands the MoS$_2$-based binary components. The results of digital/SEM images in Figure 9.2b demonstrated that the MoS$_2$-Bi$_2$Se$_3$ (MB) hybrids displayed acceleration in the char-forming procedure during the thermal decomposition of the EP. Decomposed compounds were catalyzed into char residues in the condensed phase and released less in the gas phase. Therefore, the smoke release processes were restrained. The possible mechanisms of smoke suppression and flame retardancy can be explained in Figure 9.2c: decomposed products of EP were catalyzed into char in the condensed phase, resulting in less generation in the gas phase, and thus, smoke emission was restrained.

Moreover, LDH/MoS$_2$ hybrids were facilely prepared by the self-assembly of exfoliated MoS$_2$ nanosheets and LDH via electrostatic force by Zhou, Gao et al. (2017). Incorporation of 2 wt% LDH/MoS$_2$ hybrids into EP led to an increase in char yield, and the PHRR and THR values of EP systems filled with NiFe-LDH/MoS$_2$ hybrids were remarkably reduced by 66% and 34%, respectively. In addition, the number of organic volatiles and toxic CO was inhibited after introducing LDH/ MoS_2 hybrids, suggesting that LDH and MoS_2 exhibited a combined effect on improving fire safety.

9.2.1.4   Bio-Based and Hybrid MoS_2 Nanosheet Applications

Chitosan (CS) is a sugar that comes from the outer skeleton of shellfish, including crab, lobster, and shrimp. CS has been reported to fabricate CS-MoS_2 nanosheets. Wang et al. exfoliated and non-covalently modified bulk MoS_2 to obtain ultrathin MoS_2 nanosheets by ultrasonication in a chitosan solvent, as shown in Figure 9.3 (Wang, Song et al. 2015). The nano-barrier effects of MoS_2 nanosheets suppressed the movement of EP molecules and the effusion of pyrolytic products, which also restricted the permeation of extra heat as well as oxygen, and constrained the release of volatile toxic substances, dramatically enhancing the flame retardance of EP nanocomposites and weakening smoke toxicity. Zhou et al. prepared EVA/AHP/ MCA/2 wt% MoS_2 composites to study the synergistic flame retardant effects of the applied materials. The LOI value increased to the maximum value of 38.5% and passed the V-0 rating in the UL-94 test. The PHRR and THR values were further reduced to 55.4% and 29.3%, respectively. The TGA results of 6.2% char residue suggested that MoS_2 nanosheets acted as an effective mass transport barrier to constrain mass loss during thermal degradation. As a result, the extension of MoS_2 further enhanced the thermal stability of the EVA/AHP/MCA composites.

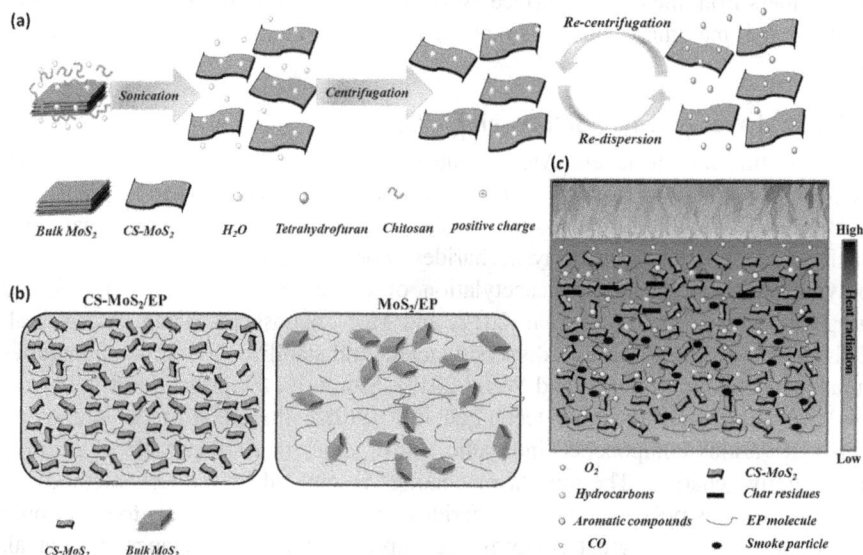

FIGURE 9.3   (a) Schematic demonstration of the MoS_2 nanosheets exfoliation process by ultrasonication; (b) Dispersion sketches of CS-MoS_2 and bulk MoS_2 nanosheets in EP matrices; (c) Nano-barrier effects of the flaming CS-MoS_2/EP nanocomposites.

Source: Reprinted from Wang, D., Song, L., Zhou, K., Yu, X., Hu, Y., and Wang, J., J. Mater. Chem. A, 3(27), 14307–14317, 2015.

In summary, the reviewed previous papers illustrate that exfoliated MoS_2 nanosheets obtained from bulk MoS_2 via either physical or chemical methods can be dispersed into polymers to reinforce the physical properties of polymer matrices. All the flame retardant properties obtained from the TGA, LOI, UL94, MCC, cone calorimeter and FTIR tests are summarized in Tables 9.1a, b. With a similar 2D structure to these layered compounds, MoS_2 has demonstrated its performance as a flame retardant additive, considering two factors the barrier effect and the catalytic effect. Furthermore, functionalized MoS_2 materials, combined with other flame retardants to generate a synergistic effect, can also improve the flame retardant efficiency of pristine MoS_2.

9.2.2   MoS₂-Based Flame Retardant Coatings

As introduced and discussed in the previous section, MoS_2 nanosheets play critical roles in polymer composites as nanofillers that physically withstand the surface migration of oxygen, pyrolyzed products and heat, thereby slowing the thermal degradation process of polymers. Moreover, the modification of MoS_2 also imparts the raw nanosheets with additional properties such as catalytic effects, carbon sources, and hierarchical physical barrier effects, which could also significantly enhance the fire safety of polymer composites. In addition to being applied as nanofillers inside a polymer matrix, MoS_2 nanosheets are also widely used as critical components in flame retardant coatings that can improve the fire retardancy of polymers from the outside surface. Similarly, the related fire safety techniques of MoS_2-based materials could also be employed in flame retardant membranes and films.

9.2.2.1   Layer-by-Layer Self-Assembly Coatings

Like montmorillonite layers, MoS_2 nanosheets are negatively charged and show great dispersion in an aqueous solution. Based on this feature, MoS_2 nanosheets possess the capability of electrostatic interaction with positively charged materials, such as nanoparticles and polysaccharides. For example, chitosan, as a soluble polysaccharide obtained by deacetylation of chitin, was employed to exfoliate graphene oxides (Guo, Sun et al. 2012) and MoS_2 nanosheets (Feng, Wang et al. 2014), and was also used as a dispersing agent and stabilizer for carbon nanotubes in aqueous solution (Huang and Wang 2013). Similarly, the principle of the layer-by-layer (LbL) self-assembly technique is to utilize the electrostatic interaction between various components (inorganic and organic) that are capable of positive and negative charges. The multilayer coatings assembled by the LbL method can integrate various polymers, polysaccharides, and inorganic materials together on a microscale and show great performance improvements. For instance, Pan et al. reported the LbL fabrication process by employing positively discharged chitosan and negatively charged MoS_2 nanosheets to form a chitosan/MoS_2 nanosheet flame retardant multilayer coating on the surface of flexible polyurethane foam (FPUF) (Pan, Shen et al. 2018). Through this method, the chitosan/MoS_2 flame retardant multilayer coating made only an 8.5% weight gain but could significantly improve the fire safety of FPUF with a 70% reduction in peak heat release rate, 62.4%

reduction in peak smoke production rate, and 33.3% reduction in total smoke release compared to the pristine sample. This work verified that negatively charged MoS_2 nanosheets showed obvious potential in LbL flame retardant coatings combined with positively charged polymers and polysaccharides.

Additionally, as mentioned above, MoS_2 nanosheets could also integrate with inorganic materials by the LbL self-assembly method to form a flame retardant coating on the surface of the polymer matrix. As reported by Zhi, Liu et al.(2019), by constructing a three-bilayer nanocoating system via the LbL assembly method, a coating consisting of MoS_2 nanosheets and C_{60} could obviously improve the fire retardancy of FPUF. The integration of MoS_2 nanosheets and C_{60} in the multilayer coating (with 5.24% weight gain) effectively enhanced the fire safety of FPUF, achieving remarkable reductions in peak heat release rate (47.5%), total heat release (51.1%), and total smoke production (33.1%) compared to control FPUF.

In Pan's work, the flame retardant mechanism of the multilayer coating was mainly the physical barrier effect contributed by MoS_2 nanosheets, which could retard the diffusion of the heat, oxygen, and flammable pyrolyzed products between the combustion zone and pyrolysis zone. Different from the mechanism introduced in Pan's work, Zhi et al. also introduced another flame retardant mechanism provided by C_{60} that could enable the free radical-trapping ability and carbonization promotion effect, which played a critical role in flame retardancy. Therefore, based on these reported works, it can be concluded that MoS_2 nanosheets with high specific surface area and negative charge characteristics show broad application potential in LbL-assembled flame retardant coatings.

9.2.2.2 MoS_2 Nanosheet-Based Polymer Composite Coatings

In addition to LbL self-assembled coatings, in the last two years, MoS_2 nanosheets have been employed in polymer composite coatings capable of delivering outstanding flame retardancy, which protects flammable substrates. As shown in Figure 9.4, Peng et al. introduced a nacre-mimetic MoS_2-based bio-nanocomposite coating by using tannic acid (TA)-modified cellulose nanofibers (CNFs) and MoS_2 nanosheets (Peng, Wang et al. 2021). The interfacial chelation between tannic acid and Fe^{3+} facilitated the alternative assembly of a "brick-and-mortar" TA-CNFs/TA-MoS_2 nanocoating covering the surface of a polyacrylonitrile (PAN) textile. Through this biologically interfacial chelating crosslinking method, this hierarchical nanocoating can be chemically and conformally coated on PAN and impart excellent thermal stability, intumescent flame retardant characteristics and washing durability. Specifically, the peak heat release rate, total heat release, smoke production rate, total smoke production and carbon oxide production rate of the coated PAN textile are significantly decreased by 68%, 65%, 79%, 79%, and 85%, respectively.

Besides acting as fillers used in polymer composite coatings, MoS_2 nanosheets may also be utilized to modify flame retardant materials and form multifunctional polymer composite coatings. The incorporation of MoS_2 nanosheets endows the nanosheet's decorated coating systems with enhanced the flame retardant efficiency. For instance, Chen et al. prepared a core-shell structural carbon microsphere (CMS) @MoS_2 system by uniformly fabricating MoS_2 nanosheets on the surface of CMS via a simple hydrothermal reaction (Chen, Xiao et al. 2022). The illustration

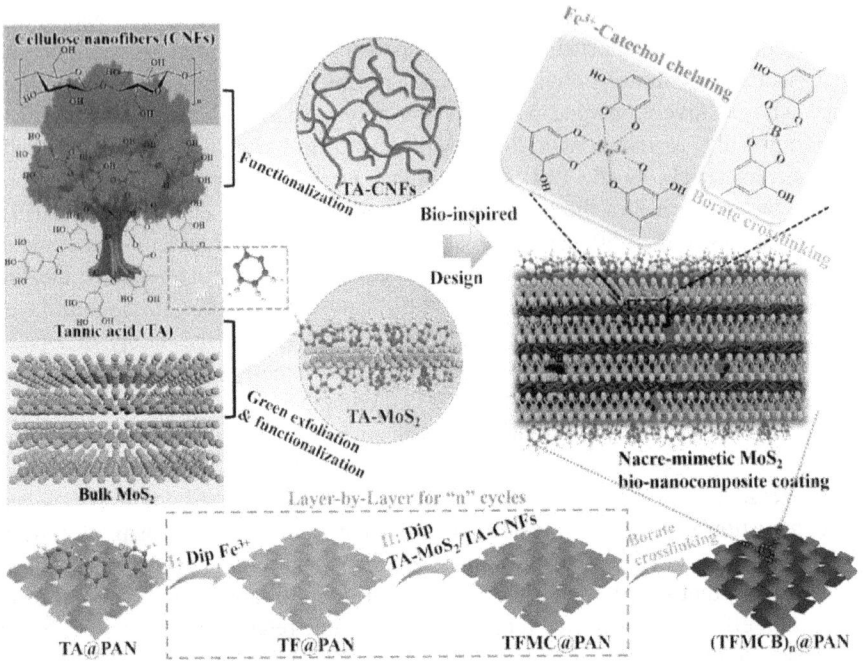

FIGURE 9.4   Scheme for preparing the TA-CNFs/TA-MoS$_2$ bio-nanocomposite coating on PAN fabric.

Source: Reprinted from Peng, H., Wang, D., and Fu, S., *Compos. B. Eng.*, 215, 108742, 2021.

diagram is shown in Figure 9.5. The CMS@MoS$_2$ flame retardant system showed an outstanding synergistic barrier effect that endowed the EP/CMS@MoS$_2$ coating with outstanding thermal insulation capability. Moreover, the modification also combined both the advantages of CMS and MoS$_2$ nanosheets, including the physical barrier effect and synergistic effect, and could significantly improve the smoke suppression property, thermal stability, and fire retardancy of the EP/CMS@ MoS$_2$ coating.

Xiao et al. also introduced a g-C$_3$N$_4$@polydopamine/MoS$_2$ nanosheet flame retardant by integrating MoS$_2$ with inorganic materials and bio-inspired poly-dopamine using the hydrothermal method (Xiao, Yang et al. 2021). The above flame retardant was incorporated into the intumescent flame retardant coating as a synergist to improve the flame retardancy efficiency, in which polydopamine could improve the dispersion stability of g-C$_3$N$_4$ while capturing free radicals and thus improve the oxidation resistance of the char layer. Furthermore, the presence of MoS$_2$ could provide the catalytic effect and facilitate the formation of the char layer.

Based on the analysis of the above works, MoS$_2$ nanosheets could be used to improve the flame retardancy of polymer composite coatings by different approaches. On the one hand, MoS$_2$ can be directly used as a flame retardant or with

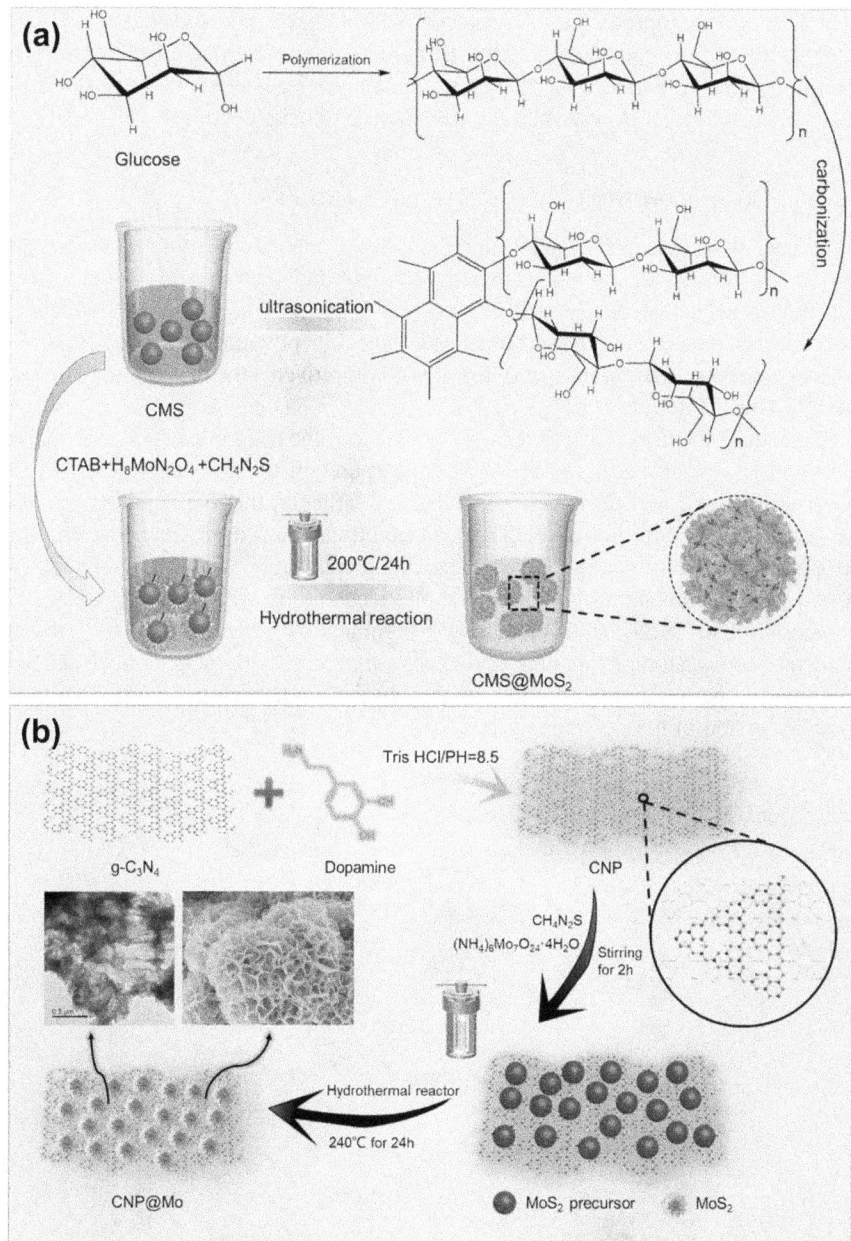

FIGURE 9.5 (a) Illustration of the preparation of CMS@MoS$_2$ hybrids); (b) Schematic preparation processes of CNP@MoS$_2$ hybrids.

Source: (a) Reprinted from Chen, C., Xiao, G., Dong, S., Zhong, F., Yang, Z., Chen, C., … and Zou, R., *Colloids Surf. A Physicochem. Eng. Asp.*, **633**, 127836, 2022. (b) Reprinted from Xiao, G., Yang, Z., Chen, C., Chen, C., Zhong, F., Wang, M., and Zou, R., *Colloids Surf. A Physicochem. Eng. Asp.*, **630**, 127575, 2021.

other additives to improve the fire resistance of polymer composite coatings. On the other hand, MoS_2 could also be used to modify flame retardants by contributing specific features to enhance the flame retardancy efficiency, such as catalytic effect and synergistic effect, to enhance the polymer composite coatings.

9.2.3 MoS_2 Nanosheet-Based Membranes and Films

MoS_2 nanosheets also show wide applications in polymer membranes and films. Due to interfacial interactions, such as hydrogen bonds between MoS_2 nanosheets and polymer chains, MoS_2 nanosheets could be modified and functionalized according to the detailed requirements and specific polymer chains and used to enhance polymer membranes and films with improved tensile strength, thermal stability, and flame retardancy.

Zhang et al. confirmed that hydrogen bonding was the main interaction mechanism between MoS_2 nanosheets and PVA. The incorporation of MoS_2 nanosheets with a low loading could realize improved mechanical strength, thermal conductivity, and fire resistance (Zhang, Lei et al. 2019). In addition, Zhou et al. also revealed that exfoliated MoS_2 nanosheets possessed a strong interaction with PVA chains and exhibited great compatibility with poly (vinyl alcohol), as shown in Figure 9.6. The incorporation of MoS_2 nanosheets could significantly improve thermal stability, flame retardancy, storage modulus, and tensile strength (Zhou, Jiang et al. 2012). The improved properties were attributed to the physical barrier effect of MoS_2 and its good dispersion in the polymer matrix.

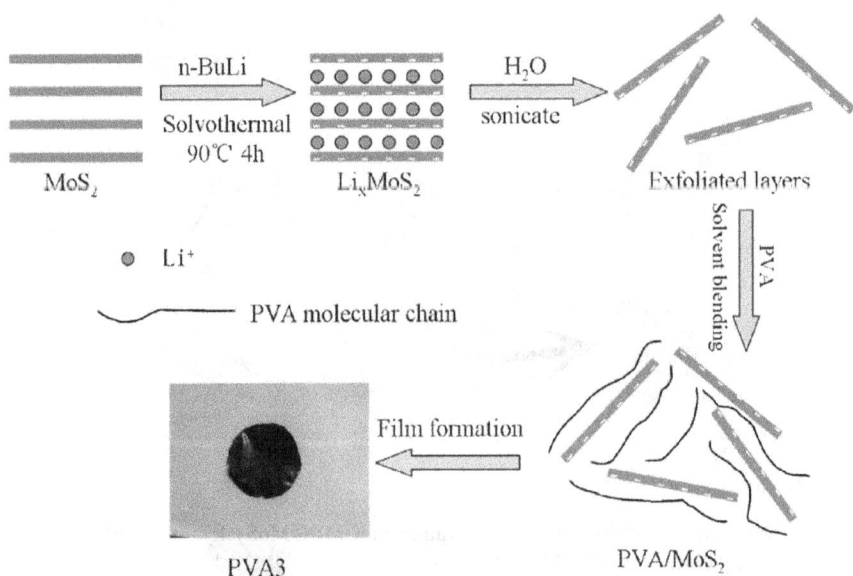

FIGURE 9.6 Illustration of the preparation process of the MoS_2/PVA film.

Source: Reprinted from Zhou, K., Jiang, S., Bao, C., Song, L., Wang, B., Tang, G., ... and Gui, Z., *RSC Adv.*, 2(31), 11695–11703, 2012.

Furthermore, to improve the performance of polymer films or membranes, MoS_2 nanosheets could also be functionalized by inorganic or organic materials to have more specific properties and thereby further enhance the properties of polymer membranes. As depicted in Figure 9.7a, Zhou et al. introduced an LDH/MoS_2 system driven by electrostatic interactions between LDH and MoS_2 nanosheets to enhance the flame retardancy of PVA matrices (Zhou, Hu et al. 2016). In addition,

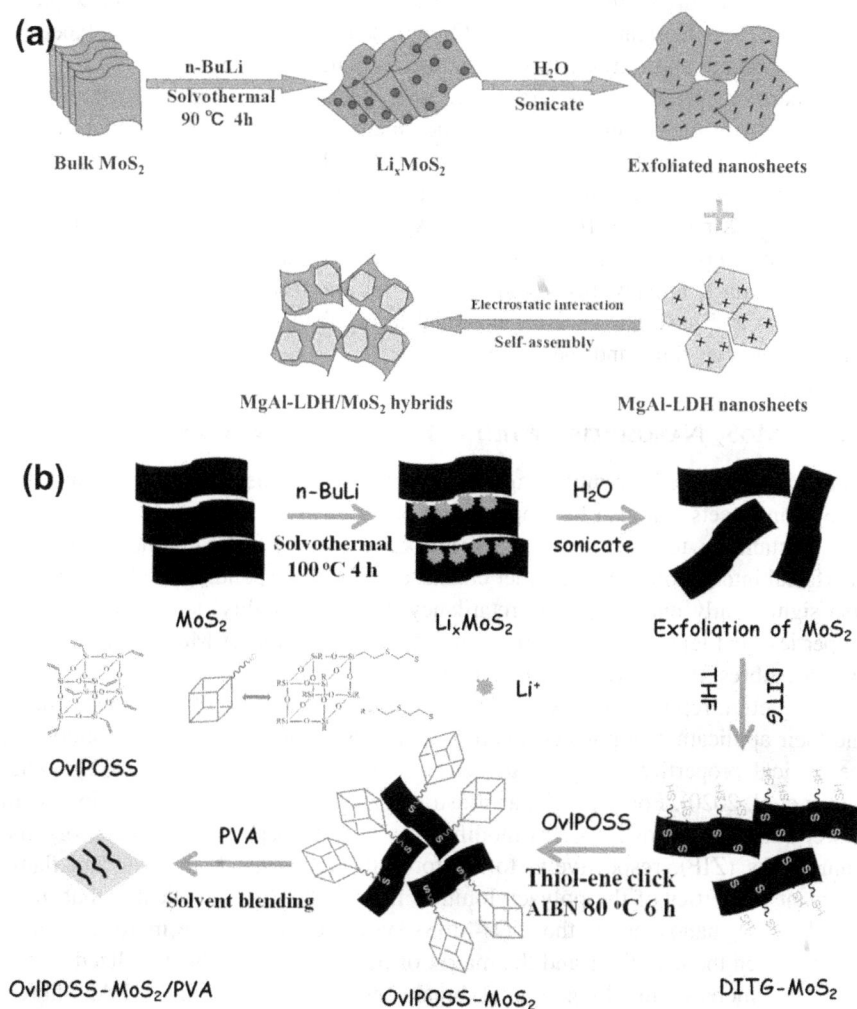

FIGURE 9.7 (a) Illustration for the preparation of MoS_2-LDH nanohybrids by self-assembly method); (b) Illustration for the functionalization of OvlPOSS-MoS_2 nanosheets and PVA nanocomposites.

Source: (a) Reprinted from Zhou, K., Hu, Y., Liu, J., Gui, Z., Jiang, S., and Tang, G., *Mater. Chem. Phys.*, 178, 1–5, 2016. Reprinted from Jiang, S. D., Tang, G., Bai, Z. M., Wang, Y. Y., Hu, Y., and Song, L., *RSC Adv.*, 4(7), 3253–3262, 2014.

as illustrated in Figure 9.7b, Jiang et al. reported another approach using a simple reflux reaction to functionalized MoS_2 nanosheets to form an OvlPOSS-MoS_2 system to improve the thermal stability and flame retardancy of PVA (Jiang, Tang et al. 2014). Meanwhile, due to the strong interfacial interactions between the matrix and OvlPOSS-MoS_2 system, the tensile strength of the membrane was increased by 57% upon the addition of 2 wt.% fillers. Moreover, MoS_2 nanosheets could also be modified by other methods to widen their applications. Qu et al. reported a ball milling method to modify MoS_2 nanosheets by using black phosphorene (BP) to fabricate BP-MoS_2/GO composite films, which showed excellent flame retardancy and ultrasensitive response for smart fire alarms.

Furthermore, properly modifying MoS_2 nanosheets is also regarded as an efficient way to improve their compatibility with polymers. Feng et al. used cetylpyridinium chloride (CPC) to modify MoS_2 nanosheets, in which the alkyl chain from CPC on MoS_2 nanosheets showed great affinity with a polyurethane acrylate (PUA) membrane (Feng, Xing et al. 2016). The good dispersion of MoS_2 nanosheets in the PUA matrix can improve flame retardancy and tensile strength. In general, MoS_2 nanosheets have already been widely used in fire retardant polymer films and membranes and have shown great potential in the field of flame retardant and fire warning polymer films and membranes.

9.2.4 MoS₂ Nanosheets Applied in Flame Retardant Polymer Fibres

Similar to being used for enhancing the properties of polymer films and membranes, MoS_2 nanosheets can also be employed in polymer fibres for better fire resistance and outstanding tensile strength. As introduced above, MoS_2 nanosheets have interfacial interactions with polymer chains, and the corresponding modification can also significantly enhance flame retardancy, thermal stability, smoke suppression properties, and tensile strength. In terms of the application of MoS_2 nanosheets in polymer fibres, the above mechanism still works.

Wang et al. reported a series of works on the modification of MoS_2 nanosheets and their application in polyacrylonitrile (PAN) fibres for improved fire safety and mechanical properties (Peng, Wang et al. 2019, Peng, Wang et al. 2020, Peng, Zhang et al. 2020). For example, as illustrated in Figures 9.8a and 9.8b, MoS_2 with flower balls and nanosheets was modified by SiO_2 nanospheres and metal-organic frameworks (ZIF), respectively, to reinforce the mechanical, thermal and flame retardant properties of the polyacrylonitrile fibres. The results indicated that modified by SiO_2 nanospheres, the SiO_2-MoS_2 system could achieve more interfacial area between the nanofiller and the matrix of polymer fibres, which realized a 35% and 42% increase in elongation at break and tensile strength of PAN fibres. Similarly, after the incorporation of Co-Zn ZIF nanosheets on the surface of MoS_2 nanosheets, the interfacial interaction between Co-Zn ZIF/MoS_2 and the fibre matrix was enhanced and facilitated the mechanical performance of the PAN fibre. In addition, this hybrid system exhibited a physical barrier effect and catalytic charring effect that significantly reinforced the thermal stability and fire resistance of PAN fibres.

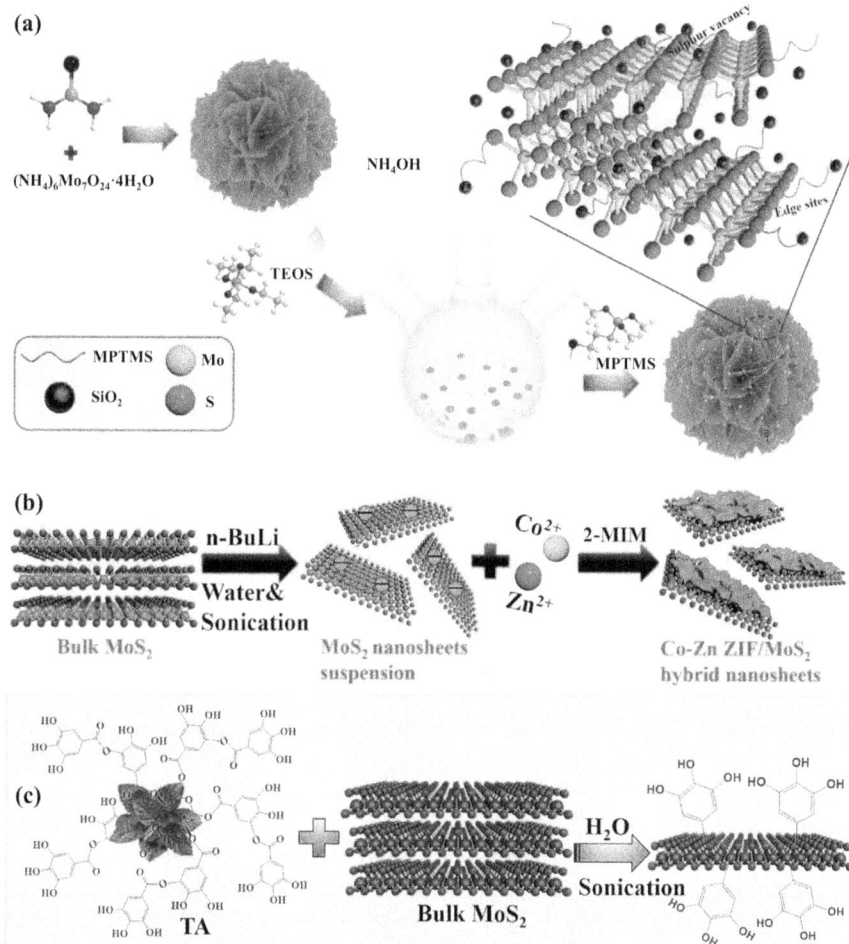

FIGURE 9.8   (a) Schematic routes of flower-like MoS_2–SiO_2 hybrids); (b) Synthetic route of Co-Zn ZIF/MoS_2 hybrid nanosheets; (c) Schematic of TA-assisted exfoliation of MoS_2 nanosheets.

Sources: (a) Reprinted from Peng, H., Wang, D., Li, M., Zhang, L., Liu, M., and Fu, S., *Compos. B. Eng.*, 174, 107037, 2019. (b) Reprinted from Peng, H., Zhang, L., Li, M., Liu, M., Wang, C., Wang, D., and Fu, S., *Chem. Eng. J.*, 397, 125410, 2020. (c) Reprinted from Peng, H., Wang, D., & Fu, S., *Chem. Eng. J.*, 384, 123288, 2020.

In addition, as shown in Figure 9.8c, Wang et al. also integrated tannic acid (TA, a plant-derived polyphenolic biomolecule with rich catechol/pyrogallol groups) with MoS_2 nanosheets to construct an organic-modified MoS_2 nanosheet system that showed strong hydrogen bond interactions with nitrile groups in PAN chains. Hence, the addition of TA/MoS_2 hybrids could significantly enhance the mechanical properties of PAN fibres. What's more. TA, as a bio-based flame retardant, brought the hybrids outstanding charring and oxygen radicals scavenging capability, which significantly improved the flame retardancy of PAN fibres. Wang's works indicated that MoS$_2$ nanosheets presented high potential for enhancing polymer fibres. The proper modification of MoS$_2$ nanosheets plays a critical role in multifunctional MoS$_2$-based nanofillers that can significantly improve the mechanical properties and flame retardancy of polymer fibres.

9.2.5 OTHER APPLICATIONS

In addition to the applications mentioned above, MoS$_2$ nanosheets also exhibited high application potential in other flame retardant materials, such as aerogels (Yang, Mukhopadhyay et al. 2017). As reported by Yang et al., the surface cations of metallic MoS$_2$ can link with carboxyl (-COOH) and hydroxyl (-OH) groups in the cellulose chains to prepare ultra-light, highly porous aerogels, which are capable of crosslinked metal-carboxylate complexes. The aerogel constructed by strong bonding between the MoS$_2$ nanosheets and nanofibers showed high porosity near 97% and a low density of approximately 4.73 mg cm^{-3} and exhibited excellent thermal insulation and fire retardancy. This kind of aerogel provided a new strategy to apply MoS$_2$ nanosheets in other application scenarios associated with fire safety materials.

9.3 SUMMARY AND FUTURE PERSPECTIVES

In recent decades, two-dimensional inorganic materials have been extensively explored for application in flame retardant materials. Among them, MoS$_2$ nanosheets are regarded as multifunctional reinforcing additives that enhance the mechanical, thermal, electrical, and fire-resistant properties of polymers. The excellent performance of MoS$_2$-based polymers is attributed to the fantastic characteristics and great modification potential of MoS$_2$. The monolayer and few-layered MoS$_2$ nanosheets are proven to significantly enhance the properties of polymer composites by a relatively low loading due to extraordinary characteristics such as high specific surface area, transparency, high Young's modulus, low thermal conductivity, and high electrical conductivities. Hence, approaches to preparing exfoliated MoS$_2$ and modification methods are critical to fabricating high-quality MoS$_2$ nanosheets.

In terms of exfoliation approaches, there are three mainstream strategies, including micromechanical exfoliation, ion-intercalation, and liquid phase exfoliation, that are widely employed to obtain monolayer and few-layered MoS$_2$ nanosheets. Generally, the employment of these methods is capable of obtaining monolayer or few-layered MoS$_2$ nanosheets, but the scale-up mass manufacturing of such processes is still challenging, which restricts the practical applications of MoS$_2$.

A good dispersion of two-dimensional nanosheets in the polymer matrix can further enhance the effectiveness of flame retardancy and other functionalities of the polymer (Hong and Chen 2014). With regard to polymer composites, the addition of inorganic fillers may cause aggregation inside the polymer matrix, which can deteriorate the mechanical and flame retardant properties. Regarding the flame retardancy of polymers, many efforts have been made to achieve ultrathin-exfoliated two-dimensional nanosheets to improve the compatibility and dispersion in a polymer matrix. Fortunately, the exfoliated MoS_2 nanosheets can realize a stable suspension in water, which provides a wonderful physical and chemical environment for further modification and application.

To improve the properties of MoS_2 nanosheets and simultaneously realize good dispersion in a polymer matrix, suitable modifications by using inorganic or/and organic materials via electrostatic interactions or chemical bonds are used to enhance the fire resistance of polymers. In detail, MoS_2 nanosheets may act as a physical barrier to retard the transfer of heat, oxygen, and flammable polymerized products and show a catalytic effect on the formation of char residue. The modification of MoS_2 nanosheets using metal oxides, bio-based polymers, and other functional materials may endow the hybrid system with other functions, such as the ability to capture free radicals, smoke suppression effects, and better charring efficiency. These approaches can be applied to enhance the fire safety of polymer composites, polymer coatings, fibres, films, and membranes. Additionally, due to the existence of negative charges on the surface, MoS_2 nanosheets can be used to construct multilayer coatings to improve the fire resistance of polymers by self-assembling positively charged polymer electrolytes and inorganic/organic materials. Similarly, this principle may also be applied in flame retardant polymer composites by preparing MoS_2/inorganic/organic hybrid systems via electrostatic interactions.

There is no doubt that the functionalization and hybridization of MoS_2 nanosheets play a critical role in improving the fire safety and other performance of polymers. However, from the aspect of practical application, the scale-up of MoS_2-based flame retardant polymers still faces challenges. The currently used materials with high costs increase the production burden, which limits further commercialization. Moreover, in addition to the merits of the raw materials, the exfoliation and corresponding functionalization of the MoS_2 layer is still at the laboratory level with a low yield. Hence, to realize the practical application of MoS_2-based flame retardant polymer composites in the future, the focus should be on optimizing the functionalization methods, yield scale-up, and cost reduction.

REFERENCES

Biby, A. H., B. A. Ali and N. K. Allam (2021). "Interplay of quantum capacitance with van der Waals forces, intercalation, co-intercalation, and the number of MoS_2 layers." *Materials Today Energy* **20**: 100677.

Cai, W., J. Wang, Y. Pan, W. Guo, X. Mu, X. Feng, B. Yuan, X. Wang and Y. Hu (2018). "Mussel-inspired functionalization of electrochemically exfoliated graphene: Based on self-polymerization of dopamine and its suppression effect on the fire hazards and smoke toxicity of thermoplastic polyurethane." *Journal of Hazardous Materials* **352**: 57–69.

Cai, W., J. Zhan, X. Feng, B. Yuan, J. Liu, W. Hu and Y. Hu (2017). "Facile construction of flame-retardant-wrapped molybdenum disulfide nanosheets for properties enhancement of thermoplastic polyurethane." *Industrial & Engineering Chemistry Research* **56**(25): 7229–7238.

Cai, W., Y. Zhu, X. Mu, Z. Li, J. Wang, Y. Hu, X. Wang and L. Song (2021). "Heterolayered boron nitride/polyaniline/molybdenum disulfide nanosheets for flame-retardant epoxy resins." *ACS Applied Nano Materials* **4**(8): 8162–8172.

Chen, C., G. Xiao, S. Dong, F. Zhong, Z. Yang, C. Chen, M. Wang and R. Zou (2022). "Construction of CMS@ MoS2 core-shell structure to enhance flame retardancy and smoke suppression performance of waterborne epoxy fireproof coatings for steel structures." *Colloids and Surfaces A: Physicochemical and Engineering Aspects* **633**: 127836.

Cheng, H., N. Dong, T. Bai, Y. Song, J. Wang, Y. Qin, B. Zhang and Y. Chen (2016). "Covalent modification of MoS_2 with poly (N-vinylcarbazole) for solid-state broadband optical limiters." *Chemistry–A European Journal* **22**(13): 4500–4507.

Choi, W., N. Choudhary, G. H. Han, J. Park, D. Akinwande and Y. H. Lee (2017). "Recent development of two-dimensional transition metal dichalcogenides and their applications." *Materials Today* **20**(3): 116–130.

Fan, B.-B., H.-N. Fan, X.-H. Chen, X.-W. Gao, S. Chen, Q.-L. Tang, W.-B. Luo, Y. Deng, A.-P. Hu and W. Hu (2021). "Metallic-state MoS_2 nanosheets with atomic modification for sodium ion batteries with a high rate capability and long lifespan." *ACS Applied Materials & Interfaces* **13**(17): 19894–19903.

Feng, X., W. Cai, X. Wang and Y. Hu (2022). "Hierarchical MoS_2/polyaniline binary hybrids with high performance for improving fire safety of epoxy resin." *Polymers for Advanced Technologies* **33**(1): 163–172.

Feng, X., B. Wang, X. Wang, P. Wen, W. Cai, Y. Hu and K. M. Liew (2016). "Molybdenum disulfide nanosheets as barrier enhancing nanofillers in thermal decomposition of polypropylene composites." *Chemical Engineering Journal* **295**: 278–287.

Feng, X., X. Wang, W. Cai, N. Hong, Y. Hu and K. M. Liew (2016). "Integrated effect of supramolecular self-assembled sandwich-like melamine cyanurate/MoS_2 hybrid sheets on reducing fire hazards of polyamide 6 composites." *Journal of hazardous materials* **320**: 252–264.

Feng, X., X. Wang, W. Xing, K. Zhou, L. Song and Y. Hu (2014). "Liquid-exfoliated MoS_2 by chitosan and enhanced mechanical and thermal properties of chitosan/MoS2 composites." *Composites Science and Technology* **93**: 76–82.

Feng, X., W. Xing, J. Liu, S. Qiu, Y. Hu and K. M. Liew (2016). "Reinforcement of organo-modified molybdenum disulfide nanosheets on the mechanical and thermal properties of polyurethane acrylate films." *Composites Science and Technology* **137**: 188–195.

Feng, X., W. Xing, L. Song and Y. Hu (2014). "In situ synthesis of a MoS_2/CoOOH hybrid by a facile wet chemical method and the catalytic oxidation of CO in epoxy resin during decomposition." *Journal of Materials Chemistry A* **2**(33): 13299–13308.

Feng, X., W. Xing, H. Yang, B. Yuan, L. Song, Y. Hu and K. M. Liew (2015). "High performance poly (ethylene oxide)/molybdenum disulfide nanocomposite films: reinforcement of properties based on the gradient interface effect." *ACS Applied Materials & Interfaces* **7**(24): 13164–13173.

Ganatra, R. and Q. Zhang (2014). "Few-layer MoS_2: a promising layered semiconductor." *ACS Nano* **8**(5): 4074–4099.

Guo, Y., X. Sun, Y. Liu, W. Wang, H. Qiu and J. Gao (2012). "One pot preparation of reduced graphene oxide (RGO) or Au (Ag) nanoparticle-RGO hybrids using chitosan as a reducing and stabilizing agent and their use in methanol electrooxidation." *Carbon* **50**(7): 2513–2523.

Guo, Y., Y. Xue, X. Zuo, L. Zhang, Z. Yang, Y. Zhou, C. Marmorat, S. He and M. Rafailovich (2017). "Capitalizing on the molybdenum disulfide/graphene synergy to produce mechanical enhanced flame retardant ethylene-vinyl acetate composites with low aluminum hydroxide loading." *Polymer Degradation and Stability* **144**: 155–166.

Gurarslan, A., S. Jiao, T. D. Li, G. Li, Y. Yu, Y. Gao, E. Riedo, Z. Xu and L. Cao (2016). "Van der waals force isolation of monolayer MoS_2." *Advanced Materials* **28**(45): 10055–10060.

Han, S., J. Sun, S. He, M. Tang and R. Chai (2019). "The application of graphene-based biomaterials in biomedicine." *American Journal of Translational Research* **11**(6): 3246.

Homa, P., K. Wenelska and E. Mijowska (2020). "Enhanced thermal properties of poly(lactic acid)/MoS$_2$/carbon nanotubes composites." *Scientific Reports* **10**(1): 1–11.

Hong, R. and Q. Chen (2014). "Dispersion of inorganic nanoparticles in polymer matrices: challenges and solutions." *Organic-Inorganic Hybrid Nanomaterials*: 1–38.

Hou, Y., Y. Hu, S. Qiu, L. Liu, W. Xing and W. Hu (2019). "Bi2Se3 decorated recyclable liquid-exfoliated MoS$_2$ nanosheets: towards suppress smoke emission and improve mechanical properties of epoxy resin." *Journal of Hazardous Materials* **364**: 720–732.

Huang, D. and A. Wang (2013). "Non-covalently functionalized multiwalled carbon nanotubes by chitosan and their synergistic reinforcing effects in PVA films." *RSC Advances* **3**(4): 1210–1216.

Huang, J., J. Chu, Z. Wang, J. Zhang, A. Yang, X. Li, C. Gao, H. Huang, X. Wang and Y. Cheng (2019). "Chemisorption of NO$_2$ to MoS$_2$ nanostructures and its effects for MoS$_2$ sensors." *ChemNanoMat* **5**(9): 1123–1130.

Huang, J., X. Deng, H. Wan, F. Chen, Y. Lin, X. Xu, R. Ma and T. Sasaki (2018). "Liquid phase exfoliation of MoS2 assisted by formamide solvothermal treatment and enhanced electrocatalytic activity based on (H$_3$Mo$_{12}$O$_{40}$P/MoS$_2$)n multilayer structure." *ACS Sustainable Chemistry & Engineering* **6**(4): 5227–5237.

Jiang, S.-D., G. Tang, Z.-M. Bai, Y.-Y. Wang, Y. Hu and L. Song (2014). "Surface functionalization of MoS$_2$ with POSS for enhancing thermal, flame-retardant and mechanical properties in PVA composites." *RSC Advances* **4**(7): 3253–3262.

Jiao, Y., A. M. Hafez, D. Cao, A. Mukhopadhyay, Y. Ma and H. Zhu (2018). "Metallic MoS$_2$ for high performance energy storage and energy conversion." *Small* **14**(36): 1800640.

Joensen, P., R. Frindt and S. R. Morrison (1986). "Single-layer MoS$_2$." *Materials Research Bulletin* **21**(4): 457–461.

Kuila, T., A. K. Mishra, P. Khanra, N. H. Kim and J. H. Lee (2013). "Recent advances in the efficient reduction of graphene oxide and its application as energy storage electrode materials." *Nanoscale* **5**(1): 52–71.

Kumar, R., W. Zheng, X. Liu, J. Zhang and M. Kumar (2020). "MoS$_2$-based nanomaterials for room-temperature gas sensors." *Advanced Materials Technologies* **5**(5): 1901062.

Lee, C., H. Yan, L. E. Brus, T. F. Heinz, J. Hone and S. Ryu (2010). "Anomalous lattice vibrations of single-and few-layer MoS$_2$." *ACS Nano* **4**(5): 2695–2700.

Li, H., J. Wu, Z. Yin and H. Zhang (2014). "Preparation and applications of mechanically exfoliated single-layer and multilayer MoS$_2$ and WSe$_2$ nanosheets." *Accounts of Chemical Research* **47**(4): 1067–1075.

Li, H., Q. Zhang, C. C. R. Yap, B. K. Tay, T. H. T. Edwin, A. Olivier and D. Baillargeat (2012). "From bulk to monolayer MoS$_2$: evolution of Raman scattering." *Advanced Functional Materials* **22**(7): 1385–1390.

Li, X. L., T. C. Li, S. Huang, J. Zhang, M. E. Pam and H. Y. Yang (2020). "Controllable synthesis of two-dimensional molybdenum disulfide (MoS$_2$) for energy-storage applications." *ChemSusChem* **13**(6): 1379–1391.

Liu, Y., H. Nan, X. Wu, W. Pan, W. Wang, J. Bai, W. Zhao, L. Sun, X. Wang and Z. Ni (2013). "Layer-by-layer thinning of MoS$_2$ by plasma." *ACS Nano* **7**(5): 4202–4209.

Luo, Z., J. Li, Y. Li, D. Wu, L. Zhang, X. Ren, C. He, Q. Zhang, M. Gu and X. Sun (2022). "Band engineering induced conducting 2H-phase MoS$_2$ by Pd—S—Re sites modification for hydrogen evolution reaction." *Advanced Energy Materials* **12**(12): 2103823.

Mahmoudi, T., Y. Wang and Y.-B. Hahn (2018). "Graphene and its derivatives for solar cells application." *Nano Energy* **47**: 51–65.

Malkappa, K., S. S. Ray and N. Kumar (2019). "Enhanced thermo-mechanical stiffness, thermal stability, and fire retardant performance of surface-modified 2D MoS$_2$ nanosheet-reinforced polyurethane composites." *Macromolecular Materials and Engineering* **304**(1): 1800562.

Manzeli, S., D. Ovchinnikov, D. Pasquier, O. V. Yazyev and A. Kis (2017). "2D transition metal dichalcogenides." *Nature Reviews Materials* **2**(8): 1–15.

Matusinovic, Z., R. Shukla, E. Manias, C. G. Hogshead and C. A. Wilkie (2012). "Polystyrene/molybdenum disulfide and poly(methyl methacrylate)/molybdenum disulfide nanocomposites with enhanced thermal stability." *Polymer Degradation and Stability* **97**(12): 2481–2486.

Pan, H., Q. Shen, Z. Zhang, B. Yu and Y. Lu (2018). "MoS$_2$-filled coating on flexible polyurethane foam via layer-by-layer assembly technique: flame-retardant and smoke suppression properties." *Journal of Materials Science* **53**(12): 9340–9349.

Pang, Y., Z. Yang, Y. Yang and T. L. Ren (2020). "Wearable electronics based on 2D materials for human physiological information detection." *Small* **16**(15): 1901124.

Peng, H., D. Wang and S. Fu (2020). "Tannic acid-assisted green exfoliation and functionalization of MoS$_2$ nanosheets: significantly improve the mechanical and flame-retardant properties of polyacrylonitrile composite fibers." *Chemical Engineering Journal* **384**: 123288.

Peng, H., D. Wang and S. Fu (2021). "Biomimetic construction of highly durable nacre-like MoS$_2$ bio-nanocomposite coatings on polyacrylonitrile textile for intumescent flame retardation and sustainable solar-thermal-electricity conversion." *Composites Part B: Engineering* **215**: 108742.

Peng, H., D. Wang, M. Li, L. Zhang, M. Liu and S. Fu (2019). "Ultra-small SiO$_2$ nanospheres self-pollinated on flower-like MoS$_2$ for simultaneously reinforcing mechanical, thermal and flame-retardant properties of polyacrylonitrile fiber." *Composites Part B: Engineering* **174**: 107037.

Peng, H., L. Zhang, M. Li, M. Liu, C. Wang, D. Wang and S. Fu (2020). "Interfacial growth of 2D bimetallic metal-organic frameworks on MoS2 nanosheet for reinforcements of polyacrylonitrile fiber: from efficient flame-retardant fiber to recyclable photothermal materials." *Chemical Engineering Journal* **397**: 125410.

Perkins, F. K., A. L. Friedman, E. Cobas, P. Campbell, G. Jernigan and B. T. Jonker (2013). "Chemical vapor sensing with monolayer MoS$_2$." *Nano Letters* **13**(2): 668–673.

Phiri, J., P. Gane and T. C. Maloney (2017). "General overview of graphene: production, properties and application in polymer composites." *Materials Science and Engineering: B* **215**: 9–28.

Presolski, S. and M. Pumera (2016). "Covalent functionalization of MoS$_2$." *Materials Today* **19**(3): 140–145

Qian, Y., W. Su, L. Li, H. Fu, J. Li and Y. Zhang (2022). "Synthesis of 3D hollow layered double hydroxide-molybdenum disulfide hybrid materials and their application in flame retardant thermoplastic polyurethane." *Polymers* **14**(8): 1506.

Qiao, Y., X. Li, T. Hirtz, G. Deng, Y. Wei, M. Li, S. Ji, Q. Wu, J. Jian and F. Wu (2019). "Graphene-based wearable sensors." *Nanoscale* **11**(41): 18923–18945.

Qiu, S., Y. Hu, Y. Shi, Y. Hou, Y. Kan, F. Chu, H. Sheng, R. K. Yuen and W. Xing (2018). "In situ growth of polyphosphazene particles on molybdenum disulfide nanosheets for flame retardant and friction application." *Composites Part A: Applied Science and Manufacturing* **114**: 407–417.

Radisavljevic, B., A. Radenovic, J. Brivio, V. Giacometti and A. Kis (2011). "Single-layer MoS$_2$ transistors." *Nature Nanotechnology* **6**(3): 147–150.

Sethulekshmi, A., J. S. Jayan, S. Appukuttan and K. Joseph (2021). "MoS$_2$: advanced nanofiller for reinforcing polymer matrix." *Physica E: Low-Dimensional Systems and Nanostructures* **132**: 114716.

Sharma, A., M. A. Verheijen, L. Wu, S. Karwal, V. Vandalon, H. C. Knoops, R. S. Sundaram, J. P. Hofmann, W. E. Kessels and A. A. Bol (2018). "Low-temperature plasma-enhanced atomic layer deposition of 2-D MoS$_2$: Large area, thickness control and tuneable morphology." *Nanoscale* **10**(18): 8615–8627.

Sui, Y., P. Li, X. Dai and C. Zhang (2021). "Green self-assembly of h-BN@ PDA@ MoS2 nanosheets by polydopamine as fire hazard suppression materials." *Reactive and Functional Polymers* **165**: 104965.

Varrla, E., C. Backes, K. R. Paton, A. Harvey, Z. Gholamvand, J. McCauley and J. N. Coleman (2015). "Large-scale production of size-controlled MoS_2 nanosheets by shear exfoliation." *Chemistry of Materials* **27**(3): 1129–1139.

Wang, D., W. Xing, L. Song and Y. Hu (2016). "Space-confined growth of defect-rich molybdenum disulfide nanosheets within graphene: application in the removal of smoke particles and toxic volatiles." *ACS Applied Materials & Interfaces* **8**(50): 34735–34743.

Wang, D., L. Song, K. Zhou, X. Yu, Y. Hu and J. Wang (2015). "Anomalous nano-barrier effects of ultrathin molybdenum disulfide nanosheets for improving the flame retardance of polymer nanocomposites." *Journal of Materials Chemistry A* **3**(27): 14307–14317.

Wang, D., P. Wen, J. Wang, L. Song and Y. Hu (2017). "The effect of defect-rich molybdenum disulfide nanosheets with phosphorus, nitrogen and silicon elements on mechanical, thermal, and fire behaviors of unsaturated polyester composites." *Chemical Engineering Journal* **313**: 238–249.

Wang, D., K. Zhou, W. Yang, W. Xing, Y. Hu and X. Gong (2013). "Surface modification of graphene with layered molybdenum disulfide and their synergistic reinforcement on reducing fire hazards of epoxy resins." *Industrial & Engineering Chemistry Research* **52**(50): 17882–17890.

Wang, P., X. Fu, Y. Kan, X. Wang and Y. Hu (2019). "Two high-efficient DOPO-based phosphonamidate flame retardants for transparent epoxy resin." *High Performance Polymers* **31**(3): 249–260.

Wang, S., B. Yu, K. Zhou, L. Yin, Y. Zhong and X. Ma (2019). "A novel phosphorus-containing MoS_2 hybrid: towards improving the fire safety of epoxy resin." *Journal of Colloid and Interface Science* **550**: 210–219.

Wang, Y., M. Qiu, M. Won, E. Jung, T. Fan, N. Xie, S.-G. Chi, H. Zhang and J. S. Kim (2019). "Emerging 2D material-based nanocarrier for cancer therapy beyond graphene." *Coordination Chemistry Reviews* **400**: 213041.

Wu, X., K. Gong, G. Zhao, W. Lou, X. Wang and W. Liu (2018). "Surface modification of MoS_2 nanosheets as effective lubricant additives for reducing friction and wear in poly-α-olefin." *Industrial & Engineering Chemistry Research* **57**(23): 8105–8114.

Xiao, G., Z. Yang, C. Chen, C. Chen, F. Zhong, M. Wang and R. Zou (2021). "Novel carbon nitride@ polydopamine/molybdenum disulfide nanoflame retardant improves fire performance of composite coatings." *Colloids and Surfaces A: Physicochemical and Engineering Aspects* **630**: 127575.

Xu, L., X. Tan, R. Xu, J. Xie and C. Lei (2019). "Influence of functionalized molybdenum disulfide (MoS_2) with triazine derivatives on the thermal stability and flame retardancy of intumescent poly(lactic acid) system." *Polymer Composites* **40**(6): 2244–2257.

Xu, T., Z. Zhang and L. Qu (2020). "Graphene-based fibers: recent advances in preparation and application." *Advanced Materials* **32**(5): 1901979.

Xu, W. Z., A. J. Li, Y. C. Liu, R. Chen and W. Li (2018). "$CuMoO4$@hexagonal boron nitride hybrid: an ecofriendly flame retardant for polyurethane elastomer." *Journal of Materials Science* **53**(16): 11265–11279.

Yang, L., A. Mukhopadhyay, Y. Jiao, Q. Yong, L. Chen, Y. Xing, J. Hamel and H. Zhu (2017). "Ultralight, highly thermally insulating and fire resistant aerogel by encapsulating cellulose nanofibers with two-dimensional MoS_2." *Nanoscale* **9**(32): 11452–11462.

Yuan, Y., W. Wang, Y. Shi, L. Song, C. Ma and Y. Hu (2020). "The influence of highly dispersed Cu_2O-anchored MoS_2 hybrids on reducing smoke toxicity and fire hazards for rigid polyurethane foam." *Journal of Hazardous Materials* **382**: 121028.

Zhang, J., W. Lei, J. Schutz, D. Liu, B. Tang, C. H. Wang and X. Wang (2019). "Improving the gas barrier, mechanical and thermal properties of poly(vinyl alcohol) with molybdenum disulfide nanosheets." *Journal of Polymer Science Part B: Polymer Physics* **57**(7): 406–414.

Zhang, X., L. Hou, A. Ciesielski and P. Samorì (2016). "2D materials beyond graphene for high-performance energy storage applications." *Advanced Energy Materials* **6**(23): 1600671.

Zhao, S., J. Yin, K. Zhou, Y. Cheng and B. Yu (2019). "In situ fabrication of molybdenum disulfide based nanohybrids for reducing fire hazards of epoxy." *Composites Part A: Applied Science and Manufacturing* **122**: 77–84.

Zhi, M., Q. Liu, S. Gao, Y. Zhao, X. Zhu and X. Jia (2019). "Layer-by-layer assembled nanocoating containing MoS_2 nanosheets and C60 for enhancing flame retardancy properties of flexible polyurethane foam." *Materials Research Express* **6**(12): 125312.

Zhong, Y., M. Li, L. Zhang, X. Zhang, S. Zhu and W. Wu (2015). "Adding the combination of CNTs and MoS_2 into halogen-free flame retarding TPEE with enhanced the anti-dripping behavior and char forming properties." *Thermochimica Acta* **613**: 87–93.

Zhou, K., R. Gao, Z. Gui and Y. Hu (2017). "The effective reinforcements of functionalized MoS_2 nanosheets in polymer hybrid composites by sol-gel technique." *Composites Part A: Applied Science and Manufacturing* **94**: 1–9.

Zhou, K., R. Gao and X. Qian (2017). "Self-assembly of exfoliated molybdenum disulfide (MoS_2) nanosheets and layered double hydroxide (LDH): towards reducing fire hazards of epoxy." *Journal of Hazardous Materials* **338**: 343–355.

Zhou, K., Y. Hu, J. Liu, Z. Gui, S. Jiang and G. Tang (2016). "Facile preparation of layered double hydroxide/MoS_2/poly(vinyl alcohol) composites." *Materials Chemistry and Physics* **178**: 1–5.

Zhou, K., S. Jiang, C. Bao, L. Song, B. Wang, G. Tang, Y. Hu and Z. Gui (2012). "Preparation of poly(vinyl alcohol) nanocomposites with molybdenum disulfide (MoS_2): structural characteristics and markedly enhanced properties." *RSC Advances* **2**(31): 11695–11703.

Zhou, K., S. Jiang, Y. Shi, J. Liu, B. Wang, Y. Hu and Z. Gui (2014). "Multigram-scale fabrication of organic modified MoS_2 nanosheets dispersed in polystyrene with improved thermal stability, fire resistance, and smoke suppression properties." *RSC Advances* **4**(76): 40170–40180.

Zhou, K., C. Liu and R. Gao (2018). "Polyaniline: a novel bridge to reduce the fire hazards of epoxy composites." *Composites Part A: Applied Science and Manufacturing* **112**: 432–443.

Zhou, K., J. Liu, Z. Gui, Y. Hu and S. Jiang (2016). "The influence of melamine phosphate modified MoS_2 on the thermal and flammability of poly(butylene succinate) composites." *Polymers for Advanced Technologies* **27**(10): 1397–1400.

Zhou, K., J. Liu, Y. Shi, S. Jiang, D. Wang, Y. Hu and Z. Gui (2015). "MoS_2 nanolayers grown on carbon nanotubes: an advanced reinforcement for epoxy composites." *ACS Applied Materials & Interfaces* **7**(11): 6070–6081.

Zhou, K., J. Liu, P. Wen, Y. Hu and Z. Gui (2014). "A noncovalent functionalization approach to improve the dispersibility and properties of polymer/MoS_2 composites." *Applied Surface Science* **316**: 237–244.

Zhou, K., J. Liu, W. Zeng, Y. Hu and Z. Gui (2015). "In situ synthesis, morphology, and fundamental properties of polymer/MoS_2 nanocomposites." *Composites Science and Technology* **107**: 120–128.

Zhou, K., G. Tang, R. Gao and H. Guo (2018). "Constructing hierarchical polymer@ MoS_2 core-shell structures for regulating thermal and fire safety properties of polystyrene nanocomposites." *Composites Part A: Applied Science and Manufacturing* **107**: 144–154.

Zhou, K., G. Tang, R. Gao and S. Jiang (2018). "In situ growth of 0D silica nanospheres on 2D molybdenum disulfide nanosheets: towards reducing fire hazards of epoxy resin." *Journal of Hazardous Materials* **344**: 1078–1089.

Zhou, K., G. Tang, S. Jiang, Z. Gui and Y. Hu (2016). "Combination effect of MoS_2 with aluminum hypophosphite in flame retardant ethylene-vinyl acetate composites." *RSC Advances* **6**(44): 37672–37680.

Zhou, K., W. Yang, G. Tang, B. Wang, S. Jiang, Y. Hu and Z. Gui (2013). "Comparative study on the thermal stability, flame retardancy and smoke suppression properties of polystyrene composites containing molybdenum disulfide and graphene." *RSC Advances* **3**(47): 25030–25040.

Zhou, K., Q. Zhang, J. Liu, B. Wang, S. Jiang, Y. Shi, Y. Hu and Z. Gui (2014). "Synergetic effect of ferrocene and MoS_2 in polystyrene composites with enhanced thermal stability, flame retardant and smoke suppression properties." *RSC Advances* **4**(26): 13205–13214.

Zhou, X., S. Qiu, W. Cai, L. Liu, Y. Hou, W. Wang, L. Song, X. Wang and Y. Hu (2019). "Construction of hierarchical $MoS_2@TiO_2$ structure for the high performance bimaleimide system with excellent fire safety and mechanical properties." *Chemical Engineering Journal* **369**: 451–462.

Zhou, X., S. Qiu, W. Xing, C. S. R. Gangireddy, Z. Gui and Y. Hu (2017). "Hierarchical polyphosphazene@ molybdenum disulfide hybrid structure for enhancing the flame retardancy and mechanical property of epoxy resins." *ACS Applied Materials & Interfaces* **9**(34): 29147–29156.

Zou, B., S. Qiu, Z. Qian, J. Wang, Y. Zhou, Z. Xu, W. Yang and W. Xing (2021). "Phosphorus/nitrogen-codoped molybdenum disulfide/cobalt borate nanostructures for flame-retardant and tribological applications." *ACS Applied Nano Materials* **4**(10): 10495–10504.

10 Black Phosphorus and Its Derivatives as a Novel Class of Flame Retardants for Fire-Safe Polymers

Wei Cai

State Key Laboratory of Fire Science, University of Science and Technology of China, Hefei, Anhui, PR China

Zhaoxin Li

State Key Laboratory of Fire Science, University of Science and Technology of China, Hefei, Anhui, PR China

Hong Kong University of Science and Technology, Kowloon, Hong Kong, China

Yuan Hu

State Key Laboratory of Fire Science, University of Science and Technology of China, Hefei, Anhui, PR China

CONTENTS

DOI: 10.1201/9781003327158-10

10.1 INTRODUCTION: BACKGROUND AND DRIVING FORCES

Due to their high phosphorus content, layered structure, and large specific surface area, black phosphorus (BP) nanosheets have become an emerging nanofiller for enhancing the comprehensive properties of polymer resins. In the initial period of polymer combustion, the layered structure of BP nanosheets is able to present a barrier function to shield heat and pyrolysis products (Qu, Wu, Jiao, et al. 2020). Then, BP nanosheets will be decomposed into phosphorus-containing acids and free radicals, thus presenting gaseous and condensed flame retardancy mechanisms (Cai, Li, Mu, et al. 2021). As a result, the flame retardancy effect of BP nanosheets has been investigated in various polymer nanocomposites, including epoxy, polyurethane, polycarbonate, and polylactic acid (Cai, Li, Mu, et al. 2021, Zou, Qiu, Wang, et al. 2022). However, due to the different interfacial compatibilities and char-forming abilities, the flame retardancy of polymer/BP nanocomposites is correspondingly different. Therefore, it is necessary to comprehensively review the flame retardancy effect of BP nanosheets in different polymer resins, thus providing a reliable reference for further research on the flame retardancy of BP nanosheets.

In addition, as a novel inorganic nanomaterial, there are many challenges that need to be overcome to realize the practical application of BP nanosheets. First, the preparation method of BP crystals usually requires high temperature, high pressure, and a long time. Even though the high-temperature mineralization route provides a relatively easy approach to prepare BP crystals, a high temperature and long heating time are still needed (Lange, Schmidt, and Nilges 2007). Second, the exfoliation efficiency and scale are still at a low level (Ta Na et al. 2018). Recent studies have mainly adopted mechanical ball milling and sonication in the liquid phase to prepare BP nanosheets. High energy consumption and long operation time are unavoidable in the above two methods. Finally, due to the incompatible interface, BP nanosheets are easily restacked into aggregates (Qu, Xu, et al. 2020). Accordingly, the flame retardancy and other effects of BP nanosheets will be suppressed. Therefore, in this chapter, the preparation, exfoliation, and functionalization of BP nanosheets are also reviewed.

10.2 PREPARATION, EXFOLIATION, AND FUNCTIONALIZATION METHOD OF BLACK PHOSPHORUS

10.2.1 PREPARATION OF BLACK PHOSPHORUS CRYSTALS

As early as 1914, bulk BP crystals were prepared by Bridgman through an extremely high-pressure method, which is difficult to generalize in some common laboratories (Brown and Rundqvist 1965, Krebs, Weitz, and Worms 1955). Then, mercury-participating high pressure and bismuth-flux routes are also put forward consecutively (Baba et al. 1989). However, the high reaction pressure, toxic reagents, and complex synthesis process have limited the preparation of BP crystals against the investigation and development of BP-based materials. Until 2007, the emergence of the mechanical ball-milling method and high-temperature mineralization route changed this predicament (Lange, Schmidt, and Nilges 2007, Park and Sohn 2007).

10.2.1.1 Mechanical Ball-Milling Method

As reported, the ball-milling method has been confirmed as an effective approach to convert amorphous red phosphorus into crystallized black phosphorus. During the collision process of ball-mill beads, high temperature and high pressure are spontaneously produced and thus provide the phase transformation condition for red phosphorus to black phosphorus. Meanwhile, the ball-milling method is not limited to a special form, and planetary milling, shake milling, and plasma-assisted ball milling are viable (Zhou et al. 2019). The first application of the ball-milling method was reported by Park et al. in 2007 (Park and Sohn 2007). With the process of high-energy mechanical milling, Zhou et al. found that the characteristic peaks corresponding to BP crystals were presented in the X-ray diffraction (XRD) pattern compared with red phosphorus (Figure 10.1a) (Zhou et al. 2019). However, the visualized peak signals did not present a highly ordered feature, indicating that the amorphous region still existed in the bulk structure of black phosphorus. The high-resolution transmission electron microscope (TEM) photograph of black phosphorus and its electron diffraction pattern further indicate the presence of an amorphous region. As reported by Jaques et al., the color evolution indicates that increasing the milling time can promote the formation of black phosphorus (Figure 10.1b) (Pedersen et al. 2020). Meanwhile, the characteristic peaks corresponding to the BP crystal can be presented in 30 min. However, the red color is still predominant in the mixture of amorphous RP and crystalline BP. In addition, ball milling for more than 4 h is not able to eliminate the characteristic peaks of RP. As the most important evidence to investigate the crystal structure, the XRD patterns of BP crystals fabricated by mechanical ball milling in previous literature all report broadly diffused peaks attributed to amorphous red phosphorus. Therefore, it is speculated that the

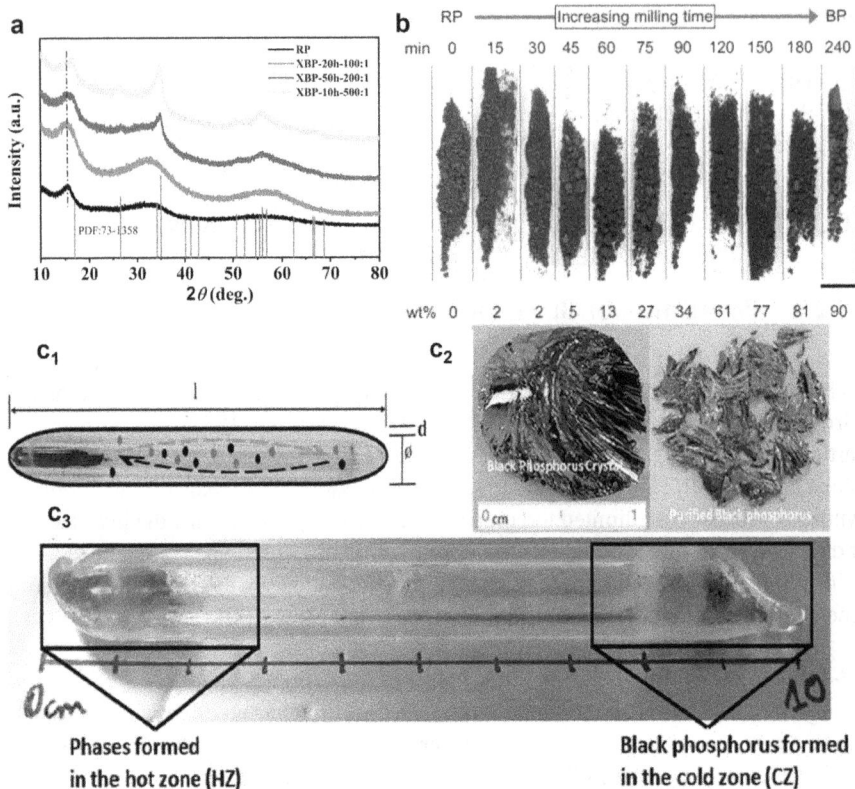

FIGURE 10.1 (a) XRD curves of red phosphorus and black phosphorus with the mechanical ball-milling method (Reproduced with permission from Ref. (Zhou et al. 2019), Copyright (2007) Wiley-VCH); (b) the morphology evolution process of black phosphorus with the mechanical ball-milling method (Reproduced with Open Access permission from Ref. (Pedersen et al. 2020), Copyright (2020) Springer Nature); schematic of high-temperature mineralization route (c₁) and digital photos of black phosphorus crystals (c₂ and c₃) (Reproduced with permission from Ref. (El Manjli et al. 2022), Copyright (2022) Elsevier).

complete formation of black phosphorus from red phosphorus is difficult to realize by the mechanical ball-milling method (Chen et al. 2017).

10.2.1.2   High-Temperature Mineralization Route

Compared to the mechanical ball-milling method, the high-temperature mineralization route has attracted more attention because of its high conversion ratio and good crystallinity. Nilges et al. first reported that black phosphorus can be prepared at a low-pressure condition of 873 K using gold, tin, and tin (IV) iodide as raw materials (Lange, Schmidt, and Nilges 2007). During this reaction, gold, tin, and tin (IV) iodide were used as mineralizers, which promoted crystallization. However, the employment of gold significantly increases the production cost of black phosphorus. Meanwhile, Au and Sn atoms also react with the P element to form

Au_3SnP_7, $AuSn$, and Sn_4P_3, mixing into the crystal structure of bulk BP. Therefore, Nilges et al. further optimized the mineralization route by using Sn and SnI_4 to replace the Au-containing mineralizer (Koepf et al. 2014). Presented by an in situ neutron diffraction study, a BP crystal is produced in the gas phase, and the whole formation process only takes several minutes. Based on the high yield and good crystal structure, more researchers have been devoted to the development of a high-temperature mineralization route. As presented in Figure 10.1c, El Manjli et al. improved the mineralization route, and the prepared BP presents good crystal luster and large crystalline grains (~10 mm), different from the powder state demonstrated by the ball-milling method (El Manjli et al. 2022). Zhao et al. sealed a mixture of red phosphorus, tins, and stannic iodides into a vacuum tube composed of silica glass (Zhao, Qian, et al. 2016). After being heated to 650 °C for 2 h, well BP ribbons are obtained. Due to its high crystal quality and high purity as well as convenient craft, the high-temperature mineralization process has great potential for large-scale production. It is believed that high-quality BP crystals can provide high electronic mobility. However, there is still a long way to go to realize this purpose. Recently, the high-temperature mineralization process has mainly been used on a laboratory scale to produce a few grams of BP crystals. Pilot plant test experiments and optimal conditions need to be further explored.

10.2.1.3   Others

In fact, in addition to the above two methods, the high-pressure method and bismuth-flux route were reported earlier. In 1914, a high-pressure method was proposed to prepare BP crystals in 1.2 GPa and 200 °C. However, such high-pressure conditions are difficult to realize in some common laboratories. Therefore, even though the feasibility of the preparation of BP crystals has been confirmed, this high-pressure method has not been further expanded. Then, mercury was used to decrease the reaction pressure by Worms et al. at a reaction temperature of 470 °C (Krebs, Weitz, and Worms 1955). There is no doubt that highly toxic mercury would cause a huge risk to research workers. White phosphorus can be converted to black phosphorus during the cooling process of a mixture of white phosphorus and bismuth previously heated to 400 °C (Brown and Rundqvist 1965). The formed BP crystal presents a needle- or rod-like structure. The bismuth-flux route effectively avoids the requirement of high pressure and toxic mercury. However, the separation and purification of BP crystals from solid-state bismuth presents a new challenge.

10.2.2   EXFOLIATION

Due to the accumulation of van der Waals forces among the multilayer, the electron mobility, phonon vibration, and other properties within the BP crystal are limited, thus influencing the application performance. Therefore, the exfoliation of BP nanosheets has become a hot and important topic to realize its huge potential.

10.2.2.1   Ball-Milling Exfoliation

Among the mechanical exfoliation methods, ball-milling exfoliation is capable of providing a strong mechanical force to overcome the interlaminar van der Waals

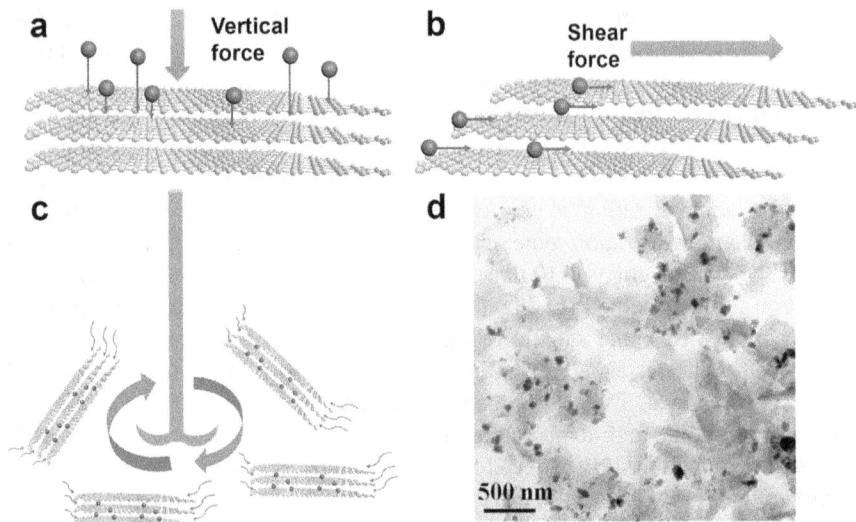

FIGURE 10.2 (a and b) Schematic for the vertical and shear forces in the ball-milling process; (c) schematic for sonication in the liquid phase; (d) TEM of BP nanosheets exfoliated by sonication in the liquid phase (Reproduced with permission from Ref. (Late 2016), Copyright (2016) Elsevier).

force, presenting the advantages of high yield and easy crafting. However, not all ball-milling methods can be used as exfoliation methods. For instance, the vertical force provided by the high-energy ball-milling method only continuously has a strong impact on the vertical direction of bulk BP, thus decreasing the scale of crystalline grains to produce fragmentation of BP crystals (Figure 10.2a). However, the interlaminar van der Waals force is not eliminated by the vertical force. Therefore, the high-energy ball-milling method is not applicable for the exfoliation of BP nanosheets. In general, research workers mainly adopt the planctary ball mill to prepare BP nanosheets, which can provide shear force to peel the bulk BP layer by layer (Figure 10.2b). The shear force impacts the bulk BP along the layer structure and pushes the movement of the BP sheets. During movement, the forming force effectively overcame the van der Waals attraction between adjacent BP flakes. In addition to the ball-milling method, micromechanical cleavage realized by Scotch tape also obtains high-quality BP nanosheets. However, the time-consuming process limits the further development of this method.

10.2.2.2 Liquid-Phase Exfoliation

Liquid-phase exfoliation methods are mainly based on sonication or shear-induced forces. During the sonication process, the fluctuating pressure will continuously cause the formation and collapse of nano/micrometer-sized bubbles or voids (Figure 10.2c). Therefore, the force caused by the collapse of bubbles or voids located in the interlamination effectively overcame the van der Waals attraction between adjacent BP flakes. In addition, to further promote the exfoliation of BP nanosheets, shear force is usually employed. However, without the assistance of sonication, a high shear force is required to exfoliate BP nanosheets. Meanwhile, during liquid-phase exfoliation, the appropriate solvents and surfactants are extremely important to disperse the BP nanosheets, avoiding the restacking process (Figure 10.2d) (Late 2016). The ideal solvents and surfactants should have a matched surface energy to BP nanosheets, thus decreasing the contact area of dispersed BP nanosheets.

10.2.2.3   Others

Recently, few-layer BP nanosheets have been obtained by the supercritical method. As reported by Yan et al., BP powder was first added to N-methylpyrrolidone (NMP) solution and ultrasonicated for 1 h (Yan et al. 2016). Then, the BP/NMP solution is sealed into a stainless steel reactor filled with CO_2 gas. The pressure of the reactor is up to 15 MPa, thus reaching supercritical conditions. Despite the high liquidity, supercritical CO_2 is capable of penetrating the BP interlayer. Once the high pressure is withdrawn, supercritical CO_2 located in the BP interlayer will be rapidly converted into gaseous CO_2 and expand the interlayer spacing to exfoliate BP nanosheets. In addition to ball-milling exfoliation and liquid-phase exfoliation, the supercritical exfoliation method undoubtedly puts forward higher operating equipment requirements. Therefore, supercritical exfoliation is not widely applied in the preparation of BP nanosheets. An ice-promoted exfoliation method was also reported by Ma et al. to prepare BP nanosheets (Zhang et al. 2019). Bulk BP powder was first mixed with NMP solution and cooled in a liquid nitrogen bath. As the melting point of NMP is −24 °C, the dispersion is frozen, and the interlayer spacing of BP crystals will be strutted by NMP ice crystals, largely reducing the interlayer van der Waals interactions. Subsequently, the frozen dispersion undergoes ultrasonication, and the BP nanosheets are successfully exfoliated from the bulk BP.

10.2.3   FUNCTIONALIZATION

The direct application of BP nanosheets in the flame retardancy field of polymer nanocomposites suffers from many restrictions, such as weak interfacial compatibility with the polymer matrix, high smoke production, and serious restacking phenomena. Surface functionalization can impart or improve the targeted properties to BP nanosheets, thus playing a vital role in promoting practical applications. In addition, surface functionalization has also been demonstrated to be advantageous in the effective passivation and ambient stability of BP.

10.2.3.1   Functionalization Based on the Coordination Bond

Each P atom within the BP crystal bonds with three P atoms to form a P-P bond in two planes, thus presenting puckered layers. The P atom (s^2p^3) has five electrons in the outermost layer. Therefore, the bonding mode of the BP crystal will produce a lone electron pair. Recently, the empty orbital in transition metal ions has been used to interact with the lone pair electrons of P atoms, thus realizing surface functionalization. For example, Yu and Chu et al. reported that different BP-based nanostructures, including quantum dots, nanosheets, and microflakes lanthanide

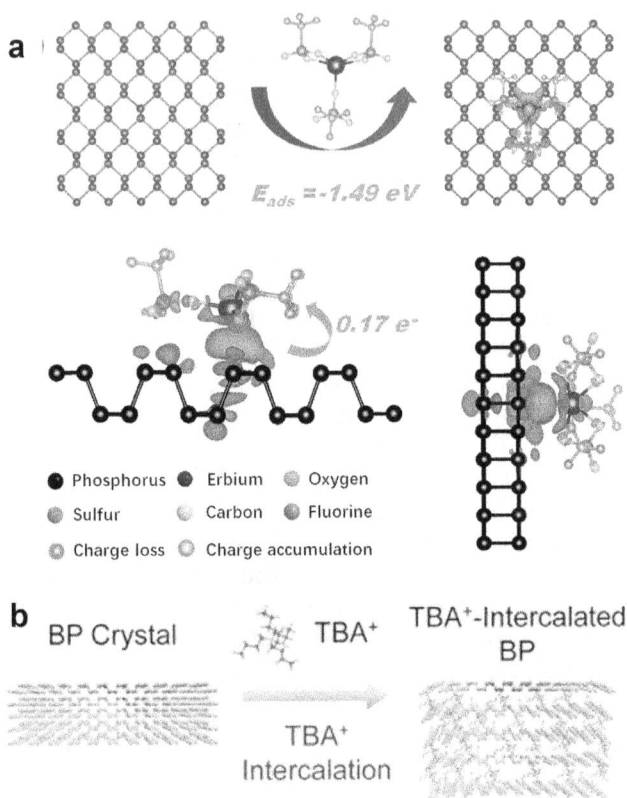

FIGURE 10.3 Schematic for functionalization based on the coordination bond (a) and electrostatic interactions (b). (Reproduced with permission from Ref.(Qu, Xu, et al. 2020), Copyright (2020) Elsevier; Ref. (Yang et al. 2021), Copyright (2021) Elsevier).

(Ln), have been modified by sulfonate complexes and titanium sulfonate ligands (Wu et al. 2018, Zhao, Wang, et al. 2016, Sun et al. 2017). The oxidation process of BP nanosheets is prevented through the modification of LnL$_3$ by using Ln-P coordination to occupy the lone-pair electrons of phosphorus, further exhibiting excellent stability in both air and water (Figure 10.3a) (Qu, Xu, et al. 2020). To further demonstrate the bonding mechanism between P and metal ions, first principle calculations (FPCs) were used to describe the theoretical models based on density functional theory (Jiang, Jin, and Gui 2020). Gui et al. found that the adsorption energies of metal ions on a BP surface via the bonding mode are much lower than those of metal ions on a BP surface via the nonbonding mode (Jiang, Jin, and Gui 2020). Meanwhile, it was demonstrated that the adsorption energy of Zn^{2+} (−2.04 eV) on BP is lower than that of other metal ions. In addition to Zn^{2+} ions, a series of metal ions, including Ag^+, Fe^{3+}, Mg^{2+}, and Hg^{2+}, have also been confirmed to be spontaneously adsorbed on the BP surface via lone pair electron–empty orbital interactions to render BP more stable in air (Guo et al. 2017). Among the functionalization methods of BP nanosheets, coordination modification is relatively simple and easy to operate, only with sonication in the liquid phase. Therefore, in some works, the exfoliation and coordination functionalization of BP nanosheets is carried out simultaneously (Jia et al. 2019).

10.2.3.2   Functionalization Based on Electron Transfer

10.2.3.2.1   Diazonium Chemistry Modification

Due to the presence of lone pair electrons, electron transfer easily occurs between BP nanosheets and electron-defect groups. Therefore, Hersam et al. first utilized the electron-defect diazonium group to modify BP nanosheets, with the formation of P-C bonds (Ryder et al. 2016). The lone pair electron of phosphorus first interacts with the positively charged nitrogen within the diazonium group of 4-nitrobenzene-diazonium. During this process, an electron is transferred to the diazonium group, thus forming a nitrogen molecule that is removed from this reaction system. Then, the positive charge is transferred to the phosphorus of BP nanosheets, and the aryl radical is formed accordingly. Finally, this free radical is introduced to the electron orbit of phosphorus, thus forming a P-C bond.

10.2.3.2.2   Alklyl Halide Modification

In addition to diazonium chemistry, the nucleophilic substitution of hydroxyl groups with alkyl halides is also effective for the covalent functionalization of BP nanosheets (Sofer et al. 2017). The lone pair electrons in the oxygen and phosphorus atoms of BP nanosheets attack the center of the positive charge in alkyl halides. The transition chemical bond is first formed between carbon and phosphorus/oxygen. Then, P-C bonds could be formed via the elimination of hydrogen iodide.

10.2.3.3   Functionalization Based on Electrostatic Interactions

Similar to other layered nanomaterials, such as graphene, MoS_2, and montmorillonite (MMT), black phosphorus nanosheets also present surface characteristics of negative charge. Due to its unique superiority, the electrostatic interaction-driven self-assembly process is frequently used to prepare BP-based nanohybrids without the destruction of crystal quality. For example, tetra-n-butyl-ammonium (TBA) ions can interact with BP nanosheets with electrostatic interactions (Figure 10.3b) (Yang et al. 2021). Through the modification of TBA ions, BP nanosheets are positively charged and can further interact with negatively charged MXene. As a result, a novel heterostructure composed of black phosphorus and MXene is presented. Some positively charged ions are also used to functionalize the surface of BP nanosheets, including 1-butyl-3-methylimidazolium tetrafluoroborate, 1-butyl-3-methylimidazolium hexa-fluorophosphate, 1-butyl-3-methylimidazolium tetrafluoroborate, and so on (Walia et al. 2017, Abellan et al. 2017).

10.2.3.4   Functionalization Based on the In Situ Hydrothermal/ Solvothermal

The presence of lone pair electrons and phosphoric acid groups located on the surface of BP nanosheets is able to adsorb the kinds of metal ions. Therefore, with the assistance of the hydrothermal/solvothermal process, some metal-based nanomaterials can be formed in situ onto the surface of BP nanosheets. As reported by Liu et al., Ag nanoparticles are stabilized on the surface of BP nanosheets by a simple hydrothermal reaction (Lei et al. 2016). Based on first-principles calculations, the interfacial interaction at the Ag/BP interface is demonstrated by the formation of a covalent P-Ag bond. In addition, Zhu et al. also reported the synthesis of Ag-loaded BP nanohybrids with a solvothermal reaction at 165 °C. Compared with the TEM photograph in the research work of Liu et al., the use of polyvinyl pyrrolidone (PVP) effectively promoted the in situ formation of Ag nanoparticles. However, due to the high temperature and water-containing environment, the degradation process of BP nanosheets easily occurs. Therefore, to maintain good crystal quality, hydrothermal/solvothermal applications should be considered prudently.

10.2.3.5 Functionalization Based on the Mechanical Force

Recently, the high-energy mechanical ball-milling method has been used to introduce novel atoms and nanoparticles to the surface of BP nanosheets. For instance, black phosphorus, TiO_2, and expandable graphite were milled for 5 h with high-energy shake milling equipment in an argon atmosphere at different ratios (Zhou, Ouyang, et al. 2020). The new characteristic peaks near 532.6 eV and 531.3 eV are attributed to C-O-P and Ti-O-P bonds, indicating that black phosphorus, TiO_2, and expandable graphite are covalently bonded. Compared to other methods, the high-energy mechanical ball-milling method is easier to operate, without the consumption of organic solvents. In addition, Kim et al. also confirmed the stable formation of P-O-C bonds between BP nanosheets and multiwall carbon nanotubes with the ball-milling method (Haghighat-Shishavan et al. 2018). These novel chemical bonds strongly confirm that the ball-milling method can promote the formation of black phosphorus-based heterostructures with covalent bonds.

10.3 THE DISPERSION STATE OF BLACK PHOSPHORUS NANOSHEETS

As a nanofiller, the dispersion state and interfacial interaction of BP nanosheets within a polymer matrix are of significant importance to improve the performance of polymer nanocomposites. Therefore, some tests, including XRD, scanning electron microscope (SEM), TEM, and torque rheometer, have been developed to investigate the dispersion state and interfacial interaction. For example, as reported by our group, the fracture surface of polyvinyl alcohol (PVA)/BP nanocomposites is observed by SEM photographs (Qiu, Zhou, et al. 2020). Increasing the addition amount of BP nanosheets promotes the formation of rough morphologies, indicating strong interfacial interactions (Figure 10.4a) (Cai, Cai, et al. 2020). However, the results from SEM photographs need some subjective judgments. The direct data provided by the torque rheometer test can present a quantitative composite viscosity, revealing the interfacial interaction. As presented by Hu et al., modification with tannic acid (TA) can significantly increase the composite viscosity of thermoplastic polyurethane (TPU)/BP nanocomposites due to the regulation of the hydroxyl group (Figure 10.4b) (Cai, Cai, et al. 2020). Meanwhile, the peak value of

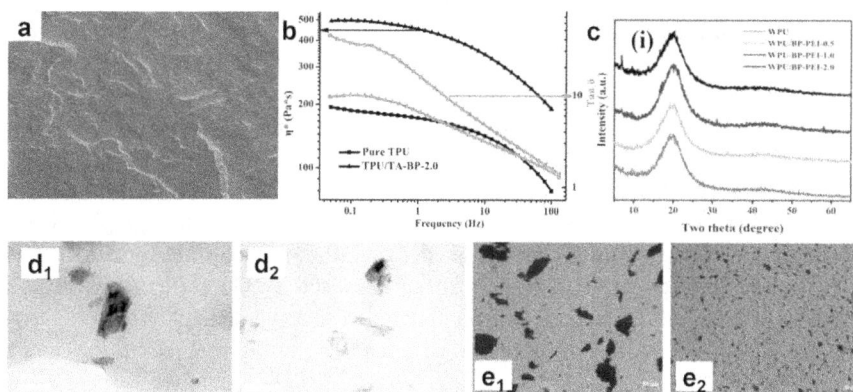

FIGURE 10.4 (a) SEM photograph of fracture surface of PU/BP nanocomposites (reproduced with permission from Ref. (Cai, Cai, et al. 2020), Copyright (2020) Elsevier); (b) the result of composite viscosity and tan δ of PU/BP nanocomposites (reproduced with permission from Ref. (Cai, Cai, et al. 2020), Copyright (2020) Elsevier); (c) XRD curves of waterborne polyurethane (WPU)/BP nanocomposites (reproduced with permission from Ref. (He et al. 2020), Copyright (2020) Elsevier); TEM photographs of cross section of WPU/pure BP nanocomposite (d_1) and WPU/functionalized BP nanocomposite (d_2) (reproduced with permission from Ref. (He et al. 2020a), Copyright (2020) Elsevier); optical images of EP/pure BP nanocomposites (e_1) and EP/functionalized BP nanocomposites (e_2) (reproduced with permission from Ref. (Qu, Wu, Jiao, et al. 2020), Copyright (2020) Elsevier).

tan δ moves to 0.13 Hz from 0.20 Hz, indicating that the molecular chains of TPU/TA-BP-2.0 need more relaxation time.

The above two characteristic methods only present information on interfacial interactions, without the investigation of the dispersion state. Therefore, XRD curves and TEM photographs are also further suggested to determine the dispersion state. As one kind of two-dimensional nanomaterial, the restacking behavior of BP nanosheets within the polymer matrix should obviously increase the intensity of the diffraction peak in the XRD results. However, recent literature does not employ this method to study the dispersion state of BP nanosheets. Meanwhile, direct TEM photographs of the ultrathin cross-section of polymer nanocomposites have not been reported widely. There are more than thirty titles reporting polymer/BP nanocomposites. However, the presence of TEM photographs of the ultrathin cross-section is less than five titles. Reported by He et al., pure BP nanosheets present bulk and thick structures within a polymer matrix (He et al. 2020). The characteristic peaks corresponding to BP nanosheets are also demonstrated in the XRD curves of the polymer/BP nanocomposites (Figure 10.4c). After being modified by polyethyleneimine (PEI), a more uniform dispersion is presented in the TEM photograph, and the characteristic peaks nearly disappear in the XRD curves. Compared to pure BP nanosheets, a good dispersion state of PEI-BP nanosheets is clearly presented in TEM photographs of the ultrathin cross-section (Figures $10.4d_1$ and d_2).

In addition, Wu and his co-operators employed optical microscope images to analyze the dispersion state of BP nanosheets (Qu, Wu, Jiao, et al. 2020). However, as presented in Figure 10.4e, the scale of BP nanosheets is at the micron scale, and the restacked BP nanosheets are more than 100 μm (Figure 10.4e$_1$). Although the functionalization of melamine-formaldehyde improves the dispersion state, the grain size of BP nanosheets is still larger than 1.0 μm. Meanwhile, BP nanosheets of 10 μm are also presented (Figure 10.4e$_2$). In another work, restacked BP nanosheets of more than 1.0 mm were reported by Qu and Wu et al. (Qu, Wu, Meng, et al. 2020). In sharp contrast, the scale of BP nanosheets in the TEM photograph is lower than 1.0 μm. In other words, the well-dispersed BP nanosheets are most likely the restacked BP structure. The restacked structure raised by Qu et al should be bulk and unexfoliated BP crystals. It can be concluded that the dispersion state of two-dimensional nanomaterials at the micro/nanoscale is not represented by the optical microscope images.

10.4   FLAME RETARDANCY APPLICATION OF BP NANOSHEETS

The flame retardancy effect of BP nanosheets was first reported by Mei et al. (Ren et al. 2018). Due to the high phosphorus content and layered structure, BP nanosheets have presented a significant effect on the flame retardancy improvement of polymer nanocomposites. However, despite the high flame retardancy, the high requirement for the preparation craft and equipment has seriously suppressed the development of BP nanosheets in flame retardant polymer nanocomposites. Recently, only a few research groups have reported the flame retardancy effect of BP nanosheets. In fact, the limitation of the flame retardancy application of BP nanosheets is the commercial cost rather than the technical challenge. Of course, the research value of BP nanosheets in flame retardancy applications still deserves to be explored.

10.4.1   EPOXY RESIN

Compared to other polymer materials, the preparation and molding of epoxy resin is much easier, without the requirement of high pressure. Meanwhile, epoxy is a char-forming polymer material. Therefore, the addition of BP nanosheets as an acid source can significantly improve the flame retardancy of epoxy nanocomposites. For example, Zou et al. reported that pure EP resin presents a peak heat release rate (PHRR) of 2082.4 kW/m^2 (Zou et al. 2020). Incorporated 2.0 wt% pure BP nanosheets decreased the PHRR to ~800 kW/m^2. However, in view of the PHRR value, it is found that the modification of multiwalled carbon nanotubes (MCNTs) is not capable of further decreasing the PHRR of EP nanocomposites (Figure 10.5a). In addition, the suppression effect of pure BP and MCNT-BP on the total smoke and heat release is extremely similar. This result may be due to the high intrinsic flame retardancy of BP nanosheets. The introduction of other components makes it difficult to further enhance the flame retardancy of BP nanosheets. A similar result was also reported in the modification of reduced graphene oxide (RGO) to BP nanosheets (Zhou, Chu, et al. 2020). Through the functionalization of carbon nitride, aryl diazonium, phenylboronic acid, triazine-based covalent organic frameworks (COFs), and so on, the flame retardancy effect of BP nanosheets is further

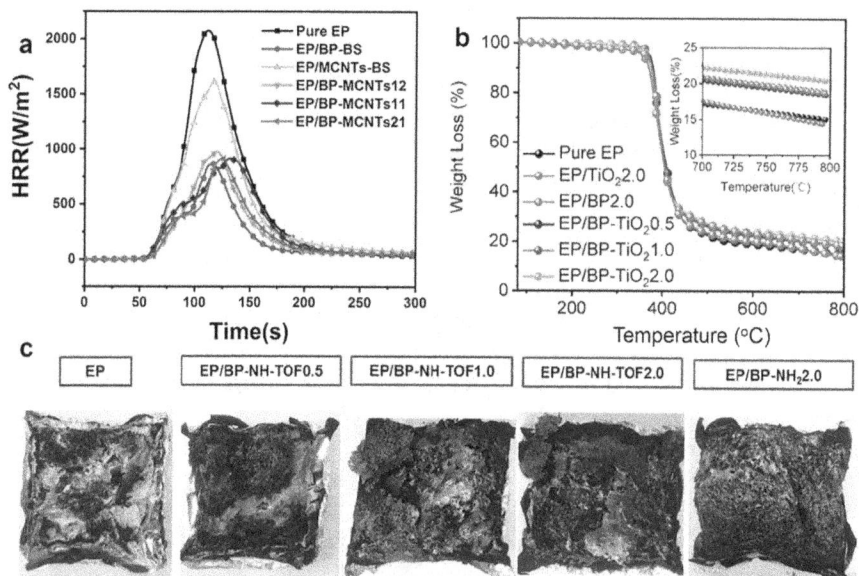

FIGURE 10.5 (a) HRR curves of EP and EP/MCNT-BP nanocomposites (reproduced with permission from Ref. (Zou et al. 2020), Copyright (2020) Elsevier); (b) Thermogravimetric Analysis (TGA) curves of EP and EP/TiO$_2$-BP nanocomposites (reproduced with permission from Ref. (Zou, Qiu, Wang, et al. 2022), Copyright (2022) Elsevier); (c) digital photos of char residue of EP and EP/BP nanocomposites (reproduced with permission from Ref. (Qiu, Zou, et al. 2020), Copyright (2020) Elsevier).

enhanced. To better compare these results, the PHRR and total heat release (THR) results of the EP/BP nanocomposites are collected and presented in Table 10.1. Obviously, the functionalized BP nanosheets can significantly decrease the PHRR of EP resin near or more than 50.0%. Meanwhile, the modification of TiO$_2$ and GO to BP nanosheets causes a reduction of more than 50.0% in the THR of EP resin (Zou, Qiu, Wang, et al. 2022, Zhou, Chu, et al. 2020). The heat release is attributed to the combustion of organic matter, i.e., epoxy resin. A reduction of more than 50.0% in THR indicates that less than 50.0% of the epoxy resin is completely burned. In other words, more char residue and smoke particles are produced. As presented in Figure 10.5b, the char residues of pure EP and EP/BP-TiO$_2$-2.0 are ~6.10 wt% and ~20.0 wt%, respectively (Zou, Qiu, Wang, et al. 2022). Although the introduction of TiO$_2$ effectively promotes char formation, the increased char residue is not able to match the reduction of burned organic matter. Lamentedly, Zou and Qiu et al did not provide the result of smoke production (Zou, Qiu, Wang, et al. 2022). However, as reported by Zhou et al., the THR and total smoke production (TSP) decreased by 54.4% and 28.5%, respectively (Zhou, Chu, et al. 2020). Obviously, the reduction in burned EP is not converted into smoke particles. Based on this phenomenon, it is believed that the char residue is responsible for the reduction in the burned EP. From the photographs of char residue, the increase in char residue does not match the decrease in THR and TSP (Figure 10.5c). Similar

TABLE 10.1

PHRR and THR Data of EP/BP Nanocomposites

Modifier Agent	Addition Amount (wt%)	Decreased PHRR (%)	Decreased THR (%)	References
TiO$_2$	2.0	58.9	50.3	(Zou, Qiu, Wang, et al. 2022)
Carbon nitride	2.0	47.72	49.6	(Ren et al. 2021)
Ferrocene oligomer	2.0	62.2	58.5	(Yang et al. 2022)
Aryl diazonium	2.0	47.0	48.1	(Wang et al. 2022)
MCNTs	2.0	55.8	41.7	(Zou et al. 2020)
GO	2.0	55.2	54.4	(Zhou, Chu, et al. 2020)
Lanthanide metal ligand	3.0	61.37	36.68	(Qu, Xu, et al. 2020)
Cobalt-based borate	2.0	48.4	41.7	(Zhou et al. 2022)
Triazine based COF	2.0	61.2	44.3	(Qiu, Zou, et al. 2020)
Hindered amine	2.0	56.0	48.2	(Qiu, Hou, et al. 2021)
MCA	2.0	47.2	42.3	(Qiu, Zhou, et al. 2021)

phenomena are also presented in the modification of melamine cyanurate (MCA), hindered amine, triazine-based COF, and so on (Qiu, Zhou, et al. 2021, Qiu, Hou, et al. 2021, Qiu, Zou, et al. 2020). Therefore, further investigation is suggested to trace the evolution of burned organic matter.

10.4.2 POLYURETHANE

Behind epoxy resin, the flame retardancy application of BP nanosheets mainly focuses on polyurethane resins, such as TPU and WPU. Due to their different molecular structures, the char-forming abilities of TPU and WPU are distinct. It is difficult to produce char residue from polyether PU during combustion. Meanwhile, due to the hydrolysis reaction, the resistance to the moist environment of polyester PU is undesirable. Therefore, waterborne polyurethane is composed of polyether groups, and polyether TPU is preferential for application in moist environments. Due to the weak char formation ability, pure WPU and polyether TPU left a spot of char residue (Figure 10.6a). The addition of BP nanosheets can promote the formation of an extremely thin char layer, as presented in Figure 10.6b. However, even with the functionalization of BN, graphene, PEI, and MXene, the char formation of WPU is not significantly promoted (He et al. 2020). The promotion effect of BP nanosheets on char formation is extremely slight. As reported by Hu and Mei et al., the suppression effect of BP nanosheets on the heat release of WPU is significant (Table 10.2) (He et al. 2020, Yin et al. 2020). There is a huge challenge that water exists in WPU/BP nanocomposites. It is found that there is still much water, as implied by the tensile curves (Figure 10.6c). The existence of water in BP nanosheets will expedite the degradation process. Therefore, the stability of WPU/BP nanocomposites in the environment is worthy of further investigation.

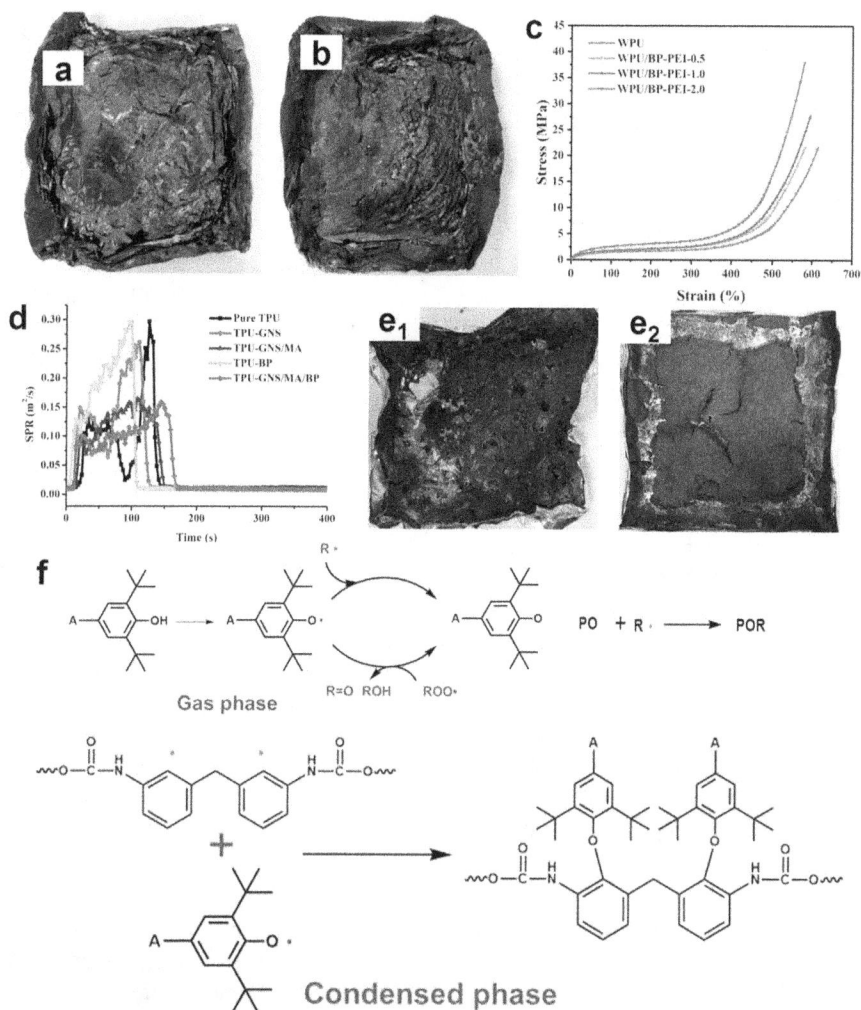

FIGURE 10.6 Digital photos of char residue of pure WPU (a) and WPU/PEI-BP nano-composites (b) (reproduced with permission from Ref. (He et al. 2020a), Copyright (2020) Elsevier); (c) tensile curves of WPU/PEI-BP nanocomposites (reproduced with permission from Ref. (He et al. 2020a), Copyright (2020) Elsevier); (d) smoke production rate curves of TPU/GNS-BP nanocomposites (reproduced with permission from Ref. (Cai, Li, Mu, et al. 2021), Copyright (2020) Elsevier); digital photos of char residue of TPU/BP nanocomposite (e$_1$) and TPU/GNS-BP nanocomposite (e$_2$) (reproduced with permission from Ref. (Cai, Li, Mu, et al. 2021), Copyright (2020) Elsevier); (f) flame retardancy mechanism of hindered phenol and BP nanosheets (reproduced with permission from Ref. (Qiu et al. 2022), Copyright (2022) Elsevier).

The 2.0 wt% addition of pure BP nanosheets decreases the PHRR of WPU from 452.5 kW/m^2 to 252.7 kW/m^2, demonstrating a reduction of 44.2% (Yin et al. 2020). Based on the slight increase in char residue, the decrease in heat release is

TABLE 10.2

PHRR and THR Data of WPU/BP Nanocomposites

Modifier	Addition Amount (wt%)	Decreased PHRR (%)	Decreased THR (%)	References
Boron nitride	0.4	50.94	23.92	(Yin et al. 2020)
Graphene	3.35	48.18	38.63	(Ren et al. 2019)
PEI	2.0	34.3	21.2	(He et al. 2020)
MoS$_2$	2.0	50.3	23.3	(He, Chu, et al. 2022)
MXene	2.0	42.2	12.9	(He, Jiang, et al. 2022)

likely attributed to the formation of smoke particles. According to the research works of Mei and his co-operators, the smoke release behavior of WPU/BP nanocomposites has not been investigated (Ren et al. 2018, Ren et al. 2019, Yin et al. 2020). Fortunately, He et al. reported the smoke release behavior of WPU/BP nanosheets in their research work (He, Jiang, et al. 2022). TSP was increased with the addition of 0.5 wt% and 1.0 wt% BP-Ti$_3$C$_2$ nanohybrids, indicating that the gaseous flame retardant mechanism of BP nanosheets will promote the incomplete combustion of WPU. However, in another work, the smoke production release of WPU/BP nanocomposites was suppressed, while the THR was only decreased by 23.3% (He, Chu, et al. 2022). The suppressed smoke release behavior is attributed to the catalytic effect of MoS$_2$, which promotes the crosslinking of pyrolysis products and the formation of char residue (He, Chu, et al. 2022). Based on the above results, it may be concluded that the THR and TSP did not simultaneously decrease without a significant increase in char residue.

As mentioned above, the flame retardancy of BP nanosheets in polyester and polyether PU is different, attributed to the char-forming ability of PU resin. The incorporation of pure BP nanosheets in polyether PU causes reductions of 47.7% and 36.9% in PHRR and THR, respectively, while total smoke release (TSR) is increased by 44.4% (Cai, Li, Mu, et al. 2021). As presented in Figure 10.6d, the smoke production process in TPU/BP nanocomposites is obviously promoted and accelerated. To address this problem, Hu et al. introduced thermally stable graphene to change the flame retardancy mechanism (Cai, Li, Mu, et al. 2021). The pyrolysis products from BP nanosheets are kept within the condensed phase to form a dense char layer (Figure 10.6e). Compared to the broken char residue in TPU/BP, the introduction of graphene nanosheets enhances the continuity and compactness of the char residue of TPU/BP-GNS (Figure 10.6e$_1$ and 10.6e$_2$). However, even with the improvement of the condensed-phase flame retardancy mechanism, the TSR of TPU/BP-GNS is still higher than that of pure polyether PU. This result indicates that the gaseous flame retardancy mechanism of BP nanosheets is difficult to avoid.

To further improve the flame retardancy of BP nanosheets in polyester TPU, hindered phenol, tannic acid, ionic liquid, and isocyanate are used to modify the surface of BP nanosheets (Cai, Cai, et al. 2020, Cai, Hu, et al. 2020, Qiu et al. 2022, Cai, Li, Liu, et al. 2021). As a result, the interfacial interaction between the BP

TABLE 10.3

PHRR and THR Data of TPU/BP Nanocomposites

Resin	Modifier	Additive Amount (wt%)	Decreased PHRR (%)	Decreased THR (%)	References
Polyester	Ionic liquid	2.0	38.2	19.7	(Cai, Hu, et al. 2020)
Polyester	Hindered phenol	2.0	49.9	49.0	(Qiu et al. 2022)
Polyester	Tannic acid	2.0	56.5	43.0	(Cai, Cai, et al. 2020)
Polyester	Isocyanate	2.0	61.6	56.7	(Cai, Li, Liu, et al. 2021)
Polyether	Graphene	2.0	54.7	23.5	(Cai, Li, Mu, et al. 2021)

nanosheets and PU resin is effectively enhanced, thus promoting the dispersion state. Compared to the obvious restacked structure of pure BP nanosheets, a similar bulk structure is not found in SEM photographs of the fracture surface of TPU nanocomposites (Cai, Cai, et al. 2020). With the aid of the torque rheometer test, the composite viscosity between tannic-modified BP nanosheets and PU resin is investigated. It is found that the composite viscosity is significantly improved by comparison with pure PU resin. The enhanced interfacial interaction is attributed to the good dispersion state. Therefore, tannic-modified BP nanosheets present a better flame retardancy effect in polyester TPU. The HRR value of TPU/TA-BP-2.0 is 562 kW/m^2, which is lower than that of TPU/BP-2.0 (640 kW/m^2). In addition to the better dispersion state, the gaseous flame retardancy effect of hindered phenol and tannic acid is capable of further suppressing the heat release. During the combustion process, hindered phenol and tannic acid will be converted into benzene free radicals and phenoxy radicals, quenching the free radicals of the combustion chain (Figure 10.6f) (Cai, Cai, et al. 2020, Qiu et al. 2022). Further comparison of the flame retardancy in PU resin of BP nanosheets is presented in Table 10.3, including WPU, polyester PU, and polyether PU.

10.4.3 OTHER POLYMER RESINS

10.4.3.1 Polyvinyl Alcohol

In addition to epoxy and polyurethane, there are a few research works reporting the flame retardancy of BP nanosheets in PVA, polycarbonate (PC), and cellulose. Inspired by mussel chemistry, polydopamine (PDA) was synthesized to encapsulate BP nanosheets by an oxidation–reduction strategy, thus isolating oxygen (Qiu, Zhou, et al. 2020). In addition, few-layer BP can strongly interact with the PVA matrix. After being ignited, no char residue was left for pure PVA. With the addition of PDA-BP nanosheets, more char residue was left in the PVA/BP nanocomposites (Figure 10.7a$_1$ and 10.7a$_2$). This result indicated that PDA-BP catalyzes the dehydration carbonization process of PVA resin (Qiu, Zhou, et al. 2020). Li et al. developed a sucrose-assisted ball-milling method to exfoliate and functionalize BP nanosheets (Guo et al. 2022). During the high-energy impact process, oxygen-containing functional groups were introduced to the BP nanosheets, thus

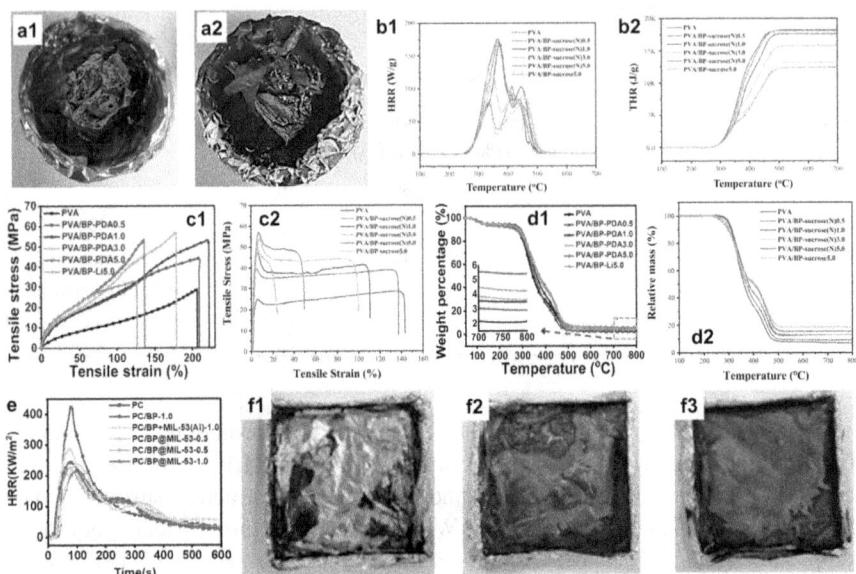

FIGURE 10.7 Digital photos of pure PVA (a_1) and PVA/PDA-BP (a_2) nanocomposites (reproduced with permission from Ref. (Qiu, Zhou, et al. 2020), Copyright (2020) Elsevier); HRR (b_1) and THR (b_2) curves of PVA/sucrose-BP nanocomposites (reproduced with permission from Ref. (Guo et al. 2022), Copyright (2022) Elsevier); tensile curves of PVA/PDA-BP (c_1) and PVA/sucrose-BP (c_2) nanocomposites (reproduced with permission from Ref. (Guo et al. 2022), Copyright (2020) Elsevier and Ref. (Qiu, Zhou, et al. 2020), Copyright (2020) Elsevier); TGA test of PVA/PDA-BP (d_1) and PVA/sucrose-BP (d_2) nanocomposites (reproduced with permission from Ref. (Qiu, Zhou, et al. 2020), Copyright (2020) Elsevier and Ref. (Guo et al. 2022), Copyright (2022) Elsevier); (e) HRR curves of PC/BP-MIL nanocomposites (reproduced with permission from Ref. (Qian et al. 2022), Copyright (2022) Elsevier); (f) digital photos of pure PLA (f_1), PLA/pure BP (f_2), and PLA/CS-BP nanocomposites (f_3) (reproduced with permission from Ref. (Zhou, Huang, et al. 2020), Copyright (2020) Elsevier).

strongly interacting with the PVA resin. The PHRR and THR of the PVA nanocomposite decreased by 52.5% and 32.8%, respectively, with the addition of 5.0 wt% BP-sucrose (Figure 10.7b_1 and 10.7b_2). The flame retardancy of BP nanosheets in PVA resin is still unknown. In the above two works, an interesting phenomenon is noticeable, presenting obviously different tensile curves (Figure 10.7c_1 and 10.7c_2). This result is due to the existence of water, as reported by Qiu, Zhou, et al. (2020). Meanwhile, a significant mass loss before 150 °C is also presented, attributed to water loss. The TGA results strongly confirmed the existence of water (Figure 10.7d_1 and 10.7d_2). Therefore, the long-term stability of PVA/BP nanocomposites is questionable.

10.4.3.2   Polycarbonate

As an important transparent optical material, polycarbonate resin has been widely applied in windows, lenses, screens, and other fields. However, the addition of BP nanosheets will seriously decrease the transmittance of polycarbonate resin due to the high absorption of visible light. Therefore, the application fields of PC/BP nanocomposites are limited. However, due to the high flame retardancy, there are still some research works reporting the combustion behavior of PC/BP nano-composites. For example, Qian and Zou et al. introduced BP nanosheets into PC resin to improve fire safety performance (Zou, Qiu, Zhou, et al. 2022). With a facile ball-milling method, BP powder and boron (B) nanoflakes are mixed uniformly and impacted by vertical force (Zou, Qiu, Zhou, et al. 2022, Qian et al. 2022). The strong impact successfully introduces B nanoflakes within the struc-ture of BP nanosheets. Incorporated 1.0 wt% BP-B nanohybrids can significantly suppress the combustion behavior of PC nanocomposites. The time corresponding to the PHRR of PC/BP-B-1.0 is delayed to ~230 s, much longer than those of PC/B-1.0 and PC/BP-1.0. Meanwhile, a reduction ratio of more than 60% is also presented in the PHRR value of PC/BP-B-1.0. Compared to the UL94 V2 rating of pure PC resin, incorporating 1.0 wt% BP-B nanohybrids enabled the PC na-nocomposite to pass the UL94 V0 rating successfully. The limit oxygen index (LOI) value of the corresponding sample also increased to 33.0 from 25.0 (pure PC). In addition, with the modification of a metal organic framework (MOF), a novel BP-based nanohybrid was also prepared by employing the typical hydro-thermal reaction (Qian et al. 2022). Qian et al. demonstrated that the PHRR and THR of PC nanocomposites were decreased by 49.4% and 19.3%, respectively (Figure 10.7e). The introduction of Al-based MOFs can present a synergistic effect to promote the formation of protective char with BP nanosheets. Despite the high flame retardancy effect, the significantly decreased transmittance has become a huge challenge for the application of PC/BP nanocomposites.

10.4.3.3   Polylactic Acid

Compared to char-forming polymer resin, the flame retardancy of polylactic acid (PLA) with low char formation ability is hard to improve by the incorporated BP nanosheets. Due to the presence of lone pair electrons and phosphoric acid groups, BP nanosheets are negatively charged. Therefore, cationic cetyl-trimethyl ammo-nium bromide (CTAB) can interact with BP nanosheets reported by Zhou and Hu et al. (Zhou, Huang, et al. 2020). As we know, this long aliphatic hydrocarbon chain in CTAB presents a nonpolar characteristic, which is adverse to improving the interfacial interaction of BP nanosheets with the polar PLA chain. Even though the introduced CTAB-BP effectively decreased the PHRR of PLA nanocomposites, the result of the LOI test is still undesirable. It was found that the LOI value slightly increased to 23.0 from 21.0. In addition, char residue isn't left nearly (Figure 10.7f). In other words, the flame retardancy of BP nanosheets in non-charring polymers requires the synergistic effect of the carbon source.

10.5   FLAME RETARDANCY MECHANISM OF BLACK PHOSPHORUS

10.5.1   Gaseous and Condensed Flame Retardancy Mechanism

As mentioned above, BP nanosheets will be pyrolyzed at temperatures greater than 500 °C and release phosphorus-containing free radicals and acids. Therefore, gaseous and condensed flame retardancy mechanisms simultaneously exist in BP nanosheets. At an early stage of the combustion process, thermally stable BP nanosheets with a layered structure can present an effective barrier function to suppress the delivery of heat and pyrolysis products. However, with the combustion process, the temperature imposed on BP nanosheets is continuously increased to more than 500 °C. As a result, BP nanosheets will decompose thermally. The released P-containing compounds can react with the pyrolysis products of polymers, i.e., water or oxygen, thus producing phosphorus-containing acids. It has been confirmed that phosphorus-containing acids not only shield fire and oxygen but also present a dehydration effect on polymer resin (Liu et al. 2015). Therefore, phosphorus-containing acids effectively promote the formation of a protective char layer. Of course, there is a precondition that the polymer resin has good char-forming ability. In addition, the released P-containing free radicals can interact with the free radicals of combustion chains, thus quenching the chain growth reaction (Braun and Schartel 2004). It can be concluded that there are three flame retardancy mechanisms for BP nanosheets during the combustion process, including barrier function, dehydration effect, and free radical quenching.

10.5.2 THE RELATION BETWEEN FUNCTIONALIZATION AND THE FLAME RETARDANCY MECHANISM

In view of the thermal stability of P-P bonds, the barrier function of BP nanosheets disappears during the combustion process. The functionalization methods could not influence the barrier function. However, the pyrolysis products of BP nanosheets need to react with water or oxygen molecules to produce metaphosphoric acid (HPO_3) and phosphorus pentoxide (P_2O_5), thus presenting a dehydration effect. Other P-containing acids only present a shield effect to fire and pyrolysis products. Therefore, to further enhance the formation of HPO_3 and P_2O_5, metal hydroxides have been employed to enhance the condensed flame retardancy mechanism of red phosphorus (Liu et al. 2015, Liu et al. 2018, Braun and Schartel 2004). It is believed that this research work can be referenced in BP nanosheets. The produced P_2O_5 and P-containing free radicals are volatilized into the gaseous phase, thus releasing more smoke particles. To suppress the escape phenomenon of pyrolysis products of BP nanosheets and polymer resin, thermally stable graphene nanosheets were introduced into BP-based nanohybrids by Hu et al. (Cai, Li, Mu, et al. 2021). It was found that the results of smoke production and char residue were obviously changed. The remarkable release peak corresponding to smoke production in the TPU/BP/GNS system disappeared, while a continuous and dense char layer was formed. It can be concluded that the functionalization methods can influence the kinds of pyrolysis products of BP nanosheets and confine the pyrolysis products within the condensed phase.

10.6 MECHANICAL PROPERTIES

In general, the significant difference in tensile modulus between BP nanosheets and polymer resins contributes to the improvement of mechanical properties. In addition, the improvement effect will be influenced by many factors, including the dispersion state, interfacial interaction, and intrinsic tensile modulus of the polymer resin. For example, the addition of pure BP nanosheets to TPU resin can significantly increase the tensile strength to ~25 MPa from 18 MPa at an elongation of 600%. The increased scale is close to 38.9%. However, incorporating pure BP nanosheets did not lead to an effective improvement in the mechanical performance of PLA (Zhou, Huang, et al. 2020). This result may be attributed to the high tensile modulus of PLA. A similar phenomenon is also observed in epoxy resin. Therefore, it can be concluded that the high tensile modulus of the polymer resin will weaken the enhancement effect of BP nanosheets on the mechanical properties. In addition, the interfacial interaction has also become the key factor for improving mechanical properties. Surface functionalization is usually employed to enhance interfacial interactions, including organic and inorganic modification. For instance, polar ionic liquid, tannic acid, COF, and metal organic frame (MOF) have been used to modify the surface of BP nanosheets (Cai, Cai, et al. 2020, Cai, Hu, et al. 2020, Qiu et al. 2022). Compared to pure TPU (29.0 MPa) and TPU/pure BP nanocomposites (17.9 MPa), the break strength of TPU/TA-BP-2.0 is effectively increased to 37.8 MPa (Cai, Hu, et al. 2020). As presented in the SEM photographs, a serious restacking phenomenon occurs in the TPU/pure BP nanocomposites. With the modification of polymerized ionic liquid, BP nanosheets are uniformly dispersed. Therefore, to improve the mechanical properties of polymer/BP nanocomposites, surface functionalization of BP nanosheets is vital.

10.7   CONCLUSION

The emerging BP nanosheets successfully integrate the respective advantages of inorganic two-dimensional nanomaterials and organophosphorus flame retardants in enhancing the fire safety of polymer nanocomposites. Due to their large specific surface area and high Young's modulus, incorporated BP nanosheets effectively improve the mechanical performance of polymer nanocomposites. Meanwhile, the layer structure presents a barrier function to hinder the delivery of mass and heat in the fire. With increasing temperature, BP nanosheets will be pyrolyzed into phosphorous compounds, such as metaphosphoric acid and phosphorus-oxygen radicals. The metaphosphoric acid contributes to the dehydration carbonization process of polymer resin, thus promoting the formation of a protective char layer. In addition, the presence of phosphoric acid also hinders the delivery process of mass and heat. The phosphorus-oxygen radicals are able to eliminate the chain growth radicals, playing a gaseous flame retardancy mechanism. Therefore, an increasing number of studies are devoted to developing the potential of BP nanosheets, thus achieving simultaneous improvement in mechanical performance and flame retardancy. However, their weak stability in moist environments and high smoke toxicity have seriously limited the application of BP nanosheets in polymer nanocomposites, especially for water-based polymers. The isolation of water and oxygen is necessary for the long-term use of BP nanosheets. The gaseous flame retardancy mechanism of BP nanosheets will unavoidably promote the formation of smoke particles. The introduction of thermally stable nanomaterials may be an effective approach to solve this problem, restricting the diffusion of smoke particles into the gaseous phase and promoting the production of char residue. According to research reports on the functionalization of BP nanosheets, a uniform dispersion state can be achieved. The real issue in the application of BP nanosheets is the extremely high preparation cost. The high cost has limited the wide application of BP-based flame retardants in common polymer nanocomposites. Meanwhile, the flame retardancy effect of BP nanosheets does not present a significant advantage to red phosphorus. Therefore, the commercial value of BP nanosheets needs to be further considered. Developing a low-cost method for the preparation of BP nanosheets has become a primary issue for the practical application of polymer/BP nanocomposites.

ACKNOWLEDGEMENT

The authors acknowledge financial supports from the National Natural Science Foundation of China (22205228), Fellowship of China Postdoctoral Science Foundation (2021M703054 and 2022T150613).

REFERENCES

Abellan, Gonzalo, Stefan Wild, Vicent Lloret, Nils Scheuschner, Roland Gillen, Udo Mundloch, Janina Maultzsch, Maria Varela, Frank Hauke, and Andreas Hirsch. 2017. "Fundamental insights into the degradation and stabilization of thin layer black phosphorus." *Journal of the American Chemical Society* no. 139 (30):10432–10440. doi: 10.1021/jacs.7b04971.

Baba, Mamoru, Fukunori Izumida, Yuji Takeda, and Akira Morita. 1989. "Preparation of black phosphorus single-crystals by a completely closed bismuth-flux method and their crystal morphology." *Japanese Journal of Applied Physics Part 1-Regular Papers Short Notes & Review Papers* no. 28 (6):1019–1022. doi: Doi 10.1143/Jjap.28.1019.

Braun, Ulrike, and Bernhard Schartel. 2004. "Flame retardant mechanisms of red phosphorus and magnesium hydroxide in high impact polystyrene." *Macromolecular Chemistry and Physics* no. 205 (16):2185–2196.

Brown, Alan, and Stig Rundqvist. 1965. "Refinement of crystal structure of black phosphorous." *Acta Crystallographica* no. 19:684-+. doi: 10.1107/s0365110x65004140.

Cai, Wei, Tongmin Cai, Lingxin He, Fukai Chu, Xiaowei Mu, Longfei Han, Yuan Hu, Bibo Wang, and Weizhao Hu. 2020. "Natural antioxidant functionalization for fabricating ambient-stable black phosphorus nanosheets toward enhancing flame retardancy and toxic gases suppression of polyurethane." *Journal of Hazardous Materials* no. 387. doi: 10.1016/j.jhazmat.2019.121971.

Cai, Wei, Yixin Hu, Ying Pan, Xia Zhou, Fukai Chu, Longfei Han, Xiaowei Mu, Zeyuan Zhuang, Xin Wang, and Weiyi Xing. 2020. "Self-assembly followed by radical polymerization of ionic liquid for interfacial engineering of black phosphorus nanosheets: Enhancing flame retardancy, toxic gas suppression and mechanical performance of polyurethane." *Journal of Colloid and Interface Science* no. 561:32–45. doi: 10.1016/j.jcis.2019.11.114.

Cai, Wei, Zhaoxin Li, Jiajia Liu, Shuilai Qiu, Ying Pan, Zhoumei Xu, Chao Ma, and Yuan Hu. 2021. "Recyclable and removable functionalization based on Diels-Alder reaction of black phosphorous nanosheets and its dehydration carbonization in fire safety improvement of polymer composites." *Composites Part a-Applied Science and Manufacturing* no. 140. doi: 10.1016/j.compositesa.2020.106157.

Cai, Wei, Zhaoxin Li, Xiaowei Mu, Lingxin He, Xia Zhou, Wenwen Guo, Lei Song, and Yuan Hu. 2021. "Barrier function of graphene for suppressing the smoke toxicity of polymer/black phosphorous nanocomposites with mechanism change." *Journal of Hazardous Materials* no. 404. doi: 10.1016/j.jhazmat.2020.124106.

Chen, Zhiyan, Yabo Zhu, Jia Lei, Wanying Liu, Yunke Xu, and Peizhong Feng. 2017. "A stage-by-stage phase-induction and nucleation of black phosphorus from red phosphorus under low-pressure mineralization." *Crystengcomm* no. 19 (47):7207–7212. doi: 10.1039/c7ce01492a.

El Manjli, Faissal, Omar Mounkachi, Ghassane Tiouitchi, Khadija El Maalam, Abdelilah Benyoussef, Frederic Boschini, Abdelfattah Mahmoud, Mohammed Hamedoun, and Mustapha Ait Ali. 2022. "From amorphous red phosphorus to black phosphorus crystal: An optimization, controllable and highest yield synthesis process." *Journal of Crystal Growth* no. 577. doi: 10.1016/j.jcrysgro.2021.126408.

Guo, Jiachen, Liu Yang, Ling Zhang, and Chunzhong Li. 2022. "Simultaneous exfoliation and functionalization of black phosphorus by sucrose-assisted ball milling with NMP intercalating and preparation of flame retardant polyvinyl alcohol film." *Polymer* no. 255. doi: 10.1016/j.polymer.2022.125036.

Guo, Zhinan, Si Chen, Zhongzheng Wang, Zhenyu Yang, Fei Liu, Yanhua Xu, Jiahong Wang, Ya Yi, Han Zhang, and Lei Liao. 2017. "Metal-ion-modified black phosphorus with enhanced stability and transistor performance." *Advanced Materials* no. 29 (42):1703811.

Haghighat-Shishavan, Safa, Masoud Nazarian-Samani, Mahboobeh Nazarian-Samani, Ha-Kyung Roh, Kyung-Yoon Chung, Byung-Won Cho, Seyed Farshid Kashani-Bozorg, and Kwang-Bum Kim. 2018. "Strong, persistent superficial oxidation-assisted chemical bonding of black phosphorus with multiwall carbon nanotubes for high-capacity ultradurable storage of lithium and sodium." *Journal of Materials Chemistry A* no. 6 (21):10121–10134. doi: 10.1039/c8ta02590h.

He, Lingxin, Fukai Chu, Xia Zhou, Lei Song, and Yuan Hu. 2022. "Cactus-like structure of BP@MoS2 hybrids: An effective mechanical reinforcement and flame retardant additive for waterborne polyurethane." *Polymer Degradation and Stability* no. 202. doi: 10.1016/j.polymdegradstab.2022.110027.

He, Lingxin, Xin Jiang, Xia Zhou, Zhaoxin Li, Fukai Chu, Xin Wang, Wei Cai, Lei Song, and Yuan Hu. 2022. "Integration of black phosphorene and MXene to improve fire safety and mechanical properties of waterborne polyurethane." *Applied Surface Science* no. 581. doi: 10.1016/j.apsusc.2021.152386.

He, Lingxin, Xia Zhou, Wei Cai, Yuling Xiao, Fukai Chu, Xiaowei Mu, Xianling Fu, Yuan Hu, and Lei Song. 2020. "Electrochemical exfoliation and functionalization of black phosphorene to enhance mechanical properties and flame retardancy of waterborne polyurethane." *Composites Part B-Engineering* no. 202. doi: 10.1016/j.compositesb.2020.108446.

Jia, Caihong, Lei Zhao, Mengsi Cui, Feng Yang, Gang Cheng, Guanghong Yang, and Zaiping Zeng. 2019. "Surface coordination modification and electrical properties of few-layer black phosphorus exfoliated by the liquid-phase method." *Journal of Alloys and Compounds* no. 799:99–107. doi: 10.1016/j.jallcom.2019.05.346.

Jiang, Xiaowen, Hui Jin, and Rijun Gui. 2020. "Emerging metal ion-coordinated black phosphorus nanosheets and black phosphorus quantum dots with excellent stabilities." *Dalton Transactions* no. 49 (34):11911–11920. doi: 10.1039/d0dt02272a.

Koepf, Marianne, Nadine Eckstein, Daniela Pfister, Carolin Grotz, Ilona Krueger, Magnus Greiwe, Thomas Hansen, Holger Kohlmann, and Tom Nilges. 2014. "Access and in situ growth of phosphorene-precursor black phosphorus." *Journal of Crystal Growth* no. 405:6–10. doi: 10.1016/j.jcrysgro.2014.07.029.

Krebs, H., H. Weitz, and K. H. Worms. 1955. "UBER DIE STRUKTUR UND EIGENS-CHAFTEN DER HALBMETALLE.8. DIE KATALYTISCHE DARSTELLUNG DES SCHWARZEN PHOSPHORS." *Zeitschrift Fur Anorganische Und Allgemeine Chemie* no. 280 (1–3):119–133. doi: 10.1002/zaac.19552800110.

Lange, Stefan, Peer Schmidt, and Tom Nilges. 2007. "Au3SnP7@ black phosphorus: An easy access to black phosphorus." *Inorganic chemistry* no. 46 (10):4028–4035.

Late, Dattatray J. 2016. "Liquid exfoliation of black phosphorus nanosheets and its application as humidity sensor." *Microporous and Mesoporous Materials* no. 225:494–503. doi: 10.1016/j.micromeso.2016.01.031.

Lei, Wanying, Tingting Zhang, Ping Liu, Jose A. Rodriguez, Gang Liu, and Minghua Liu. 2016. "Bandgap- and local field-dependent photoactivity of Ag/black phosphorus nanohybrids." *ACS Catalysis* no. 6 (12):8009–8020. doi: 10.1021/acscatal.6b02520.

Liu, Jichun, Yingbin Guo, Haibo Chang, Hang Li, Airong Xu, and Bingli Pan. 2018. "Interaction between magnesium hydroxide and microencapsulated red phosphorus in flame-retarded high-impact polystyrene composite." *Fire and Materials* no. 42 (8):958–966. doi: 10.1002/fam.2650.

Liu, Jichun, Shuge Peng, Yanbin Zhang, Haibo Chang, Zhuoli Yu, Bingli Pan, Chang Lu, Junying Ma, and Qingshan Niu. 2015. "Influence of microencapsulated red phosphorus on the flame retardancy of high impact polystyrene/magnesium hydroxide composite and its mode of action." *Polymer Degradation and Stability* no. 121:208–221. doi: 10.1016/j.polymdegradstab.2015.09.011.

Park, Cheol-Min, and Hun-Joon Sohn. 2007. "Black phosphorus and its composite for lithium rechargeable batteries." *Advanced Materials* no. 19 (18):2465-+. doi: 10.1002/adma.200602592.

Pedersen, Samuel V., Florent Muramutsa, Joshua D. Wood, Chad Husko, David Estrada, and Brian J. Jaques. 2020. "Mechanochemical conversion kinetics of red to black phosphorus and scaling parameters for high volume synthesis." *Npj 2D Materials and Applications* no. 4 (1). doi: 10.1038/s41699-020-00170-4.

Qian, Ziyan, Bin Zou, Yuling Xiao, Shuilai Qiu, Zhoumei Xu, Yuting Yang, Guangyong Jiang, Zixuan Zhang, Lei Song, and Yuan Hu. 2022. "Targeted modification of black phosphorus by MIL-53(Al) inspired by "Cannikin's Law" to achieve high thermal stability of flame retardant polycarbonate at ultra-low additions." *Composites Part B-Engineering* no. 238. doi: 10.1016/j.compositesb.2022.109943.

Qiu, Shuilai, Yanbei Hou, Xia Zhou, Yifan Zhou, Jingwen Wang, Bin Zou, Wenhao Yang, Lei Song, and Yuan Hu. 2021. "Light stabilizer and diazo passivation of black phosphorus nanosheets: Covalent functionalization endows air stability and flame retadancy enhancements." *Chemical Engineering Journal* no. 425. doi: 10.1016/j.cej.2021.131532.

Qiu, Shuilai, Jing Liang, Yanbei Hou, Xia Zhou, Yifan Zhou, Jingwen Wang, Bin Zou, Weiyi Xing, and Yuan Hu. 2022. "Hindered phenolic antioxidant passivation of black phosphorus affords air stability and free radical quenching." *Journal of Colloid and Interface Science* no. 606:1395–1409. doi: 10.1016/j.jcis.2021.08.098.

Qiu, Shuilai, Yifan Zhou, Xiyun Ren, Bin Zou, Wenwen Guo, Lei Song, and Yuan Hu. 2020. "Construction of hierarchical functionalized black phosphorus with polydopamine: A novel strategy for enhancing flame retardancy and mechanical properties of polyvinyl alcohol." *Chemical Engineering Journal* no. 402. doi: 10.1016/j.cej.2020.126212.

Qiu, Shuilai, Yifan Zhou, Weiyi Xing, Xiyun Ren, Bin Zou, and Yuan Hu. 2021. "Conceptually Novel Few-Layer Black Phosphorus/Supramolecular Coalition: Noncovalent Functionalization Toward Fire Safety Enhancement." *Industrial & Engineering Chemistry Research* no. 60 (34):12579–12591. doi: 10.1021/acs.iecr.1c02313.

Qiu, Shuilai, Bin Zou, Tao Zhang, Xiyun Ren, Bin Yu, Yifan Zhou, Yongchun Kan, and Yuan Hu. 2020. "Integrated effect of NH2-functionalized/triazine based covalent organic framework black phosphorus on reducing fire hazards of epoxy nanocomposites." *Chemical Engineering Journal* no. 401. doi: 10.1016/j.cej.2020.126058.

Qu, Zhencai, Kun Wu, Enxiang Jiao, Weilong Chen, Zhuorong Hu, Changan Xu, Jun Shi, Shan Wang, and Zhiyou Tan. 2020. "Surface functionalization of few-layer black phosphorene and its flame retardancy in epoxy resin." *Chemical Engineering Journal* no. 382. doi: 10.1016/j.cej.2019.122991.

Qu, Zhencai, Kun Wu, Weihua Meng, Bingfei Nan, Zhuorong Hu, Chang-an Xu, Zhiyou Tan, Qian Zhang, Huifa Meng, and Jun Shi. 2020. "Surface coordination of black phosphorene for excellent retardancy and thermal conductivity in epoxy resin stability, flame." *Chemical Engineering Journal* no. 397. doi: 10.1016/j.cej.2020.125416.

Qu, Zhencai, Chang-an Xu, Zhuorong Hu, Yue Li, Huifa Meng, Zhiyou Tan, Jun Shi, and Kun Wu. 2020. "(CF3SO3) Er-decorated black phosphorene for robust ambient stability and excellent flame retardancy in epoxy resin." *Composites Part B-Engineering* no. 202. doi: 10.1016/j.compositesb.2020.108440.

Ren, Xinlin, Yi Mei, Peichao Lian, Delong Xie, Weibin Deng, Yaling Wen, and Yong Luo. 2019. "Fabrication and application of black phosphorene/graphene composite material as a flame retardant." *Polymers* no. 11 (2). doi: 10.3390/polym11020193.

Ren, Xinlin, Yi Mei, Peichao Lian, Delong Xie, Yunyan Yang, Yongzhao Wang, and Zirui Wang. 2018. "A novel application of phosphorene as a flame retardant." *Polymers* no. 10 (3). doi: 10.3390/polym10030227.

Ren, Xiyun, Bin Zou, Yifan Zhou, Zhixin Zhao, Shuilai Qiu, and Lei Song. 2021. "Construction of few-layered black phosphorus/graphite-like carbon nitride binary hybrid nanostructure for reducing the fire hazards of epoxy resin." *Journal of Colloid and Interface Science* no. 586:692–707. doi: 10.1016/j.jcis.2020.10.139.

Ryder, Christopher R., Joshua D. Wood, Spencer A. Wells, Yang Yang, Deep Jariwala, Tobin J. Marks, George C. Schatz, and Mark C. Hersam. 2016. "Covalent functionalization and passivation of exfoliated black phosphorus via aryl diazonium chemistry." *Nature Chemistry* no. 8 (6):597–602. doi: 10.1038/nchem.2505.

Sofer, Zdenek, Jan Luxa, Daniel Bousa, David Sedmidubsky, Petr Lazar, Tomas Hartman, Hilde Hardtdegen, and Martin Pumera. 2017. "The Covalent Functionalization of Layered Black Phosphorus by Nucleophilic Reagents." *Angewandte Chemie-International Edition* no. 56 (33):9891–9896. doi: 10.1002/anie.201705722.

Sun, Zhengbo, Yuetao Zhao, Zhibin Li, Haodong Cui, Yayan Zhou, Weihao Li, Wei Tao, Han Zhang, Huaiyu Wang, Paul K. Chu, and Xue-Feng Yu. 2017. "TiL4-coordinated black phosphorus quantum dots as an efficient contrast agent for in vivo photoacoustic imaging of cancer." *Small* no. 13 (11). doi: 10.1002/smll.201602896.

Ta Na, Bao, O. Tegus Hasichaolu, Ning Jun, and Narengerile. 2018. "Preparation of Black Phosphorus by the Mechanical Ball Milling Method and its Characterization." *Solid State Phenomena* no. 271:18–22. doi: 10.4028/www.scientific.net/SSP.271.18.

Walia, Sumeet, Sivacarendran Balendhran, Taimur Ahmed, Mandeep Singh, Christopher El-Badawi, Mathew D. Brennan, Pabudi Weerathunge, Md Nurul Karim, Fahmida Rahman, Andrea Rassell, Jonathan Duckworth, Rajesh Ramanathan, Gavin E. Collis, Charlene J. Lobo, Milos Toth, Jimmy Christopher Kotsakidis, Bent Weber, Michael Fuhrer, Jose M. Dominguez-Vera, Michelle J. S. Spencer, Igor Aharonovich, Sharath Sriram, Madhu Bhaskaran, and Vipul Bansal. 2017. "Ambient protection of few-layer black phosphorus via sequestration of reactive oxygen species." *Advanced Materials* no. 29 (27). doi: 10.1002/adma.201700152.

Wang, Jingwen, Shuilai Qiu, Liang Cheng, Weijian Chen, Yifan Zhou, Bin Zou, Longfei Han, Zhoumei Xu, Wenhao Yang, Yuan Hu, and Lei Song. 2022. "Synergistic effects of aryl diazonium modified Few-Layer black Phosphorus/Ultrafine rare earth yttrium oxide with enhancing flame retardancy and catalytic smoke toxicity suppression of epoxy resin." *Applied Surface Science* no. 571. doi: 10.1016/j.apsusc.2021.151356.

Wu, Lie, Jiahong Wang, Jiang Lu, Danni Liu, Na Yang, Hao Huang, Paul K Chu, and Xue-Feng Yu. 2018. "Lanthanide-Coordinated Black Phosphorus." *Small* no. 14 (29):1801405.

Yan, Shancheng, Bojun Wang, Zhulan Wang, Dong Hu, Xin Xu, Junzhuan Wang, and Yi Shi. 2016. "Supercritical carbon dioxide-assisted rapid synthesis of few-layer black phosphorus for hydrogen peroxide sensing." *Biosensors & Bioelectronics* no. 80:34–38. doi: 10.1016/j.bios.2016.01.043.

Yang, Jie, Zhenghui Pan, Jun Zhong, Shuo Li, John Wang, and Po-Yen Chen. 2021. "Electrostatic self-assembly of heterostructured black phosphorus-MXene nano-composites for flexible microsupercapacitors with high rate performance." *Energy Storage Materials* no. 36:257–264. doi: 10.1016/j.ensm.2020.12.025.

Yang, Wenhao, Shuilai Qiu, Yifan Zhou, Jingwen Wang, Bin Zou, and Lei Song. 2022. "Covalent grafting diazotized black phosphorus with ferrocene oligomer towards smoke suppression and toxicity reduction." *Chemosphere* no. 303. doi: 10.1016/j.chemosphere.2022.135012.

Yin, Sihao, Xinlin Ren, Peichao Lian, Yuanzhi Zhu, and Yi Mei. 2020. "Synergistic Effects of Black Phosphorus/Boron Nitride Nanosheets on Enhancing the Flame-Retardant Properties of Waterborne Polyurethane and Its Flame-Retardant Mechanism." *Polymers* no. 12 (7). doi: 10.3390/polym12071487.

Zhang, Qingzhe, Shengyun Huang, Jiujun Deng, Deepak Thrithamarassery Gangadharan, Fan Yang, Zhenhe Xu, Giacomo Giorgi, Maurizia Palummo, Mohamed Chaker, and Dongling Ma. 2019. "Ice-Assisted Synthesis of Black Phosphorus Nanosheets as a Metal-Free Photocatalyst: 2D/2D Heterostructure for Broadband H-2 Evolution." *Advanced Functional Materials* no. 29 (28). doi: 10.1002/adfm.201902486.

Zhao, Ming, Haolei Qian, Xinyue Niu, Wei Wang, Liao Guan, Jian Sha, and Yewu Wang. 2016. "Growth Mechanism and Enhanced Yield of Black Phosphorus Microribbons." *Crystal Growth & Design* no. 16 (2):1096–1103. doi: 10.1021/acs.cgd.5b01709.

Zhao, Yuetao, Huaiyu Wang, Hao Huang, Quanlan Xiao, Yanhua Xu, Zhinan Guo, Hanhan Xie, Jundong Shao, Zhengbo Sun, Weijia Han, Xue-Feng Yu, Penghui Li, and Paul K. Chu. 2016. "Surface Coordination of Black Phosphorus for Robust Air and Water Stability." *Angewandte Chemie-International Edition* no. 55 (16):5003–5007. doi: 10.1002/anie.201512038.

Zhou, Fengchen, Liuzhang Ouyang, Jiangwen Liu, Xu-Sheng Yang, and Min Zhu. 2020. "Chemical bonding black phosphorus with TiO2 and carbon toward high-performance lithium storage." *Journal of Power Sources* no. 449. doi: 10.1016/j.jpowsour.2019.227549.

Zhou, Fengchen, Liuzhang Ouyang, Meiqin Zeng, Jiangwen Liu, Hui Wang, Huaiyu Shao, and Min Zhu. 2019. "Growth mechanism of black phosphorus synthesized by different ball milling techniques." *Journal of Alloys and Compounds* no. 784:339–346. doi: 10.1016/j.jallcom.2019.01.023.

Zhou, Yifan, Fukai Chu, Shuilai Qiu, Wenwen Guo, Shenghe Zhang, Zhoumei Xu, Weizhao Hu, and Yuan Hu. 2020. "Construction of graphite oxide modified black phosphorus through covalent linkage: An efficient strategy for smoke toxicity and fire hazard suppression of epoxy resin." *Journal of Hazardous Materials* no. 399. doi: 10.1016/j.jhazmat.2020.123015.

Zhou, Yifan, Fukai Chu, Wenhao Yang, Shuilai Qiu, and Yuan Hu. 2022. "Innovative design of hierarchical cobalt-based borate functionalized black phosphorus structure with petal-like wrinkle: Enhancing the fire safety and mechanical properties of epoxy resin." *Composites Part B-Engineering* no. 238. doi: 10.1016/j.compositesb.2022.109886.

Zhou, Yifan, Jiali Huang, Junling Wang, Fukai Chu, Zhoumei Xu, Weizhao Hu, and Yuan Hu. 2020. "Rationally designed functionalized black phosphorus nanosheets as new fire hazard suppression material for polylactic acid." *Polymer Degradation and Stability* no. 178. doi: 10.1016/j.polymdegradstab.2020.109194.

Zou, Bin, Shuilai Qiu, Xiyun Ren, Yifan Zhou, Feng Zhou, Zhoumei Xu, Zhixin Zhao, Lei Song, Yuan Hu, and Xinglong Gong. 2020. "Combination of black phosphorus nanosheets and MCNTs via phosphoruscarbon bonds for reducing the flammability of air stable epoxy resin nanocomposites." *Journal of Hazardous Materials* no. 383:121069. doi: 10.1016/j.jhazmat.2019.121069.

Zou, Bin, Shuilai Qiu, Jingwen Wang, Wenhao Yang, Can Liao, Heng Yu, Xingjun Li, Liangyuan Qi, Zhou Gui, and Yuan Hu. 2022. "Revealing and modeling of fire products in gas-phase for epoxy/black phosphorus-based nanocomposites." *Chemosphere* no. 305. doi: 10.1016/j.chemosphere.2022.135504.

Zou, Bin, Shuilai Qiu, Yifan Zhou, Ziyan Qian, Zhoumei Xu, Jingwen Wang, Yuling Xiao, Can Liao, Wenhao Yang, Longfei Han, Fukai Chu, Lei Song, Zhou Gui, and Yuan Hu. 2022. "Photothermal-healing, and record thermal stability and fire safety black phosphorus-boron hybrid nanocomposites: Mechanism of phosphorus fixation effects and charring inspired by cell walls." *Journal of Materials Chemistry A* no. 10 (27):14423–14434. doi: 10.1039/d2ta02430f.

11 Layer-by-Layer Assembly of 2D Nanomaterials into Flame-Retardant Coatings for Fire-Safe Polymers

Giulio Malucelli

Department of Applied Science and Technology and Local INSTM Unit, Politecnico di Torino, Alessandria, Italy

CONTENTS

11.1 BACKGROUND

Polymeric materials are well known for their beneficial features, such as light weight, low temperature processing, dielectric behavior, and optical properties. Conversely, when exposed to an irradiative heat flux or to a flame, most polymers effortlessly burn if not protected with flame retardants. In addition, during combustion, melt dripping may occur, which drives flame propagation toward other flammable materials. This behavior is very crucial and represents a strong limitation for the effective exploitation of polymeric materials for various applications, especially when their flame retardancy is mandatory (Vahabi, Saeb, and Malucelli 2022; Hilado 1998; Horrocks 2000; Hu 2020; Morgan 2022).

For this reason, physical or chemical processes are usually employed for making polymers flame retardant (i.e., delaying or blocking flame propagation). Various flame retardant treatments have been designed, synthesized, and applied to several polymer systems. The history of flame retardants dates to approximately the 1950s, when the first products, based on halogenated molecules (mainly chlorinated and brominated), were synthesized. From that time forward, significant progress has been made, moving toward the use of inorganic fillers (such as magnesium hydroxide or

DOI: 10.1201/9781003327158-11

aluminum trihydrate) and the currently employed non-halogenated flame retardant systems, which are mainly based on phosphorus (also in combination with nitrogen) chemistry. In parallel, several products based on different nano-objects (nano-particles, nanoclays, carbon nanotubes, graphenes, etc.) have been proposed, showing interesting and effective flame retardant features. In particular, the utilization of either top-down (employing preformed nano-object suspensions) or bottom-up (through the creation of single nano-objects or nano-objects assemblies) methods has been thoroughly investigated (Fink 2020; Morgan 2019).

Different strategies can be successfully exploited for obtaining flame retarded polymer systems. Based on the chemical structure, architecture, and morphology of the selected polymer, it is possible to utilize melt compounding (that becomes spinning, when utilized for synthetic fibers), simple mechanical dispersion (often carried out at room temperature) into thermosetting resins, or chemical reactions (namely, copolymerization or grafting), which involve covalent bonding of the flame retardant to the polymer chains (ensuring higher stability and durability).

A very effective strategy for providing flame retardancy to polymer systems is based on so-called surface engineering, which involves the chemical or physical deposition of the flame retardants just on the surface of the polymer: this gives rise to a coating that is able to limit the heat and mass transfer from and toward the underlying polymer substrate. Being able to place the flame retardant exclusively on the polymer surface, the critical interface between the condensed phase and the gas phase is important for quickly suppressing fire without altering bulk polymer behavior. Indeed, the flame retardant is placed only where it is truly needed. It has been demonstrated that the homogeneous dispersion of the flame retardant within the polymer bulk usually requires higher FR loadings with respect to the surface approach, as most of the additive incorporated into the bulk does not activate during the application of a flame or the exposure to an irradiative heat source, so the protective function is not fulfilled (Malucelli et al. 2014).

Among the different surface engineering approaches (i.e., plasma treatments (Akovali and Takrouri 1991), (nano)particle absorption (Carosio, Alongi, and Frache 2011), and sol-gel processes (Alongi and Malucelli 2012)), layer-by-layer (LbL) methods are currently gathering much interest from both academic and industrial communities. In fact, they are simple to perform, limiting the environmental impact, as most of their components are environmentally friendly. Additionally, LbL treatments can be adapted to any polymer substrate, irrespective of shape, size, and complexity. Further, apart from flame retardancy, it is possible to impart additional properties (including antibacterial, hydrophobicity, and electrical/thermal conductivity) by simply selecting the appropriate components of the layers (Qiu et al. 2018). For all these reasons, layer-by-layer assembly has been widely applied to different polymeric materials, including bulk polymers, fabrics, polymer composites, and foams. This review aims to provide a clear overview of this surface-engineered approach, specifically focused on the use of 2D nanostructures, highlighting the current state of the art, the main limitations, and challenging issues, and discussing some possible further advances for the future.

11.2 LAYER-BY-LAYER ASSEMBLY

To find the birth of the LbL method, it is necessary to return to the late 1960s, when Iler, in a pioneering work, proved the potential of this technique, demonstrating that a molecularly organized process is suitable for the buildup of nanostructured assemblies (Iler 1966). Then, it took approximately 25 years to discover a practical LbL methodology, as demonstrated by Decher and Hong (1991), suitable for its direct exploitation to obtain alternating polycations/polyanion layers in the form of ultrathin multilayered films.

At present, LbL is gathering high attractiveness and is inspiring academic and industrial researchers toward the disclosure of new potentialities, derived from the unlimited possibility of combining a vast number of components (such as biomacromolecules, micro- and nanoparticles, and polyelectrolytes) to obtain assembled coatings that even show multifunctional features (Richardson, Björnmalm and Caruso 2015; Zhang et al. 2019; Richardson et al. 2016).

Although LbL coatings are usually obtained by exploiting electrostatic interactions occurring between positively and negatively charged components, the scientific literature has clearly demonstrated the possibility of using other types of interactions, namely, the formation of covalent bonds (Fang et al. 1997; Ichinose, Kawakami, and Kunitake 1998), donor/acceptor exchanges (Shimazaki et al. 1997; Benten et al. 2008) and hydrogen bonding (Stockton and Rubner 1997; Wang et al. 2000).

A general scheme of the layer-by-layer process exploiting different types of components (bio(macro)molecules, polyelectrolytes, micro/nanoparticles, sheets, or rods/tubes) is depicted in Figure 11.1.

By using multiple dipping cycles, it is possible to obtain bilayered assemblies.

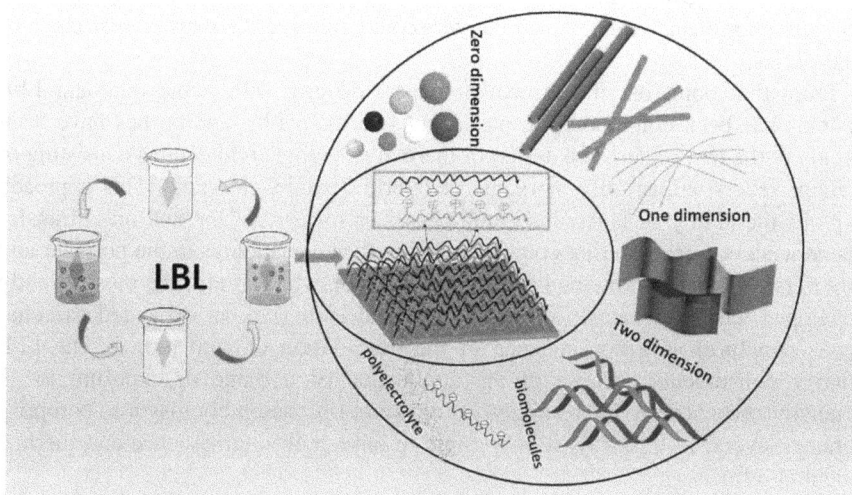

FIGURE 11.1 Scheme of LbL: the polymer substrate is successively dipped in baths containing polyelectrolytes, micro/nanoparticles, bio(macro)molecules, sheets or rods/tubes. Reprinted with permission from (Qiu et al. 2018). Copyright 2018 Elsevier.

In the buildup of a multilayered assembly based on electrostatic forces, the polymer substrate (which generally bears a weak negative electrical charge) needs to be put in contact (by means of either dipping or spray techniques) with waterborne polyelectrolyte solutions or (nano)particle suspensions. The dipping method is preferred for lab-scale processes, while spraying is more suitable for semipilot/ pilot/industrial plants. The duration of each single spraying or dipping stage is quite variable, depending on several factors, namely: the type of polymer substrate and layer components and the utilization of rinsing and drying steps, although rinsing is not mandatory but strongly suggested for avoiding probable contaminations taking place between the oppositely charged solutions/suspensions, thus extending the duration of the utilized baths. In addition, rinsing allows the removal of the improperly adhered layer components. By following this procedure, it is possible to build multilayered assemblies, exploiting the reversal of the surface charges. Deposition procedures can proceed until the envisioned number of deposited layers is reached; commonly, 10 to 50 (or even beyond) layers can be deposited, according to the designed LbL assemblies.

It is worth highlighting that both the overall deposition process and the morphology of the obtained assemblies are significantly affected by different experimental factors; the most important include the temperature, at which deposition occurs (Tan et al. 2003), the pH of the employed baths (Chang et al. 2008; Shiratori and Rubner 2000), the chemical features (i.e., structure and molecular weight) of the polyelectrolytes or bio(macro)molecules (Mermut and Barrett 2003; Sui, Salloum, and Schlenoff 2003), the adsorption time (Fujita and Shiratori 2006) and the ionic strength (McAloney et al. 2001).

Specifically, referring to flame retardancy, the utilization of layer-by-layer methods was proposed approximately 15 years ago by Srikulkit et al. (Srikulkit, Iamsamai, and Dubas 2006), who demonstrated that an LbL coating made of polyphosphoric acid and chitosan bilayers could enhance the thermal resistance of silk fabrics.

From that point on, the growth of papers dealing with flame retardant LbL systems has been continuous. In particular, two main LbL approaches have been explored: the first regards the design of thermal shielding architectures consisting of ceramic layers, either fully inorganic or hybrid organic–inorganic. This approach exploits the ability of layered ceramic phases to lower the heat and mass transfer phenomena occurring during combustion from the surroundings to the polymer and vice versa. The second method (i.e., intumescence), which is also the most recently developed, takes advantage of the formation of *char* (i.e., an expanded carbonaceous structure) as a consequence of the degradation or activation of the LbL coating constituents because of the application of a flame or exposure to an irradiative heat source. Interestingly, intumescent LbL assembly may also comprise ceramic layers: in this way, it is possible to have both intumescence and thermal shielding effects.

Specifically, referring to 2D nanomaterials (such as nanoclays, graphene, and double layer hydroxides, among a few to mention) assembled in LbL flame retardant coatings, the scientific literature contains plenty of good examples that will be presented in the next paragraphs.

11.3   FLAME RETARDANT LbL COATINGS CONTAINING NANOCLAYS

Lamellar structures possess a high specific surface area and can be successfully exploited for designing effective flame retardant coatings, which act as a barrier (i.e., thermal shield) on the surface of the polymeric substrate, hence blocking dripping phenomena during flame spread tests and remarkably lowering the peak of the heat release rate in forced-combustion (i.e., cone calorimetry) tests.

Lamellar silicates (particularly Montmorillonites and Laponites) are one of the 2D nanostructures that were first utilized in LbL flame retardant architectures. One of the first papers dates back to 2009 (Li, Schulz, and Grunlan 2009) and describes the preparation and characterization of flame-retarded bilayered assemblies made of branched polyethyleneimine and synthetic Laponite clay; up to 30 bilayers were assembled and then successfully deposited on cotton fabrics. A general scheme of the performed LbL deposition is shown in Figure 11.2; the typical morphology of a 30 bilayered coating, as assessed by transmission electron microscopy, is presented in Figure 11.3. The LbL deposition on cotton resulted in a well-covered substrate, where each individual cellulosic fiber was uniformly coated: this morphology, together with the high ceramic content of the assemblies (up to approximately 83 wt.%), accounted for a certain protection on the underlying substrate, as revealed by the significant increase in the char residues after vertical flame spread tests and the improved thermal stability (as assessed by thermogravimetric analyses).

In a further research effort, the same group prepared other flame retardant bilayered assemblies replacing Laponite with Montmorillonite (Li et al. 2010). In particular, two pH values of the branched polyethyleneimine bath (namely, 7 and 10) and two concentrations of the nanoclay suspensions (0.2 and 1.0 wt.%) were employed. As assessed by flammability tests carried out in the vertical configuration, although

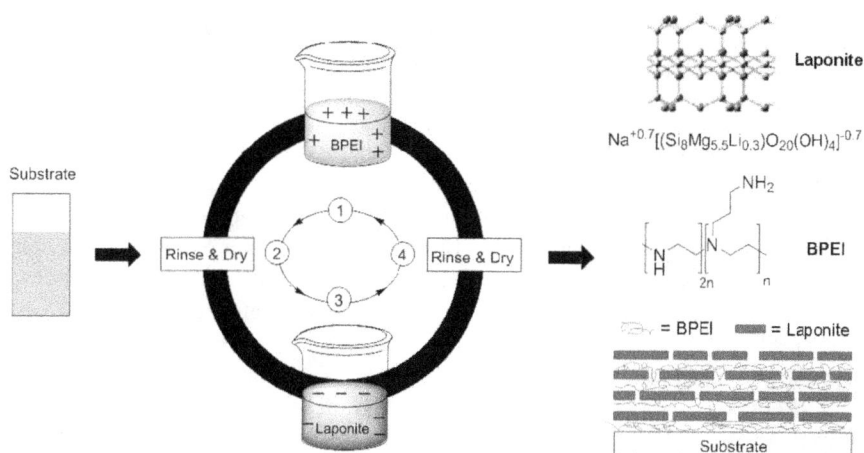

FIGURE 11.2   Scheme of the LbL process for obtaining bilayered assemblies made of branched polyethyleneimine (BPEI) and synthetic Laponite clay. Reprinted with permission from (Li, Schulz, and Grunlan 2009). Copyright 2009 American Chemical Society.

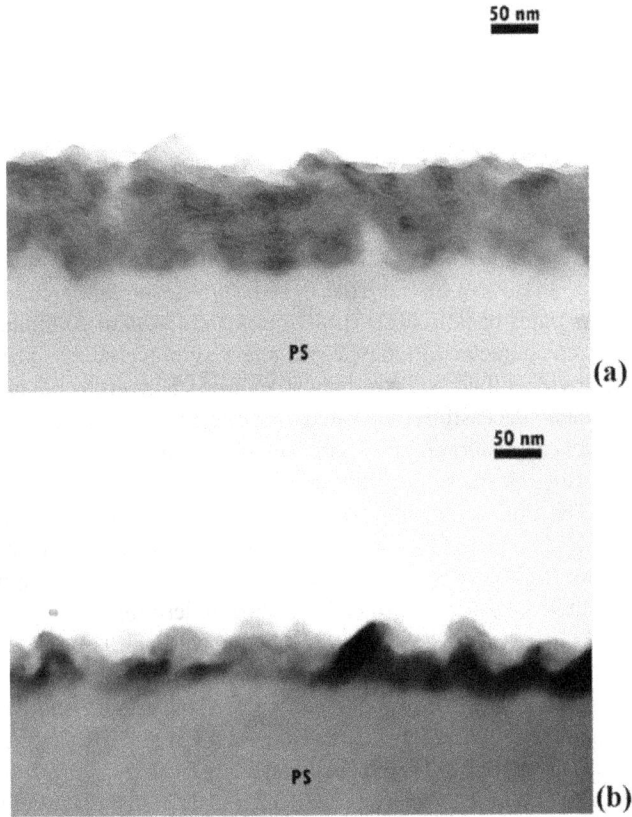

FIGURE 11.3 TEM cross-sectional images of 30 bilayered assemblies (on a polystyrene substrate) made of Laponite and BPEI at two different pH values in a Laponite dispersion bath, namely, pH 10 (a) and pH 8 (b). Adapted with permission from (Li, Schulz, and Grunlan 2009). Copyright 2009 American Chemical Society.

self-extinction was not achieved, 20 bilayer coatings obtained with the branched polyethyleneimine bath at pH 10 and with 1 wt.% nanoclay suspension significantly increased the residue at the end of the tests, hence indicating a high char-forming character of the designed assembly. In addition, the same system, tested with microscale combustion calorimetry, showed an important decrease in the heat release capacity, from 274 (untreated fabric) to approximately 221 J/(gK) (treated cotton). Overall, the paper highlighted the importance of pH in determining the structure, compactness, and crosslinking density of the obtained LbL assemblies because of the dependence of the ionization degree of the branched poly-ethyleneimine on the pH.

Further improvements were obtained by using a continuous LbL deposition apparatus (Chang et al. 2014), which allowed the coating of three different types of cotton (untreated print cloth, mercerized print cloth, and twill cotton fabrics) with up to 50 bilayers made of branched polyethylenimine with urea and diammonium

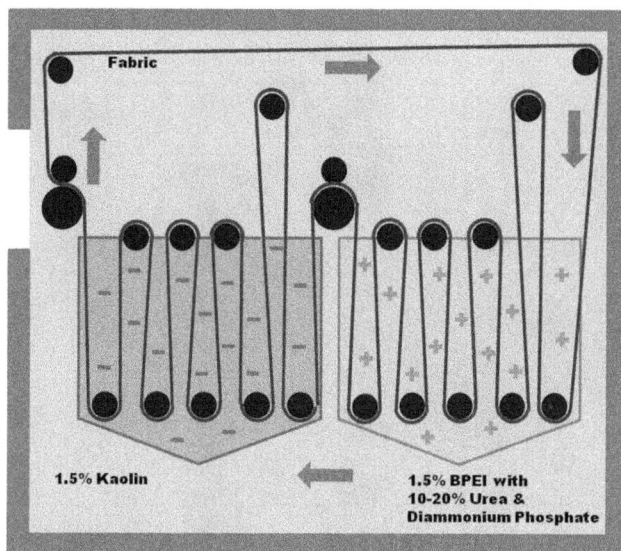

FIGURE 11.4 Schematic of the continuous LbL deposition process used to apply 50 bilayered branched polyethyleneimine-urea/clay-diammonium phosphate coatings. Arrows indicate the directional flow of the fabric as it is processed. Reprinted with permission from (Chang et al. 2014). Copyright 2014 American Chemical Society.

phosphate (positively charged bath) and clay nanoparticles (Kaolin, negatively charged bath); a general scheme is shown in Figure 11.4.

The combination of branched polyethyleneimine, urea, diammonium phosphate and nanoclay accounted for self-extinction, as revealed by vertical flame spread tests, as well as for a remarkable increase in the limiting oxygen index (LOI) values, as shown in Figure 11.5.

New and coworkers (New et al. 2016) applied 10 or 20 bilayers made of sodium montmorillonite (negatively charged) and branched polyethyleneimine (positively charged) to polyester fabrics. The designed treatments showed some significant enhancements in flame retardancy, as revealed by microscale combustion calorimetry tests: in particular, the peak of heat release rate markedly decreased from 629 (untreated polyester) to 393 (fabric treated with 10 bilayers) and 368 W/g (fabric treated with 20 bilayers), as displayed in Figure 11.6. However, the main novelty of the paper was the investigation of the fabric physiological properties (namely, vertical wicking, moisture management and air permeability), which were correlated with the number of deposited bilayers.

A copolymer derived from the reaction between acrylic acid and N-(2-(5,5-dimethyl-1,3,2-dioxaphosphinyl-2-ylamino)-ethylacetamide-2-propenyl) acid was exploited in combination with a montmorillonite functionalized with 3-aminopropyltrimethoxysilane to provide cotton fabrics with flame retardant features (Huang et al. 2012). To this aim, 5 and 20 bilayered assemblies were deposited on the cellulosic substrate; their fire behavior was compared with that exhibited by architectures made of functionalized montmorillonite and pristine acrylic acid. As assessed by

	Untreated	Clay/ BPEI only (Wt%)	Clay/ BPEI / diammonium phosphate/urea (Wt%)		
	0	2.9	10.0	17.8	20.2
Print cloth				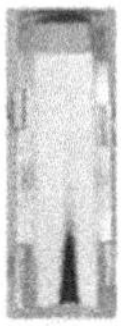	
After-flame time (s)	32	20	4	0	0
After-glow time (s)	9	4	2	0	0
Char length (cm)	NA	NA	15.0	8.0	8.5
LOI (%)	18±5	24±5	28±5	38±5	39±5
	0	2.6	9.4	16.1	18.2
Mercerized print cloth				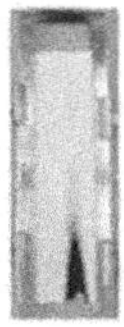	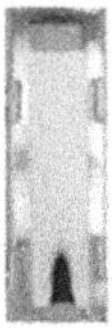
After-flame time (s)	33	21	12	0	0
After-glow time (s)	10	5	14	0	0
Char length (cm)	NA	NA	>30.0	7.5	6.0
LOI (%)	18±5	23±5	29±5	38±5	40±5
	0	2.2	5.8	12.7	19.5
Twill					
After-flame time (s)	106	55	2	0	0
After-glow time (s)	27	16	0	0	0
Char length (cm)	NA	NA	9.5	6.0	6.0
LOI (%)	19±5	27±5	27±5	36±5	38±5

FIGURE 11.5 Results of vertical flame spread and limiting oxygen index tests of untreated and treated cotton fabrics. Reprinted with permission from (Chang et al. 2014). Copyright 2014 American Chemical Society.

forced-combustion tests performed under 35 kW/m^2 irradiative heat flux, the LbL-treated cotton showed decreased values of total heat release, average mass loss rate and peak of heat release rate with respect to both neat cotton and fabric treated with functionalized montmorillonite and pristine acrylic acid bilayers (Table 11.1). These findings clearly proved the enhanced flame retardancy provided by the designed assemblies, in which the 2D nanoparticles exerted a good thermal shielding effect.

Only a few works report on the durability (i.e., washing fastness) of LbL-treated fabrics containing 2D nanoparticles. In particular, Xu and coworkers (Xu et al. 2014) treated polyimide fabrics with 5 to 20 bilayers of poly(N-benzyloxycarbonyl-

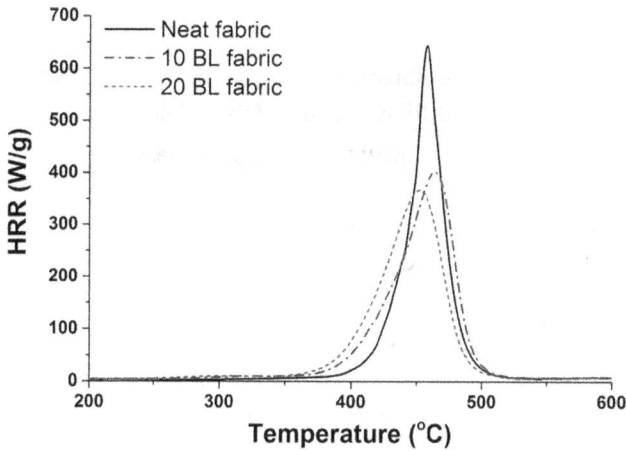

FIGURE 11.6 Results from microscale combustion calorimetry tests for untreated polyester and polyester fabrics treated with 10 or 20 bilayers. Reprinted with permission from (New et al. 2016). Copyright 2016 Elsevier.

3,4-dihydroxyphenylalanine) and sodium montmorillonite. The LbL-treated fabrics achieved self-extinction in vertical flame spread tests, which was maintained even after 20 washing cycles carried out according to the ISO105-C03 standard.

In an effort to make LbL assemblies greener, Wei and coworkers (Wei et al. 2015) succeeded in depositing 5 and 10 bilayers of chitosan and sodium montmorillonite on sisal fiber cellulose microcrystals. The effect provided by the LbL treatment on the fire behavior was assessed through microscale combustion

TABLE 11.1

Results from Cone Calorimetry Tests for Cotton Fabrics, Neat, or LbL Treated with 5 and 20 Bilayers

Fabric System	TTI (s)	pkHRR (kW/m^2)	THR (MJ/m^2)	AMLR (g/s)
Cotton	42	144	25.1	0.057
Cotton treated with 5 BL of acrylic acid and modified montmorillonite	44	124	23.4	0.053
Cotton treated with 20 BL of acrylic acid and modified montmorillonite	48	109	22.4	0.045
Cotton treated with 5 BL of FR acrylic acid and modified montmorillonite	48	105	22.1	0.044
Cotton treated with 20 BL of FR acrylic acid and modified montmorillonite	59	77	20.7	0.032

Legend: BL: bilayer; FR acrylic acid: copolymer derived from the reaction between acrylic acid and N-(2-(5,5-dimethyl-1,3,2-dioxaphosphinyl-2-ylamino)-ethylacetamide-2-propenyl acid; TTI: time to ignition; pkHRR: peak of heat release rate; AMLR: average mass loss rate.

TABLE 11.2

Results from Microscale Combustion Calorimetry Tests for Neat and LbL-Treated (with 5 and 10 Bilayers) Sisal Fiber Cellulose Microcrystals

System	pkHRR (W/g)	ΔpkHRR (%)	THR (kJ/g)	ΔTHR (%)
Neat sisal fiber cellulose microcrystals	296	–	15	–
Sisal fiber cellulose microcrystals treated with 5 bilayers of chitosan and sodium montmorillonite	282	–4.7	11.3	–24.6
Sisal fiber cellulose microcrystals treated with 10 bilayers of chitosan and sodium montmorillonite	249	–15.9	11.7	–22.0

calorimetry tests: increasing the number of deposited bilayers significantly decreased both the peak of heat release rate and total heat release (Table 11.2).

A similar approach was exploited by Deng and coworkers (Deng et al. 2016), who deposited 7, 14, and 21 bilayers of chitosan and sodium montmorillonite on a silicone foam: 7 bilayers, as assessed by cone calorimetry tests, were sufficient for significantly reducing both peaks of the heat release rate (ca. −19% with respect to the untreated foam) and total smoke production (by approximately 18%); in addition, the limiting oxygen index increased from 20.2 (untreated foam) to 23.4% (foam treated with 7 bilayers).

Another type of polymer substrate to which research on LbL coatings containing 2D nanoparticles has been addressed is polyurethane foam because of its wide flame-retarding applications. In this context, Kim and coworkers (Kim, Harris, and Davis 2012) deposited either a trilayered assembly (up to 8 trilayers) based on cationic branched polyethyleneimine, anionic sodium montmorillonite, and anionic poly(acrylic acid) or a bilayered coating (up to 20 bilayers) based on cationic branched polyethyleneimine and anionic sodium montmorillonite, as shown in Figure 11.7, on a polyurethane foam.

The trilayered assembly performed better than the bilayered assembly, as assessed by forced-combustion tests carried out under a 35 kW/m^2 irradiative heat flux. The comparison between untreated polyurethane foam and the counterpart LbL-treated with 8 trilayers is shown in Figure 11.8. Both curves show two peaks: the former is attributed to the pyrolysis of isocyanate, while the second is attributed to the pyrolysis of polyol. It is noteworthy that the presence of the trilayered coating decreased the peaks of the heat release rate and total heat release by 17 and 6%, respectively, with respect to untreated polyurethane foam while increasing the time to peak.

In a further research effort, the same group (Li et al. 2013) studied the effects of the concentration of the three components (branched-polyethyleneimine, sodium montmorillonite and poly(acrylic acid)) on the fire behavior of treated polyurethane foams. In particular, it was demonstrated that the overall flame retardancy is more affected by the increase in nanoclay loading in the assembled coating than by the increase in both polymer layers.

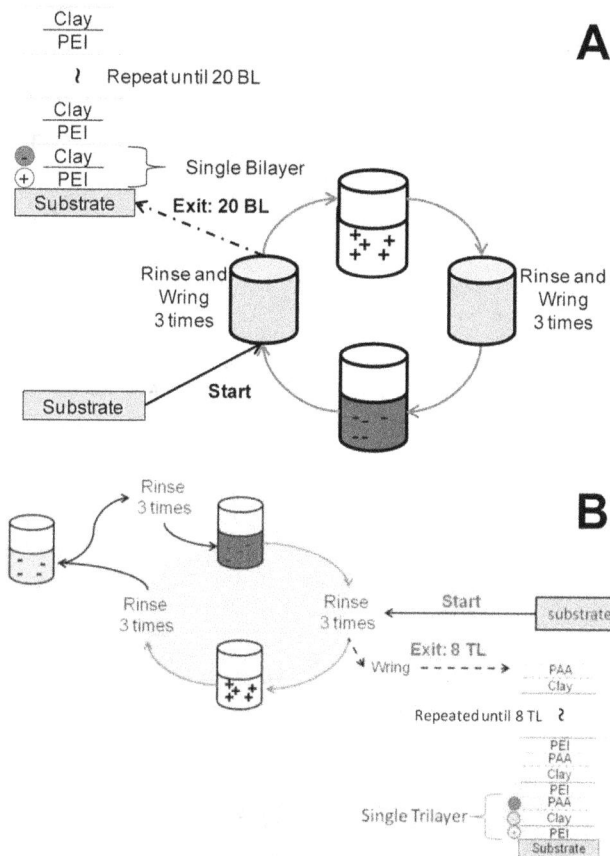

FIGURE 11.7 Scheme of LbL assembly on polyurethane foam. Reprinted with permission from (Kim, Harris, and Davis 2012). Copyright 2012 American Chemical Society.

Pursuing this research, sodium montmorillonite was replaced with hydrotalcite (i.e., layered double hydroxide) in the former trilayered structure (Li et al. 2015). As revealed by SEM analyses, the polyurethane foam was entirely covered by the LbL assemblies, irrespective of the number of deposited trilayers, and showed fracture-free, homogeneous coverage (Figure 11.9). Furthermore, just 2 trilayered assemblies deposited on the polyurethane foam were enough to remarkably lower the peak of heat release rate and the average heat release rate (by approximately 40 and 80%, respectively) with respect to the untreated polymer substrate, as assessed by forced-combustion tests. The protection exerted by the deposited coatings was further supported by the incomplete combustion of polyurethane, as well as by the absence of dimensional changes between the pre- and post-tests.

Finally, sodium montmorillonite, hydrotalcite, branched-polyethylenimine, and poly(acrylic acid) were deposited on a polyurethane foam in the form of bi, tri- and quad-layered architectures (Yang et al. 2015). Each specific type of assembly exhibited different flame retardant features because of its diverse morphology and

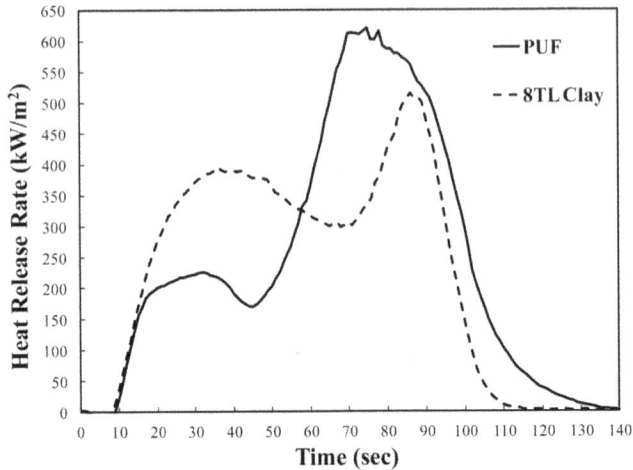

FIGURE 11.8   Heat release rate vs. time for polyurethane foam untreated and treated with 8 trilayers of branched polyethyleneimine, sodium montmorillonite, and poly(acrylic acid). Reprinted with permission from (Kim, Harris, and Davis 2012). Copyright 2012 American Chemical Society.

coating growth rate; in particular, hybrid organic–inorganic bilayers combining one of the two polymers with one of the two inorganic nanofillers performed better than the other complex assemblies, also considering the ease of processing of the former.

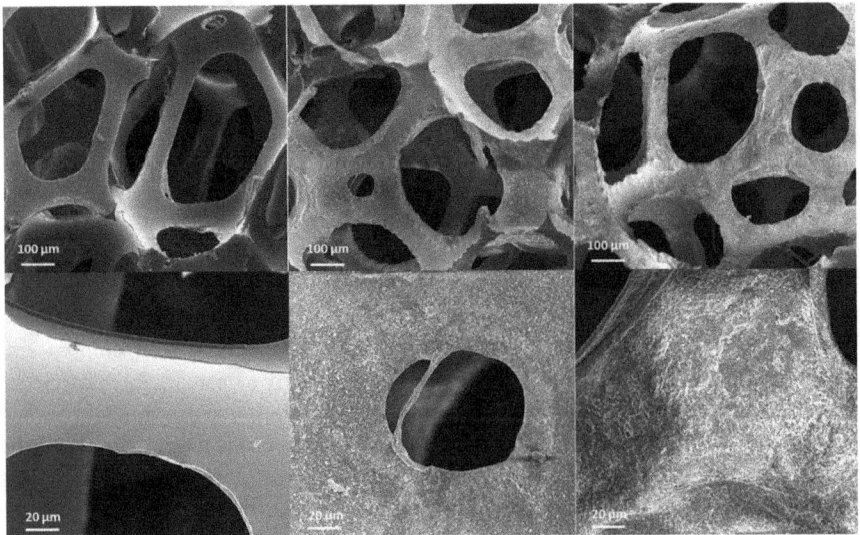

FIGURE 11.9   SEM micrographs of untreated polyurethane foam (left), polyurethane foam treated with 1 tri-layer (middle) and polyurethane foam treated with 2 tri-layers (right) at 100× magnification (top) and 500× magnification (bottom). Reprinted with permission from (Li et al. 2015). Copyright 2015 Elsevier.

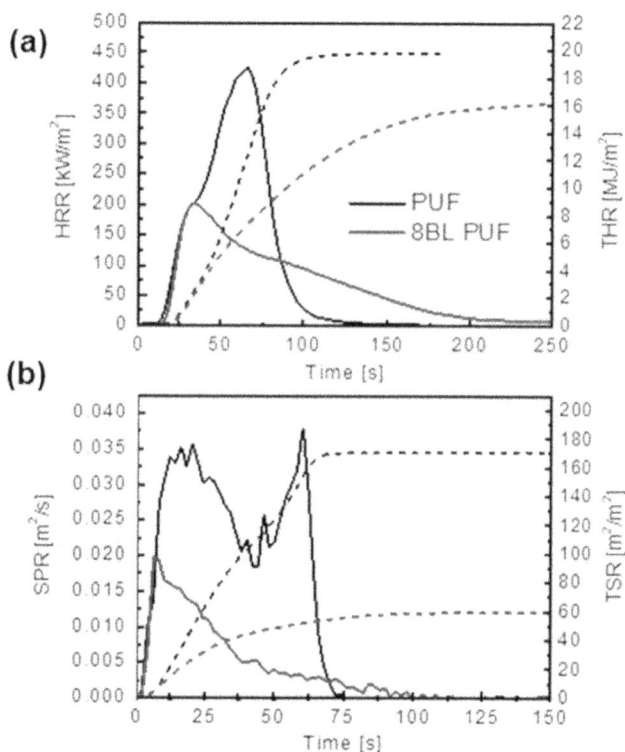

FIGURE 11.10   Cone calorimetry data for flexible polyurethane foams, neat or treated with 8 bilayers of chitosan and vermiculite clay: (a) heat release rate (HRR) and total heat release (THR), (b) smoke production rate (SPR) and total smoke release (TSR). Reprinted with permission from (Lazar et al. 2018). Copyright 2018 American Chemical Society.

Quite recently, Lazar and coworkers (Lazar et al. 2018) demonstrated that a few bilayers (namely, 8) made of chitosan and vermiculite clay are capable of strongly enhancing the flame retardancy of a flexible polyurethane foam. Indeed, as assessed by forced-combustion tests carried out at 35 kW/m^2, a significant decrease in the peak heat release rate (by 53%) and total smoke release (−63%) was observed after the deposition of the LbL coating (Figure 11.10). These findings were ascribed to an efficient flame retardant action, occurring in the condensed phase only, provided by the combination of the char forming character of chitosan with the thermal shielding effect of the clay, which promoted the formation of a stable aromatic char. This flame retardant mechanism justified the only partial degradation the treated polyurethane foam underwent at the backside of the sample after being exposed to an 1100°C flame (i.e., after a burn-through test).

In a further research effort aimed at designing a biosourced LbL assembly, Ur Rehman and coworkers (Ur Rehman et al. 2021) deposited 5, 7, 10, and 15 bilayers consisting of potato starch (positively charged) and vermiculite/titania (negatively charged) on cotton fabrics. As assessed by vertical flame spread tests, the deposited

TABLE 11.3

Results from Microscale Combustion Calorimetry Tests for Untreated and LbL-Treated (with 5, 7, 10, and 15 bilayers) Cotton Fabrics

System	pkHRR (W/g)	ΔpkHRR (%)	HRC (J/gK)	ΔHRC (%)	Final Residue (%)
Untreated cotton	242	–	521	–	17.5
Cotton treated with 5 bilayers	206	−14	434	−17	13.2
Cotton treated with 7 bilayers	193	−20	390	−25	30.6
Cotton treated with 10 bilayers	206	−14	382	−27	20.8
Cotton treated with 15 bilayers	210	−13	384	−26	7.4

assemblies were capable of slowing flame propagation, although self-extinction was not achieved. In addition, the LOI values increased with increasing the number of deposited bilayers, from 19.7% (untreated cotton) to 22.4% (cotton treated with 15 bilayers). Furthermore, microscale combustion calorimetry analyses revealed that the deposition of 7 bilayers provided the best FR performance (in terms of the reduced peak of the heat release rate and increased final residue), as shown in Table 11.3.

Recently, Zhou and coworkers (Zhou et al. 2021) exploited the layer-by-layer technique for conferring flame retardant features to wood plastic composites. To this aim, 80 bilayered assemblies consisting of sodium montmorillonite and polyethyleneimine/graphitized multiwalled carbon nanotubes were deposited by spraying. The treated substrates achieved self-extinction in vertical flame spread tests; in addition, as assessed by forced-combustion tests (irradiative heat flux: 35 kW/m^2), the time to ignition was delayed by approximately 400% compared to the untreated substrate. Furthermore, the thermal shielding effect of the nanoclay accounted for a significant decrease in the peak of heat release rate (−43%), total heat release (-45%), smoke production rate (−65%) and total smoke production (−33%).

The LbL method also demonstrated high potential for making cotton hydrophobic and flame retardant (Pan et al. 2021). For this purpose, 10 bilayers made of polyethyleneimine and hypophosphorous acid-modified chitosan/sepiolite were deposited on cellulosic fabrics that were then coated with polydimethylsiloxane. As assessed by microscale combustion calorimetry tests, the treated fabrics exhibited a remarkable decrease in the peak heat release rate and total heat release (−56 and −55%, respectively) compared to untreated cotton. In addition, treatment with polydimethylsiloxane allowed superhydrophobic surfaces to be achieved, with water contact angles as high as 133°.

A similar approach was exploited by Zhu and coworkers (Zhu et al. 2022), who deposited 3, 5, and 8 bilayers of chitosan and sodium montmorillonite on a flexible polyurethane foam; after deposition, the foams were further treated with polydimethylsiloxane. Forced-combustion tests revealed a significant decrease in the

TABLE 11.4

Results from Cone Calorimetry Tests for Untreated and LbL-Treated (with 5, 7, 10, and 15 bilayers) Cotton Fabrics

System	Time to Ignition (s)	pkHRR (kW/m^2)	THR (MJ/m^2)
Untreated polyurethane foam	1	455	14.3
Polyurethane foam treated with 3 bilayers	4	381	13.3
Polyurethane foam treated with 5 bilayers	2	344	12.8
Polyurethane foam treated with 8 bilayers	2	311	12.8

peak of heat release rate, as shown in Table 11.4. In addition, the presence of the low-surface tension polymer accounted for a significant decrease in the foam wettability (water contact angles up to 120°) and a high selectivity for oil-water separation.

Kumar Kundu and coworkers designed and deposited biosourced flame retardant LbL assemblies made of chitosan, sodium alginate, phytic acid and sodium montmorillonite on polyamide 66 fabrics; then, the effectiveness as a cross-linking agent, i.e., either citric acid or oxidized sucrose (i.e., a carbohydrate molecule derived from plant species), was assessed (Kumar Kundu, Song, and Hu 2021). In particular, quad- or hexa-layered assemblies (Figure 11.11) were prepared: the former

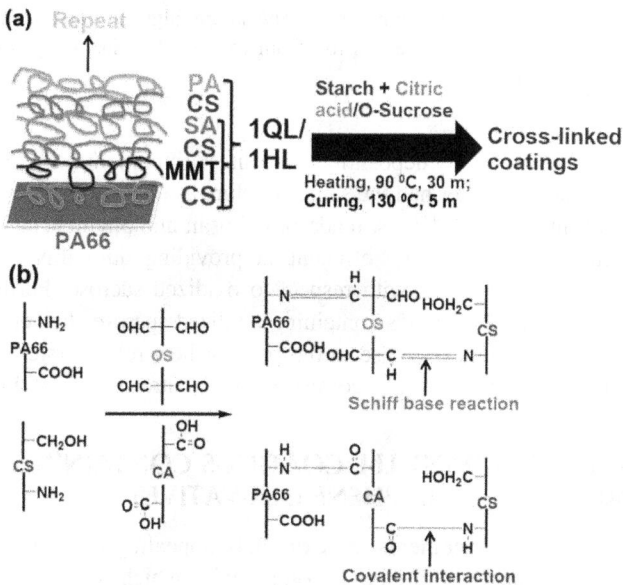

FIGURE 11.11 Scheme of (a) the structure of LbL assemblies deposited on polyamide 66 (PA66) and (b) cross-linking reactions. Copyright 2021 Elsevier. CA: citric acid; CS: chitosan; MMT: sodium montmorillonite; SA: sodium alginate; PA: phytic acid; OS: oxidized sucrose; QL: quad-layer; HL: hexa-layer. Reprinted with permission from (Kumar Kundu, Song, and Hu 2021). Copyright 2021 Elsevier.

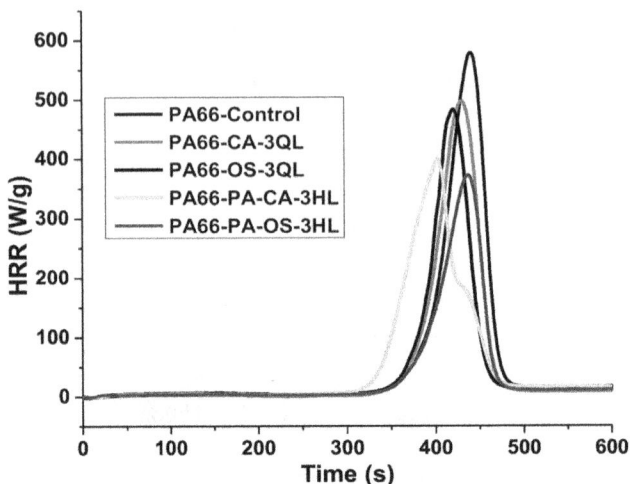

FIGURE 11.12 Heat release rate vs. time of PA66 fabric samples from microscale combustion calorimetry tests. PA66-Control: untreated fabric; PA66-CA-3QL: fabric treated with 3 quad-layers (chitosan/sodium montmorillonite/chitosan/sodium alginate) and crosslinked with citric acid; PA66-OS-3QL: fabric treated with 3 quad-layers (chitosan/sodium montmorillonite/chitosan/sodium alginate) and crosslinked with oxidized sucrose; PA66-PA-CA-3HL: fabric treated with 3 hexa-layers (chitosan/sodium montmorillonite/chitosan/sodium alginate) and crosslinked with citric acid; PA66-PA-OS-3HL: fabric treated with 3 hexa-layers (chitosan/sodium montmorillonite/chitosan/sodium alginate) and crosslinked with oxidized sucrose. Reprinted with permission from (Kumar Kundu, Song, and Hu 2021). Copyright 2021 Elsevier.

were obtained by subsequently depositing chitosan, sodium montmorillonite, chitosan and sodium alginate; the latter had the same structure as the quad-layered assemblies, apart from the fifth and sixth layers, made of chitosan and phytic acid, respectively. Citric acid was found to be less efficient in providing durability (i.e., washing fastness) to LbL-treated fabrics with respect to oxidized sucrose. Furthermore, the quad- and hexa-layered assemblies containing oxidized sucrose showed the best fire behavior, with an important decrease in the peak of heat release rate and total heat release (−36 and −12%, respectively) compared to the untreated fabric (Figure 11.12).

11.4 FLAME RETARDANT LbL COATINGS CONTAINING GRAPHENE OR GRAPHENE DERIVATIVES

The scientific literature documents the extremely appealing characteristics of graphene, one of the most up-to-date nanomaterials, which shows a peculiar 2D structure made of one-atom-thick honeycomb layers of sp^2-bonded carbon atoms (Novoselov et al. 2012). This nanofiller, together with its main derivatives (such as graphene oxide and reduced graphene oxide), has been investigated as a potential component in LbL flame retardant assemblies. In particular, significant attention has been given to graphene oxide, as it bears different oxygen-containing functionalities

TABLE 11.5

Results from Cone Calorimetry Tests for Neat and LbL-Treated (with 2, 5, and 10 trilayers) Flexible Polyurethane Foams

System	pHRR (kW/m^2)	ΔpHRR (%)	TSR (m^2/ m^2)	ΔTSR (%)	Residue (%)
Untreated polyurethane foam	794	–	203	–	0.8
Polyurethane foam treated with 2 tri-layer	667	−16	197	−3	2.5
Polyurethane foam treated with 5 tri-layers	527	−34	174	−14	4.1
Polyurethane foam treated with 10 tri-layers	318	−60	141	−30	7.2

(comprising epoxy, hydroxyl, and carboxyl groups) and intrinsically possesses a weak negative electrical charge, exploitable for establishing electrostatic interactions with other positively charged layers.

In a pioneering study, Zhang and coworkers (Zhang et al. 2016) flame retarded a flexible polyurethane foam through the deposition of trilayered assemblies made of chitosan, graphene oxide and alginate; in particular, 2, 5, and 10 trilayers were successfully deposited. As assessed by forced combustion tests carried out at an irradiative heat flux of 35 kW/m^2 (Table 11.5), the presence of the assemblies significantly decreased the peak of heat release rate (pHRR) and total smoke release (TSR), highlighting the remarkable barrier effect exerted by the 2D nanofiller toward heat and mass transfer from the polymer to the surroundings and vice versa.

Then, hybrid bilayered assemblies consisting of sodium alginate/graphene oxide (mixed together in a single deposition bath) and polyethylenimine were deposited on a flexible polyurethane foam, obtaining 3, 7, and 12 bilayered architectures (Pan et al. 2016a). In addition, half of the LbL-treated foams were subjected to a thermal reduction carried out at 180°C for 3 h; this way, it was possible to obtain coated foams containing reduced graphene oxide. The latter showed higher thermal stability within 430 and 600°C, but at the same time lower fire performance with respect to the counterparts treated with graphene oxide (Table 11.6) because of the loss of higher amounts of reduced graphene oxide nanosheets during combustion, as well as to the lowered thickness of the LbL assemblies containing the reduced nanofiller.

In further research works, the same group prepared either graphene oxide/amino-terminated silica nanosphere assemblies (made of 2 or 5 bilayers, (Pan et al. 2016b)) or graphene oxide nanosheet/β-FeOOH nanorod (only 5 bilayers) assemblies (Pan et al. 2016c) on flexible polyurethane foams. The use of functionalized silica nanospheres with graphene oxide was responsible for a remarkable lowering of the peak of heat release rate (by approximately 51%), as assessed by forced-combustion tests performed under 35 kW/m^2 irradiative heat flux; in addition, the 2nd peak in the HRR vs. time curves disappeared, unlike the foams treated with assemblies made of either graphene oxide or unmodified silica nanospheres only. These findings were

TABLE 11.6

Results from Cone Calorimetry Tests for Neat and LbL-Treated (with 3, 7, and 12 bilayers) Flexible Polyurethane (PU) Foams

System	pHRR (kW/m^2)	ΔpHRR (%)	TSR (m^2/ m^2)	ΔTSR (%)
Untreated PU foam	714	–	313	–
PU foam treated with 3 bilayers containing graphene oxide	378	–47	193	–38
PU foam treated with 7 bilayers containing graphene oxide	231	–68	191	–39
PU foam treated with 12 bilayers containing graphene oxide	203	–72	136	–56
PU foam treated with 3 bilayers containing reduced graphene oxide	594	–17	198	–37
PU foam treated with 7 bilayers containing reduced graphene oxide	337	–53	217	–31
PU foam treated with 12 bilayers containing reduced graphene oxide	252	–65	269	–14

attributed to the barrier action exerted by the hybrid assemblies, which were able to limit the heat and mass transfer between the burning polymer and the flame. In addition, 5 bilayered assemblies increased the tensile strength (ca. +23%), decreasing the ductility (–33% elongation at break) with respect to the untreated foam.

Regarding the LbL architectures made of graphene oxide nanosheets and β-FeOOH nanorods, 5 bilayered coatings in cone calorimetry tests remarkably lowered the peak of heat release rate of the polyurethane foam from 730 to 370 kW/m^2; again, the 2nd peak in the HRR vs. time curves disappeared, unlike the foams treated with assemblies made with the single components (Figure 11.13).

Very recently, aiming at minimizing the deposition steps required by the LbL methodology, Maddalena, Carosio, and Fina (2021) succeeded in obtaining fireproof open cell polyurethane foams by exploiting a water-based single-step deposition of coatings made of graphene oxide nanosheets and deoxyribonucleic acid (DNA, in the form of oligonucleotides). In particular, as shown in Figure 11.14, the surface of the foams was first anionized through activation in a diluted polyacrylic acid solution; then, the activated foams were dipped into a suspension containing graphene oxide platelets and DNA.

The hybrid coating was responsible for self-extinction in both vertical and horizontal flame spread tests (Figure 11.15), as well as for an important decrease in the peak of heat release rate and total smoke release (–75 and –30%, respectively, in comparison with the untreated polyurethane foam) in cone calorimetry tests performed under 35 kW/m^2 irradiative heat flux. All these findings were attributed to the concurrent barrier effect exerted by the graphene oxide platelets, together with the inherent intumescent behavior of deoxyribonucleic acid.

2, 5, 10, 15, and 20 quad-layered complex film assemblies made of poly(vinyl alcohol), graphene oxide nanosheets, phosphorylated chitin, and laponite were

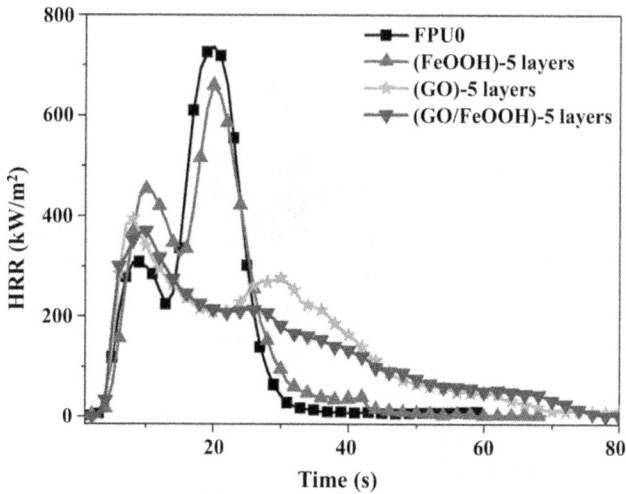

FIGURE 11.13 Cone calorimetry data for flexible polyurethane foams, neat or LbL-treated. FPU0: untreated polyurethane foam; (GO/FeOOH)-5 layers: polyurethane foam treated with 5 bilayers made of graphene oxide nanosheets and β-FeOOH nanorods; (FeOOH)-5 layers: LbL-treated foam (5 trilayers), alternately dipped into branched polyethyleneimine solution (cationic), alginate solution (anionic) and β-FeOOH nanorod suspension (cationic); (GO)-5 layers: LbL-treated foam (5 bilayers), alternately dipped into branched polyethyleneimine solution (cationic), alginate solution/graphene oxide nanosheet suspension (anionic). Reprinted with permission from (Pan et al. 2016c). Copyright 2016 Elsevier.

designed and prepared by Batool et al. (2022). Prior to deposition, quartz or silicon, used as model substrates for preparing the multilayered coatings, were dipped in a polyethyleneimine solution to cationize their surface. As assessed by vertical flame spread tests, all the quad-layered assemblies achieved a V-0 rating (Figure 11.16).

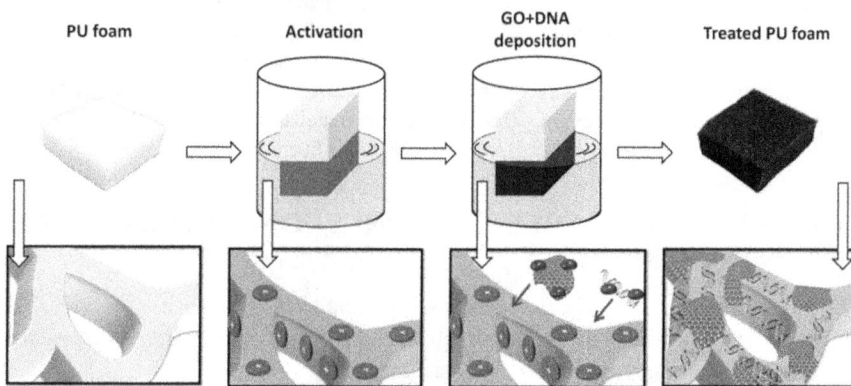

FIGURE 11.14 Scheme of the single-step deposition. Polyurethane foams are first activated in polyacrylic acid solution and then dipped in anionic graphene oxide platelet/deoxyribonucleic acid suspension. Reprinted from (Maddalena, Carosio and Fina 2021) under CC BY 4.0 license.

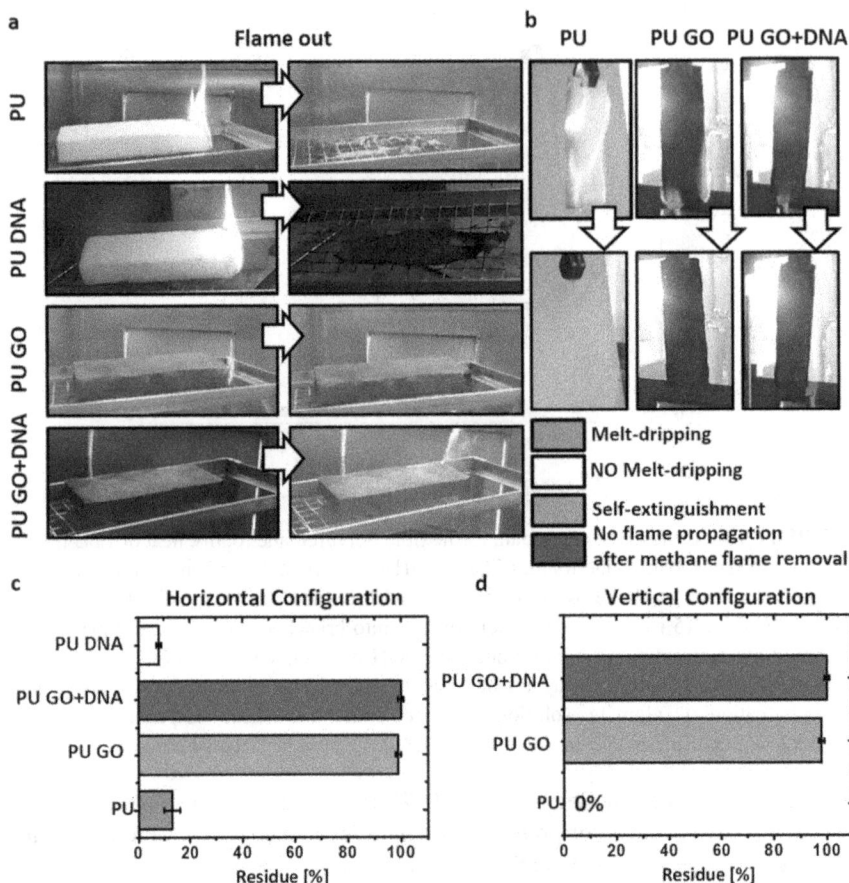

FIGURE 11.15 Snapshots from flame spread tests carried out in (a) horizontal and (b) vertical configuration of untreated vs. coated foams. End-of-test residues and flammability behavior in the (c) horizontal and (d) vertical configurations. PU: untreated foam; PU DNA: foam treated with DNA only; PU GO: foam treated with graphene oxide platelets only; PU GO+DNA: foam treated with graphene oxide platelets and DNA. Reprinted from (Maddalena, Carosio, and Fina 2021) under CC BY 4.0 license.

11.5   FLAME RETARDANT LbL COATINGS CONTAINING OTHER 2D NANOPARTICLES

The scientific literature reports some nice examples about the use of other 2D nanoparticles for the buildup of flame retardant LbL assemblies. The main research outcomes are summarized in the next paragraphs.

Pan and coworkers (Pan et al. 2015) constructed trilayered assemblies (chitosan/α-zirconium phosphate/sodium alginate) on a flexible polyurethane foam; in particular, as observed in forced-combustion tests (irradiative heat flux: 35 kW/m^2), 9 deposited trilayers markedly decreased the peak of heat release rate (−71%) with respect to untreated polyurethane.

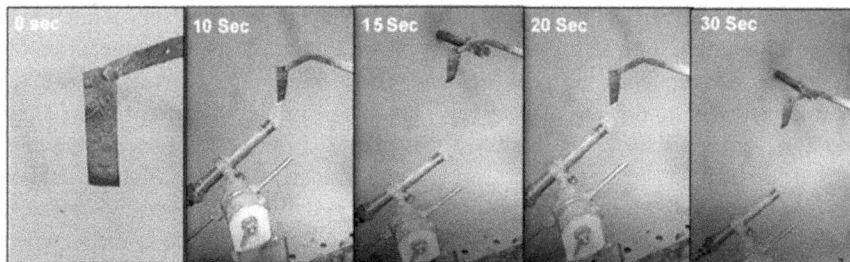

FIGURE 11.16 Snapshots of vertical flame spread tests carried out on 20 quad-layers made of poly(vinyl alcohol), graphene oxide nanosheets, phosphorylated chitin, and laponite. Reprinted with permission from (Batool et al. 2022). Copyright 2022 Elsevier.

MnO_2 nanosheets were combined in a trilayered assembly with polyethyleneimine and sodium alginate deposited on a flexible polyurethane foam (Wang et al. 2016). Two, four, and six trilayers were assembled. As revealed by forced combustion tests (under 35 kW/m^2 irradiative heat flux), the physical insulation provided by MnO_2 nanosheets accounted for a significant decrease in the peak of heat release rate (Figure 11.17).

In a further research effort, the same group designed bilayered flame retardant assemblies based on sodium alginate in combination with either Mg-Al layered double hydroxide (Pan et al. 2016d) or α-cobalt hydroxide (Mu et al. 2017). The first LbL architecture (made of 5, 10 or 20 sodium alginate/Mg-Al layered double hydroxide bilayers) was deposited on cotton fabrics, while the second (made of 1, 2, 3, or 4 sodium alginate/α-cobalt hydroxide bilayers) was deposited on a flexible polyurethane foam.

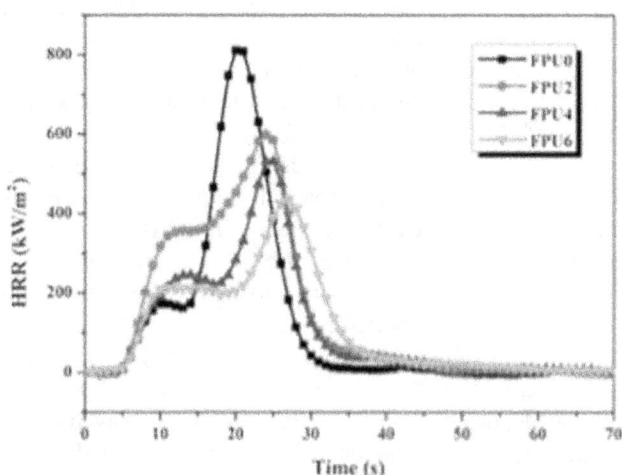

FIGURE 11.17 Heat release rate vs. time curves for flexible polyurethane foam samples. FPU0: untreated foam; FPUx: LbL-treated foam, where x represents the number of deposited MnO_2 nanosheets/polyethyleneimine/sodium alginate trilayers. Reprinted with permission from Wang et al. (2016). Copyright 2016 Elsevier.

As revealed by vertical flame spread tests, 20 bilayers provided self-extinction to cotton. In addition, in microscale combustion calorimetry tests, the same assembly noticeably lowered both the peak of heat release rate and total heat release (by approximately 35 and 26%, respectively) compared to the untreated fabric. These findings were ascribed to the barrier effect provided by the 2D nanofiller, which slowed the heat and mass transfer during the degradation of the cellulosic fabric.

As evaluated by forced-combustion tests (irradiative heat flux: 35 kW/m^2), replacing Mg-Al layered double hydroxide with α-cobalt hydroxide in the bilayered promoted resulted in, with just one bilayer applied on the polyurethane foam, an outstanding decrease in the peak of heat release rate (−59%) with respect to the untreated substrate. The obtained results were attributed to the char-forming character of α-cobalt hydroxide nanosheets that contributed to the graphitization of the carbonaceous residues during combustion, hence forming a stable protective layer on the underlying degrading polymer.

11.6 CONCLUSION AND PERSPECTIVES

It is undeniable that the layer-by-layer approach that exploits 2D nanostructures is a valuable method not only for nanostructuring different types of surfaces but also for enhancing the flame retardancy of the treated substrates, also, in some cases, providing multifunctional features.

Several advantages are offered by LbL, such as the practically unrestricted choice in the design of the assemblies, which may represent the most important added value. In particular, specifically referring to flame retardancy, 2D nanostructure-containing assemblies have played and are still playing an important role in limiting or even stopping the flame propagation or the degradation of a polymer substrate exposed to an irradiative heat flux. Their main flame retardant mechanism relies on the formation of a continuous and coherent ceramic layer in the condensed phase, which protects the underlying polymer (i.e., through a thermal shielding effect), limiting, at the same time, the mass transfer phenomena involved during combustion (i.e., barrier effect).

In addition, the flame retardant mechanisms involved may be multiple, as they strictly rely on the structure (i.e., number and morphology) and chemical composition of the deposited assemblies. Therefore, in some cases, synergistic effects may take place between the components of the LbL flame retardant architectures.

Furthermore, the possibility of selecting biosourced constituents for the assemblies implements the current circular economy concept (Kalmykova, Sadagopan and Rosado 2018), with significant improvements, especially considering the low environmental impact of some of the designed LbL flame retardant architectures.

In addition, the deposited flame-retarded assemblies may provide the treated polymer substrates with multifunctional features (such as hydrophobicity, selectivity for oil-water separation, thermal and/or electrical conductivity, and antibacterial activity), hence widening the possible application fields.

However, despite the very high potential of this surface engineering technique, several drawbacks limit its full exploitation. First, scalability as a minimum from the lab to the semipilot or pilot scale is a current challenging issue that is not easy to overcome. Dipping-assisted LbL (which is the most common procedure) is tedious and time consuming and has to be performed with a large manual practice, although some automatic robots are currently available on the market. At variance, spray-assisted LbL appears more appropriate for industrial exploitation, as it is much faster than dipping and prevents cross-contamination of the suspension/solution baths.

A further important limit of the current LbL assemblies for flame retardant purposes, especially those specifically designed for fabrics, refers to the poor durability (i.e., washing fastness) of the deposited layers. In fact, the latter are usually derived from aqueous suspension/solution baths and are mainly based on electrostatic interactions. Therefore, if covalent bonds among the deposited layers cannot be formed, it is very likely that the assemblies will be destroyed during laundry cycles, often carried out in the presence of surfactants. In this way, the flame retardant (or even multifunctional) properties previously conferred to the fabrics are usually partially or fully lost during laundry occasions because of the (partial) removal of the assemblies.

Generally, durability also relates to the stability of LbL coatings, which may undergo physical friction or may have to withstand bacterial erosion: again, in these circumstances, the chance to design covalently bonded assemblies may significantly improve their weatherability. In this context, the LbL deposition of flame retardant superhydrophobic assemblies could represent a possible solution.

In brief, the potential of the LbL approach for designing novel, efficient and "green" flame retardant assemblies surely explains its rapid and continuous growth over diverse polymer substrates; it is likely to expect some flame retardant LbL systems "ready to the market" within the forthcoming years.

Furthermore, the integration of some fabrication procedures within the LbL process will help to remarkably reduce the process times and confer particular properties to the obtained assemblies.

Finally, it would be desirable to design proper and consistent methods to predict the properties and performances of LbL flame retardant assemblies, even before their deposition on polymer substrates, simply on the basis of the experimental parameters chosen for deposition (namely, assembly procedure, chemical composition of the selected layers, and deposition conditions); undoubtedly, this would reduce the gap between LbL design and industrial exploitation.

REFERENCES

Akovali G. and Takrouri F. 1991. "Studies on modification of some flammability characteristics by plasma. II. Polyester fabric." *Journal of Applied Polymer Science* 42:2717–2725. doi: 10.1002/app.1991.070421010

Alongi J. and Malucelli G. 2012. "State of the art and perspectives on sol–gel derived hybrid architectures for flame retardancy of textiles." *Journal of Materials Chemistry*, 22:21805–21809. doi: 10.1039/C2JM32513F

Batool S., Guo W., Gill R., Xin W., and Hu Y. 2022. "Chitin based multilayered coatings with flame retardancy an approach to mimic nacre: Synthesis, characterization and mechanical properties." *Carbohydrate Polymers* 291:119488. doi: 10.1016/j.carbpol.2022.119488

Benten H., Ogawa M., Ohkita H., and Ito S. 2008. "Design of multilayered nanostructures and donor-acceptor interfaces in solution-processed thin-film organic solar cells." *Advanced Functional Materials* 18:1563–1572. doi: 10.1002/adfm.200701167

Carosio, F., Alongi, J., and Frache A. 2011. "Influence of surface activation by plasma and nanoparticle adsorption on the morphology, thermal stability and combustion behavior of PET fabrics." *European Polymer Journal* 47:893–902. doi: 10.1016/j.eurpolymj.2011.01.009

Chang L., Kong X., Wang F., Wang L., and Shen J. 2008. "Layer-by-layer assembly of poly (N-acryloyl-N'-propylpiperazine) and poly (acrylic acid): Effect of pH and temperature". *Thin Solid Films* 516:2125–2129. doi: 10.1016/j.tsf.2007.07.188

Chang S.C., Slopek R.P., Condon B., and Grunlan J.C. 2014. "Surface Coating for Flame-Retardant Behavior of Cotton Fabric Using a Continuous Layer-by-Layer Process". *Industrial & Engineering Chemistry Research* 53:3805–3812. doi: 10.1021/ie403992x

Decher G., and Hong J. D. 1991. "Buildup of Ultrathin Multilayer Films by a Self-assembly Process. 1. Consecutive Adsorption of Anionic and Cationic Bipolar Amphiphiles on Charged Surfaces." *Makromolekulare Chemie. Macromolecular Symposia* 46:321–327. doi: 10.1002/masy.19910460145

Deng S.B., Liao W., Yang J.C., Cao Z.J., and Wang Y.Z. 2016. "Flame-retardant and smoke suppressed silicone foams with chitosan-based nanocoatings." *Industrial & Engineering Chemistry Research* 55:7239–7248. doi: 10.1021/acs.iecr.6b00532

Fang M., Kaschak D.M., Sutorik A.C., and Mallouk T.E. 1997. "A 'mix and match' ionic-covalent strategy for self-assembly of inorganic multilayer films." *Journal of the American Chemical Society* 119:12184–12191. doi: 10.1021/ja972569e

Fink J.K. 2020. *Flame retardants. Materials and applications.* Hoboken, USA: Wiley Scrivener Publishing LLC.

Fujita S., and Shiratori S. 2006. "The optical properties of ultrathin films fabricated by layer-by-layer adsorption process depending on dipping time." *Thin Solid Films*, 499:54–60. doi: 10.1016/j.tsf.2005.07.039

Hilado, C. J. 1998. *Flammability handbook for plastics.* 5th Edition, Boca Raton, USA: CRC Press.

Horrocks A.R., and Price D. (2000). *Fire retardant materials.* Amsterdam, Holland: Woodhead Publishing Limited.

Hu Y., and Wang X. (2020). *Flame retardant polymeric materials. A Handbook.* Boca Raton, USA: CRC Press.

Huang G., Liang H., Wang X., and Gao J. 2012. "Poly(acrylic acid)/clay thin films assembled by Layer-by-Layer deposition for improving the flame retardancy properties of cotton." *Industrial & Engineering Chemistry Research* 51:12299–12309. doi: 10.1021/ie300820k

Iler R. K. 1966. "Multilayers of colloidal particles." *Journal of Colloid and Interface Science* 21:569–594. doi: 10.1016/0095-8522(66)90018-3

Ichinose I., Kawakami T., and Kunitake T. 1998. "Alternate molecular Layers of metal oxides and hydroxyl polymers prepared by the surface sol-gel process." *Advanced Materials* 10:535–539. doi: (SICI)1521-4095(199805)10:7<535::AID-ADMA535>3.0.CO;2-Q

Kalmykova Y., Sadagopan M., and Rosado L. 2018. "Circular Economy – from Review of Theories and Practices to Development of Implementation Tools." *Resources, Conservation and Recycling* 135:190–201. doi: 10.1016/j.resconrec.2017.10.034

Kim Y.S., Harris R., and Davis R. 2012. "Innovative approach to rapid growth of highly clay filled coatings on porous polyurethane foam." *ACS Macro Letters* 1:820–824. doi: 10.1021/mz300102h

Kumar Kundu C., Song L., and Hu Y. 2021. "Sucrose derivative as a cross-linking agent in enhancing coating stability and flame retardancy of polyamide 66 textiles." *Progress in Organic Coatings* 159:106438. doi: 10.1016/j.porgcoat.2021.106438

Lazar S., Carosio F., Davesne A.-L., Jimenez M., Bourbigot S., and Grunlan J. 2018. "Extreme Heat Shielding of Clay/Chitosan Nanobrick Wall on Flexible Foam." *ACS Applied Materials & Interfaces* 10:31686–31696. doi: 10.1021/acsami.8b10227

Li Y.C., Schulz J., and Grunlan J.C. 2009. "Polyelectrolyte/nanosilicate thin-film assemblies: influence of pH on growth, mechanical behavior, and flammability." *ACS Applied Materials & Interfaces* 1:2338–2347. doi: 10.1021/am900484q

Li Y.C., Schulz J., Mannen S., Delhom C., Condon B., Chang S., Zammarano M., and Grunlan J.C. 2010. "Flame retardant behavior of polyelectrolyte–clay thin film assemblies on cotton fabric." *ACS Nano* 4:3325–3337. doi: 10.1021/nn100467e

Li Y.C., Kim Y.S., Shields J.R., and Davis R.D. 2013. "Controlling polyurethane foam flammability and mechanical behavior by tailoring the composition of clay-based multilayer nanocoatings." *Journal of Materials Chemistry A* 1:12987–12997. doi: 10.1039/C3TA11936J

Li Y.C., Yang Y.H., Shields J.R., and Davis R.D. 2015. "Layered double hydroxide-based fire resistant coatings for flexible polyurethane foam." *Polymer* 56:284–292. doi: 10.1016/j.polymer.2014.11.023

Maddalena L., Carosio F., and Fina A. 2021. "In Situ Assembly of DNA/Graphene Oxide Nanoplates to Reduce the Fire Threat of Flexible Foams." *Advanced Materials Interfaces* 8:2101083. doi: 10.1002/admi.202101083

Malucelli G., Carosio F., Alongi J., Fina A., Frache A., and Camino G. 2014. "Materials engineering for surface-confined flame retardancy." *Materials Science Engineering R: Reports* 84:1–20. doi: 10.1016/j.mser.2014.08.001

Mermut O., and Barrett, C.J. 2003. "Effects of charge density and counterions on the assembly of polyelectrolyte multilayers." *The Journal of Physical Chemistry B* 107:2525–2530. doi: 10.1021/jp027278t

McAloney R.A., Sinyor M., Dudnik V., and Goh M.C. 2001. "Atomic force microscopy studies of salt effects on polyelectrolyte multilayer film morphology." *Langmuir* 17:6655–6663. doi: 10.1021/la010136q

Morgan A.B. 2019. "The Future of Flame Retardant Polymers – Unmet Needs and Likely New Approaches." *Polymer Reviews* 59:25–54. doi: 10.1080/15583724.2018.1454948

Morgan A.B. 2022. *Nonhalogenated flame retardant Handbook*, 2nd Edition. Hoboken, USA: Wiley Scrivener Publishing LLC.

Mu X.W., Yuan B.H., Pan Y., Feng X.M., Duan L.J., Zong R.W., and Hu Y. 2017. "A single alpha-cobalt hydroxide/sodium alginate bilayer layer-by-layer assembly for conferring flame retardancy to flexible polyurethane foams." *Materials Chemistry and Physics* 191:52–61. doi: 10.1016/j.matchemphys.2017.01.023

New J., Zope I.S., Abdul Rahman S.N., Yap X.L.W., and Dasari A. 2016. "Physiological comfort and flame retardancy of fabrics with electrostatic self-assembled coatings." *Materials & Design* 89:413–420. doi: 10.1016/j.matdes.2015.10.006

Novoselov K.S., Falko V.I., Colombo L., Gellert P.R., Schwab M.G., and Kim K. 2012. "A roadmap for graphene." *Nature* 490:192–200. doi: 10.1038/nature11458

Pan Y., Pan H., Yuan B., Hong N., Zhan J., Wang B., Song L., and Hu Y. 2015. "Construction of organic–inorganic hybrid nanocoatings containing α-zirconium phosphate with high efficiency for reducing fire hazards of flexible polyurethane foam." *Materials Chemistry and Physics* 163:107–115. doi: 10.1016/j.matchemphys.2015.07.020

Pan H., Yu B., Wang W., Pan Y., Song L., and Hu Y. 2016a. "Comparative study of layer by layer assembled multilayer films based on graphene oxide and reduced graphene oxide on flexible polyurethane foam: flame retardant and smoke suppression properties." *RSC Advances* 6:114304–114312. doi: 10.1039/C6RA15522G

Pan H., Lu Y., Song L., Zhang X., and Hu Y. 2016b. "Fabrication of binary hybrid-filled layer by layer coatings on flexible polyurethane foams and studies on their flame-retardant and thermal properties." *RSC Advances* 6:78286–78295. doi: 10.1039/C6RA03760G

Pan H., Lu Y., Song L., Zhang X., and Hu Y. 2016c. "Construction of layer-by-layer coating based on graphene oxide/β-FeOOH nanorods and its synergistic effect on improving flame retardancy of flexible polyurethane foam." *Composite Science and Technology* 129:116–122. doi: 10.1016/j.compscitech.2016.04.018

Pan H., Wang W., Shen Q., Pan Y., Song L., Hu Y., and Lu Y. 2016d. "Fabrication of flame retardant coating on cotton fabric by alternate assembly of exfoliated layered double hydroxides and alginate." *RSC Advances* 6:111950–111958. doi: 10.1039/C6RA21804K

Pan Y., Liang Q., Song L., and Zhao H. 2021. "Fabrication of layer-by-layer self-assembled coating modified cotton fabric with flame retardancy and hydrophobicity based on sepiolite." *Polymer-Plastics Technology and Materials* 60 (12):1368–1376. doi: 10.1080/25740881.2021.1904982.

Qiu X., Li Z., Li X., and Zhang Z. 2018. "Flame retardant coatings prepared using layer by layer assembly: A review." *Chemical Engineering Journal* 334:108–122. doi: 10.1016/j.cej.2017.09.194

Richardson J. J., Björnmalm M., and Caruso F. 2015. "Technology-driven layer-by-layer assembly of nanofilms." *Science* 348:aaa2491-1/11. doi:10.1126/science.aaa2491

Richardson J. J., Cui J., Björnmalm M., Braunger J.A., Ejima H., and Caruso F. 2016. "Innovation in Layer-by-Layer Assembly." *Chemical Reviews* 116:14828–14867. doi: 10.1021/acs.chemrev.6b00627

Shimazaki Y., Mitsuishi M., Ito S., and Yamamoto M. 1997. "Preparation of the Layer-by-Layer deposited ultrathin film based on the charge-transfer interaction." *Langmuir* 13:1385–1387. doi: 10.1021/la9609579

Shiratori S.S., and Rubner M.F. 2000. "pH-dependent thickness behavior of sequentially adsorbed layers of weak polyelectrolytes." *Macromolecules* 33:4213–4219. doi: 10.1021/ma991645q

Srikulkit K., Iamsamai C., and Dubas S.T. 2006. "Development of flame retardant poly-phosphoric acid coating based on the polyelectrolyte multilayers technique." *International Journal of Minerals, Metallurgy and Materials* 16:41–45.

Stockton W.B., and Rubner M.F. 1997. "Molecular-level processing of conjugated polymers. Layer-by-layer manipulation of polyaniline via hydrogen-bonding interactions." *Macromolecules* 30:2717–2725. doi: 10.1021/ma9700486

Sui Z.J., Salloum D., and Schlenoff J.B. 2003. "Effect of molecular weight on the construction of polyelectrolyte multilayers: stripping versus sticking." *Langmuir* 19:2491–2495. doi: 10.1021/la026531d

Tan H. L., McMurdo M.J., Pan G.Q., and Van Patten P.G. 2003. "Temperature dependence of polyelectrolyte multilayer assembly." *Langmuir* 19:9311–9314. doi: 10.1021/la035094f

Ur Rehman Z., Huh S.-H., Ullah Z., Pan Y.-T., Churchill D.G., and Koo B.H. 2021. "LBL generated fire retardant nanocomposites on cotton fabric using cationized starch-clay-nanoparticles matrix." *Carbohydrate Polymers* 274:118626. doi: 10.1016/j.carbpol.2021.118626

Vahabi H., Saeb M.R. & Malucelli G. (Eds.) (2022). *Analysis of flame retardancy in polymer science*. New York, USA: Elsevier.

Wang W., Pan Y., Pan H., Yang W., Liew K.M., Song L., and Hu Y. 2016. "Synthesis and characterization of MnO2 nanosheets based multilayer coating and applications as a flame retardant for flexible polyurethane foam." *Composite Science and Technology* 123:212–221. doi: 10.1016/j.compscitech.2015.12.014

Wang L., Cui S., Wang Z., Zhang X., Jiang M., Chi L., and Fuchs H. 2000. "Multilayer assemblies of copolymer PSOH and PVP on the basis of hydrogen bonding." *Langmuir* 16:10490–10494. doi: 10.1021/la000733x

Wei C., Zeng S., Tan Y., Wang W., Lv J., and Liu H. 2015. "Impact of Layer-by-Layer self-assembly clay-based nanocoating on flame retardant properties of sisal fiber cellulose microcrystals." *Advances in Materials Science and Engineering* 3:1–7. doi: 10.1155/2 015/691290

Xu T., Zhang L., Zhong Y., and Mao Z. 2014. "Fire retardancy and durability of poly(N-benzyloxycarbonyl-3,4-dihydroxyphenylalanine)-montmorillonite composite film coated polyimide fabric." *Journal of Applied Polymer Science* 131:39608–39615. doi: 10.1002/APP.39608

Yang Y.H., Li Y.C., Shields J.R., and Davis R.D. 2015. "Layer double hydroxide and sodium montmorillonite multilayer coatings for the flammability reduction of flexible poly-urethane foams." *Journal of Applied Polymer Science* 132:41767–41774. doi: 10.1002/app.41767

Zhang X., Shen Q., Zhang X., Pan H., Lu Y. 2016. "Graphene oxide-filled multilayer coating to improve flame-retardant and smoke suppression properties of flexible polyurethane foam." *Journal of Materials Science* 51:10361–10374. doi: 10.1007/s10853-016-0247-3

Zhang X., Xu Y., Zhang X., Wu H., Shen J., Chen R., Xiong Y., Li J., and Guo S. 2019. "Progress on the layer-by-layer assembly of multilayered polymer composites: Strategy, structural control and applications." *Progress in Polymer Science* 89:76–107. doi: 10.1016/j.progpolymsci.2018.10.002

Zhou X., Fu Q., Zhang Z., Fang Y., Wang Y., Wang F., Song Y., Pittman Jr. C.U., and Wang Q. 2021. "Efficient flame-retardant hybrid coatings on wood plastic composites by layer-by-layer assembly." *Journal of Cleaner Production* 321:128949. doi: 10.1016/j.jclepro.2021.128949

Zhu G., Wang J., Yuan X., and Yuan B. 2022. "Hydrophobic and fire safe polyurethane foam coated with chitosan and nanomontmorillonite via layer-by-layer assembly for emer-gency absorption of oil spill." *Materials Letters* 316:132009. doi: 10.1016/j.matlet.2022.132009

12 Challenges and Perspectives of Two-Dimensional Nanomaterials for Fire-Safe Polymers

Yu-Ting Yang and Xin Wang
State Key Laboratory of Fire Science, University of Science and Technology of China, Hefei, Anhui, PR China

CONTENTS

12.1 INTRODUCTION

Two-dimensional materials are a new pattern of sheet-like nanomaterials, including layered double hydroxide (LDHs), graphene (oxide) (GO), montmorillonite (MMT), molybdenum disulphide (MoS_2), black phosphorene (BP), boron nitride, MXene and carbon nitride, and have been extensively studied as fillers for polymer-based composites in recent years. These materials have a good interaction with the polymer matrix and can greatly enhance the properties of polymers at even very little loading because of their large specific area with lateral sizes from hundreds of nanometres to a few micrometres and small average thickness (Liu, Ullah et al. 2019). Due to their unique layered structure and high thermal stability, two-dimensional nanomaterials have been widely applied in the field of fire-safe polymers (Wang, Pan et al. 2016). In addition, the uniform dispersion of a small amount of layered two-dimensional nanomaterials in a polymer matrix can not only greatly improve the mechanical properties (Cai, Zhang et al. 2018) and thermal stability (Kaul, Samson et al. 2017) of polymers, but can also further improve the barrier effect of residual char in heat and mass transfer (Kashiwagi, Du et al. 2005, Zare 2015, Zhou, Gao et al. 2017). Different two-dimensional nanomaterials have potential electrical conductivity, excellent catalytic activity, thermal conductivity, and UV resistance, which can meet the flame retardant demands of polymer composites in various applications. These attractive physical properties make two-dimensional nanomaterials ideal candidate materials for producing fire-safe polymer nanocomposites.

Figure 12.1 shows the growth trend of the literature on two-dimensional nano-materials for fire-safe polymers over the last decade. The number of publications using ((flame retardant OR fire safety OR flammability) AND (clay OR montmo-rillonite OR layered double hydroxide OR MXenes OR graphene (oxide) OR carbon nitride OR layered boron nitride OR molybdenum disulphide OR black phosphorus)) as keywords in ISI Web of Science has increased significantly, which indicates that fire-safe polymer nanocomposites are a rising research hotspot and frontier field.

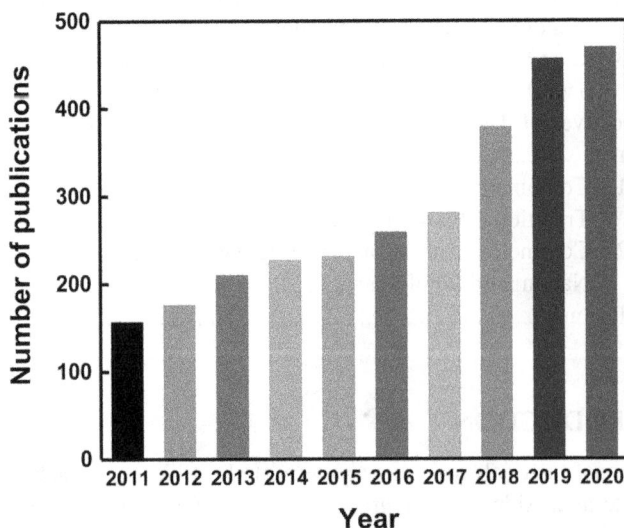

FIGURE 12.1 The number of publications involving two-dimensional nano-materials for fire-safe polymer nanocomposites.

12.2   CHALLENGES OF TWO-DIMENSIONAL NANOMATERIALS FOR FIRE-SAFE POLYMERS

12.2.1   Dispersion and Distribution of Two-Dimensional Nanomaterials

Due to their unique layered structure, two-dimensional nanomaterials have been widely studied in the field of flame retardant polymeric materials. The mechanism of flame retardancy involves the formation of physical barriers and continuous char layers (Schartel, Weiss et al. 2011). Two-dimensional nanomaterials can be used as effective physical barriers to reduce the mass loss rate during the thermal decomposition process. Additionally, two-dimensional nanomaterials are conducive to forming a protective char layer on the burning material surface (Wu, Schartel et al. 2012). The protective layer can isolate oxygen from burning materials to reduce the transfer of combustible volatiles and heat to the flame zone. However, the mode of action of fire-safe polymers containing two-dimensional nanomaterials is considerably dependent upon the dispersion state of two-dimensional nanomaterials. Tang's group investigated the influence of the dispersion state of octadecylammonium modified montmorillonite (OMMT) on the flammability of polypropylene (PP) nanocomposites (Song, Wang et al. 2007). Two kinds of OMMT containing different loadings (80% and 150%) of organo-modifier were studied, which were denoted as OMMT-80 and OMMT-150. OMMT-80 or OMMT-150 was prone to aggregate within the PP matrix (Figure 12.2a and 12.2c), while the presence of maleic anhydride-grafted-polypropylene copolymer (PP-MA) (serving as a compatibilizer) was conducive to forming OMMT-intercalated PP nanocomposites (Figure 12.2b and 12.2d). The better dispersion state of OMMT in the PP/PP-MA matrix was more effective in decreasing the peak heat release rate (PHRR) during combustion, as demonstrated by the comparison of the heat release rate curves of PP/OMMT-80 versus PP/PP-MA/OMMT-80 (Figure 12.3a) and PP/OMMT-150 *versus* PP/PP-MA/ OMMT-150 (Figure 12.3b). Figure 12.4 illustrates the mode of action of the dispersion state of two-dimensional nanomaterials on the flammability of polymers (Wang, Kalali et al. 2015). The good dispersion state of two-dimensional nanomaterials is beneficial to form a continuous and dense char layer that covers the surface of burning polymers to block heat and mass exchange, while the poor dispersion state of two-dimensional nanomaterials is apt to form char residues with cracks and holes that cannot inhibit the diffusion of flammable volatiles and heat irradiation.

Until now, meticulous control of the dispersion of two-dimensional nanomaterials has remained the main problem in effectively improving the fire safety and other properties of polymer-based nanocomposites (Bartholmai and Schartel 2004). For the sake of solving the problem of uniform dispersion of two-dimensional nanomaterials in a polymer matrix, surface functionalization, covalent grafting and hybridization have been proven to be effective and simple methods to prepare polymer-based nanocomposites with good dispersion and enhanced performance. The type and loading of modifiers for two-dimensional nanomaterials also have a considerable influence on flammability. Table 12.1 lists the comparison of the effect of unmodified and modified two-dimensional nanomaterials on the fire-safe characteristics of polymers. Modified two-dimensional nanomaterials are more efficient in suppressing heat

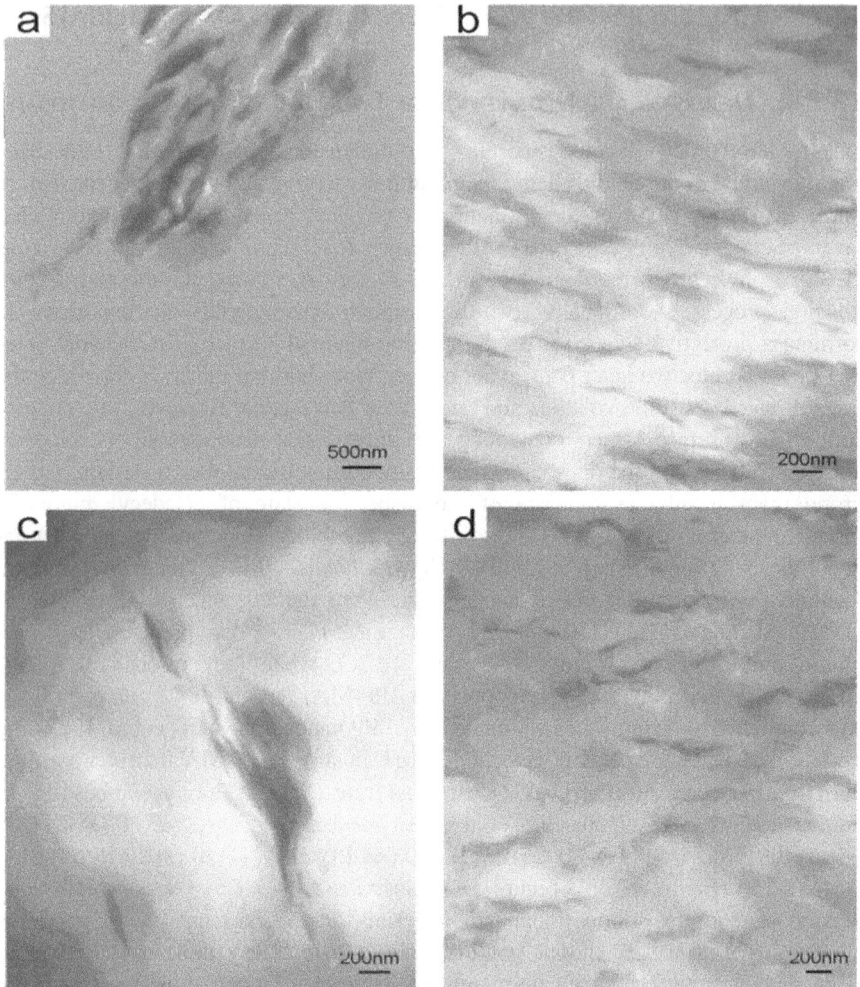

FIGURE 12.2 TEM images of (a) PP/OMMT-80, (b) PP/PP-MA/OMMT-80, (c) PP/OMMT-150 and (d) PP/PP-MA/OMMT-150.

Source: Reprinted from Song, R. et al., *J. Appl. Polym. Sci.*, 106, 3488–3494, 2007. With permission.

release and smoke emission than unmodified ones during the combustion process, owing to the more uniform dispersion state of modified two-dimensional nanomaterials and synergistic effect between two-dimensional nanomaterials and modifiers.

12.2.2 LIFE CYCLE ASSESSMENT OF FIRE-SAFE TWO-DIMENSIONAL NANOMATERIAL/POLYMER COMPOSITES

The increasing exploitation of two-dimensional nanomaterials requires a comprehensive assessment of the latent impact of two-dimensional nanomaterials on

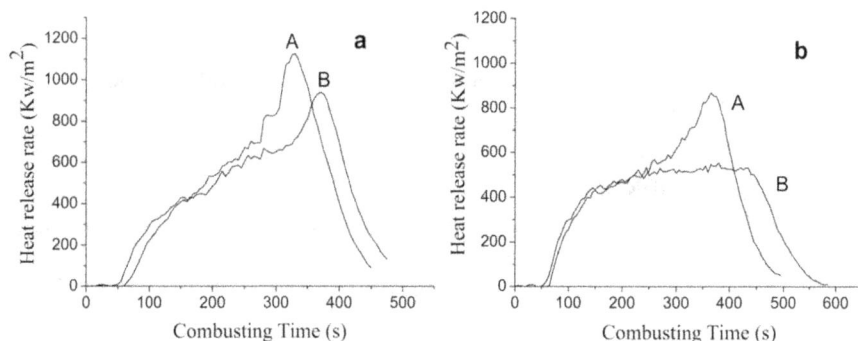

FIGURE 12.3 Comparison of heat release rates for PP/OMMT-80, PP/PP-MA/OMMT-80, PP/OMMT-150 and PP/PP-MA/OMMT-150 at 35 kW/m^2 heat flux. (a) A: PP/OMMT-80, B: PP/PP-MA/OMMT-80; (b) A: PP/OMMT-150, B: PP/PP-MA/OMMT-150.

Source: Reprinted from Song, R. et al., *J. Appl. Polym. Sci.*, **106**, 3488–3494, 2007. With permission.

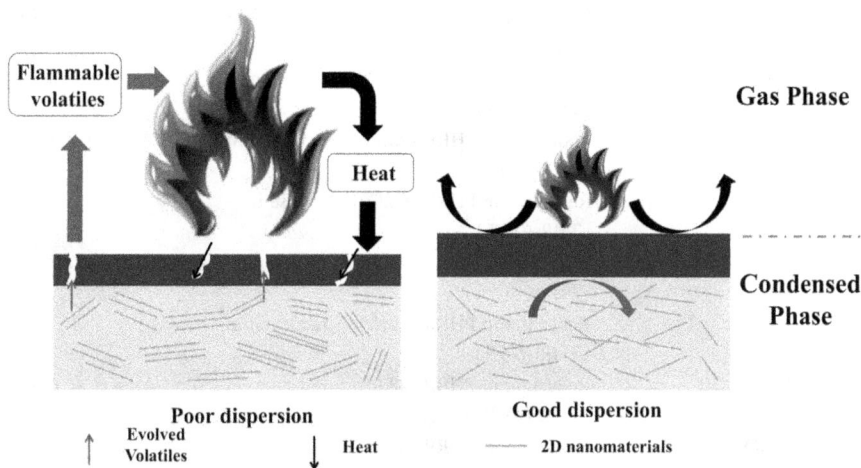

FIGURE 12.4 Schematic illustration of the mode of action of the dispersion state of two-dimensional nanomaterials on flammability of polymers.

Source: Reprinted from Wang, X. et al., *ACS Sustainable Chem. Eng.*, **3**, 3281–3290, 2015. With permission.

human health and the environment. The development of new nanomaterials and their biomedical applications is increasing, which has caused people to pay more attention to their biological safety and potential health effects (Tsukanov, Turk et al. 2022). Additionally, some studies have noted the risk of occupational exposure with further commercialization of two-dimensional nanomaterials (Wazalwar, Sahu et al. 2021). The most common situation of human occupational

TABLE 12.1

The Influence of Modified Two-Dimensional Nanomaterials on the Flammability of Polymer Nanocomposites

Polymer Matrix	Types of 2D Nanomaterials	Achieved Properties	References
PS	PANI-BNO	Reduction in PHRR (31.3%), reduction in THR (1.5%)	Zhi et al. (2018)
	BNO	Reduction in PHRR (28.9%), increase in THR (1.4%)	
TPU	PANI-BNO	Reduction in PHRR (32.6%), reduction in THR (10.8%)	Zhi et al. (2018)
	BNO	Reduction in PHRR (22.8%), reduction in THR (7.8%)	
EP	MoS_2@PPN	Reduction in PHRR (30.7%), reduction in THR (23.6%)	Qiu et al. (2018)
	MoS_2	Reduction in PHRR (19.6%), reduction in THR (21.7%)	
UPR	MoS_2-HPA	Reduction in PHRR (43.2%), reduction in THR (39.6%)	Wang et al. (2017)
	MoS_2	Reduction in PHRR (12.7%), reduction in THR (4.6%)	
EP	LDH-β-FeOOH	Reduction in PHRR (28.7%), reduction in THR (8.9%)	Wang et al. (2016)
	LDH	Reduction in PHRR (6.2%), increase in THR (0.8%)	
EVA	PAHPA-LDHs	Reduction in PHRR (42.7%), reduction in THR (18.5%)	Huang et al. (2012)
	LDHs	Reduction in PHRR (30.4%), reduction in THR (6.4%)	
ABS	TiO_2/GO	Reduction in PHRR (48.0%), reduction in THR (48.3%)	Attia et al. (2016)
	GO	Reduction in PHRR (41.4%), reduction in THR (37.2%)	
EP	Ce–MnO_2–GNS	Reduction in PHRR (53.7%), reduction in THR (35.5%)	Jiang et al. (2014)
	GNS	Reduction in PHRR (30.1%), reduction in THR (17.0%)	

exposure to two-dimensional materials is in the process of industrial production or laboratory/industrial waste through inhalation, eye, skin or mouth (Figure 12.5). Once settled in the lungs, these substances can cause cytotoxicity and initiate cell damage.

Regarding the environmental hazard assessment of two-dimensional nanomaterials, a variety of organisms in ecosystems have been investigated, such as bacteria,

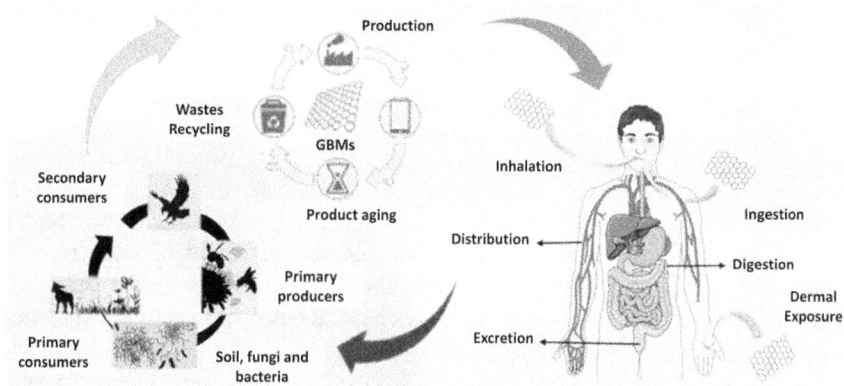

FIGURE 12.5   Schematic indicating health and environmental hazard of graphene exposure.

Source: Reprinted from Fadeel, B. et al., *ACS Nano*, 12, 10582–10620, 2018. With permission.

algae, seed plants, invertebrates, and vertebrates. For example, GO can significantly enhance the phytotoxicity of arsenic in wheat (Hu, Kang et al. 2014), cadmium in freshwater cyanobacteria, and Microcystis aeruginosa (Tang, Tian et al. 2015). It is vital to ascertain how nanomaterials interact with other pollutants co-existing in the environment with respect to sorption, conveyance, and the subsequent influences on pollutant poisonousness and biodegradability.

12.2.3   COST-EFFECTIVENESS AND SCALABILITY OF FIRE-SAFE TWO-DIMENSIONAL NANOMATERIAL/POLYMER COMPOSITES

Although two-dimensional nanomaterials show attractive prospects, their disadvantages are also obvious: low monolayer rate, wide distribution of lamellar size, and low yield. The existence of these shortcomings requires more research and breakthroughs in proposing advanced stripping methods. In the study of the morphology and structure of two-dimensional nanomaterial/polymer nanocomposites, improving the compatibility between nanoparticles and polymer matrix will be an important research direction. Most two-dimensional nanomaterials have the tendency of agglomeration and restacking. In addition, their hydrophilicity (compatibility with hydrophobic polymers) also leads to poor dispersion in polymer matrices (Yue, Li et al. 2019). Considering the economic problems of large-scale production and the restrictions on the application of fire-safe two-dimensional nanomaterial/polymer composites in terms of upgrading and reforming equipment, how to prepare high-quality and large-aspect-ratio layered nanomaterials in a large-scale manner and improve the processability of two-dimensional nanomaterials and polymer composites remain challenging tasks.

12.3 PERSPECTIVES OF TWO-DIMENSIONAL NANOMATERIALS FOR FIRE-SAFE POLYMERS

12.3.1 COMBINATION OF TWO-DIMENSIONAL NANOMATERIALS WITH TRADITIONAL FLAME RETARDANTS

The flammability characteristics of the obtained nanocomposite are the major areas of concern. In recent years, inorganic nanoparticles have attracted increasing attention as important flame retardants. The flame retardancy of each type of nanoparticle on polymer materials is different and mainly hinges on the geometric characteristics of nanomaterials, such as spherical, tubular and layered. Compared with zero-dimensional nanomaterials (such as spherical SiO_2 nanoparticles) and one-dimensional nanomaterials (such as MWNTs), two-dimensional nanomaterials show better flame retardancy. Under the same addition content, two-dimensional nanomaterials render polymer nanocomposites more reduction in HRR and MLR than tubular structures and spherical structures (Dittrich, Wartig et al. 2015, Yan, Xu et al. 2017).

In order to study the effect of carbon nanomaterials with different particle sizes and shapes on the flame retardant behaviours of PP, Dittrich, Wartig et al. (2013) compared layered thermally reduced graphite oxide (TRGO) with spherical carbon black (CB), multiwall carbon nanotubes (MWNTs), expanded graphite (EG) platelets and multilayered graphene (MLG). As shown in Figure 12.6, TRGO with good dispersibility has the most considerable reduction in the PHRR of PP composites, which is superior over EG, tubular structure nanoparticle MWNT and spherical structure nanoparticle CB. In Figure 12.6b, two-dimensional nanomaterials (TRGO and MLG) impart much better anti-flammability to PP than EG. Specifically, PP/5.0MLG250 and PP/5.0TRGO exhibited a notable reduction in PHRR of 72% and 74%, respectively.

Although two-dimensional nanomaterials are very effective in reducing the heat release rate and smoke emission, it is difficult to enable polymers to meet industrial

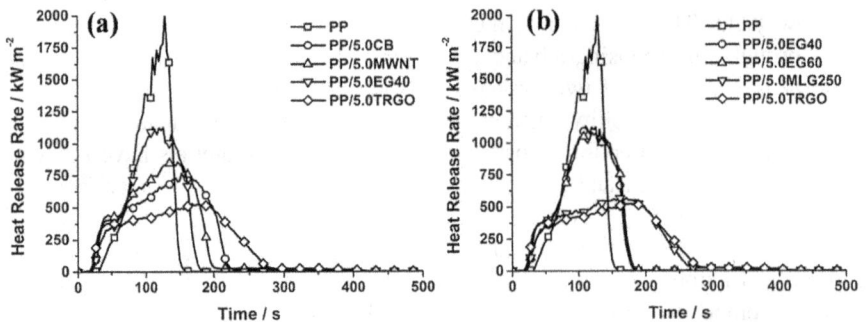

FIGURE 12.6 Heat release rate curves of PP and its carbon nanocomposites with (a) different morphologies and (b) different surface areas of carbon nanofillers.

Source: Reprinted from Dittrich, B. et al., *Polym. Degrad. Stab.*, 98, 1495–1505, 2013. With permission.

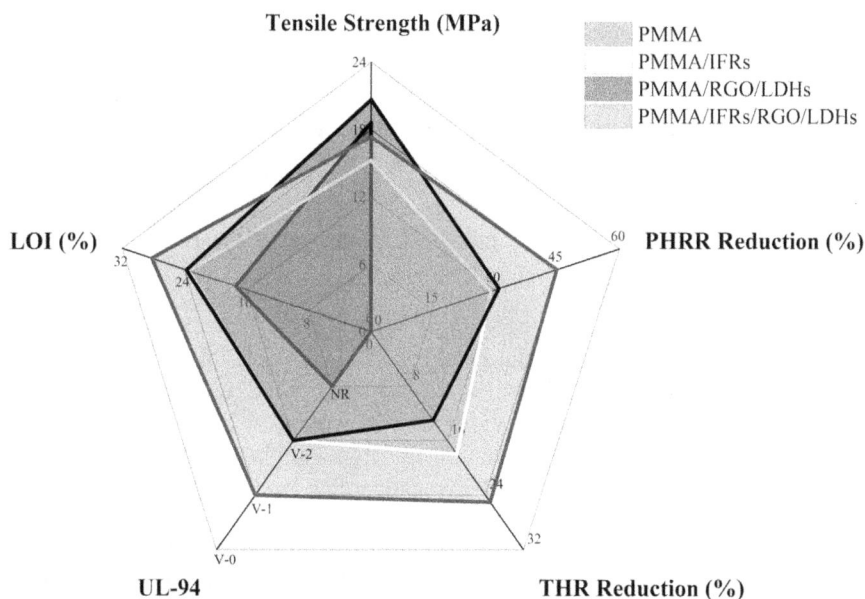

FIGURE 12.7 Scheme illustrating comparison of flame retardant and mechanical properties of polymers based on traditional flame retardants and combination of two-dimensional nanomaterials with traditional flame retardants.

flame retardant requirements, such as the LOI and UL-94 vertical combustion tests. Therefore, two-dimensional nanomaterials are usually combined with conventional flame retardants (such as compounds containing phosphorus, nitrogen or silicon elements) through physical blending or chemical modification to create a synergistic flame retardant effect (Liao, Liu et al. 2012, Bai, Jiang et al. 2014, Attia, Elashery et al. 2021). For example, Huang et al. (Huang, Chen et al. 2014) prepared an intumescent flame retardant poly(methyl methacrylate) (PMMA) nanocomposite through in situ polymerization by adding graphene, layered double hydroxides (LDHs), and intumescent flame retardants (IFRs). The fire resistance and mechanical properties of pure PMMA and its nanocomposites are shown in Figure 12.7. Compared to pure PMMA, the PHRR and THR of the PMMA/IFR composite were reduced by 29% and 18%, respectively. For the PMMA/RGO/LDH composite, the PHRR and THR exhibit 31% and 13% reductions, respectively, indicating that the addition of graphene and LDHs has an effective reduction effect on the flammability of the PMMA matrix. The PHRR and THR of the PMMA/IFR/RGO/LDH composite were reduced by 45% and 25%, respectively. Then, LOI and UL-94 tests are used to assess the flame retardant grade of the PMMA nanocomposites. The LOI value of PMMA is 17.4% and fails to pass the UL-94 test. Compared with pure PMMA, the LOI values of PMMA/IFR and PMMA/RGO/LDHs are increased to 23.4% and 23.7%, respectively, and reach the V-2 rating. The LOI value of PMMA/IFR/RGO/LDH is markedly increased to 28.2% and can reach a V-1 rating. The results above demonstrate the enhanced flame retardancy of the IFR system in combination with RGO and LDHs. It is worth noting that the tensile strength of PMMA almost does not decrease when

IFRs, LDHs and graphene coexist, which indicates that the PMMA/IFR/RGO/LDH composites generally achieve prospective mechanical properties during the application process. The combination of two-dimensional nanomaterials with traditional flame retardants is the most promising flame retardant technology for fabricating fire-safe polymers with excellent comprehensive properties.

12.3.2 COMMERCIALIZATION OF FIRE-SAFE TWO-DIMENSIONAL NANOMATERIAL/POLYMER COMPOSITES

It has been more than two decades since the rise of flame retardant polymer nanocomposites. The rise of various kinds of two-dimensional nanomaterials has spawned considerable interest in extending their applications as flame retardants for polymers; however, most studies are still in the laboratory stage. There is increasing demand that fire-safe two-dimensional nanomaterial/polymer composites be commercialized to replace conventional flame retardant polymers with high smoke emission or deteriorated mechanical/thermal properties.

Toyota Motor Center R&D Lab developed and commercialized nylon/clay hybrid materials. A small amount of inorganic filler can greatly improve the stiffness, heat resistance and barrier properties between gas and automobile fuel (Shelley, Mather et al. 2001).

The State Key Laboratory of Fire Science (SKLFS) at the University of Science and Technology of China cooperated with a renowned wire and cable company, BaoSheng Group, to apply OMMT to produce fire-safe cable materials with a low heat release rate, low smoke and low toxicity (Li, Cheng et al. 2021), which have successfully achieved large-scale commercial applications in the fields of subways and mines.

The Korea Institute of Machinery and Materials (KIMM) has developed a rolling-based non-destructive transfer technology that can transfer two-dimensional nanomaterials to the wafer scale without damage (Kim, Yoon et al. 2021). This technology has many applications, such as transparent indicators, semiconductors and autonomous vehicle indicators.

Although there are several kinds of commercialized fire-safe polymer products, the commercialization of fire-safe polymer/two-dimensional nanomaterial composites is still in the starting stage. More efforts should be made to strengthen the collaboration between scientific and industrial communities to increase the commercialization of fire-safe two-dimensional nanomaterial/polymer composites.

12.4 CONCLUSIONS

Polymer nanocomposites reinforced by two-dimensional nanomaterials are very attractive due to their enhanced thermal and mechanical properties and fire safety. Since two-dimensional nanomaterials have been used as flame retardants, their remarkable advantages have attracted the attention of researchers.

First, the dispersibility of two-dimensional nanomaterials in the polymer matrix is the primary factor in maximizing the barrier effect. Therefore, surface functionalization, covalent bond grafting and hybridization are expected to enhance the fire safety of polymer composites, including polystyrene, thermoplastic polyurethane and epoxy resin. Second, these materials may cause cytotoxicity, initiate cell damage and cause human hazards or environmental pollution. Finally, the mass production and high cost of two-dimensional nanomaterials flame retardant polymer materials have also hampered their commercialization.

It is concluded that not all two-dimensional materials are suitable for industrialization, so it is necessary to comprehensively consider the price, processability, flame retardant efficiency and other factors to screen out the optimal formula. Additionally, improving the dispersibility of two-dimensional nanomaterials in a polymer matrix maximizes the performance enhancement of composite materials. There is no doubt that fire-safe two-dimensional nanomaterial/polymer composites will exhibit broad application prospects.

REFERENCES

Attia, N. F., N. S. Abd El-Aal and M. A. Hassan (2016). "Facile synthesis of graphene sheets decorated nanoparticles and flammability of their polymer nanocomposites." *Polymer Degradation and Stability* **126**: 65–74.

Attia, N. F., S. E. A. Elashery, A. M. Zakria, A. S. Eltaweil and H. Oh (2021). "Recent advances in graphene sheets as new generation of flame retardant materials." *Materials Science and Engineering B-Advanced Functional Solid-State Materials* **274**: 115460.

Bai, Z. M., S. D. Jiang, G. Tang, Y. Hu, L. Song and R. K. K. Yuen (2014). "Enhanced thermal properties and flame retardancy of unsaturated polyester-based hybrid materials containing phosphorus and silicon." *Polymers for Advanced Technologies* **25**(2): 223–232.

Bartholmai, M. and B. Schartel (2004). "Layered silicate polymer nanocomposites: new approach or illusion for fire retardancy? Investigations of the potentials and the tasks using a model system." *Polymers for Advanced Technologies* **15**(7): 355–364.

Cai, W., D. Zhang, B. Wang, Y. Shi, Y. Pan, J. Wang, W. Hu and Y. Hu (2018). "Scalable one-step synthesis of hydroxylated boron nitride nanosheets for obtaining multifunctional polyvinyl alcohol nanocomposite films: multi-azimuth properties improvement." *Composites Science and Technology* **168**: 74–80.

Dittrich, B., K.-A. Wartig, D. Hofmann, R. Muelhaupt and B. Schartel (2013). "Flame retardancy through carbon nanomaterials: carbon black, multiwall nanotubes, expanded graphite, multi-layer graphene and graphene in polypropylene." *Polymer Degradation and Stability* **98**(8): 1495–1505.

Dittrich, B., K.-A. Wartig, D. Hofmann, R. Muelhaupt and B. Schartel (2015). "The influence of layered, spherical, and tubular carbon nanomaterials' concentration on the flame retardancy of polypropylene." *Polymer Composites* **36**(7): 1230–1241.

Hu, X., J. Kang, K. Lu, R. Zhou, L. Mu and Q. Zhou (2014). "Graphene oxide amplifies the phytotoxicity of arsenic in wheat." *Scientific Reports* **4**: 6122.

Huang, G., S. Chen, P. Song, P. Lu, C. Wu and H. Liang (2014). "Combination effects of graphene and layered double hydroxides on intumescent flame-retardant poly(methyl methacrylate) nanocomposites." *Applied Clay Science* **88–89**: 78–85.

Huang, G., Z. Fei, X. Chen, F. Qiu, X. Wang and J. Gao (2012). "Functionalization of layered double hydroxides by intumescent flame retardant: preparation, characterization, and application in ethylene vinyl acetate copolymer." *Applied Surface Science* **258**(24): 10115–10122.

Jiang, S.-D., Z.-M. Bai, G. Tang, L. Song, A. A. Stec, T. R. Hull, J. Zhan and Y. Hu (2014). "Fabrication of Ce-doped MnO$_2$ decorated graphene sheets for fire safety applications of epoxy composites: flame retardancy, smoke suppression and mechanism." *Journal of Materials Chemistry A* **2**(41): 17341–17351.

Kashiwagi, T., F. M. Du, K. I. Winey, K. A. Groth, J. R. Shields, S. P. Bellayer, H. Kim and J. F. Douglas (2005). "Flammability properties of polymer nanocomposites with single-walled carbon nanotubes: effects of nanotube dispersion and concentration." *Polymer* **46**(2): 471–481.

Kaul, P. K., A. J. Samson, G. T. Selvan, I. V. M. V. Enoch and P. M. Selvakumar (2017). "Synergistic effect of LDH in the presence of organophosphate on thermal and flammable properties of an epoxy nanocomposite." *Applied Clay Science* **135**: 234–243.

Kim, C., M.-A. Yoon, B. Jang, H.-D. Kim, J.-H. Kim, A. T. Hoang, J.-H. Ahn, H.-J. Jung, H.-J. Lee and K.-S. Kim (2021). "Damage-free transfer mechanics of 2-dimensional materials: competition between adhesion instability and tensile strain." *NPG Asia Materials* **13**(1): 44.

Li, Z., W. Cheng, C. Dong, B. Wang, Z. Jin, Q. Fang and Y. Hu (2021). "Synergistic properties of microencapsulated intumescent flame retardant-organically modified montmorillonite/ethylene-vinyl acetate copolymer composites." *Acta Materiae Compositae Sinica* **38**(8): 2546–2553.

Liao, S.-H., P.-L. Liu, M.-C. Hsiao, C.-C. Teng, C.-A. Wang, M.-D. Ger and C.-L. Chiang (2012). "One-step reduction and functionalization of graphene oxide with phosphorus-based compound to produce flame-retardant epoxy nanocomposite." *Industrial & Engineering Chemistry Research* **51**(12): 4573–4581.

Liu, W., B. Ullah, C.-C. Kuo and X. Cai (2019). "Two-dimensional nanomaterials-based polymer composites: fabrication and energy storage applications." *Advances in Polymer Technology* **2019**: 4294306.

Qiu, S., Y. Hu, Y. Shi, Y. Hou, Y. Kan, F. Chu, H. Sheng, R. K. K. Yuen and W. Xing (2018). "In situ growth of polyphosphazene particles on molybdenum disulfide nanosheets for flame retardant and friction application." *Composites Part A – Applied Science and Manufacturing* **114**: 407–417.

Schartel, B., A. Weiss, H. Sturm, M. Kleemeier, A. Hartwig, C. Vogt and R. X. Fischer (2011). "Layered silicate epoxy nanocomposites: formation of the inorganic-carbonaceous fire protection layer." *Polymers for Advanced Technologies* **22**(12): 1581–1592.

Shelley, J. S., P. T. Mather and K. L. DeVries (2001). "Reinforcement and environmental degradation of nylon-6/clay nanocomposites." *Polymer* **42**(13): 5849–5858.

Song, R. J., Z. Wang, X. Y. Meng, B. Y. Zhang and T. Tang (2007). "Influences of catalysis and dispersion of organically modified montmorillonite on flame retardancy of polypropylene nanocomposites." *Journal of Applied Polymer Science* **106**(5): 3488–3494.

Tang, Y., J. Tian, S. Li, C. Xue, Z. Xue, D. Yin and S. Yu (2015). "Combined effects of graphene oxide and Cd on the photosynthetic capacity and survival of *Microcystis aeruginosa.*" *Science of the Total Environment* **532**: 154–161.

Tsukanov, A. A., B. Turk, O. Vasiljeva and S. G. Psakhie (2022). "Computational indicator approach for assessment of nanotoxicity of two-dimensional nanomaterials." *Nanomaterials* **12**(4): 650.

Wang, D., P. Wen, J. Wang, L. Song and Y. Hu (2017). "The effect of defect-rich molybdenum disulfide nanosheets with phosphorus, nitrogen and silicon elements on mechanical, thermal, and fire behaviors of unsaturated polyester composites." *Chemical Engineering Journal* **313**: 238–249.

Wang, W., H. Pan, Y. Shi, Y. Pan, W. Yang, K. M. Liew, L. Song and Y. Hu (2016). "Fabrication of LDH nanosheets on beta-FeOOH rods and applications for improving the fire safety of epoxy resin." *Composites Part A – Applied Science and Manufacturing* **80**: 259–269.

Wang, X., E. N. Kalali and D.-Y. Wang (2015). "Renewable cardanol-based surfactant modified layered double hydroxide as a flame retardant for epoxy resin." *ACS Sustainable Chemistry & Engineering* **3**(12): 3281–3290.

Wazalwar, R., M. Sahu and A. M. Raichur (2021). "Mechanical properties of aerospace epoxy composites reinforced with 2D nano-fillers: current status and road to industrialization." *Nanoscale Advances* **3**(10): 2741–2776.

Wu, G. M., B. Schartel, H. Bahr, M. Kleemeier, D. Yu and A. Hartwig (2012). "Experimental and quantitative assessment of flame retardancy by the shielding effect in layered silicate epoxy nanocomposites." *Combustion and Flame* **159**(12): 3616–3623.

Yan, L., Z. Xu and J. Zhang (2017). "Influence of nanoparticle geometry on the thermal stability and flame retardancy of high-impact polystyrene nanocomposites." *Journal of Thermal Analysis and Calorimetry* **130**(3): 1987–1996.

Yue, X., C. Li, Y. Ni, Y. Xu and J. Wang (2019). "Flame retardant nanocomposites based on 2D layered nanomaterials: a review." *Journal of Materials Science* **54**(20): 13070–13105.

Zare, Y. (2015). "New models for yield strength of polymer/clay nanocomposites." *Composites Part B –Engineering* **73**: 111–117.

Zhi, Y.-R., B. Yu, A. C. Y. Yuen, J. Liang, L.-Q. Wang, W. Yang, H.-D. Lu and G.-H. Yeoh (2018). "Surface manipulation of thermal-exfoliated hexagonal boron nitride with polyaniline for improving thermal stability and fire safety performance of polymeric materials." *ACS Omega* **3**(11): 14942–14952.

Zhou, K., R. Gao and X. Qian (2017). "Self-assembly of exfoliated molybdenum disulfide (MoS$_2$) nanosheets and layered double hydroxide (LDH): towards reducing fire hazards of epoxy." *Journal of Hazardous Materials* **338**: 343–355.

Index

Note: Page numbers in *italic* and **bold** refer to figures and tables, respectively.

For Product Safety Concerns and Information please contact our EU
representative GPSR@taylorandfrancis.com
Taylor & Francis Verlag GmbH, Kaufingerstraße 24, 80331 München, Germany

www.ingramcontent.com/pod-product-compliance
Lightning Source LLC
Chambersburg PA
CBHW060805220326
41598CB00022B/2543